$A + B = B + A$ 4-3

$A \cdot B = B \cdot A$ 4

$A + (B + C) = (A + B) + C$ 5

$A(B \cdot C) = (A \cdot B)C$ 6

$A(B + C) = AB + AC$ 7

$A + 0 = A$ 8

$A \cdot 0 = 0$ 9

$A + 1 = 1$ 10

$A \cdot 1 = A$ 11

$A + A = A$ 12

$A \cdot A = A$ 13

$A + \overline{A} = 1$ 14

$A \cdot \overline{A} = 0$ 15

$\overline{\overline{A}} = A$ 16

de Morgan's:

$\overline{A + B} = \overline{A} \cdot \overline{B}$ 17

$\overline{A \cdot B} = \overline{A} + \overline{B}$ 18

$A + AB = A$ 19

$A(A + B) = A$ 20

$(A + B)(A + C) = A + BC$ 21

$A + \overline{A}B = A + B$ 22

$A(\overline{A} + B) = AB$ 23

$(A + B)(\overline{A} + C) = AC + \overline{A}B$ 24

$AB + \overline{A}C = (A + C)(\overline{A} + B)$ 25

5, 6, 7, 8, 9.1-9.3

74141	p. 60, 127
7443	61
7444	61
74145	61
7447	61, 127, 184
7400	98
7410	98
7420	98
7430	98
7483	127
7486	127
7490	127, 175, 177, 184
7493	127, 175, 177
7421	138
7432	138
7475	182, 184

digital principles and applications

digital principles and applications

second edition

Albert Paul Malvino, Ph.D.

Donald P. Leach, Ph.D.
Chairman, Engineering and Technology Division
Foothill College
Los Altos Hills, California

McGraw-Hill Book Company

New York
St. Louis
Dallas
San Francisco
Auckland
Düsseldorf
Johannesburg
Kuala Lumpur
London
Mexico
Montreal
New Delhi
Panama
Paris
São Paulo
Singapore
Sydney
Tokyo
Toronto

Library of Congress Cataloging in Publication Data

Malvino, Albert Paul.
Digital principles and applications.

1. Digital electronics. 2. Electronic digital computers—Circuits. I. Leach, Donald P., joint author. II. Title.
TK7868.D5M3 1975 621.3815 74-19172
ISBN 0-07-039837-2

Digital Principles and Applications

2 3 4 5 6 7 8 9 0 KPKP 7 8 3 2 1 0 9 8 7 6 5

The editors for this book were *Gordon Rockmaker* and *Susan Schwartz,* the designer was *Tracy Glasner,* and the production supervisor was *Laurence Charnow.* It was set in Optima by Progressive Typographers. Printed and bound by Kingsport Press, Inc.

Contents

Preface

The use of digital devices and systems has grown so fantastically that it is now mandatory for every technician and engineer to know at least the fundamentals of digital electronics.

This new edition maintains our original approach. Rather than focus on digital computers, computer architecture, and programming, we prefer to emphasize those principles that apply not only to computers but to automobiles, communication equipment, data processing, industrial automation, medicine, process control, transportation, etc. This general introduction to digital electronics provides a broader base for study in special areas.

To strengthen and modernize the book, we have included

1. New chapters on Karnaugh mapping and computer organization
2. Discussions of TTL, ECL, and CMOS logic
3. Material on discrete, bar-matrix, and dot-matrix displays
4. Standard SSI, MSI, and LSI circuits such as gates, decoders, latches, adders, D/A converters, and ROMs
5. Industry-wide logic symbols in all illustrations

The main prerequisite for this book is an understanding of how semiconductor diodes and transistors work. The length and level make the book suitable for a beginning course in digital electronics. Behaviorial or learning objectives are stated at the beginning of each chapter; summaries, glossaries, review questions, and problems appear at the end of each chapter.

A. P. Malvino
D. P. Leach

Introduction

1

Digital electronics, how did it all start? What is a computer? What are logic circuits? These are a few of the questions answered in this first chapter.

After studying this chapter, you should be able to

1. Name and describe the main parts of a digital computer.
2. Differentiate between saturated and nonsaturated logic.
3. Name five types of bipolar ICs and one type of MOS IC.

1-1 JACQUARD, BABBAGE, AND BOOLE

Digital electronics started in caves, thousands of years before written history. It began when the first man learned to count, when he learned to associate number names with objects in a group. Most counting was done on the fingers (digits), and for this reason the basic number names (one, two, three, . . .) are known as *digits*.

The invention of numbers led to arithmetic and all kinds of calculating devices like the abacus, Napier's bones (the first slide rule), and Pascal's calculator (the first adding machine). But the really crucial inventions in the evolution of digital electronics were made in the nineteenth century.

To begin with, Jacquard (1801) invented an automatic loom whose main feature was the use of *punched cards*. Figure 1-1 shows an example. In Jacquard's loom, needles passed through the holes in such a card and stitched a pattern onto cloth. By using cards with different holes, Jacquard could produce all kinds of figures easily and reliably.

In 1833, Babbage visualized the first *computer,* a machine that used punched cards to carry out arithmetic calculations automatically. By a prearranged code, certain groups of holes in these cards were to represent either numbers or instructions. The key idea in Babbage's computer was to enter all numbers and instructions *before* the calculation began; then, on command, the computer was to carry out all the steps in the calculation without human intervention. (This is the crucial difference between a calculator and a computer. A calculator depends on human intervention because someone has to enter numbers and instructions while the calculation is in progress.)

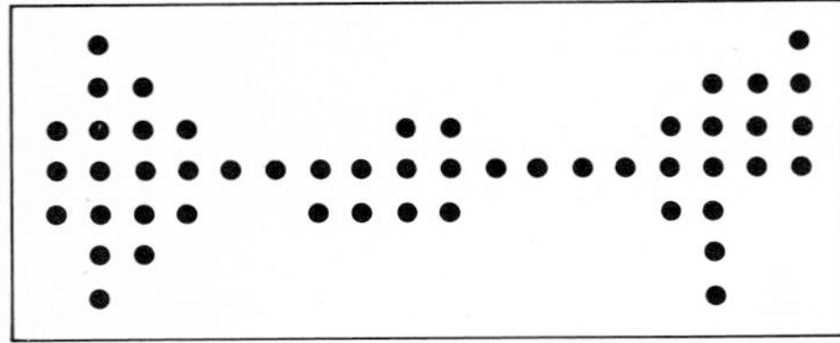

Fig. 1-1. Punched card.

In 1854 Boole found a new way of thinking, a new way to reason things out. He decided to use symbols instead of words to reach logical conclusions. Boole saw a pattern in the way we think that allowed him to invent *symbolic logic*, a method of reasoning based on the manipulation of letters and symbols. In many ways, symbolic logic resembles ordinary algebra. This is why it's called *Boolean algebra*.

Although originally intended for solving logic problems, Boolean algebra now finds its greatest use in the design of digital computers. By a coincidence, the rules of symbolic logic apply to the electronic circuits in computers and other digital systems.

1-2 COMPUTERS

Babbage's work in 1833 laid the foundation for modern digital computers (those whose operation is based on the digits 0, 1, 2, etc.). Let's take a closer look at what Babbage did. As he saw it, a computer had to have the functions shown in Fig. 1-2. Here's what each block does:

Input This transfers into the computer a set of instructions called a *program* and a set of numbers known as the *data*.

Memory Here is where the program and data are stored before a calculation begins. Also, during a calculation the memory stores intermediate answers, similar to the way we write intermediate answers on a scratch pad.

Control This part of the computer eliminates the human operator. The control section directs other parts of the computer to manipulate the data as prescribed in the program.

Fig. 1-2. Main parts of a computer.

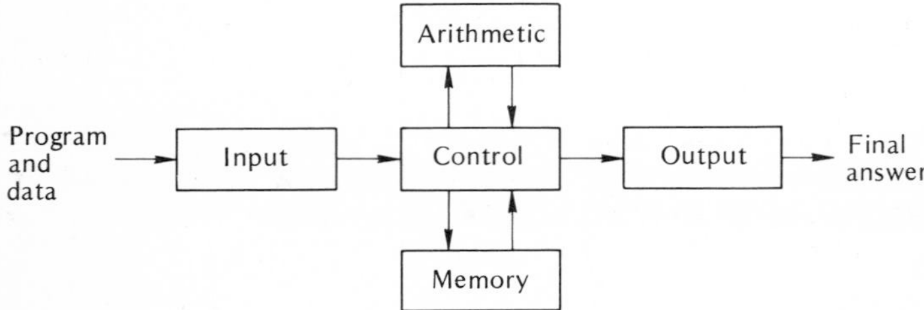

Arithmetic This section adds, subtracts, multiplies, and divides. This may not sound like much, but it's adequate for more advanced mathematical problems. Why? Because even the toughest problems can usually be reduced to an *arithmetic algorithm,* a series of simple arithmetic operations where the result of each step is used in the next step.

Output This transfers the final answer from the computer to the outside world.

Babbage never built a working model of a digital computer, but his notes prove he knew how to go about it. His ideas opened up a whole new world and led to today's modern computers.

The first electronic computers based on Babbage's ideas appeared in the early 1950s. These *first-generation* computers used vacuum tubes. Toward the end of the same decade, *second-generation* computers were developed. (They used transistors.) In the early 1960s *third-generation* computers evolved; these used transistors and some integrated circuits. We're now in the *fourth generation* of computers; these make extensive use of integrated circuits.

Computers are having an enormous impact on all areas of electronics. Some of the digital circuits developed for computers are now being used in automobiles, communications, data processing, industrial automation, manufacturing controls, medicine, transportation, etc.

1-3 SATURATED AND NONSATURATED LOGIC

The circuits in digital computers duplicate logical processes of the mind. Because of this, Boolean algebra can be used in the analysis and design of digital computers. Any circuit that can be analyzed with Boolean algebra is known as a *logic circuit.*

The output of a logic circuit is either a low or a high voltage. For instance, Fig. 1-3 shows a simple logic circuit using a transistor with a β_{dc} greater than 10. Before point *A* in time, the input voltage is zero and the transistor is cut off; therefore, the output voltage is +5 V. At point *A* in time, the input voltage switches from 0 V to +5 V, enough to drive the transistor into saturation; ideally, the output voltage

Fig. 1-3. Saturated logic circuit.

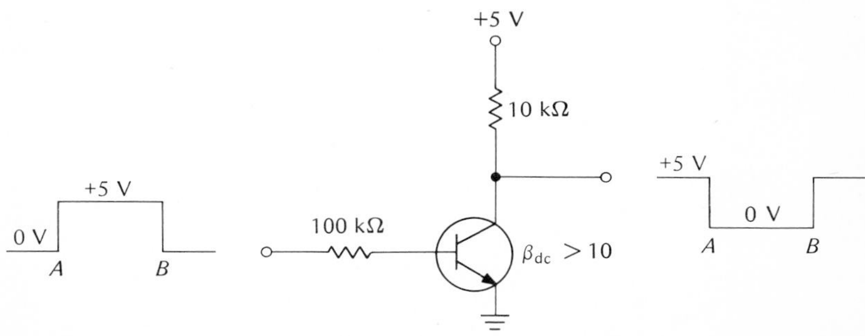

drops immediately from +5 V to 0 V. Later, at point B in time, the input voltage drops back to 0 V, and the output voltage goes back to +5 V. An important thing to remember is this: the output of a logic circuit is either a low or a high voltage.

When logic circuits use saturated transistors, the circuits are classified as *saturated logic circuits,* or simply saturated logic. The majority of logic circuits fall into this category. But there are some logic circuits in which saturation is deliberately avoided; these circuits are referred to as *nonsaturated logic*.

The disadvantage of saturated logic is the delay that occurs when a transistor is brought out of saturation. When a transistor is heavily saturated, a large number of extra carriers flood the base region. Even though the base voltage is suddenly switched off, the transistor remains saturated until all carriers leave the base region. The time required for this departure is called the *saturation delay time* (designated t_s).

For instance, a transistor may have a t_s of 40 ns. In Fig. 1-3 this means the output voltage does not rise immediately at point B in time; rather, a delay of 40 ns occurs before the transistor comes out of saturation.

In most applications the saturation delay time is too small to matter, but there are some applications where the fastest possible switching speed is required. This is where nonsaturated logic comes in. By designing circuits to avoid transistor saturation, we can eliminate saturation delay time. The resulting circuits can then switch on and off very rapidly. This is the advantage of nonsaturated logic over saturated logic.

1-4 BIPOLAR INTEGRATED CIRCUITS

As already mentioned, we're now in the fourth generation of computers, where integrated circuits are almost exclusively used. The two basic kinds of integrated circuits (ICs) are the *bipolar* and the *metal-oxide semiconductor* (MOS). The remainder of this section discusses bipolar ICs; the next section covers MOS ICs.

The earliest bipolar ICs used only resistors and transistors. As a result, they became known as *resistor-transistor logic,* abbreviated RTL. Figure 1-4 shows two

Fig. 1-4. Two examples of RTL.

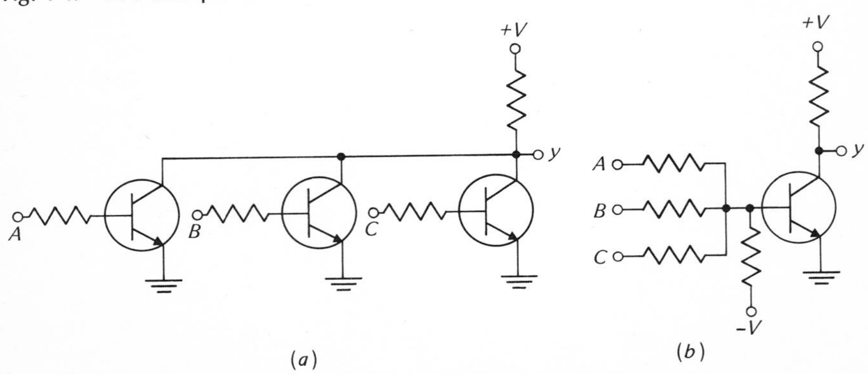

+V
y
A
B
C
(a)
+V
y
A
B
C
−V
(b)

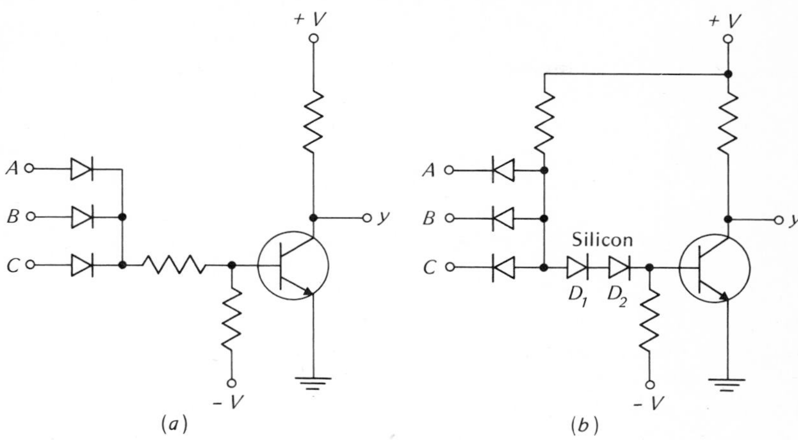

Fig. 1-5. Two examples of DTL.

examples of RTL. During normal operation the transistors switch between cutoff and saturation; therefore, RTL is a saturated logic. RTL is of historical interest only; it has so many disadvantages that it's now virtually obsolete.

After RTL came *diode-transistor logic* (DTL). Figure 1-5 shows two of many possible examples. With DTL, the active integrated components are diodes and transistors. Again, the transistors operate between cutoff and saturation; so DTL falls into the category of saturated logic. Although still used, DTL is being replaced by newer types.

Transistor-transistor logic (TTL) became commercially available in 1964. Since then, it has become the most popular bipolar type of IC. Figure 1-6 is an example of a TTL circuit. Notice the *multiple-emitter* input transistor and the *totem-pole* output transistors. The action of this and other TTL circuits is discussed later. For

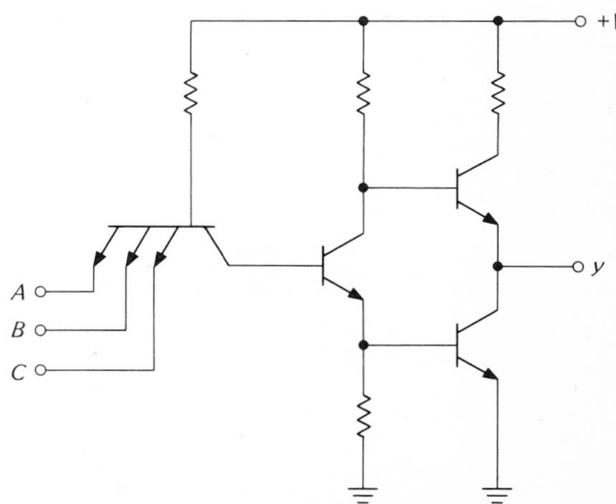

Fig. 1-6. Example of TTL.

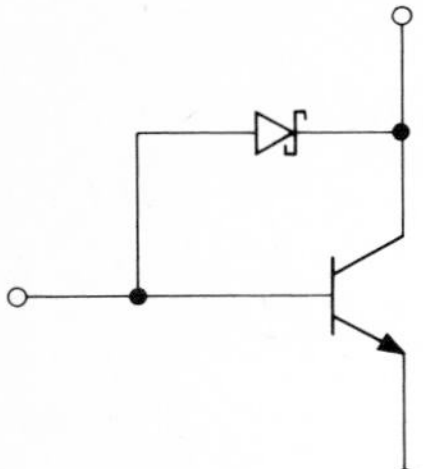

Fig. 1-7. Schottky TTL transistor.

the moment, it's enough to know TTL is important; more than 65 percent of all bipolar ICs are in the TTL family.

The transistors of Fig. 1-6 operate between cutoff and saturation. This makes TTL a saturated logic. As discussed earlier, saturation delay time slows down the switching action of a logic circuit. For this reason, *standard* TTL (the kind using saturated transistors) is not fast enough for some applications.

Schottky TTL is different. In this variation of TTL, a Schottky diode is integrated along with a bipolar transistor as shown in Fig. 1-7. Because the Schottky diode has a forward voltage of about 0.4 V, it prevents the transistor from fully saturating. Schottky TTL is therefore a nonsaturated logic that finds use in high-speed applications.

Emitter-coupled logic (ECL) is even faster. An ECL circuit uses differential-amplifier stages whose transistors stay well out of saturation. Speed is ECL's main advantage. Roughly, 20 to 25 percent of all bipolar ICs fall into this nonsaturated-logic family. Figure 1-8 shows one of many possible ECL circuits.

Fig. 1-8. Example of ECL.

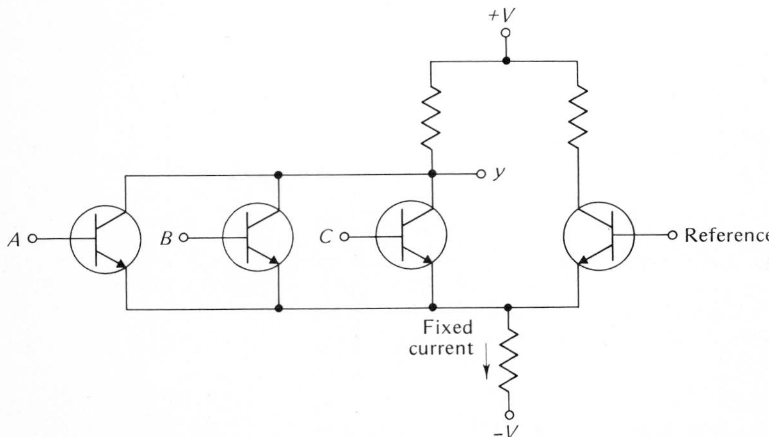

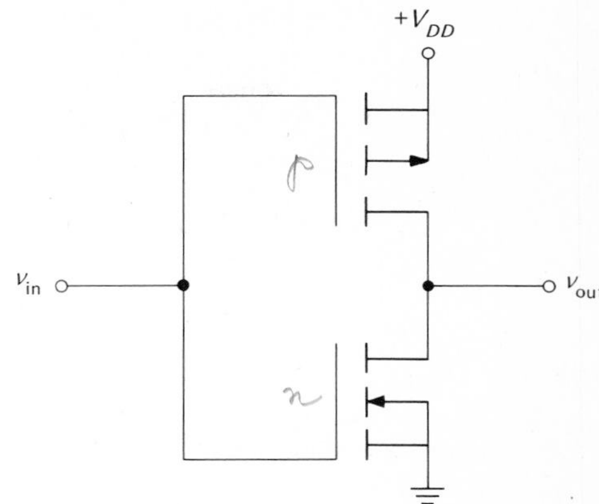

Fig. 1-9. Example of CMOS.

1-5 MOS INTEGRATED CIRCUITS

The main integrated component in MOS ICs is the enhancement MOSFET (metal-oxide semiconductor field-effect transistor). An enhancement MOSFET is normally off; to turn it on, you have to apply a sufficiently large gate voltage.[1]

An important variation of MOS logic circuits is *complementary* MOS (CMOS). Figure 1-9 shows an example of CMOS. A *p*-channel enhancement MOSFET (upper device) is in series with an *n*-channel enhancement MOSFET. When the input voltage v_{in} is low, the *p*-channel MOSFET is on and the *n*-channel MOSFET is off. Conversely, when v_{in} is high, the *p*-channel device turns off and the *n*-channel device turns on.

In either case, the series current through both MOSFETs is very small (equal to the leakage current of the off device). As a result, the total power dissipation is extremely low (power-supply voltage times leakage current). This is the main reason for the popularity of CMOS; the very low power consumption is a tremendous advantage in applications such as electronic wristwatches, space vehicles, and pocket calculators.

STUDY AIDS

Summary

Babbage was the one who first visualized all the key ideas behind a digital computer. He intended to use punched cards to enter the program (set of instructions) and data (set of numbers) into his computer.

Boole invented a new kind of algebra that simplified logic problems. Without

[1] For a detailed explanation, see A. P. Malvino, "Electronic Principles," pp. 355–358, McGraw-Hill Book Company, New York, 1973.

knowing it, he was also inventing the most popular tool used in the analysis and design of digital computers.

Electronic digital computers began appearing in the middle of this century. Successive generations have used vacuum tubes, transistors, and integrated circuits. We're now in the fourth generation of computers, where integrated circuits are used almost exclusively.

Standard TTL is the most important bipolar family of ICs. For high-speed applications, Schottky TTL or ECL is used. When low power consumption is vital, CMOS is the way to go.

Glossary

algorithm A series of operations in which the result of each step is used in the next step.

Boolean algebra Also known as symbolic logic. Invented by Boole, this algebra is the main tool used in analyzing and designing digital circuits.

calculator A device for working out arithmetic problems. It differs from a computer in that a human operator is needed to enter numbers and instructions while a calculation is in progress.

data A set of numbers that goes into the computer before the calculation begins.

digit One of the basic symbols used in a number system. In the familiar decimal number system, ten digits are used:

0, 1, 2, 3, 4, 5, 6, 7, 8, 9

logic circuit Any digital circuit that can be analyzed with Boolean algebra.

program A set of instructions entered into a computer before the start of a calculation.

Review Questions

1. What's the difference between a calculator and a computer?
2. Name the five main parts of a computer.
3. In what section of a computer are the program and data stored? Which part of a computer performs the job of the human operator of a calculator?
4. What's the difference between saturated and nonsaturated logic? Which would you use to get faster switching speed?
5. Which bipolar family is the most widely used? For the fastest possible speed, which family is used? To get the lowest power dissipation, which family is used?

Problems

1-1. Idealize the transistor of Fig. 1-3 by neglecting its V_{BE} drop. In this case, what is the base current when the transistor is conducting? The collector current?

1-2. If the V_{BE} drop in Fig. 1-3 is 0.7 V, what does the base current equal when the transistor is on? If the transistor is saturated, what does the collector current equal? What is the minimum β_{dc} needed to ensure a saturated transistor?

1-3. V_{DD} equals 10 V in Fig. 1-9. The upper MOSFET is cutoff. If its data sheet lists a cutoff current of 2 nA, what is the total power dissipation of the circuit?

Numbers

1-22-77

On hearing the word *number,* most of us immediately think of the familiar decimal number system with its ten digits:

0, 1, 2, 3, 4, 5, 6, 7, 8, 9

It may seem strange and useless to discuss number systems with two digits, or eight digits, or sixteen digits. Nevertheless, we shall spend this entire chapter discussing such number systems because they are used in computers and other digital systems.

On completion of this chapter, you ought to be able to

1. Add, subtract, multiply, and divide using binary numbers.
2. Convert from octal to binary numbers, and vice versa.
3. Convert from hexadecimal to binary numbers, and vice versa.

2-1 BINARY NUMBERS

A number system is nothing more than a code. For each distinct quantity, there is an assigned symbol. After memorizing the code, we can count, and this leads to arithmetic and higher forms of mathematics.

The most familiar number-system is the decimal system, whose digits are symbolized in Table 2-1. The black circles indicate pebbles which we shall use to represent different quantities. In the table are ten basic symbols or digits: 0 through 9. Each of these symbols stands for a certain number of pebbles. Since we long ago memorized the meaning of each symbol, we immediately know that 4 represents ●●●●, 7 denotes ●●●●●●●, and so on. But these symbols have the meaning they do only because we have *memorized* them. Other symbols could just as easily be used. For instance, instead of 0, 1, 2, . . . , 9, we can use *A, B, C,* . . . , *J*. In this case, *C* would mean ●●, *D* would denote ●●●, and so on.

The use of 0 through 9 is unnecessary. After all, since a number system is only a code, we can use any number of code symbols we want. A *binary* number system is a code that uses only two basic symbols. The digits can be any two distinct characters like *A* and *B*, · and -, or the customary 0 and 1. Table 2-2 shows the basic symbols of the binary number system.

Table 2-1
THE DECIMAL DIGITS

Pebbles	Symbol
None	0
•	1
••	2
•••	3
••••	4
•••••	5
••••••	6
•••••••	7
••••••••	8
•••••••••	9

Probably the first question that comes to mind is how do we represent quantities that are greater than • ? What can we use for ••, •••, and so on?

After reaching 9 in the decimal number system, we form combinations of decimal digits to get 10, 11, 12, etc. In other words, the next decimal number after 9 is obtained by using *the second digit followed by the first* to get 10. The decimal number after 10 is obtained by using the second digit followed by the second to get 11, and so forth.

In the binary number system we use the same approach. After reaching 1, we have run out of binary digits (there is no 2, 3, . . . in the binary number system). To represent ••, use the second binary digit followed by the first to get 10. To represent •••, use 11. Therefore, we count in binary as follows: 0, 1, 10, 11. To avoid confusion with decimal numbers, it helps to read these binary numbers as zero, one, one-zero, and one-one.[1]

What is the next binary number after 11? It is not 12 because 2 is not a binary digit. After reaching 99 in the decimal number system, we have exhausted all the two-digit numbers; we then use three digits at a time to get 100, 101, 102, 103, etc. By following this same pattern with binary numbers, we count as follows: 0, 1, 10, 11, 100, 101, 110, 111. Table 2-3 summarizes these numbers.

[1] A tribe of the Torres Straits counts in binary as follows: urapun, okosa, okosa-urapun, okosa-okosa.

Table 2-2
THE BINARY DIGITS

Pebbles	Symbol
None	0
•	1

Table 2-3
THREE WAYS OF COUNTING TO SEVEN

Quantity	Binary number	Decimal number
None	0	0
●	1	1
●●	10	2
●●●	11	3
●●●●	100	4
●●●●●	101	5
●●●●●●	110	6
●●●●●●●	111	7

To remember how to count with binary numbers, the following method is also useful.

1. Think of the decimal numbers.
2. Eliminate any decimal number with a digit greater than 1.
3. The numbers that remain are the binary numbers.

For instance, if you think of the decimal numbers and cross out all numbers with a digit greater than 1, you get

0, 1, ~~2~~, ~~3~~, . . . , ~~9~~, 10, 11, ~~12~~, ~~13~~, . . . , ~~99~~, 100, 101, ~~102~~, ~~103~~, . . . , ~~109~~, 110, 111, ...

Collecting the numbers that remain gives the binary number system:

0, 1, 10, 11, 100, 101, 110, 111, ...

The binary number system of Table 2-3 is nothing more than a code. After some practice, it becomes almost as familiar as the decimal number system.

Of course, we can form many other number systems by selecting a different group of basic symbols or digits. The *base* or *radix* of a number system refers to the number of basic symbols used. For instance, ten is the base of the decimal number system because this system uses ten digits, 0 through 9. The binary number system has a base of two because the only digits used are 0 and 1. The octal number system, which we shall study shortly, has a base of eight.

Example 2-1

Show how to count in a number system with a base of three. Use the digits 0, 1, 2.

Solution

The first three numbers are 0, 1, 2. At this point we have run out of basic symbols, and we now form two-digit combinations. The number after 2 is 10 (second digit followed by first). The number after 10 is 11 (second followed by second), and the number after 11 is 12 (second followed by third). Up to this point, we count as follows: 0, 1, 2, 10, 11, 12. Since 2s are allowed, the next two-digit number after 12 is 20. Then come 21 and 22.

We now can form three-digit combinations, then four, five, and so forth. Therefore, here is how to count in a base-3 system:

0, 1, 2, 10, 11, 12, 20, 21, 22, 100, 101, 102, 110, 111, 112, 120, 121, 122, 200, 201, 202, 210, 211, 212, 220, 221, 222, 1000, ...

Example 2-2

Instead of using the digits 0 and 1 for the binary numbers, use *A* and *B*, respectively. Show how to count to seven.

Solution

A, B, BA, BB, BAA, BAB, BBA, BBB

Example 2-3

In Fig. 2-1, an on-lamp stands for binary number 1, whereas an off-lamp denotes binary number 0. Reading from left to right, what binary number do the lamps symbolize? What is the corresponding decimal number?

Solution

The lamps are on-off-on, which stands for 101. Table 2-3 shows that this represents decimal number 5.

Example 2-4

In Fig. 2-2*a* we have a transistor circuit. Whenever a voltage is approximately 5 V, we shall use binary number 1 to indicate this condition; whenever a voltage is zero, we shall use binary number 0 for this. What binary numbers do the input and output waveforms of Fig. 2-2*b* and *c* represent? (Read from left to right during *each* second of time.)

Solution

Since +5 V denotes binary 1, and 0 V stands for binary 0, we read the input waveform as 10110.

The output waveform indicates binary number 01001.

Incidentally, the transistor is either saturated (shorted) or cut off (open) so the output voltage is either 0 or +5 V.

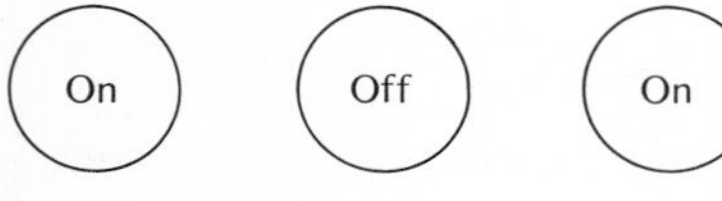

Fig. 2-1. Lamps represent binary number.

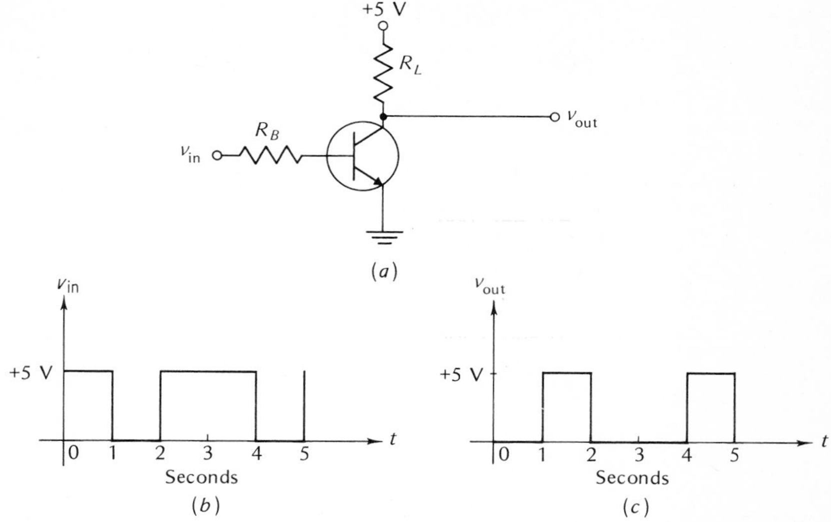

Fig. 2-2. Input and output voltages represent binary numbers.

2-2 BINARY ADDITION

Addition is a manipulation of numbers that represents the combining of physical quantities. For example, in the decimal number system, $2 + 3 = 5$ symbolizes the combining of •• with ••• to get a total of •••••.

To discover the rules for binary addition, we need to discuss four simple cases.

Case 1 When nothing is combined with nothing, we get nothing. The binary representation of this is $0 + 0 = 0$.

Case 2 When nothing is combined with •, we get •. Using binary numbers to denote this gives $0 + 1 = 1$.

Case 3 Combining • with nothing gives •. The binary equivalent of this is $1 + 0 = 1$.

Case 4 When we combine • with •, the result is ••. Using binary numbers, we symbolize this by $1 + 1 = 10$.

The last result is sometimes disturbing because of our long-time association with decimal numbers. But it is correct and makes sense because *we are using binary numbers*. Binary number 10 stands for •• and not for •••••••••• (ten). Naturally, it is important to know whether binary numbers or decimal numbers are being used. Ordinarily, it is clear from the context which kind of numbers we are talking about. But when there is doubt, we shall use a subscript to indicate the base. For instance, 11_2 denotes binary number 11. Or we can write 45_{10} to mean decimal number 45. Therefore, you should be able to understand why the following equations make sense:

$$1_{10} + 1_{10} = 2_{10}$$
$$1_2 + 1_2 = 10_2$$

To summarize our results for binary addition,

0 + 0 = 0
0 + 1 = 1
1 + 0 = 1
1 + 1 = 10 (zero, carry one)

To add larger binary numbers, carry into higher-order columns as is done with decimal numbers. As an example, add 10 to 10 as follows:

$$\begin{array}{r} 10 \\ +10 \\ \hline 100 \end{array}$$

In the first column, zero plus zero is zero. In the second column, one plus one is zero, carry a one. (In decimal numbers the foregoing addition was 2 + 2 = 4.)

As another example, take 1 + 1 + 1. Add two of the 1s to get 10 + 1. Adding again gives 11 as follows:

$$\begin{aligned} 1 + 1 + 1 &= 10 + 1 \\ &= 11 \end{aligned}$$

If four 1s are to be added, cumulatively add two numbers each time.

$$\begin{aligned} 1 + 1 + 1 + 1 &= 10 + 1 + 1 \\ &= 11 + 1 \\ &= 100 \end{aligned}$$

Example 2-5

(*a*) Add 101 and 110.
(*b*) Add 111 and 110.

Solution

(*a*)	101	first column: 1 + 0 = 1
	+110	second column: 0 + 1 = 1
	1011	third column: 1 + 1 = 10 (zero, carry one)
(*b*)	111	first column: 1 + 0 = 1
	+110	second column: 1 + 1 = 10 (zero, carry one)
	1101	third column: 1 + 1 + 1 = 10 + 1 = 11

Example 2-6

Add the following:

(*a*) 1010 + 1101
(*b*) 1011 + 1010

Solution

$$(a)\quad \begin{array}{r} 1010 \\ +1101 \\ \hline 10111 \end{array}$$

$$(b)\quad \begin{array}{r} 1011 \\ +1010 \\ \hline 10101 \end{array}$$

2-3 BINARY-TO-DECIMAL CONVERSION

How do you convert a binary number into its decimal counterpart? For instance, what decimal number does binary number 1010011 stand for? This section shows how to convert a binary number quickly and easily into its decimal equivalent.

We can express any decimal *integer* (a whole number) in units, tens, hundreds, thousands, and so on. For instance, decimal number 2945 decomposes into

$$2945 = 2000 + 900 + 40 + 5$$

In powers of 10, this becomes

$$2945 = 2(10^3) + 9(10^2) + 4(10^1) + 5(10^0)$$

In a similar way, we can partition any binary number into simpler parts. For instance, binary number 111 becomes

$$111 = 100 + 10 + 1 \qquad (2\text{-}1)$$

This decomposition is valid because the right side of the equation equals the left side.

Something interesting occurs when Eq. (2-1) is converted into its decimal counterpart. Using Table 2-3,

$$\begin{array}{ccccccccl} 111 & = & 100 & + & 10 & + & 1 & & \text{binary} \\ \downarrow & & \downarrow & & \downarrow & & \downarrow & & \quad\downarrow \\ 7 & = & 4 & + & 2 & + & 1 & & \text{decimal} \end{array}$$

Breaking a binary number into parts is equivalent to splitting its decimal equivalent into units, twos, fours, eights, and so on. In other words, each digit position in a binary number has a value or weight. The least significant digit (the one on the right) has a value of 1. The second position from the right has a value of 2, the next

4, then 8, 16, 32, etc. Diagrammatically, the digit positions of a binary number have the following decimal weights:

etc. ←	32	16	8	4	2	1

As a matter of fact, these numbers are in ascending powers of 2.

etc. ←	2^5	2^4	2^3	2^2	2^1	2^0

The important point about the weights is this: whenever you look at a binary number, you can find its decimal equivalent as follows:

1. When there is a 1 in a digit position, add the weight of that position.
2. When there is a 0 in a digit position, disregard the weight of that position.

For example, binary number 101 has a decimal equivalent of

$$4 + 0 + 1 = 5$$

As another example, binary number 1011 is equivalent to

$$8 + 0 + 2 + 1 = 11$$

Still another example is 11001, equivalent to

$$16 + 8 + 0 + 0 + 1 = 25$$

We can streamline binary-to-decimal conversion by the following procedure:

1. Write the binary number.
2. Directly under the binary number write 1, 2, 4, 8, 16, . . . , working from right to left.
3. If a zero appears in a digit position, cross out the decimal weight for that position.
4. Add the remaining weights to get the decimal equivalent.

As an example of this approach, let us convert binary 101 to its decimal equivalent.

Step 1 1 0 1
Step 2 4 2 1
Step 3 4 $\not{2}$ 1
Step 4 $4 + 1 = 5$

As another example, notice how quickly 10101 is converted to its decimal equivalent.

$$\begin{array}{ccccc} 1 & 0 & 1 & 0 & 1 \\ 16 & \not{8} & 4 & \not{2} & 1 \rightarrow 21 \end{array}$$

So far, we have discussed binary *integers* (whole numbers). How are binary fractions converted into corresponding decimal equivalents? For instance, what is the decimal equivalent of 0.101? In this case, the weights of digit positions to the right of the binary point are given by

$\frac{1}{2}$	$\frac{1}{4}$	$\frac{1}{8}$	$\frac{1}{16}$

• ↑ Binary point → etc.

Or, in powers of 2,

2^{-1}	2^{-2}	2^{-3}	2^{-4}

• ↑ Binary point → etc.

Therefore, 0.101 has a decimal equivalent of

$$\begin{array}{ccc} 0.1 & 0 & 1 \\ \frac{1}{2} & \not{\frac{1}{4}} & \frac{1}{8} \rightarrow 0.5 + 0.125 = 0.625 \end{array}$$

Another example, the decimal equivalent of 0.1101 is

$$\begin{array}{cccc} 0.1 & 1 & 0 & 1 \\ \frac{1}{2} & \frac{1}{4} & \not{\frac{1}{8}} & \frac{1}{16} \rightarrow 0.5 + 0.25 + 0.0625 = 0.8125 \end{array}$$

As far as mixed numbers are concerned (numbers that have an integer and a fractional part), handle each part according to the rules just developed. The weights for a mixed number are

2^3	2^2	2^1	2^0	2^{-1}	2^{-2}	2^{-3}

etc. ← → etc.

• ↑ Binary point (between 2^0 and 2^{-1})

With practice, the conversion of binary numbers into decimal numbers becomes a fast and easy procedure.

Example 2-7

Convert binary 110.001 to a decimal number.

Solution

$$1\ 1\ 0.\ 0\ 0\ 1$$

$$4\ 2\ 1\ \frac{1}{2}\ \frac{1}{4}\ \frac{1}{8} \rightarrow 6.125$$

Example 2-8

Solve for x in the following equation:

$$1011.11_2 = x_{10}$$

Solution

This is just another way of asking for the decimal equivalent of a binary number. (Remember that the subscript tells us the base or radix of the number system.)

$$1\ 0\ 1\ 1.1\ 1$$

$$8\ 4\ 2\ 1\ \frac{1}{2}\ \frac{1}{4} \rightarrow 11.75$$

Therefore,

$$1011.11_2 = 11.75_{10}$$

2-4 DECIMAL-TO-BINARY CONVERSION

One way to convert a decimal number into its binary equivalent is to reverse the process described in the preceding section. For instance, suppose you want to convert decimal 9 into the corresponding binary number. All you need to do is express 9 as a sum of powers of 2, and then write 1s and 0s in the appropriate digit positions:

$$9 = 8 + 1 = 8 + 0 + 0 + 1$$
$$\rightarrow 1001$$

As another example,

$$25 = 16 + 8 + 1 = 16 + 8 + 0 + 0 + 1$$
$$\rightarrow 11001$$

A more popular way to convert decimal numbers to binary numbers is the *double-dabble* method. In the double-dabble method you progressively divide the decimal number by 2, writing down the remainder after each division. The remainders, taken in reverse order, form the binary number. You'll understand this method after going through some examples. To convert 9 to a binary number, progressively divide by 2 as follows:

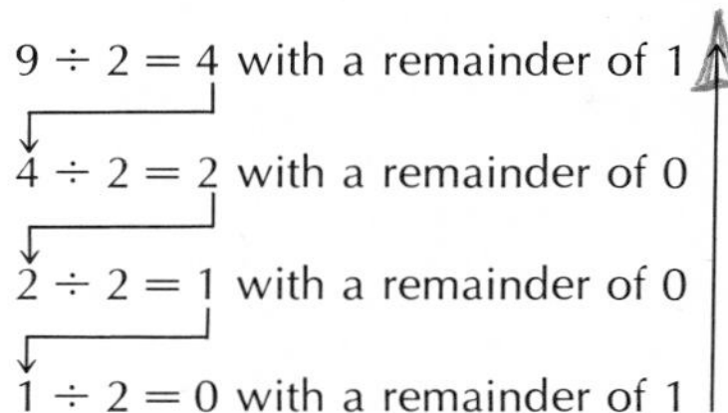

In the first step you divide 9 by 2 to get 4 with a remainder of 1. Next divide the result of the first division (4) by 2 to get 2 with a remainder of 0. You continue in this fashion, progressively dividing the result of each division by 2, and writing down the remainder after each division. In the last step, $1 \div 2 = 0$ with a remainder of 1. In other words, 2 is not contained in 1; so write 0 with a remainder of 1. The remainders taken in *reverse* order (from the bottom, up) give the answer, which is 1001.

Let us convert decimal 25 to its binary equivalent using the double-dabble method.

$25 \div 2 = 12$ with a remainder of 1
$12 \div 2 = 6$ with a remainder of 0
$6 \div 2 = 3$ with a remainder of 0
$3 \div 2 = 1$ with a remainder of 1
$1 \div 2 = 0$ with a remainder of 1

The remainders taken in reverse order (from the bottom, up) give the binary equivalent, which is 11001. The foregoing process can be simplified by writing the numbers in an orderly fashion like

2	25	
	12	1
	6	0
	3	0
	1	1
	0	1

In other words, starting at the top, divide 25 by 2 and write 12 with a remainder of 1. Then divide 12 by 2 and write 6 with a remainder of 0, and so forth. With a little practice you can quickly and easily convert decimal integers to binary integers.

As far as fractions are concerned, you *multiply* by 2 and record a carry in the integer position. The carries taken in *forward* order are the binary fraction. As an example, convert 0.625 to a binary fraction.

$$0.625 \times 2 = 1.25 = 0.25 \text{ with a carry of } 1$$

$$0.25 \times 2 = 0.5 \qquad \text{with a carry of } 0$$

$$0.5 \times 2 = 1.0 = 0 \qquad \text{with a carry of } 1$$

By taking the carries in forward order, we get 0.101, the binary equivalent of 0.625.[2]

As another example, 0.85 converts to binary as follows:

$$0.85 \times 2 = 1.7 = 0.7 \text{ with a carry of } 1$$
$$0.7 \times 2 = 1.4 = 0.4 \text{ with a carry of } 1$$
$$0.4 \times 2 = 0.8 = 0.8 \text{ with a carry of } 0$$
$$0.8 \times 2 = 1.6 = 0.6 \text{ with a carry of } 1$$
$$0.6 \times 2 = 1.2 = 0.2 \text{ with a carry of } 1$$
$$0.2 \times 2 = 0.4 = 0.4 \text{ with a carry of } 0$$

Taking the carries in forward order gives binary fraction 0.110110. In this case, we stopped the process after getting six binary digits; our answer is an approximation. If more accuracy is needed, continue multiplying by 2 until you have as many digits as necessary.

Incidentally, the word *bit* is common in digital work. It is a short way of saying "binary digit." A binary number like 1101 has four binary digits, or four bits. 111010 has six binary digits, or six bits.

Example 2-9

Convert 21.6_{10} to a binary number.

Solution

Split 21.6 into an integer of 21 and a fraction of 0.6, and apply double-dabble to each part.

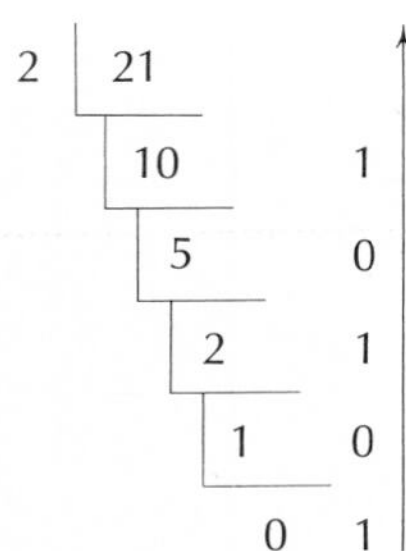

[2] Note that in this example the last product is 1.0. Whenever 1.0 is product, we can terminate the conversion because all succeeding carries must be 0s.

and

$$\begin{array}{ll} 0.6 \times 2 = 1.2 = 0.2 & \text{with a carry of 1} \\ 0.2 \times 2 = 0.4 & \text{with a carry of 0} \\ 0.4 \times 2 = 0.8 & \text{with a carry of 0} \\ 0.8 \times 2 = 1.6 = 0.6 & \text{with a carry of 1} \\ 0.6 \times 2 = 1.2 = 0.2 & \text{with a carry of 1} \downarrow \end{array}$$

The binary number is 10101.10011. This 10-bit number is an approximation of 21.6_{10} because we terminated the conversion of the fractional part after five bits.

2-5 BINARY SUBTRACTION

To subtract binary numbers, we first need to discuss four simple cases.

Case 1	$0 - 0 = 0$
Case 2	$1 - 0 = 1$
Case 3	$1 - 1 = 0$
Case 4	$10 - 1 = 1$

The last result represents

●● − ● = ●

which makes sense.

To subtract larger binary numbers, subtract column by column, borrowing from the adjacent column when necessary. For instance, in subtracting 101 from 111, proceed as follows (all examples will also show the equivalent decimal subtraction):

$$\begin{array}{r} 7 \\ -5 \\ \hline 2 \end{array} \qquad \begin{array}{r} 111 \\ -101 \\ \hline 010 \end{array} \qquad \begin{array}{ll} \text{first column:} & 1 - 1 = 0 \\ \text{second column:} & 1 - 0 = 1 \\ \text{third column:} & 1 - 1 = 0 \end{array}$$

Here is another example: subtract 1010 from 1101.

$$\begin{array}{r} 13 \\ -10 \\ \hline 3 \end{array} \qquad \begin{array}{r} 1101 \\ -1010 \\ \hline 0011 \end{array} \qquad \begin{array}{ll} \text{first column:} & 1 - 0 = 1 \\ \text{second column:} & 10 \text{ (after borrow)} - 1 = 1 \\ \text{third column:} & 0 \text{ (after borrow)} - 0 = 0 \\ \text{fourth column:} & 1 - 1 = 0 \end{array}$$

Binary numbers can also be negative, just like decimal numbers. As an illustration, subtract 111 from 100 in this way:

$$\begin{array}{r} 4 \\ -7 \\ \hline -3 \end{array} \qquad \begin{array}{r} 100 \\ -111 \\ \hline -011 \end{array} \qquad \begin{array}{ll} \text{first column:} & 1 - 0 = 1 \\ \text{second column:} & 1 - 0 = 1 \\ \text{third column:} & 1 - 1 = 0 \end{array}$$

Note that we subtracted the *smaller magnitude from the larger,* and prefixed the *sign of the larger magnitude;* this is identical to the procedure used with decimal numbers.

For another example, subtract 11011 from 10111.

23	10111
−27	−11011
− 4	−00100

Again note that the smaller magnitude is subtracted from the larger.

The foregoing examples illustrate one method of subtraction. Many digital computers subtract in this way. However, there are other methods that require less circuitry. Before we describe these other ways to subtract, we need to define the 1's and 2's *complements.*

The 1's complement of a binary number is the number that results when we change each 0 to a 1, and each 1 to a 0. In other words, the 1's complement of 100 is 011 (each 1 is changed to 0, and each 0 to 1). The 1's complement of 1001 is 0110. More examples of 1's complements are

1010 has a 1's complement of 0101
1110 has a 1's complement of 0001
0011 has a 1's complement of 1100

The 2's complement is the binary number that results when we *add* 1 to the 1's complement. That is,

$$\text{2's complement} = \text{1's complement} + 1$$

For instance, to find the 2's complement of 1011, first get the 1's complement, which is 0100. Next, add 1 to 0100 to get 0101—the 2's complement of 1011. For another example, find the 2's complement of 11001. First, get the 1's complement, which is 00110. After adding 1 to this you have 00111. Therefore, 00111 is the 2's complement of 11001. Here are some more worked-out 2's complements.

Number	*2's complement*
1110	0001 + 1 = 0010
0001	1110 + 1 = 1111
10110	01001 + 1 = 01010

How can we use complements to do subtraction? Instead of subtracting a number, we can add its 2's complement, and disregard the last carry. For instance, to subtract 101 from 111, proceed as follows:

Decimal	*Conventional*	*2's complement*	
7	111	111	
−5	−101	+011	(2's complement of 101)
2	010	~~1~~010	

Note that we added the 2's complement, and crossed out the final carry. What remains is the answer obtained by conventional subtraction.

Take another example: subtract 1010 from 1101.

```
  13      1101      1101
 -10     -1010     +0110
   3      0011     /10011
```

Again, by forming the 2's complement and adding, you get the same answer as with conventional subtraction, provided you disregard the final carry.[3]

Some digital computers subtract by the 2's-complement method. The advantage is a reduction in hardware; instead of having digital circuits that directly add and subtract, only adding-type circuits are needed.

Still another approach is the 1's-complement method of subtraction. Instead of subtracting a number, we add the 1's complement of the number. The last carry is then added to get the final answer. The method is best understood by an example. To subtract 101 from 111, we proceed as follows:

```
    111
   +010     (the 1's complement of 101)
  ┌1001
  │ 001
  └→ +1
    010
```

Here is what we did: (1) We formed the 1's complement of 101 to get 010. (2) We added 010 to 111 to get 1001. (3) The last carry is 1. Whenever there is a 1 carry in the last position, *remove* it and *add* it onto the remainder as shown. This is called "end-around carry."

Take another example. Subtract 1010 from 1101.

```
  13      1101      1101
 -10     -1010     +0101     (the 1's complement of 1010)
   3      0011    ┌10010
                  │ 0010
                  └→  +1
                    0011
```

Whenever there is an end-around carry, the answer is positive and is in binary form. But when there is no end-around carry, the answer is negative and is in 1's-complement form. For instance, when we subtract 1101 from 1010, we get

[3] A thorough study of the 2's complement shows that the carry indicates whether the answer is positive or negative. When the carry is 1, the answer is positive. When there is no carry, however, the answer is negative and is the 2's complement of the actual magnitude.

```
  10      1010      1010
 −13     −1101     +0010     (the 1's complement of 1101)
 − 3     −0011      1100 → −0011
                    ↑
              (There is no carry.)
```

Whenever there is no end-around carry, as in this case, the answer is negative and in 1's-complement form. Therefore, we take the 1's complement of 1100 to get 0011 and prefix a minus sign for a final answer of −0011.

The general proofs for using the 1's and 2's complements must be left to advanced books in mathematics.

The use of the 1's complement to subtract is popular in digital computers because only adders are needed; also, it is easy with digital circuits to get the 1's complement of a number (this will be shown in a later chapter).

Example 2-10

Subtract 01101 from 11011. Use the 1's complement.

Solution

```
   11011
  +10010     (the 1's complement of 01101)
 ┌101101
 │ 01101
 └→   +1
   01110
```

Example 2-11

Subtract 11011 from 01101. Use the 1's complement. Also, show the decimal subtraction and the conventional binary subtraction.

Solution

```
  13     01101      01101
 −27    −11011     +00100     (1's complement)
 −14    −01110  ┌→  10001 → −01110
                └─(No carry)
```

Since there is no final carry, we take the 1's complement and prefix a minus sign.

2-6 THE 9'S AND 10'S COMPLEMENTS

There also are complements in the decimal system. The 9's complement (similar to the 1's complement) is found by subtracting each decimal digit from 9. For instance, the 9's complement of 25 is obtained as follows:

```
 99
-25
 74      the 9's complement of 25
```

We subtracted each decimal digit from 9. As another example, the 9's complement of 6291 is

```
 9999
-6291
 3708      the 9's complement of 6291
```

Again, we subtracted each decimal digit from 9.

The 10's complement (similar to the 2's complement) of a decimal integer is one greater than the 9's complement, that is,

$$10\text{'s complement} = 9\text{'s complement} + 1$$

As an example, the 10's complement of 25 is found as follows:

```
 99
-25
 74 + 1 = 75      the 10's complement of 25
```

You get the 9's complement by subtracting 25 from 99; you then add 1 to get the 10's complement. As another example, the 10's complement of 6291 is

```
 9999
-6291
 3708 + 1 = 3709      the 10's complement of 6291
```

We can use the 9's and 10's complements to subtract decimal numbers. For instance, to subtract 25 from 83 using the 9's complement, proceed as follows:

```
 83        83
-25       +74      the 9's complement of 25
 58     ┌─157
        │  57
        └→  1      end-around carry
           58
```

We can get the same answer with the 10's complement:

```
 83
+75      the 10's complement of 25
 1̸58
```

(The use of the 9's and 10's complements is similar to the use of 1's and 2's complements, respectively.)

2-7 BINARY MULTIPLICATION AND DIVISION

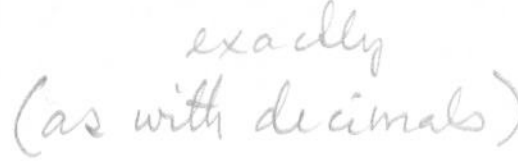

Binary multiplication is much easier than decimal multiplication because the multiplication table for binary has only four cases.

Case 1	$0 \times 0 = 0$
Case 2	$0 \times 1 = 0$
Case 3	$1 \times 0 = 0$
Case 4	$1 \times 1 = 1$

To handle larger numbers, that is, numbers with several bits, follow the pattern used with decimal numbers—form partial products and add. For instance, in multiplying 111 by 101, proceed as follows:

Decimal	Binary
7	111
×5	×101
35	111
	000
	111
	100011

Consider another example: 10110 times 110.

Decimal	Binary
22	10110
×6	×110
132	00000
	10110
	10110
	10000100

Division follows the same pattern in binary as in decimal. To divide 1100 by 10, we proceed as follows:

Decimal	Binary
6	110
2)12	10)1100
	10
	10
	10
	00

With practice, binary multiplication and division become easier than decimal multiplication and division.

Digital circuits for multiplying and dividing are straightforward, and we shall discuss them in a later chapter.

Example 2-12

Multiply 101.1 by 11.01.

Solution

```
     5.5           101.1
   ×3.25          ×11.01
   -----          ------
     275            1011
    110            0000
   165            1011
   -----         1011
   17875         --------
   17.875        10001111
                 10001.111
```

2-8 OCTAL NUMBERS

Octal numbers are important in digital work. To begin with, the octal number system has a base of eight, meaning it has eight basic symbols. Although we can use any eight distinct symbols, it is customary to use the first eight decimal digits. In other words, the digits of the octal number system are

0, 1, 2, 3, 4, 5, 6, 7

(There is no 8 or 9.) These digits, 0 through 7, have exactly the same physical meaning as decimal symbols; that is, 2 stands for ●●, 5 symbolizes ●●●●●, and so on.

How do you count beyond 7 with octal numbers? As with binary and decimal numbers, after running out of basic symbols you form two-digit combinations, taking the second digit followed by the first digit, then the second followed by the second, and so forth. Therefore, after 7 the next octal number is 10 (second digit followed by first), then 11 (second followed by second), and so on. You count as follows with octal numbers:

0, 1, 2, 3, 4, 5, 6, 7,
10, 11, 12, 13, 14, 15, 16, 17,
20, 21, 22, 23, 24, 25, 26, 27, . . .

To remember the octal numbers, think of the decimal numbers and cancel any number with a digit greater than 7. The numbers that remain are octal numbers. In other words, after thinking of the decimal numbers and canceling numbers containing 8 or 9,

0, 1, 2, 3, 4, 5, 6, 7, ~~8~~, ~~9~~, 10, 11, 12, 13, 14, 15, 16, 17, ~~18~~, ~~19~~,
20, 21, . . . , 75, 76, 77, ~~78~~, ~~79~~, ~~80~~, . . . , 100, 101, . . .

the numbers left over are the octal numbers.

How do we convert octal numbers to decimal numbers? In the octal number system each digit corresponds to a power of 8. The weights of the digit positions in an octal number are as follows:

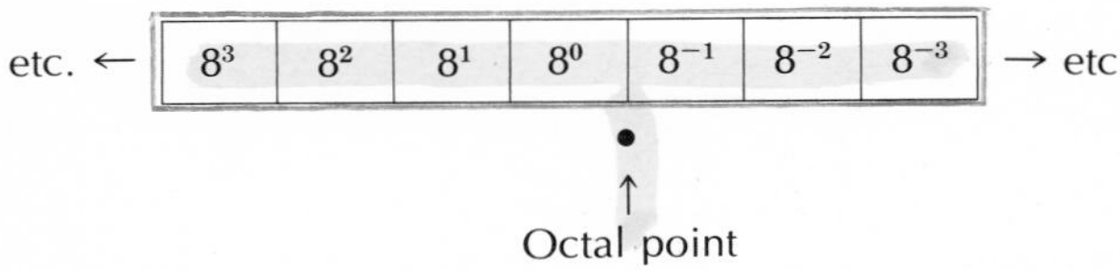

Therefore, to convert from octal to decimal, multiply each octal digit by its weight and add the resulting products. For instance, in powers of 8, octal number 23 becomes

$$2(8^1) + 3(8^0) = 16 + 3 = 19 \qquad \text{decimal}$$

What decimal number does 257_8 represent? Expressed in powers of 8, 257_8 becomes

$$2(8^2) + 5(8^1) + 7(8^0) = 128 + 40 + 7 = 175_{10}$$

How do you convert in the opposite direction, that is, from decimal to octal? A method like double-dabble is used with octal numbers. Instead of dividing by 2 (the base of binary numbers), divide by 8 (the base of octal numbers), writing down the remainders after each division. The remainders in reverse order form the octal number. As an example, we convert decimal 19 as follows:

$19 \div 8 = 2$ with a remainder of 3 ↑
$2 \div 8 = 0$ with a remainder of 2

In reverse order the remainders give 23, the octal equivalent of decimal 19. To convert 175_{10} to an octal number:

$175 \div 8 = 21$ with a remainder of 7 ↑
$21 \div 8 = 2$ with a remainder of 5
$2 \div 8 = 0$ with a remainder of 2

We can condense these steps by writing

$$
\begin{array}{r|l}
8 & 175 \\
\hline
 & 21 \quad 7 \\
 & 2 \quad 5 \\
 & 0 \quad 2
\end{array}
$$

With decimal fractions; multiply instead of divide, writing the carry into the integer position. An example of this is to convert decimal 0.23 into an octal fraction.

$$
\begin{aligned}
0.23 \times 8 &= 1.84 = 0.84 \text{ with a carry of } 1 \\
0.84 \times 8 &= 6.72 = 0.72 \text{ with a carry of } 6 \\
0.72 \times 8 &= 5.76 = 0.76 \text{ with a carry of } 5 \\
&\text{etc.}
\end{aligned}
$$

The carries taken in forward order give the octal fraction 0.165. We terminated after three places; if more accuracy were required, we would continue multiplying to obtain more octal digits.

2-9 OCTAL-BINARY CONVERSION

By far, the most important use of octal numbers lies in octal-binary conversions. The relation between octal digits and binary digits is obtained by counting to 7 in each system:

000	001	010	011	100	101	110	111
0	1	2	3	4	5	6	7

With this tabulation you can convert any octal number up to 7 into its binary equivalent.

How do you convert larger octal numbers? Because the base 8 of octal numbers is the third power of 2, the base of binary numbers, the conversion from octal to binary numbers becomes simple: convert one octal digit at a time to its binary equivalent. For instance, to change octal 23 to its binary equivalent, proceed as follows:

$$
\begin{array}{cc}
2 & 3 \\
\downarrow & \downarrow \\
010 & 011
\end{array}
$$

We converted each octal digit into its binary equivalent (2 becomes 010, and 3 becomes 011). The binary equivalent of octal 23 is 010 011, or 010011. Often, a space is left between groups of three bits; this makes it easier to read the binary number. In other words, we can leave 010011 in the form of 010 011.

As another example, convert octal 3574 to its binary equivalent.

$$\begin{array}{cccc} 3 & 5 & 7 & 4 \\ \downarrow & \downarrow & \downarrow & \downarrow \\ 011 & 101 & 111 & 100 \end{array}$$

Hence, binary 011101111100 is equivalent to octal 3574. The binary number is easier to read if we leave a space between groups of three bits.

Mixed octal numbers are no problem. Convert each octal digit to its equivalent binary value. Octal 34.562 becomes 011 100.101 110 010 in binary.

Converting from binary to octal is a reversal of the foregoing procedures. Simply remember to group the bits in threes, starting at the binary point; then convert each group of three to its octal equivalent (0s are added at each end, if necessary). For instance, binary number 1011.01101 converts as follows:

$$\begin{array}{lcccc} 1011.01101 \rightarrow & 001 & 011 & .011 & 010 \\ & \downarrow & \downarrow & \downarrow & \downarrow \\ & 1 & 3 & .\ 3 & 2 \end{array}$$

You start at the binary point and, working both ways, separate the bits into groups of three. When necessary, as in this case, add 0s to complete the outside groups. Then convert each group of three into its binary equivalent. Therefore;

$$1011.01101_2 = 13.32_8$$

The simplicity of converting octal to binary and vice versa has many advantages in digital work. For one thing, getting information into and out of a digital system requires less circuitry because it is easier to read, record, and print out octal numbers than binary numbers. Another advantage is that large decimal numbers are more easily converted to binary if we first convert to octal and then to binary; the reason is that a direct decimal-to-binary conversion requires many more divisions than a decimal-to-octal-binary conversion.

Example 2-13

What is the binary equivalent of decimal number 363? Convert to octal, then to binary.

Solution

$$\begin{array}{r|ll} 8 & 363 & \\ \hline & 45 & 3 \\ & 5 & 5 \\ & 0 & 5 \end{array}$$

Next, convert octal 553 to its binary equivalent, which is

101 101 011 (the binary equivalent of 363_{10})

We could have converted 363 directly to binary by progressively dividing by 2, but this would have been tedious, requiring 9 divisions. The decimal-octal-binary method speeds up the conversion process.

2-10 HEXADECIMAL NUMBERS

The *hexadecimal* number system has a base of 16. Although any 16 distinct symbols may be used, it's customary to use 0 through 9 and A through F as shown in Table 2-4. After reaching 9 in the hexadecimal system, you continue counting as follows:

A, B, C, D, E, F

How do we count beyond F with hexadecimal numbers? After running out of basic symbols, we form two-digit combinations, taking the second digit followed by the first digit, then the second followed by the second, and so on. So the next number after F in hexadecimal numbers, is 10. This is followed by

11, 12, 13, 14, 15, 16, 17, 18, 19, 1A, 1B, 1C, 1D, 1E, 1F, 20, 21,

and so forth.

Table 2-4
HEXADECIMAL DIGITS

Decimal	Binary	Hexadecimal
0	0000	0
1	0001	1
2	0010	2
3	0011	3
4	0100	4
5	0101	5
6	0110	6
7	0111	7
8	1000	8
9	1001	9
10	1010	A
11	1011	B
12	1100	C
13	1101	D
14	1110	E
15	1111	F

Mixing decimal digits and letters may seem queer at first. But like any code, the hexadecimal system is easy after some practice. You can actually become comfortable with numbers such as A1C, F3DA, and B27A3.

The main use of hexadecimal numbers lies in binary-to-hexadecimal conversion. Because 16 is the fourth power of 2, the conversion from hexadecimal to binary is simple. With Table 2-4, convert one hexadecimal digit at a time to its binary equivalent. For instance, here's how to convert A1C to its binary equivalent:

$$\begin{array}{ccc} A & 1 & C \\ \downarrow & \downarrow & \downarrow \\ 1010 & 0001 & 1100 \end{array}$$

As you see, each hexadecimal digit is converted to its binary equivalent (A becomes 1010, 1 becomes 0001, and C becomes 1100). For another example, B27A3 converts as follows:

$$\begin{array}{ccccc} B & 2 & 7 & A & 3 \\ \downarrow & \downarrow & \downarrow & \downarrow & \downarrow \\ 1011 & 0010 & 0111 & 1010 & 0011 \end{array}$$

Many computers, especially *minicomputers* (ones that cost less than $50,000), use binary numbers processed in groups of four bits each. In discussing these computers it's convenient to use the hexadecimal equivalent of the binary number. For instance, suppose binary number 10011101 is stored in the memory section of a minicomputer. The hexadecimal equivalent is

$$\begin{array}{cc} 1001 & 1101 \\ \downarrow & \downarrow \\ 9 & D \end{array}$$

To simplify conversation, we could say the memory stores 9D; this is much easier to say than 10011101.

STUDY AIDS

Summary

A number system relates quantities and symbols. The base or radix tells us how many digits the number system uses. The decimal system has a base of ten, binary has a base of two, and octal uses a base of eight.

In binary numbers $1 + 1 = 10$ because we are combining • with • to get ••. When necessary, a subscript can be used to indicate the base. Therefore, we can write

$$1_2 + 1_2 = 10_2$$

To convert from binary to decimal numbers, add the weights of each bit position (1, 2, 4, 8, . . .) when there is a 1 in that position. To convert from decimal to binary numbers, double-dabble is useful.

Octal numbers are important in octal-binary conversions. Octal numbers also expedite decimal-binary conversions.

Hexadecimal numbers have a base of 16. These numbers are useful when binary numbers are processed in groups of four bits.

Glossary

base The number of digits or basic symbols in a number system. The decimal system has a base of ten because it uses ten digits. Octal has a base of eight, binary a base of two.

binary Refers to a number system with a base of two, that is, with two basic symbols or digits.

bit An abbreviated form of *binary digit*. Instead of saying that 10110 has five binary digits, we can say that it has five bits.

digit A basic symbol used in a number system. The decimal system has ten digits, 0 through 9.

end-around carry Used with the 1's complement in subtracting. It means we take the last carry and add it to the least significant digit (Sec. 2-5).

hexadecimal Refers to a number system with a base of 16. The hexadecimal system has digits 0 through 9, followed by A through F.

minicomputer A computer costing less than $50,000.

octal This refers to a number system with a base of eight, that is, one that uses eight basic symbols or digits. Normally, these are 0, 1, 2, 3, 4, 5, 6, and 7.

radix A synonym for the word *base*.

weight Refers to the decimal value of each digit position of a number. For decimal numbers the weights are 1, 10, 100, 1000, . . . , working from the decimal point to the left. For binary numbers the weights are 1, 2, 4, 8, . . . to the left of the binary point. And with octal numbers, the weights become 1, 8, 64, . . . to the left of the octal point.

Review Questions

1. How many digits are there in the decimal number system? What are they?
2. Count to 7 using binary numbers.
3. What is the base or radix of a number system?
4. What does 110_2 mean? And 110_{10}?
5. What are the decimal weights of the binary positions to the left of the binary point? What are the weights to the right of the binary point?
6. Describe the double-dabble method for converting a decimal integer into a binary integer.
7. What is the condensed word for *binary digit?*
8. Define the 1's and 2's complements of a binary number.
9. In the 2's-complement method of subtraction, what is done with a carry in the

last position? What is done with a last carry when the 1's complement is used?

10. What does end-around carry mean?

11. What is the radix of the octal number system? What are the digits in this system? The octal number following 17 is what?

12. Why is the octal number system so useful in digital work?

13. Count up to 16 in hexadecimal numbers. What binary number does B3D stand for?

14. What is a minicomputer?

Problems

2-1. Write the first 27 binary numbers starting with 1.

2-2. A base-3 number system is to use the symbols *X*, *Y*, and *Z* instead of 0, 1, 2. Write the first 20 numbers in this system, starting with 1.

2-3. Add the following binary numbers:
(*a*) 1011 and 1001.
(*b*) 1110001 and 1010101.
(*c*) 1111010 and 1001101.

2-4. Add the following:
(*a*) 1111 and 1011.
(*b*) 111, 111, and 111.
(*c*) 1110010111 and 1100110011.

2-5. Convert the following decimal numbers to binary numbers: 24, 65, and 106.

2-6. What binary number does decimal 268 stand for?

2-7. Convert decimal 23.45 to a binary number.

2-8. What decimal number does 110101_2 represent?

2-9. What decimal number does 11001.011_2 represent?

2-10. Subtract binary 1101 from 11110.

2-11. Subtract the following binary numbers:
(*a*) 111 − 1010.
(*b*) 110011 − 100011.
(*c*) 100011 − 111010.

2-12. What are the 1's and 2's complements of the following binary numbers: 101010, 111101, 10001000, 00000010?

2-13. What four-bit number is equal to its 2's complement? Is there any four-bit number that is equal to its 1's complement?

2-14. In Fig. 2-2*b* the waveform can be interpreted as a binary number if we let +5 V stand for binary number 1, and 0 V for binary number 0. What is the 1's complement of the input waveform, and how is this complement related to the output waveform?

2-15. Use the 1's and 2's complements to perform the following binary subtractions:

(*a*) 1111 − 1011.
(*b*) 110011 − 100101.
(*c*) 100011 − 111010.

2-16. What binary number does 1100 × 1010 equal?

2-17. Multiply the following binary numbers:

(*a*) 1.11 × 101.
(*b*) 11.110 × 100.1.
(*c*) −11101 × 100.1

2-18. What does 101.111 × 1000.101 equal in binary?

2-19. Divide 11011_2 by 100_2.

2-20. Divide these binary numbers:

(*a*) 11001 by 110.
(*b*) 110.10 by 100.1.

2-21. Convert octal 65 to a decimal number.

2-22. What is the octal equivalent of decimal number 583?

2-23. Perform the following octal-decimal conversions:

(*a*) $654_{10} = x_8$.
(*b*) $x_{10} = 327_8$.

2-24. Convert the following octal numbers to binary numbers: 34, 567, 4673.

2-25. Convert the following binary numbers to octal numbers:

(*a*) 10101111.
(*b*) 1101.0110111.
(*c*) 1010011.101101.

2-26. The following decimal numbers are to be converted to binary numbers by converting first to octal, then to binary: 352, 850, 7563.

2-27. Convert the following hexadecimal numbers into binary numbers:

(*a*) E5.
(*b*) B4D.
(*c*) 7AF4.

2-28. Convert these binary numbers into hexadecimal numbers:

(*a*) 10001100.
(*b*) 00110111.
(*c*) 111101010110.

Binary Codes

Having used decimal numbers for many years, we would like to keep using them. Digital systems, however, force us to use binary numbers. Fortunately, we can compromise by using binary-coded decimals (BCD). These codes combine features of decimal and binary numbers. There are an enormous number of BCD codes. This chapter examines some common ones.

Your main objectives in this chapter are to be able to

1. Code decimal numbers in 8421 form, and decode 8421 numbers in decimal form.
2. Code and decode with excess-3 numbers.
3. Differentiate between even and odd parity.
4. Convert binary numbers to Gray form, and vice versa.

3-1 THE 8421 CODE

The 8421 code expresses each decimal digit by its four-bit binary equivalent. For instance, decimal number 429 is changed to its binary equivalent as follows:

$$\begin{array}{ccc} 4 & 2 & 9 \\ \downarrow & \downarrow & \downarrow \\ 0100 & 0010 & 1001 \end{array}$$

Therefore, in the 8421 code, 0100 0010 1001 stands for decimal number 429. As another example, let us encode 8963.

$$\begin{array}{cccc} 8 & 9 & 6 & 3 \\ \downarrow & \downarrow & \downarrow & \downarrow \\ 1000 & 1001 & 0110 & 0011 \end{array}$$

Again, we have changed each decimal digit to its binary equivalent.

Table 3-1 shows more of the 8421 code. As you can see, each decimal digit is changed to its equivalent four-bit group. Note that 1001 is the largest four-bit group in the 8421 code. In other words, only 10 of the 16 possible four-bit groups are

Table 3-1

Decimal	8421	Binary
0	0000	0000
1	0001	0001
2	0010	0010
3	0011	0011
4	0100	0100
5	0101	0101
6	0110	0110
7	0111	0111
8	1000	1000
9	1001	1001
10	0001 0000	1010
11	0001 0001	1011
12	0001 0010	1100
13	0001 0011	1101
. . .	. . . . . .	. . .
98	1001 1000	1100010
99	1001 1001	1100011
100	0001 0000 0000	1100100
101	0001 0000 0001	1100101
102	0001 0000 0010	1100110
. . .	. . . .	. . .
578	0101 0111 1000	1001000010
. . .	. . . .	. . . . . .

used. The 8421 code does not use the numbers 1010, 1011, 1100, 1101, 1110, and 1111. (If any of these forbidden numbers appeared in a machine using the 8421 code, you would know an error had occurred.)

The 8421 code is identical to binary through decimal number 9. Because of this, it is called the 8421 code; the weights in the groups are 8, 4, 2, 1, reading from left to right—the same as for binary numbers.

Above 9, the 8421 code differs from the binary-number code. For instance, the binary number for 12 is 1100, but the 8421 number for 12 is 0001 0010. Or, decimal number 24 is 11000 in binary, but it becomes 0010 0100 in the 8421 code. Therefore, above 9 every binary number differs from the corresponding 8421 number.

The main advantage of the 8421 code is the ease of converting to and from decimal numbers; we need only remember the binary numbers for 0 through 9 because we encode only one decimal digit at a time. A disadvantage, however, of the 8421 code is that the rules for binary addition do not apply to the entire 8421 number, but only to the individual four-bit groups. For instance, to add 12 and 9 in straight binary is easy.

$$
\begin{array}{r} 12 \\ +\ 9 \\ \hline 21 \end{array} \qquad \begin{array}{r} 1100 \\ +1001 \\ \hline 10101 \end{array} \rightarrow 21
$$

If we try it in the 8421 code, we get an unacceptable answer.

```
  12      0001 0010
+  9    +      1001
----    -----------
  21      0001 1011
```

We are unable to decode 0001 1011 because 1011 does not exist in the 8421 code. Remember the largest 8421 code group is 1001 (9). Therefore, addition of 8421 numbers is not so simple as for binary numbers (discussed in Chap. 5).

The 8421 code is one of the many codes referred to as binary-coded decimals (BCD). An enormous number of such codes exist.[1] In general, a BCD code is one in which the digits of a decimal number are encoded—one at a time—into groups of binary digits. For this encoding we can use four-bit groups, or five-bit groups, or six-bit, etc.

The 8421 code is a mixed-base code; it is binary within each group of four bits, but it is decimal from group to group.

Because the 8421 BCD code is the most natural type of BCD code, it is often referred to as BCD without qualifying it. In other words, if we refer to the *BCD code,* we mean the 8421 code.

Example 3-1

Encode the following decimal numbers into 8421 numbers:

(*a*) 45
(*b*) 732
(*c*) 94,685

Solution

(*a*)

```
  4     5
  ↓     ↓
0100  0101
```

(*b*)

```
  7     3     2
  ↓     ↓     ↓
0111  0011  0010
```

(c)

```
  9     4     6     8     5
  ↓     ↓     ↓     ↓     ↓
1001  0100  0110  1000  0101
```

Example 3-2

Decode the following 8421 numbers:

(*a*) 1000 0101 0110 0011
(*b*) 0011 0101 0001 1001 0111

[1] Four-bit BCD codes use only 10 of 16 possible states; there are 16!/6! possible codes, approximately 30 billion codes.

Solution

(*a*)

1000	0101	0110	0011
↓	↓	↓	↓
8	5	6	3

(*b*)

0011	0101	0001	1001	0111
↓	↓	↓	↓	↓
3	5	1	9	7

3-2 THE EXCESS-3 CODE

The excess-3 code is another important BCD code. To encode a decimal number into its excess-3 form, add 3 to each decimal digit before converting to binary. For example, convert 12 to an excess-3 number as follows:

$$\begin{array}{cc} 1 & 2 \\ \underline{+3} & \underline{+3} \\ 4 & 5 \end{array}$$

We have added 3 to each decimal digit. Now convert each digit to its binary equivalent.

$$\begin{array}{cc} 4 & 5 \\ \downarrow & \downarrow \\ 0100 & 0101 \end{array}$$

So, 0100 0101 in the excess-3 code stands for decimal 12.

Take another example; convert 29 to an excess-3 number:

$$\begin{array}{cc} 2 & 9 \\ \underline{+3} & \underline{+\ 3} \\ 5 & 12 \\ \downarrow & \downarrow \\ 0101 & 1100 \end{array}$$

After adding 9 and 3, do *not* carry the 1 into the next column; instead leave the result intact as 12; then convert as shown. Therefore, 0101 1100 in the excess-3 code stands for decimal 29.

Table 3-2 shows the excess-3 code. 3 is added to each decimal digit before converting each group into binary.

The excess-3 code uses only 10 of the 16 possible four-bit groups. The six unused groups are 0000, 0001, 0010, 1101, 1110, and 1111. If any of these forbidden combinations turned up in an excess-3 computer, we would know an error had occurred.

The excess-3 code is a *self-complementing* code. This means the 1's complement of any excess-3 number represents the 9's complement of the decimal number. For

Table 3-2

Decimal	Excess-3
0	0011
1	0100
2	0101
3	0110
4	0111
5	1000
6	1001
7	1010
8	1011
9	1100
10	0100 0011
11	0100 0100
12	0100 0101
13	0100 0110
. . .	. . . . . .
98	1100 1011
99	1100 1100
100	0100 0011 0011
101	0100 0011 0100
102	0100 0011 0101
. . .	. . . .
578	1000 1010 1011
. . .	. . . .

instance, 0101 in excess-3 code stands for decimal 2. The 1's complement of 0101 is 1010, which stands for decimal 7. But decimal 7 is the 9's complement of 2. A check on any other excess-3 number gives the same result—the 1's complement represents the 9's complement of the encoded decimal number. (Not all BCD codes are self-complementing. The 8421 code, for instance, does not have this property.)

The excess-3 code is an unweighted code, whereas the 8421 code is weighted. In a weighted code each bit position has a fixed value. In the 8421 code the number 0101 has weights of 8, 4, 2, and 1, reading from left to right. Whenever a 1 appears, we add the weight of its position. For instance,

$$\begin{array}{l} 0 \quad 1 \quad 0 \quad 1 \\ \not{8} + 4 + \not{2} + 1 = 5 \end{array}$$

Similarly, in the 8421 code the number 1001 has a decimal equivalent of

$$\begin{array}{l} 1 \quad 0 \quad 0 \quad 1 \\ 8 + \not{4} + \not{2} + 1 = 9 \end{array}$$

The excess-3 code, however, is unweighted; there is no way to assign fixed weights or values to the bit positions.

The meaning of the term *excess-3* is clear at this point: each four-bit group is the binary equivalent of a decimal digit that is 3 larger than the encoded digit. In a similar way, an excess-6 number means each four-bit group is the binary equivalent of a decimal number that is 6 larger than the encoded digit. For instance, decimal number 2 would be 0101 (5) in excess-3 code, but would be 1000 (8) in excess-6 code.

As pointed out in the preceding section, an obstacle arises when we try to add 8421 numbers whose decimal sum exceeds 9. The excess-3 code was devised to get rid of this obstacle. To understand why, we must discuss two cases.

Case 1 In excess-3 code whenever we add two decimal digits whose sum is 9 or less, an excess-6 number results. To return to excess-3 form, we must subtract 3. For instance, for 2 + 5

2	0101	excess-3 equivalent of 2
+5	+1000	excess-3 equivalent of 5
7	1101	excess-6 equivalent of 7
	−0011	subtract 3
	1010	excess-3 equivalent of 7

We added two excess-3 numbers and got an excess-6 number. To restore the answer to excess-3 form, we subtracted 3. The final answer is 1010, the excess-3 equivalent of 7.

As another example, add 43 and 36.

43	0111 0110	excess-3 for 43
+36	+0110 1001	excess-3 for 36
79	1101 1111	excess-6 for 79
	−0011 0011	subtract 3 from each group
	1010 1100	excess-3 for 79

There was no carry in either group, so the decimal sum was 9 or less for each group. As a result, the answer is in excess-6 form and we must subtract 3 from each group to return to excess-3 code.

Case 2 Whenever the sum of decimal digits exceeds 9, there will be a carry from one group into the next. When this happens, the group that produced the carry will revert to 8421 form; this occurs because of the excess-6 and the six unused four-bit groups. To restore the answer to excess-3 code, we must add 3 to the group that produced the carry. For instance, to add 29 and 39,

	A	*B*	
29	0101	1100	excess-3 for 29
+39	+0110	1100	excess-3 for 39
68	1100	1000	first results
	−0011	+0011	subtract and add 3
	1001	1011	excess-3 for 68

Here is what happens. In column *B* we add 1100 and 1100 to get 1000 *with a* carry of 1 into column *A*. In column *A* we add 0101 and 0110 and the carry to get 1100 *with no carry*. The first result in column *B* is back in 8421 form because this column produced a carry. The first result in column *A* is still in excess-6 form because this column does not produce a carry. Therefore, we must subtract 3 from column *A*, and add 3 to column *B*. The final answer is 1001 1011, the excess-3 number for 68.

To summarize addition with excess-3 numbers:

> Add the numbers using the rules for binary addition.
> If any group produces a decimal carry, add 0011 to that group.
> If any group does not produce a decimal carry, subtract 0011 from that group.

The excess-3 code has the advantage that all operations in addition use ordinary binary addition. (The 8421 code is not like this. It requires special operations to handle decimal carries.) Because the excess-3 code is self-complementing, it also has the advantage that the 1's or 2's complements can be used to subtract excess-3 numbers. (This is not true for 8421 numbers.)

Example 3-3

What are the excess-3 numbers for the following decimal numbers:

(*a*) 35
(*b*) 569
(*c*) 2468

Solution

(*a*)

3	5
+3	+3
6	8
↓	↓
0110	1000

Therefore, 0110 1000 is the excess-3 equivalent of decimal 35.

(*b*)

5	6	9
+3	+3	+3
8	9	12
↓	↓	↓
1000	1001	1100

So, decimal 569 becomes 1000 1001 1100 in excess-3 code.

(*c*)

2	4	6	8
+3	+3	+3	+3
5	7	9	11
↓	↓	↓	↓
0101	0111	1001	1011

The excess-3 number 0101 0111 1001 1011 stands for 2468_{10}.

Example 3-4

Decode the following excess-3 numbers:

(*a*) 1100 0110

(*b*) 1010 1011 0011 1001

Solution

(*a*)

1100	0110
↓	↓
12	6
−3	−3
9	3

Decimal 93 is the decoded value of 1100 0110 (excess-3).

(*b*)

1010	1011	0011	1001
↓	↓	↓	↓
10	11	3	9
−3	−3	−3	−3
7	8	0	6

Decimal 7806 is the decoded value of the excess-3 number 1010 1011 0011 1001.

Example 3-5

Add 567 and 295 using excess-3 numbers.

Solution

567	1000	1001	1010
+295	+0101	1100	1000
862	1110	10110	10010
	−0011	+0011	+0011
	1011	1001	0101

The two least significant groups produced carries; therefore, add 3 to each group. The most significant group did not produce a carry; therefore, subtract 3 from it. The final answer is 1011 1001 0101, the excess-3 equivalent of 862.

3-3 OTHER FOUR-BIT BCD CODES

Many other four-bit codes exist. Tables 3-3 and 3-4 show some common ones. As usual, decimal numbers greater than 9 are encoded a digit at a time. For instance, decimal number 16 becomes the following in 2421 code:

1	6	decimal
↓	↓	↓
0001	1100	2421 code

Table 3-3

FOUR-BIT BCD CODES

Decimal	7421	6311	5421	5311	5211
0	0000	0000	0000	0000	0000
1	0001	0001	0001	0001	0001
2	0010	0011	0010	0011	0011
3	0011	0100	0011	0100	0101
4	0100	0101	0100	0101	0111
5	0101	0111	1000	1000	1000
6	0110	1000	1001	1001	1001
7	1000	1001	1010	1011	1011
8	1001	1011	1011	1100	1101
9	1010	1100	1100	1101	1111

Or, decimal 75 in 5421 code is

$$\begin{array}{cccc} 7 & 5 & & \text{decimal} \\ \downarrow & \downarrow & & \downarrow \\ 1010 & 1000 & & \text{5421 code} \end{array}$$

Or, decimal 693 in 6311 code is

$$\begin{array}{ccccc} 6 & 9 & 3 & & \text{decimal} \\ \downarrow & \downarrow & \downarrow & & \downarrow \\ 1000 & 1100 & 0100 & & \text{6311 code} \end{array}$$

All the codes in the tables are weighted codes.[2] All use positive weights except the last two codes of Table 3-4. These two codes use negative weights, as well as

[2] Sometimes a given set of weights may generate a number of different codes. Tables 3-3 and 3-4 show the most popular code forms.

Table 3-4

FOUR-BIT BCD CODES

Decimal	4221	3321	2421	$84\bar{2}\bar{1}$	$74\bar{2}\bar{1}$
0	0000	0000	0000	0000	0000
1	0001	0001	0001	0111	0111
2	0010	0010	0010	0110	0110
3	0011	0011	0011	0101	0101
4	1000	0101	0100	0100	0100
5	0111	1010	1011	1011	1010
6	1100	1100	1100	1010	1001
7	1101	1101	1101	1001	1000
8	1110	1110	1110	1000	1111
9	1111	1111	1111	1111	1110

positive. For instance, in the $84\bar{2}\bar{1}$ code the least significant digit has a weight of -1, and the next position has a weight of -2. Therefore, an $84\bar{2}\bar{1}$ number like 1011 is decoded as follows:

$$\begin{array}{cccc} 1 & 0 & 1 & 1 \\ 8 + & \not{4} - & 2 - & 1 = 5 \end{array}$$

The $84\bar{2}\bar{1}$ code can be derived from the excess-3 code. By complementing the 0s and 1s in the last two significant positions of the excess-3 code, we get the $84\bar{2}\bar{1}$ code. For instance, 5 in excess-3 code is 1000. If we complement the two least significant digits, we obtain 1011, the $84\bar{2}\bar{1}$ number for 5. As another example, 1100 is the excess-3 equivalent of 9. Complementing the two least significant digits, we get 1111, the $84\bar{2}\bar{1}$ number for 9.

The 2421 code uses the first five and the last five four-bit binary numbers (0000, 0001, 0010, 0011, 0100 and 1011, 1100, 1101, 1110, 1111). Electronic counters occasionally use this code.

Recall that the excess-3 code is a self-complementing code; that is, the 1's complement of any excess-3 number stands for the 9's complement of the encoded decimal number. By inspection of Tables 3-3 and 3-4, the following codes also are self-complementing: the 4221 code, the 3321, the 2421, and the $84\bar{2}\bar{1}$.

Tables 3-3 and 3-4 will be useful for later reference.

Example 3-6

Encode the following:

(*a*) Decimal number 842 into 4221 form

(*b*) Decimal 951 into 6311 form

Solution

(*a*) Table 3-4 gives the 4221 code. So,

$$\begin{array}{cccc} 8 & 4 & 2 & \text{decimal} \\ \downarrow & \downarrow & \downarrow & \downarrow \\ 1110 & 1000 & 0010 & 4221 \text{ code} \end{array}$$

(*b*) Table 3-3 shows the 6311 code.

$$\begin{array}{cccc} 9 & 5 & 1 & \text{decimal} \\ \downarrow & \downarrow & \downarrow & \downarrow \\ 1100 & 0111 & 0001 & 6311 \text{ code} \end{array}$$

Example 3-7

Decode the following BCD numbers:

(*a*) 0100 1011 1101 in 5311 code

(*b*) 1100 0101 1111 0001 in 3321 code

Solution

(*a*) Using Table 3-3, decode the 5311 number as follows:

0100	1011	1101	5311 code
↓	↓	↓	↓
3	7	9	decimal

(*b*) Using Table 3-4, decode the 3321 number as follows:

1100	0101	1111	0001	3321 code
↓	↓	↓	↓	↓
6	4	9	1	decimal

3-4 THE PARITY BIT

In a digital system a *word* is a group of bits that is treated, stored, and moved around as a unit. For instance, suppose an 8421 computer is about to add 0101 1000 0011 and 0010 0100 0110. (This would be 583 and 246.) Each of these BCD numbers is a word. The computer brings each of these words out of memory and puts them into the arithmetic unit. The sum is a new word which is then put back into memory.

While words are being moved and stored, errors can get into the words. For instance, one of the 0s in a word might accidentally be changed to a 1 by an intermittent failure, or by noise, or by a transient, and so on. Under normal operating conditions such a change is unlikely, but any error at all could be disastrous. Therefore, we need methods for detecting errors that crop up while a word is being moved or stored.

In all the BCD codes discussed so far, we have used only 10 of the 16 possible

Table 3-5
EVEN PARITY

8421 code	Added bit
0000	0
0001	1
0010	1
0011	0
0100	1
0101	0
0110	0
0111	1
1000	1
1001	0

Table 3-6
ODD PARITY

8421 code	Added bit
0000	1
0001	0
0010	0
0011	1
0100	0
0101	1
0110	1
0111	0
1000	0
1001	1

four-bit combinations. Therefore, one way to find errors is to look for forbidden combinations. If any of these forbidden combinations appear in memory, we immediately know an error has occurred.

Checking for forbidden combinations is possible, but it does not give a high enough probability for catching errors. The most widely used approach for detecting errors that arise in storing and moving words is to attach a parity bit to the word.

Even parity means attaching an extra bit to a group of bits to produce an even number of 1s. For instance, suppose we have a word like 0111. There are three 1s in this word, an odd number of 1s. We attach an extra 1 to the word to get 0111 1. Now we have an even number of 1s. This new word can be moved and stored by the computer, and can be checked for even parity at different points to assure that no errors have crept into the word.

As another illustration of even parity, Table 3-5 shows the 8421 code with an even-parity bit. The parity bit produces an even number of 1s for each code group.

There is also *odd parity*. In this case, the added parity bit makes the number of 1s odd. Table 3-6 shows the 8421 code with an odd-parity bit. Note that the odd-parity bit is always the complement of the even-parity bit.

Both types of parity are commonly used; there is no strong argument in favor of either type.

Using a parity bit to detect errors rests upon two assumptions that apply to most digital systems:

> The probability of errors is very small.
>
> In the few cases where an error does get into a word, the error is most likely to be a one-bit error. The chances of two or more bits changing accidentally are very remote unless there is a complete failure, readily detected by other means.[3]

In other words, a parity check will catch all one-bit errors but not double errors. In most digital systems the chances for a double error are extremely small.

Parity checks are especially common in storage devices such as magnetic tapes, drums, cores, and paper tapes. Magnetic tape is an inexpensive way to store large amounts of digital information, but it is more susceptible to error than other techniques. The chances for a double error in magnetic tape are higher, so that a single-parity check is not so reliable as in other parts of a digital system.

To improve matters, we can use a *double-parity check* with magnetic tape. Figure 3-1 illustrates a double-parity check, using odd parity. The information is put on the tape in what is called a "field" or "block." We have shown a block with seven horizontal rows and seven vertical columns. The black dots represent magnetized points, and the light dots stand for unmagnetized points. (Or, black could be one polarity and light the other.) To simplify matters, suppose binary numbers have

[3] For example, suppose there are five holes in the ground. The chances of a meteor falling into one of the holes are small. But the chances of two meteors falling into two holes are far smaller.

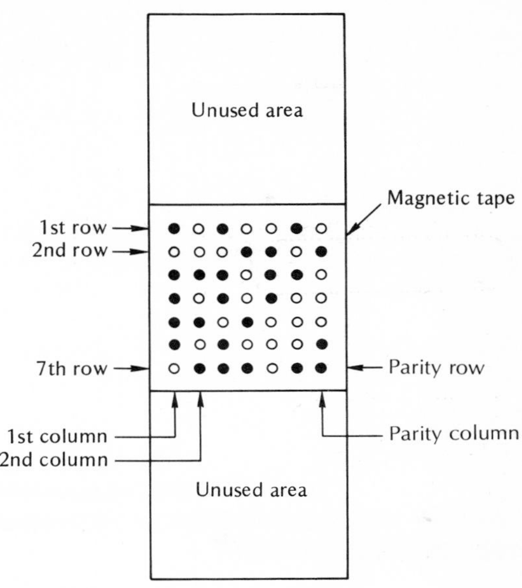

Fig. 3-1. Double-parity check of magnetic tape.

been encoded into the first six bits of each row, and the seventh bit in each row is a horizontal parity bit. In the first row we read 101001 with a parity bit of 0. In the second row we read 000110 with a parity bit of 1. Each row contains an odd-parity bit, so that when the tape is read electronically, an odd-parity check can be made.

The entire seventh row contains odd-parity bits, one for each column. The first vertical column is 101111 with a parity bit of 0 (reading from the top down.) The second column reads 001010 with a parity bit of 1. In this way, every column has an odd-parity bit. The advantage of this double parity is that double errors can be electronically detected during the reading of the tape. For instance, suppose the first row (101001 0) is somehow incorrectly read as 011001 0 (the first two bits have been complemented). A double error has occurred, so horizontal odd parity still checks. However, the vertical odd parity will not check because the first and second columns will then have an even number of 1s.

Later chapters discuss circuits for checking even and odd parity. As already indicated, checking parity is the most widespread method of error detection.

Example 3-8

Encode the following numbers into 8421 code, and attach an odd-parity bit to the end of each word:

(*a*) 592

(*b*) 8307

Solution

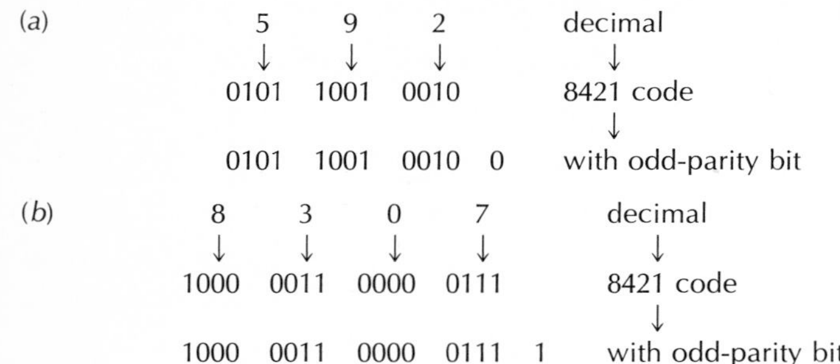

(a)

5	9	2		decimal
↓	↓	↓		↓
0101	1001	0010		8421 code
				↓
0101	1001	0010	0	with odd-parity bit

(b)

8	3	0	7		decimal
↓	↓	↓	↓		↓
1000	0011	0000	0111		8421 code
					↓
1000	0011	0000	0111	1	with odd-parity bit

3-5 FIVE-BIT CODES

Five-bit codes also exist. Although only four bits are needed to encode any decimal digit from 0 to 9, an extra bit will allow us to decode the number more easily, as well as to detect errors more readily. Table 3-7 shows some of these five-bit codes.

As usual, any decimal number greater than 9 is encoded a digit at a time. For instance, to encode decimal 847 into the shift-counter code,

8	4	7	decimal
↓	↓	↓	↓
11000	01111	11100	shift-counter code

The *2-out-of-5 code* is an unweighted code that has been used in telephone and communication work. Not only does it have even parity, it also has exactly two 1s in each code group. Because of this, errors can be detected more reliably than with just an even-parity check.

The *shift-counter code* (sometimes called the "Johnson code") is an unweighted

Table 3-7

FIVE-BIT BCD CODES

Decimal	2-out-of-5	63210	Shift-counter	86421	51111
0	00011	00110	00000	00000	00000
1	00101	00011	00001	00001	00001
2	00110	00101	00011	00010	00011
3	01001	01001	00111	00011	00111
4	01010	01010	01111	00100	01111
5	01100	01100	11111	00101	10000
6	10001	10001	11110	01000	11000
7	10010	10010	11100	01001	11100
8	10100	10100	11000	10000	11110
9	11000	11000	10000	10001	11111

for "noisy" installations

Table 3-8

MORE-THAN-FIVE-BIT CODES

Decimal	50 43210	543210	9876543210
0	01 00001	000001	0000000001
1	01 00010	000010	0000000010
2	01 00100	000100	0000000100
3	01 01000	001000	0000001000
4	01 10000	010000	0000010000
5	10 00001	100001	0000100000
6	10 00010	100010	0001000000
7	10 00100	100100	0010000000
8	10 01000	101000	0100000000
9	10 10000	110000	1000000000

code used in electronic counters. It has the advantage of being easy to decode electronically (see Chap. 9).

The 51111 code is weighted and self-complementing. It is similar to the shift-counter code, so that it is easily decoded with electronic circuitry.

The 63210 code is weighted, except for the decimal value of 0. Above decimal number 0, the weights are 6, 3, 2, 1, 0, reading from left to right. It has exactly two 1s in each code group, allowing reliable error detection. This code has been used for storage of digital data on magnetic drums.

Example 3-9

Encode 529_{10} into 2-out-of-5 code.

Solution

5	2	9	decimal
↓	↓	↓	↓
01100	00110	11000	2-out-of-5-code

3-6 CODES WITH MORE THAN FIVE BITS

We can decode and detect errors more easily by adding more bits. Table 3-8 shows some BCD codes that use more than five bits. These are all weighted codes.

The 9876543210 code, sometimes called the *ring-counter code,* uses only a single 1 in each code group; this makes it easy to decode and to detect errors. The code has the disadvantage of requiring more electronic circuitry than the simpler 4- and 5-bit codes.

The 50 43210 code, also called the "biquinary code," is occasionally used in electronic counters.[4] (Note that biquinary means two-five.) Each member of this

[4] The abacus uses the biquinary code. Its beads are arranged in twos and fives.

code contains a group of two bits and a group of five bits. The group of two bits indicates whether the number is more or less than 5. The group of five bits denotes the count. Reliable error detection is possible because a single 1 is in the group of two bits, and a single 1 is in the group of five bits.

The 543210 code is like the biquinary code, except that six bits are used instead of seven. The most significant digit (far left) tells us if the number is less than 5 or not. This code is sometimes used in electronic counters.

3-7 THE GRAY CODE

The *Gray code* is an unweighted code not suited to arithmetic operations, but useful for input-output devices, analog-to-digital converters, and other peripheral equipment.

Table 3-9 shows the Gray code, along with the corresponding binary numbers. Each Gray number differs from the preceding number by a single bit. For instance, in going from decimal 7 to 8, the Gray-code numbers change from 0100 to 1100; these numbers differ only in the most significant bit. As another example, decimal numbers 13 and 14 are represented by Gray-code numbers 1011 and 1001; these numbers differ in only one digit position (the second position from the right). So it is with the entire Gray code—every number differs by only one bit from the preceding number.

Sometimes, we have to convert binary numbers into Gray-code numbers and vice versa. Here is how to convert from *binary to Gray:*

The first Gray digit is the same as the first binary digit.
Add each pair of adjacent bits to get the next Gray digit. *Disregard* any carries.[5]

An example is the best way to describe the conversion from binary to Gray. Take binary number 1100. Here is how to find the corresponding Gray-code number:

Step 1 The first Gray digit is the same as the first binary digit. So, repeat the first digit.

```
1 1 0 0     binary
↓
1           Gray
```

Step 2 Now, add the first two bits of the binary number, disregarding any carries. The sum is the next Gray digit.

```
 ⌒
1 1 0 0     binary
  ↓
1 0         Gray
```

[5] This is formally called mod-2 addition, or exclusive-OR addition. The four rules for this kind of addition are: $0 + 0 = 0$, $0 + 1 = 1$, $1 + 0 = 1$, $1 + 1 = 0$.

Table 3-9
GRAY CODE UP TO 15_{10}

Decimal	Gray code	Binary
0	0000	0000
1	0001	0001
2	0011	0010
3	0010	0011
4	0110	0100
5	0111	0101
6	0101	0110
7	0100	0111
8	1100	1000
9	1101	1001
10	1111	1010
11	1110	1011
12	1010	1100
13	1011	1101
14	1001	1110
15	1000	1111
. . .	. . . . . .	. . .

In other words, add the first two bits of the binary number to get 1 + 1 = 0 carry a 1. Write down the 0, but discard the 1.

Step 3 Add the next two binary digits to get the next Gray digit.

```
1 1 0 0    binary
    ↓
1 0 1      Gray
```

Step 4 Add the last two binary digits to get the Gray digit.

```
1 1 0 0    binary
      ↓
1 0 1 0    Gray
```

Therefore, 1010 is the Gray-code equivalent of binary number 1100.

Take another example. To convert binary 110100110 to Gray code, proceed as follows:

Step 1 Repeat the most significant digit.

```
1 1 0 1 0 0 1 1 0    binary
↓
1                    Gray
```

Step 2 Add the first two binary digits to get the next Gray digit. (Disregard all carries.)

```
1 1 0 1 0 0 1 1 0     binary
  ↓
1 0                   Gray
```

Step 3 Continue adding each pair of adjacent bits to get

```
1 1 0 1 0 0 1 1 0     binary
    ↓ ↓
1 0 1 1 1 0 1 0 1     Gray
```

To convert *from Gray code back to binary,* we use a method that is similar, but not exactly the same. Again, an example best describes the method. Let us convert Gray-code number 101110101 back to its binary equivalent.

Step 1 Repeat the most significant digit.

```
1 0 1 1 1 0 1 0 1     Gray
↓
1                     binary
```

Step 2 Add *diagonally* as shown to get the next binary digit.

```
1 0 1 1 1 0 1 0 1     Gray
 ↗↓
1 1                   binary
```

(1 + 0 = 1.)

Step 3 Continue adding *diagonally* to get the remaining binary digits.

```
1 0 1 1 1 0 1 0 1     Gray
   ↗↓
1 1 0 1 0 0 1 1 0     binary
```

By these methods, you can convert Gray to binary and vice versa whenever the need arises.

Example 3-10

(*a*) Convert binary 1000110111 to its Gray-code equivalent.
(*b*) Convert Gray number 11100100011 to its binary equivalent.

Solution

(a) To go from binary to Gray, add pairs of adjacent bits as follows:

```
1 0 0 0 1 1 0 1 1 1    binary
↓
1 1 0 0 1 0 1 1 0 0    Gray
```

(*b*) To convert from Gray to binary, add diagonally as shown.

```
1 1 1 0 0 1 0 0 0 1 1    Gray
↗↓
1 0 1 1 1 0 0 0 0 1 0    binary
```

3-8 ALPHANUMERIC DISPLAYS

Most people don't know how to decode binary or BCD numbers, so it is essential to decode such numbers electronically and display the decimal equivalent. Furthermore, in many applications it is necessary to display letters of the alphabet. This section is about devices that display *alphanumerics* (alphabet letters, numbers, and other symbols).

The three common ways to display alphanumerics are the *discrete* method, the *bar-matrix* method, and the *dot-matrix* method. In the discrete method a single light source produces each symbol. For instance, in the well-known nixie tube there are one anode and several cathodes, each shaped like the character to be displayed. The characters are transparent until activated. As an example, Fig. 3-2 shows a typical nixie circuit. Grounding a particular cathode causes the neon gas around that cathode to ionize. Because of the resulting glow, a particular digit is displayed.

Another example of the discrete format uses incandescent lamps, one lamp for each symbol. When a lamp is lit, it projects the particular symbol on a screen.

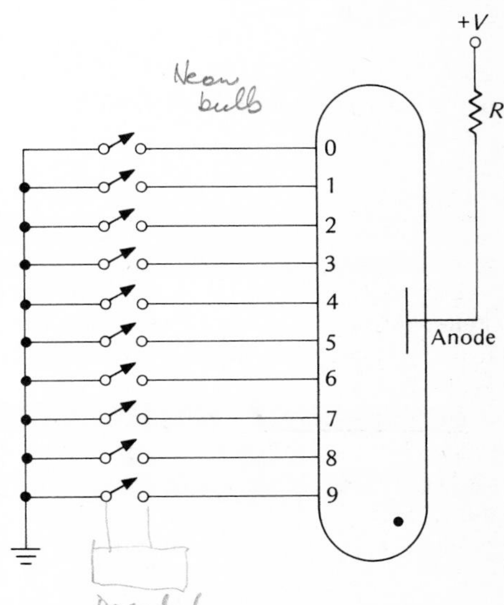

Fig. 3-2. Nixie circuit.

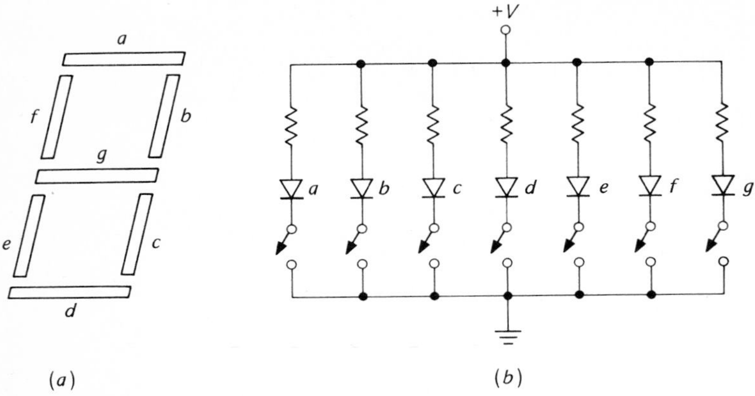

Fig. 3-3. (*a*) Seven-segment indicator. (*b*) Readout circuit.

The main thing to remember about the discrete method is this: A single light source is involved in the display of each symbol. As a result, it's necessary to activate only one pin on the display device to produce each symbol.

Bar-matrix devices are different. In this method one or more light sources may be involved in the display of a particular symbol. By far, the best known bar-matrix device is the seven-segment readout shown in Fig. 3-3*a*. To display a symbol, you have to light up the bars or segments associated with the symbol. For instance, to display a 0, you have to light up segments *a* through *f*; to get a 1, segments *b* and *c* can be lit; in a similar way, any decimal digit from 0 through 9 can be displayed.

Light-emitting diodes (LEDs) are the most common light sources in a seven-segment readout. Typically, a common supply voltage drives the anodes as shown in Fig. 3-3*b*. When a switch closes, the corresponding LED is forward biased and emits light. The series resistor is a current-limiting resistor needed to set up the current specified on the LED data sheet. (In smaller devices a single LED is adequate for each segment, but in larger devices two or more LEDs are used for each segment.)

Besides LEDs, seven-segment displays use light sources such as *liquid crystals* (materials that are transparent until a voltage is applied to them) and *fluorescent phosphors* (they glow when bombarded by electrons).

The main thing to remember about bar-matrix devices is this: One or more light sources shaped like bars or segments are involved in the display of a symbol. As a result, it is necessary to activate one or more pins on the display device to produce each symbol.

Dot-matrix devices have many individual light sources shaped like dots. A typical example is the 5 × 7 LED matrix shown in Fig. 3-4. To light up LED in this matrix, you have to apply a voltage to its anode and a ground to its cathode. In Fig. 3-4 the circled LED is lit because a voltage is applied to the fifth vertical column, and a ground is applied to the fourth horizontal row. By applying voltage to more than one column and grounding more than one row, we can display any decimal digit, any alphabet letter, and many other symbols (described in Sec. 3-11).

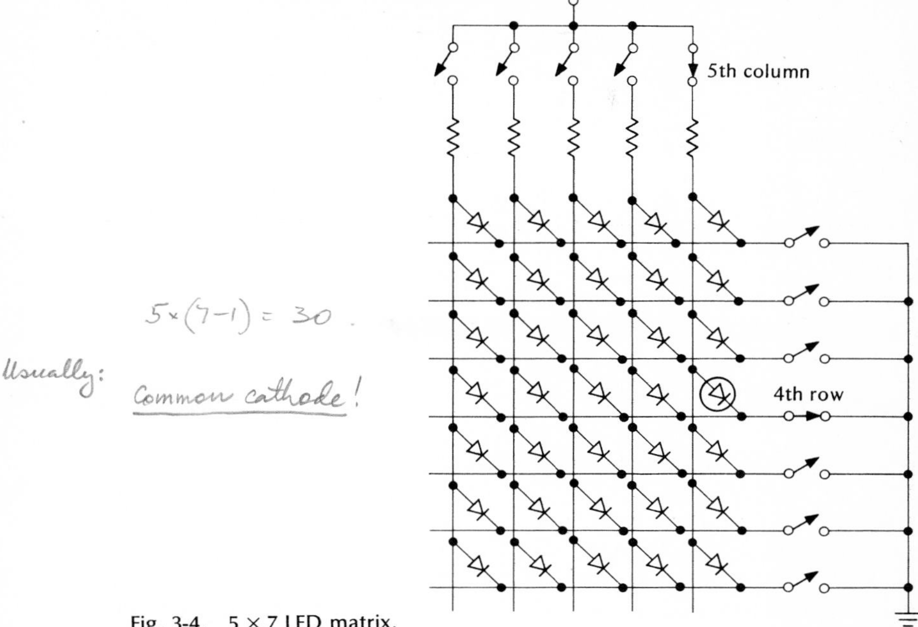

Fig. 3-4. 5 × 7 LED matrix.

3-9 DECIMAL DECODERS

The 7400 series, a line of TTL circuits introduced in 1964, has become the most widely used group of bipolar ICs. Because of this, our discussions usually include examples showing how 7400-series devices can be used. In this section, we shall look at 7400-series decimal decoders.

Before we can display a decimal digit or alphabet letter, we have to decode the

Fig. 3-5. 74141 decoder/driver.

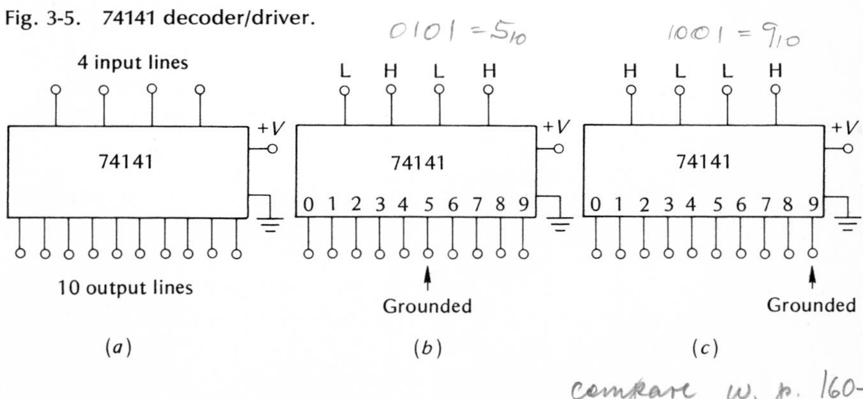

binary information into suitable voltages for driving the display device. For instance, a 74141 is a BCD-to-decimal decoder used for driving a discrete-display device like a nixie tube. In Fig. 3-5*a* the supply voltage *V* is applied to one of the pins on the right, while the other pin is grounded. When a BCD digit drives the four input lines, one of the 10 output lines will be internally grounded.

If a low voltage (L) represents a 0 and a high voltage (H) stands for a 1, then a BCD input like LHLH represents 0101. Similarly, a BCD input like HLLH stands for 1001. When inputs like these drive a 74141, the appropriate output line is grounded. For instance, if the BCD input is LHLH (see Fig. 3-5*b*), the fifth output line is grounded. Or, given a BCD input like HLLH (Fig. 3-5*c*), the 74141 automatically grounds the ninth output line. (Incidentally, the 74141 is a second-generation device that replaces the 7441, a widely used first-generation device.)

A 74141 decoder is needed for each BCD digit. For example, Fig. 3-6 shows three 74141s decoding

LLHH HLLL LHHL

which is equivalent to

0011 1000 0110

Since each 74141 acts like a BCD-to-decimal decoder, the *hundreds* decoder grounds its third output line, the *tens* decoder grounds its eighth output line, and the *units* decoder grounds its sixth output line.

To display a decimal digit on a discrete-display device, the ground on the output line of 74141 can be applied to the display device. Figure 3-7 illustrates how it's done with a nixie tube. Each BCD input to the 74141 is decoded as a ground on one of 10 output lines; this ground then causes the neon gas around a particular cathode to ionize. (A circuit like Fig. 3-7 is needed for each decimal digit.)

Incidentally, if a forbidden 8421 combination (1010 through 1111) drives the

Fig. 3-6. Decoding three BCD digits.

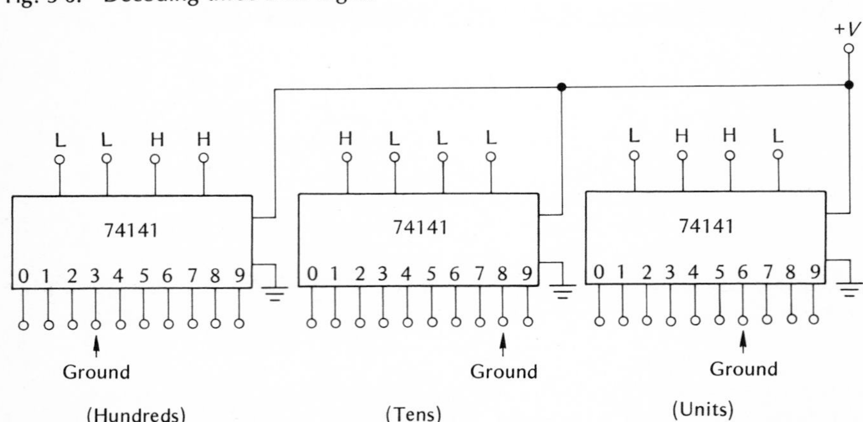

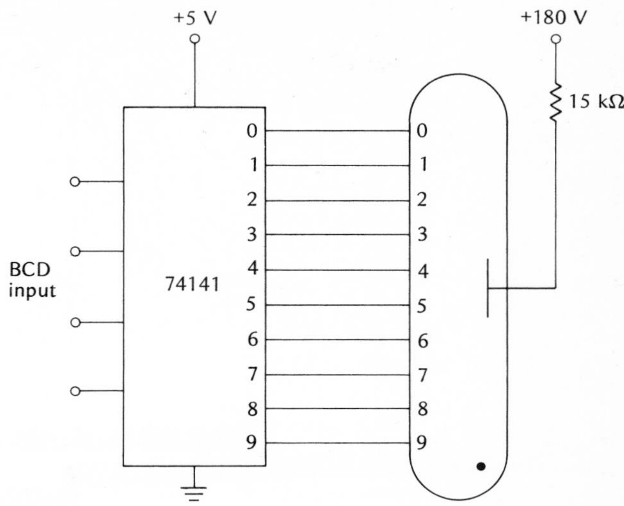

Fig. 3-7. Decoder driving nixie.

74141 of Fig. 3-7, none of the output lines is grounded. Because of this, the nixie is blank (no number displayed). This feature of the 74141 allows designers to blank leading zeros. In other words, it's possible to blank the leading zeros of 003726 to get a final display of 3726.

Manufacturers' catalogs include other decimal decoders, such as the 7443 (excess-3-to-decimal), the 7444 (excess-3 Gray-to-decimal), the 74145 (BCD-to-decimal for driving incandescent lamps, relays, etc.), and many others.[6]

3-10 SEVEN-SEGMENT DECODERS

The 7447 is an example of a BCD-to-seven-segment decoder. Given a BCD input, the 7447 provides suitable grounds for a seven-segment LED display. For instance, if the BCD input of Fig. 3-8*a* is LHHH (7), output lines *a*, *b*, and *c* are grounded. These grounds can then be used to activate segments *a*, *b*, and *c* in Fig. 3-8*b*.

The straightforward way to display a five-digit decimal number is to use a 7447 and a seven-segment LED indicator for each decimal digit. So to display a decimal number like 35842, we'd need five 7447s and five seven-segment indicators. With this approach, all decimal digits are excited simultaneously.

Alternatively, to reduce the number of decoders and the wiring, we can *multiplex* as shown in Fig. 3-9. The details are too complicated to go into here, but this is the basic idea: Instead of activating all decimal digits simultaneously, we can display them in rapid order, one after another. If this is done fast enough, the number 35842 appears to be continuously displayed. In other words, the 3 is flashed first,

[6] For a complete list of TTL decoders, see "The TTL Data Book for Design Engineers," Texas Instruments Inc., 1973.

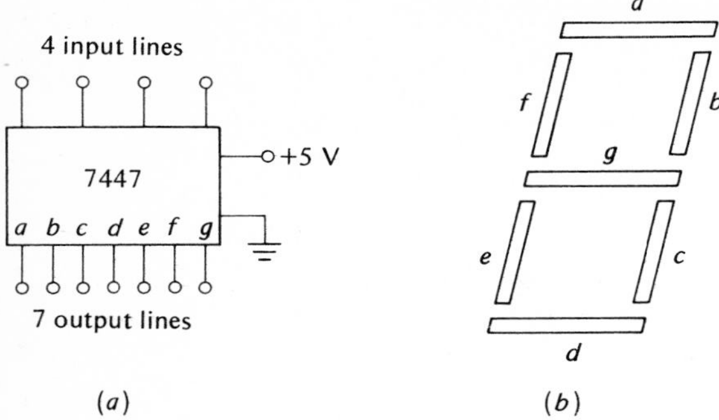

Fig. 3-8. 7447 decoder/driver.

then the 5, then the 8, and so on. When this is done at a high rate (usually greater than 50 Hz), the number 35842 appears without flicker.

Other seven-segment decoders are available. The CD4055 is a CMOS BCD-to-seven-segment decoder suitable for driving a liquid-crystal indicator. Liquid crystals have less power dissipation than LEDs. Used with CMOS logic, liquid-crystal displays are ideal for low-power applications such as electronic wristwatches.

3-11 DOT-MATRIX DECODING

To display a decimal digit with a dot-matrix readout, the required LEDs are not lit simultaneously (see Fig. 3-10). Normally, they are activated a row at a time in rapid order. For instance, in Fig. 3-10 we'd start by grounding the first horizontal row of

Fig. 3-9. Multiplexing.

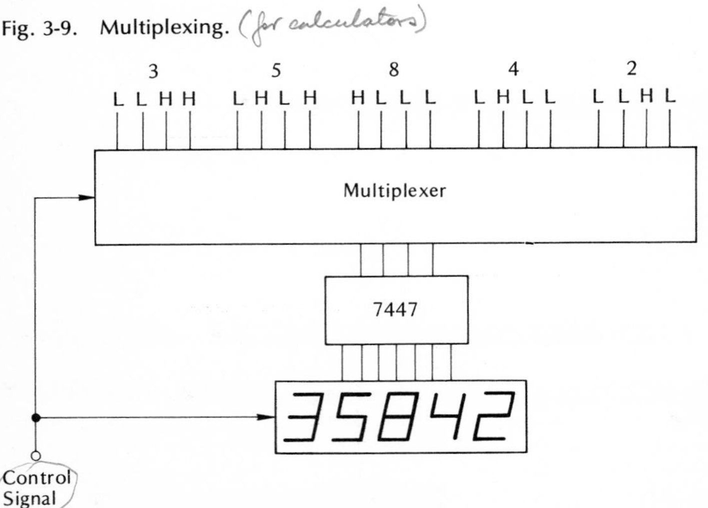

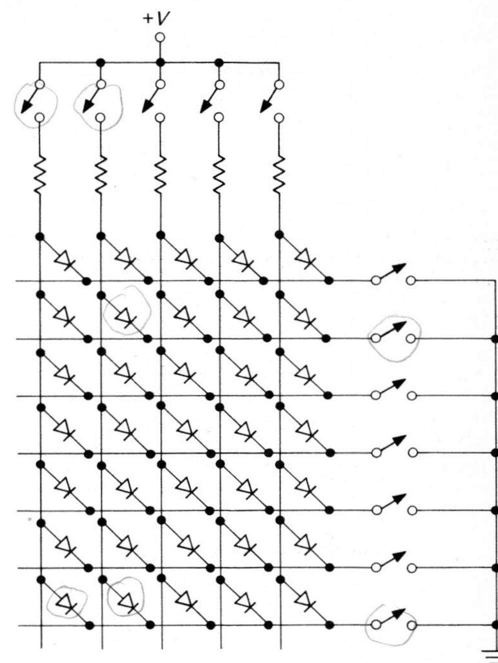

Fig. 3-10. Dot-matrix decoding.

LEDs and applying voltages to appropriate vertical columns. After a brief display, the second row of LEDs would be grounded, and voltages applied to certain columns. We would proceed this way through the remaining rows and then start over. When this process is repeated fast enough, an alphanumeric is displayed without flicker.

The circuits driving the dot matrix are advanced because they have to scan through the horizontal rows and simultaneously apply voltages to the appropriate columns. Typically, a *ring counter* (Chap. 9) scans the horizontal rows, while a *read-only memory* (Chap. 12) applies voltages to the vertical columns.

STUDY AIDS

Summary

BCD codes are a compromise between the binary and the decimal number systems.

The best-known BCD code is the 8421 code. The weights from left to right are 8, 4, 2, 1—the same as for straight binary numbers. Above 9 the 8421 code is completely different from the straight binary system. To encode in 8421 form, we change *each* decimal digit to its binary equivalent. The main advantage of the 8421 code is the ease of converting to and from decimal numbers.

To encode a decimal number into excess-3 form, add 3 to each decimal digit before converting to binary. The excess-3 code is a self-complementing code, which means that the 1's complement of any excess-3 number represents the 9's complement of the encoded decimal number.

There are many four-bit codes (about 30 billion). Tables 3-3 and 3-4 show some of the popular ones.

Sometimes, a parity bit is attached to a word or group of bits to make the total number of 1s either even or odd. By doing this, we can more easily detect errors that occasionally get into words as the words are being moved around or stored.

Five-bit codes also exist. The extra bit allows us to decode the number more easily and facilitates error detection. The shift-counter code is important in some electronic counters.

Some BCD codes have more than five bits. The 50 43210 code, also called the biquinary code, is important and is used in some electronic counters.

The Gray code has the outstanding characteristic that each Gray-code number differs from the preceding Gray-code number by only one bit.

Three common ways to display alphanumerics are the discrete, bar-matrix, and dot-matrix methods. The 74141 is a TTL decoder/driver for BCD-to-decimal conversion, specifically for use with nixies. The 7447 is an example of a TTL decoder/driver for BCD-to-seven-segment conversion.

Glossary

alphanumeric Alphabet letters, numbers, and other symbols.

bar matrix A display using bars or segments. The seven-segment indicator is an example.

biquinary Stands for two-five and refers to the 50 43210 code, whose code groups contain a group of two bits and a group of five bits.

8421 code The most popular BCD code. Each decimal digit is changed into its binary equivalent.

excess-3 A BCD code in which 3 is added to each decimal digit before converting to equivalent binary groups.

Gray code Sometimes called the reflected binary code. This code has the outstanding property that each Gray-code number differs from the preceding one by only one bit.

parity bit A bit that is deliberately attached to a group of bits to make the total number of 1s either even or odd.

self-complementing Means a BCD code has the property that the 1's complement of a code group represents the 9's complement of the encoded decimal digit. Only some of the BCD codes have this property.

weighted code A code whose bit positions carry a fixed value, usually given by the name of the code. For instance, the 8421 code has weights of 8, 4, 2, and 1 in the corresponding bit positions. Some codes like the excess-3 are unweighted; the bit positions have no fixed values.

word In a digital system this is a group of bits that is treated, stored, and moved as a unit.

Review Questions

1. What is the 8421 code? What is the 8421-code number for 395?
2. What is a BCD code?
3. How are decimal numbers changed into excess-3 numbers?
4. What does *self-complementing* mean in reference to a BCD code?
5. What is a weighted code? An unweighted code?
6. Name some of the four-bit codes. Are these weighted or unweighted?
7. What is a parity bit? What are the two kinds of parity?
8. What does *word* refer to?
9. Name some of the five-bit codes. Are these weighted or unweighted?
10. What are the weights in the biquinary code? What does *biquinary* mean?
11. What is the main characteristic of the Gray code?
12. What does alphanumeric mean? What are three ways to display alphanumerics?
13. What is a 74141? A 7447?
14. With respect to an alphanumeric display, what does *multiplexing* mean?

Problems

3-1. Encode the following decimal numbers into 8421 BCD numbers:
(*a*) 59.
(*b*) 39,584.

3-2. What are the 8421 BCD numbers for these decimal numbers:
(*a*) 649.
(*b*) 71,465.

3-3. Decode the following 8421 BCD numbers:
(*a*) 0011 1000 0111.
(*b*) 1001 0110 0111 1000 0111 0011.

3-4. Decode the following 8421 BCD numbers:
(*a*) 1001 0111 1000.
(*b*) 0101 0110 0111 0001 0000 0100.

3-5. What are the excess-3 numbers for these decimal numbers:
(*a*) 467.
(*b*) 5839.

3-6. Change the following decimal numbers into excess-3 numbers:
(*a*) 5932.
(*b*) 386,258.

3-7. Decode these excess-3 numbers:
(*a*) 1011 0111 1001.
(*b*) 1010 1000 0110 0101 0011.

3-8. What is the excess-3 number for octal 743?

3-9. Use excess-3 numbers to add these decimal numbers: 457 and 293. What is the final answer in excess-3 code?

3-10. What is the excess-3 sum of decimals 45 and 34?

3-11. What are the 5421 BCD numbers for:
(*a*) Decimal 435.
(*b*) Octal 435.

3-12. Decode the following BCD numbers:
(*a*) 1011 1001 0011 0001 (5421 code).
(*b*) 1110 1001 0101 0111 ($74\bar{2}\bar{1}$ code).

3-13. An electronic counter uses the 4221 code. Suppose that its output is given in the following form *on-on-on-on, off-on-on-on, off-off-on-off*. What decimal number does this output represent if *on* stands for 1? What does it represent if *on* stands for 0?

3-14. The 2421 code is used in an electronic counter. What do the following outputs represent?
(*a*) 0001 1011 1110 1011 0010.
(*b*) 0011 1011 1101 0100 0000.

3-15. Add an even-parity bit to each of the following words:
(*a*) 1011001.
(*b*) 110001100011.
(*c*) 10101110011100.

3-16. Determine whether the parity bit in the following words is even or odd:
(*a*) 1001100101 1.
(*b*) 1000110011 1.
(*c*) 1000011101 0.

3-17. Suppose we want to add an even-parity bit to the 3321 code. What bits should we add?

3-18. To add an odd-parity bit to the 5211 code, what bits should we add?

3-19. Encode the following decimal numbers into shift-counter BCD numbers: 84, 932, and 5482.

3-20. Decode the following 2-out-of-5 BCD numbers:
(*a*) 01010 10010 11000.
(*b*) 00011 01100 10100 00101 10100.

3-21. Write the following decimal numbers as biquinary BCD numbers: 25, 452.

3-22. Decode the following biquinary BCD numbers:
(*a*) 10 00010.
(*b*) 01 01000.

3-23. Convert the following binary numbers into Gray numbers:
(*a*) 10110.
(*b*) 100111011001.
(*c*) 10101101110011.

3-24. Convert the following Gray numbers into binary numbers:
(*a*) 10101.
(*b*) 100110011.
(*c*) 00111000010110.

3-25. The supply voltage in Fig. 3-3*b* is 10 V. If a conducting LED has 1.5 V across it, what size should the series resistor be to get 10 mA through the LED?

Boolean Algebra

4

Is an action *right* or *wrong?* A motive *good* or *bad?* A conclusion *true* or *false?* Much of our thinking and logic involves trying to find the answers to *two-valued* questions like these. The binary or two-valued nature of logic had a major influence on Aristotle, who worked out precise methods for getting to the truth, given a set of true assumptions. Logic next attracted mathematicians, who intuitively sensed some kind of algebraic process running through all thought.

De Morgan came close to capturing the connection between logic and mathematics. But it was Boole (1854) who put it all together. He invented a new kind of algebra that replaced Aristotle's methods. Boole proved binary or two-valued logic is valid for letters and symbols instead of words. The advantages of Boolean algebra are simplicity, speed, and accuracy.

Boolean algebra did not have an impact on digital electronics until almost a century later (1938) when Shannon applied the new algebra to telephone switching circuits. Because a switch is a binary device (*on* or *off*), Shannon was able to analyze and design complicated switching circuits using Boolean algebra.

This chapter familiarizes you with Boolean algebra, logic circuits, and other topics that are important in digital electronics. After finishing this chapter, you should be able to

1. Write the Boolean equation for a circuit with AND, OR, and inverter functions.
2. Apply De Morgan's theorems.
3. Recall Boolean equations (4-3) through (4-18).
4. Differentiate between totem-pole and open-collector TTL.

4-1 THE OR GATE

A *gate* is a logic circuit with one output and two or more inputs; an output signal occurs only for certain combinations of input signals.

The first kind of gate we study is the OR gate. Figure 4-1*a* shows a two-input OR gate where *A* and *B* are the inputs and *y* is the output. For the moment, let us analyze this OR gate by restricting the input voltages to either 0 or 1 V. There are only four possible cases to analyze:

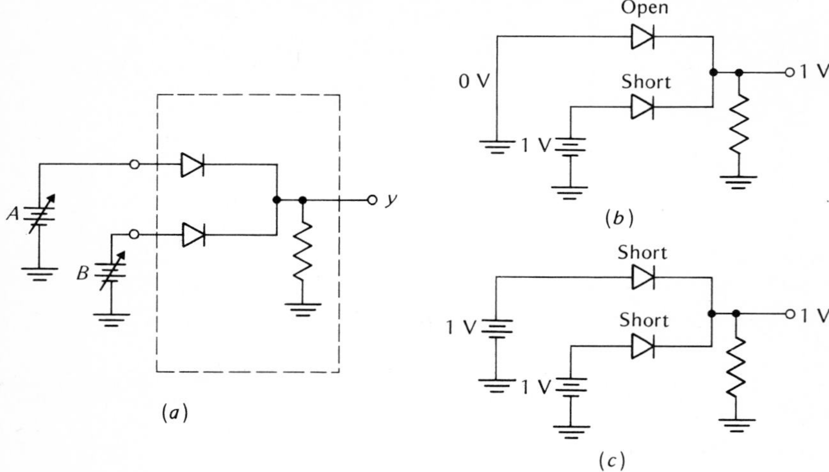

Fig. 4-1. OR gate.

Case 1 $A=0$ and $B=0$. With both input voltages at zero the output voltage must be zero because no voltage exists anywhere in the circuit. Therefore, $y=0$.

Case 2 $A=0$ and $B=1$. The B battery forward biases the lower diode, causing the output to be ideally 1 V. Since the A battery is 0 V, it looks like a short. Figure 4-1*b* shows the circuit for this condition. The upper diode is off, the lower diode is on, and the output $y=1$ V.

Case 3 $A=1$ and $B=0$. Because of the symmetry of the circuit the argument is similar to case 2. The upper diode is on, the lower diode is off, and $y=1$ V.

Case 4 $A=1$ and $B=1$. With both inputs at 1 V both diodes are forward biased. Since the voltages are in parallel, the output voltage is ideally 1 V. Therefore, $y=1$ V as shown in Fig. 4-1*c*.

Table 4-1 lists the input-output conditions of an OR gate. Examine this table carefully and memorize the following: the OR gate has a 1 output when either *A or B* or both are 1. In other words, the OR gate is an *any-or-all* gate; an output occurs when any or all of the inputs are present.

Table 4-1

THE OR-GATE TRUTH TABLE

A	*B*	*y*
0	0	0
0	1	1
1	0	1
1	1	1

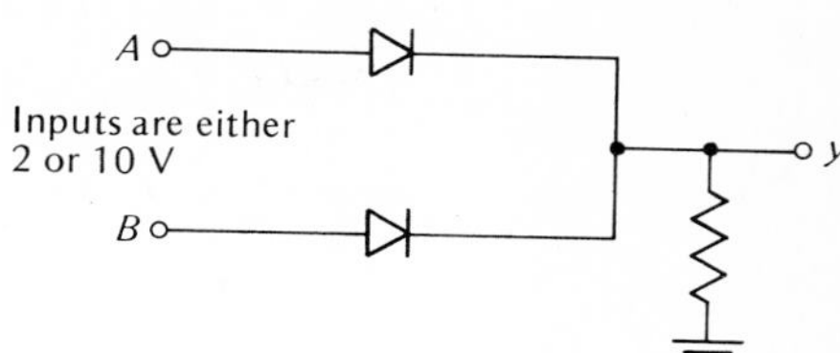

Fig. 4-2. Two-input OR gate.

Incidentally, a *truth table* (sometimes called a "table of combinations") is a table that shows all input-output possibilities for a logic circuit. Table 4-1 is one example of a truth table. All possible *AB* inputs, 00, 01, 10, and 11, are shown along with the resulting outputs.

Most digital circuits use diodes and transistors as switches to change from one voltage level to another. When we analyze digital circuits, we determine whether a voltage is low or high. The exact magnitude or value is unimportant, as long as the voltage is distinguishable as low or high.

In digital circuits low and high voltages are often represented by 0 and 1, respectively. For instance, in the OR gate of Fig. 4-2 the input levels are either 2 or 10 V—low or high. The operation of this circuit is like that of Fig. 4-1; Table 4-2 shows the truth table. If we let 0 stand for 2 V and 1 for 10 V, we can show an equivalent truth table with 0s and 1s (Table 4-3). The point is that we can use the actual voltages, or we can use 0s and 1s to represent the low and high voltages; in either case, the OR gate has a high output when any or all inputs are high.

The OR gate can have any number of inputs. Figure 4-3 shows a three-input OR gate. If *A* or *B* or *C* is high, *y* will be high because the diode associated with the high input will turn on. Letting 0 stand for low and 1 for high, we get the truth table of Table 4-4. In general, no matter how many inputs there are, the OR gate has a 1 output when any or all of the inputs are 1.

Incidentally, the number of horizontal rows in a truth table equals 2^n where n is the number of inputs. For a two-input gate, the truth table has 2^2 or 4 rows. A three-input gate will have a truth table with 2^3 or 8 rows, while a four-input gate results in 2^4 or 16 rows, and so on. To include all possible entries, it helps to list the entries in a binary-number progression. In Table 4-4 we deliberately listed the *ABC* entries as 000, 001, 010, 011, . . . , following the binary-number count; this guarantees that we will not forget any possibility.

We can also use transistors to make OR gates. Figure 4-4 shows a three-input OR gate where *A, B,* and *C* are the inputs and *y* is the output. The circuit has three

Table 4-2

A	*B*	*y*
2 V	2 V	2 V
2 V	10 V	10 V
10 V	2 V	10 V
10 V	10 V	10 V

Table 4-3

A	*B*	*y*
0	0	0
0	1	1
1	0	1
1	1	1

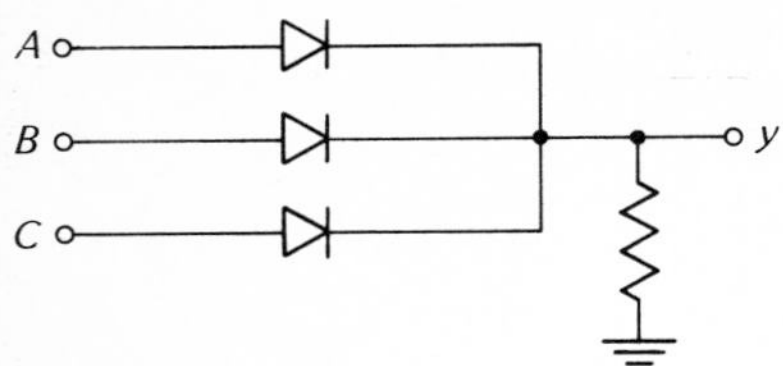

Fig. 4-3. Three-input OR gate.

n-p-n emitter followers in parallel. With all input voltages in the low state, the output must be low because the emitters follow the inputs. When any or all the inputs go high, the output follows. Therefore, we have OR-gate action. By letting 0 denote low voltage and 1 denote high voltage, we get a truth table that is the same as Table 4-4.

There are many other ways to build OR gates by using various combinations of resistors, diodes, and transistors. The important thing to remember is that the OR gate has a 1 output when any or all inputs are 1.

4-2 THE AND GATE

The AND gate is another basic kind of digital circuit—it has an output only when *all* inputs are present. Consider the two-input AND gate of Fig. 4-5*a*. Again, use ideal diodes and restrict all voltages to either 0 or 1 V. There are four cases to analyze:

Case 1 $A = 0$ and $B = 0$. Since both input batteries are at 0 V, they are like shorts (Fig. 4-5*b*). The 1-V battery forces conventional current in the direction of each diode triangle; therefore, both diodes are on or shorted. Carefully examine Fig. 4-5*b* and realize that the output is shorted to ground through the diodes and the batteries. Therefore, $y = 0$.

Case 2 $A = 0$ and $B = 1$. The upper diode is forward-biased as shown in Fig. 4-5*c*; the output is still shorted to ground through the upper diode and the battery. Therefore, $y = 0$.

Fig. 4-4. Transistor OR gate.

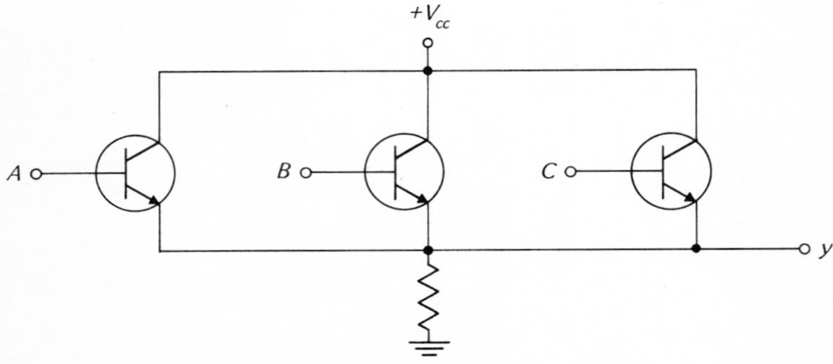

Table 4-4
THREE-INPUT OR GATE

A	B	C	y
0	0	0	0
0	0	1	1
0	1	0	1
0	1	1	1
1	0	0	1
1	0	1	1
1	1	0	1
1	1	1	1

Case 3 $A = 1$ and $B = 0$. Because of symmetry the argument is similar to case 2, and $y = 0$.
Case 4 $A = 1$ and $B = 1$. No current flows in the circuit (Fig. 4-5*d*). With no current in R there is no voltage drop across R; therefore, y must equal 1 V.

As usual, we can summarize the circuit action by a truth table (Table 4-5). Examine this table carefully and memorize the following: The AND gate has a 1 output

Fig. 4-5. AND gate.

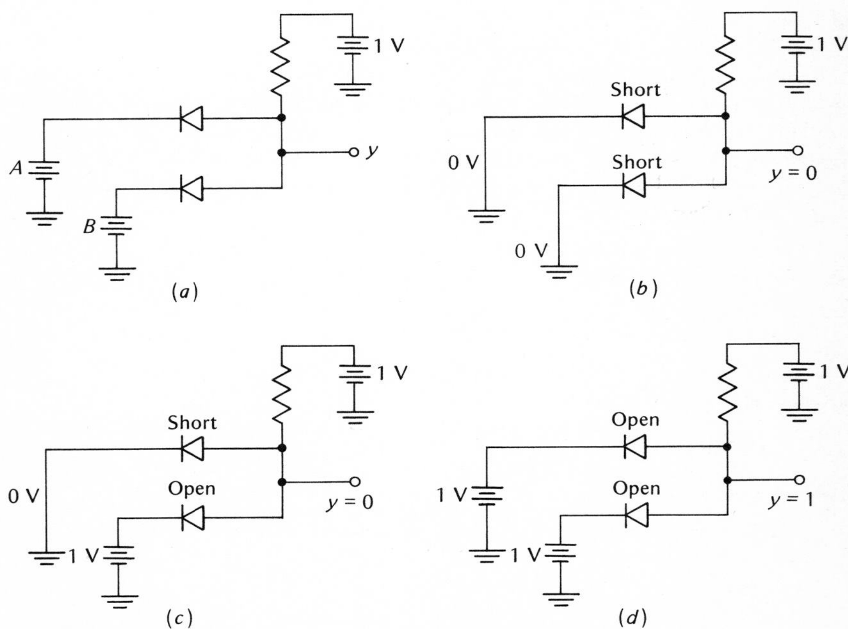

Table 4-5

AND-GATE TRUTH TABLE

A	B	y
0	0	0
0	1	0
1	0	0
1	1	1

Table 4-6

A	B	y
2 V	2 V	2 V
2 V	10 V	2 V
10 V	2 V	2 V
10 V	10 V	10 V

when *A and B* are 1. In other words, the AND gate is an *all-or-nothing* gate; an output occurs only when all inputs are present.

The use of 0 and 1 V was only a convenience for analysis. We can use any two distinct voltages. In Fig. 4-5*a* suppose the 1-V battery is changed to a 10-V battery, and the inputs are either 2 or 10 V. If we use actual voltages, we get the truth table of Table 4-6.

The AND gate can have more than two inputs. For instance, Fig. 4-6 shows a three-input AND gate. Assume the two distinct voltage levels are 0 and 10 V. If any input is at 0 V (grounded), the diode connected to that input is forward biased or shorted; therefore, the output will be shorted to ground. Therefore, $y = 0$ when any input is 0. The only way to make $y = 10$ V is to have all inputs simultaneously equal to 10 V. In this case no current flows through *R*, and the output voltage rises to the supply voltage. If we let 0 stand for low voltage and 1 for high voltage, we get the truth table of Table 4-7. In general, the AND gate is an *all-or-nothing* gate; no matter how many inputs there are, the AND gate has an output only when all inputs are present.

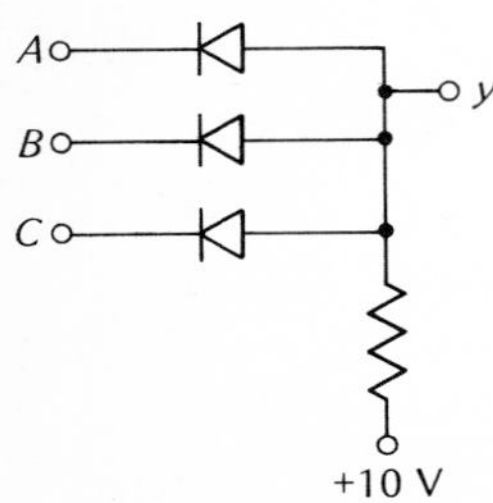

Fig. 4-6. Three-input AND gate.

Table 4-7
THREE-INPUT AND GATE

A	B	C	y
0	0	0	0
0	0	1	0
0	1	0	0
0	1	1	0
1	0	0	0
1	0	1	0
1	1	0	0
1	1	1	1

We can build AND gates with transistors instead of diodes. Figure 4-7 shows one way to make a three-input AND gate. Here is how it works. Suppose the input voltages (A, B, and C) are either 0 or 10 V. If any of the inputs is 0 V (grounded), the base-emitter diode of the associated transistor will be on (ideally, a short); therefore, the output will be held approximately at ground, so that $y = 0$. The only way to make $y = 10$ V is to raise all inputs to 10 V. Under this condition none of the base-emitter diodes is on, and the output is free to rise to 10 V. By letting 0 stand for low voltage and 1 for high voltage, we get a truth table like Table 4-7.

4-3 POSITIVE AND NEGATIVE LOGIC SYSTEMS

A *positive* logic system is a system in which a 1 represents the more positive of the two voltage levels; in a *negative* logic system the 1 stands for the more negative voltage. For instance, suppose a digital system has voltage levels of +5 and 0 V. We are free to choose the definitions for a 0 and a 1. If we say a 1 stands for +5 V and a 0 for 0 V, the system becomes a positive logic system. On the other hand, if we let 1 denote 0 V and let 0 symbolize +5 V, we have a negative logic system. (Up to now, we have been using positive logic.)

Fig. 4-7. Transistor AND gate.

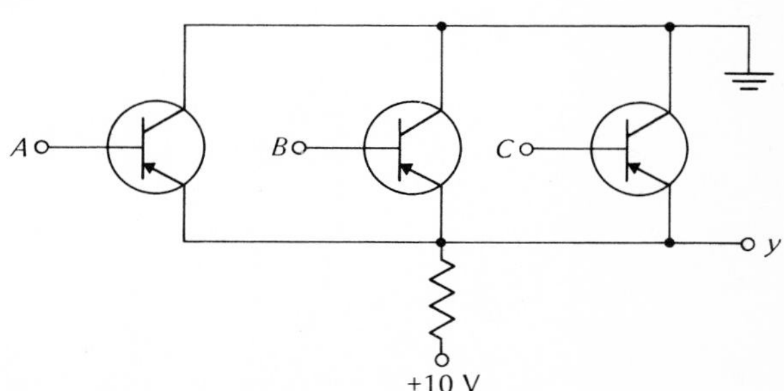

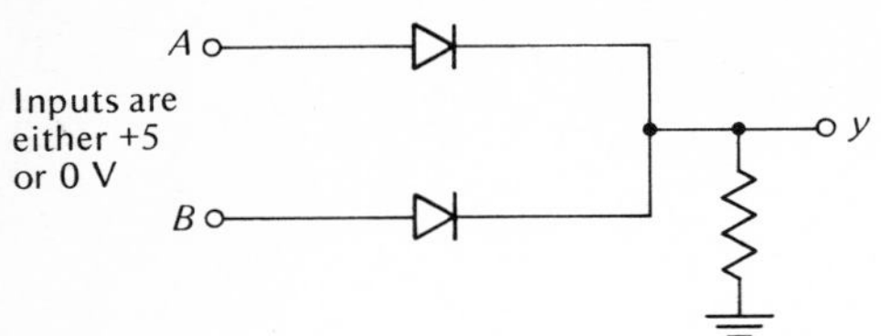

Fig. 4-8. Positive OR gate versus negative AND gate.

The distinction between positive and negative logic is important because an OR gate in a positive logic system becomes an AND gate in a negative logic system. To understand why, consider the circuit of Fig. 4-8. The truth table for this circuit (Table 4-8) gives actual voltage levels. The output is 5 V when either *A* or *B* is 5 V. Is this circuit an OR gate or an AND gate? The answer depends on how we define 0 and 1. If 1 stands for +5 V and 0 for 0 V, we have a positive logic system and can show the truth table as in Table 4-9. This is the truth table of an OR gate because there is a 1 output when *A* or *B* is 1.

On the other hand, if a 1 stands for 0 V and a 0 for +5 V, we have a negative logic system. By looking at Table 4-8 and writing a 1 for each 0-V entry and a 0 for each 5-V entry, we get the truth table of Table 4-10. By inspection of this table a 1 output occurs only when *A* and *B* are 1; therefore, the circuit is an AND gate in a negative logic system.

A digital system can be either a positive or a negative logic system; it all depends on how we define 0 and 1. The choice of which system to use is arbitrary—like defining the direction of current from + to − or from − to +. For consistency we shall use *positive logic* throughout this book; 1 always will stand for the more positive of the two voltage levels.

4-4 THE NOT CIRCUIT

Another of the basic digital circuits is the NOT circuit, also called a "complementary" circuit or an inverter." This circuit has one input and one output. All it does is invert the input signal; if the input is high, the output is low, and vice versa.

Figure 4-9 shows one way to build a NOT circuit. When the input voltage is high enough, the transistor saturates; therefore, the output is low. On the other hand,

Table 4-8

A	*B*	*y*
0 V	0 V	0 V
0 V	5 V	5 V
5 V	0 V	5 V
5 V	5 V	5 V

Table 4-9

A	*B*	*y*
0	0	0
0	1	1
1	0	1
1	1	1

Table 4-10

A	*B*	*y*
1	1	1
1	0	0
0	1	0
0	0	0

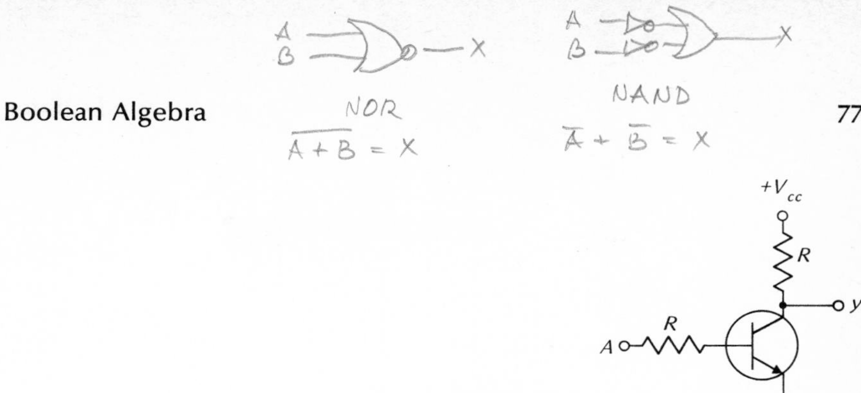

Fig. 4-9. NOT circuit.

when the input voltage is low enough, the transistor cuts off, and the output voltage is high. The truth table of the NOT circuit is

Input	Output
0	1
1	0

We call this circuit a NOT circuit because the output is *not* the same as the input.

4-5 OR ADDITION

In ordinary algebra when we solve an equation for its roots, we can get a real number—positive, negative, fractional, and so forth. In other words, the set of numbers in ordinary algebra is infinite. In Boolean algebra when we solve an equation, we get either a 0 or a 1. No other answers are possible, because the set of numbers includes only the binary digits 0 and 1.

Another startling difference about Boolean algebra is the meaning of the plus sign. To bring out this meaning, consider Fig. 4-10 which show a two-input OR gate with A and B inputs and a y output. In Boolean algebra the + sign symbolizes the action of an OR gate. In other words, we may think of the OR gate as a device that combines A with B to give a result of y. In Boolean algebra when we write

$$y = A + B$$

we mean that A and B are to be combined in the same way that an OR gate combines A and B. Read the expression $y = A + B$ as y equals A OR B. To repeat, the + sign does not stand for ordinary addition; it stands for OR addition whose rules are given by the OR truth table (Table 4-1).

To get used to OR addition, let us work out the value of $y = A + B$ for the four possible input conditions.

A, B → OR gate → y

Fig. 4-10. OR addition.

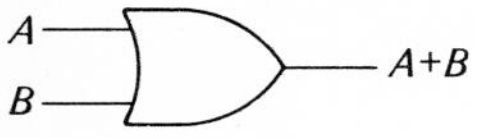

Fig. 4-11. Logic symbol for OR gate.

Case 1 $A = 0$ and $B = 0$. We have

$$y = A + B = 0 + 0 = 0$$

because an OR gate combines 0 with 0 togive 0.
Case 2 $A = 0$ and $B = 1$. This gives

$$y = A + B = 0 + 1 = 1$$

because an OR gate combines 0 with 1 to give 1.
Case 3 $A = 1$ and $B = 0$. This is like case 2.

$$y = A + B = 1 + 0 = 1$$

Case 4 $A = 1$ and $B = 1$. We get

$$y = A + B = 1 + 1 = 1$$

because an OR gate combines 1 with 1 to give 1.

It may take a while to get used to the last result because of your built-in understanding of the + sign. You simply must remember that in digital work the + sign has several meanings. With decimal numbers it means ordinary addition—the first kind you learned about. With binary numbers, however, it refers to binary addition. In Boolean algebra the + sign stands for OR addition—the kind of addition that an OR gate does. To display the three different meanings:

$1 + 1 = 2$	decimal addition
$1 + 1 = 10$	binary addition
$1 + 1 = 1$	OR addition

The meaning of the + sign is usually clear from the context. In other words, in solving decimal-arithmetic problems, we use the ordinary meaning of the + sign. In solving a Boolean equation, however, we will use the new meaning of OR addition. In case of doubt, we shall indicate what type of addition the + sign refers to.

Incidentally, Fig. 4-11 shows the symbol used in industry for the OR gate. It should be memorized.

4-6 AND MULTIPLICATION

The multiplication sign (either × or ·) has a new meaning in Boolean algebra. To understand this meaning, look at the AND gate of Fig. 4-12. Think of an AND gate as a device that combines A and B to give a result of y. So, in Boolean algebra when

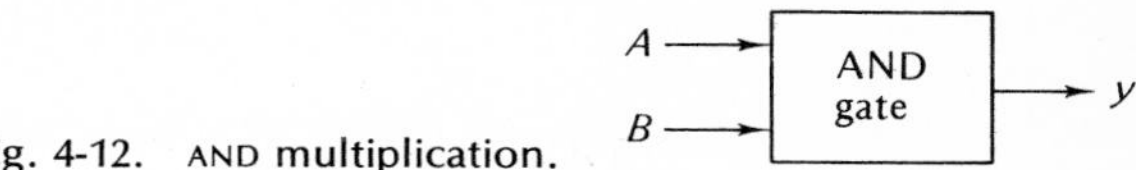

Fig. 4-12. AND multiplication.

we write

$$y = A \cdot B$$

or

$$y = AB$$

we mean A and B are to be combined in the same way that an AND gate combines A with B to give a y output. For practice, let us solve $y = AB$ for the four possible cases.

Case 1 $A = 0$ and $B = 0$. This gives

$$y = AB = \underline{0 \cdot 0 = 0}$$

because an AND gate combines 0 with 0 to give 0.
Case 2 $A = 0$ and $B = 1$. We get

$$y = AB = \underline{0 \cdot 1 = 0}$$

because an AND gate combines 0 with 1 to give 0.
Case 3 $A = 1$ and $B = 0$. This is like case 2:

$$y = AB = \underline{1 \cdot 0 = 0}$$

because an AND gate combines 1 with 0 to give 0.
Case 4 $A = 1$ and $B = 1$. This gives

$$y = AB = \underline{1 \cdot 1 = 1}$$

These four results are easy to remember. Even though the dot does not mean multiplication in the ordinary sense, the results of AND multiplication are the same as for ordinary multiplication.

Figure 4-13 shows the symbol used in industry for the AND gate. Memorize it.

Example 4-1

What is the Boolean expression for the output of the circuit shown in Fig. 4-14*a*?

Solution

The output of the AND gate is AB. This goes into the OR gate along with B. Therefore, the final output is

$$y = AB + B$$

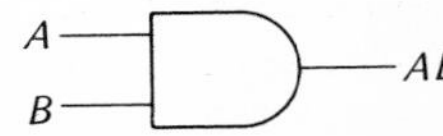

Fig. 4-13. Logic symbol for AND gate.

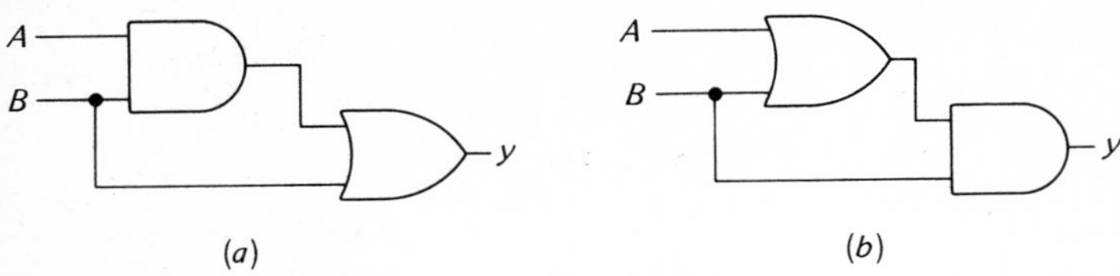

Fig. 4-14. Logic circuits.

Example 4-2

Evaluate the Boolean expression in the preceding example for
(a) $A = 0$ and $B = 1$
(b) $A = 1$ and $B = 0$

Solution

(a) To evaluate the expression $y = AB + B$, substitute the values of A and B into the expression and work out an answer.

$$y = AB + B = 0 \cdot 1 + 1 = 0 + 1 = 1$$

(Multiply before you add just as in ordinary algebra.)

(b) For $A = 1$ and $B = 0$,

$$y = AB + B = 1 \cdot 0 + 0 = 0 + 0 = 0$$

Example 4-3

Find the Boolean expression for the output in Fig. 4-14*b*, and evaluate the expression for $A = 0$ and $B = 1$.

Solution

The output of the OR gate is $A + B$. This goes into the AND gate along with the other input, which is B. Therefore, the final output is

$$y = (A + B)B$$

(Note that the parentheses are also used to denote AND multiplication.)
To evaluate this expression for $A = 0$ and $B = 1$, substitute to get

$$y = (0 + 1)1 = (1)1 = 1$$

4-7 THE NOT OPERATION

In Fig. 4-15 the A input to the NOT circuit is inverted: if a 0 goes in, a 1 comes out; if a 1 goes in, a 0 comes out. In Boolean algebra the expression

$$y = \overline{A}$$

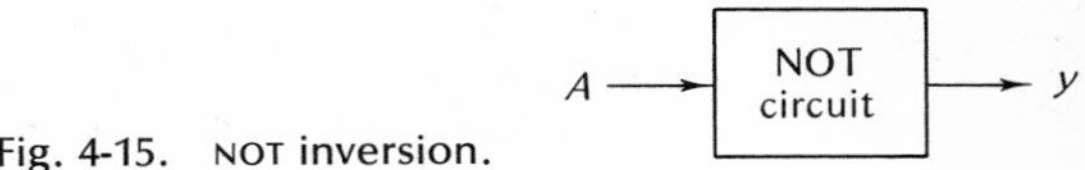

Fig. 4-15. NOT inversion.

means you are to change A in the same way a NOT circuit changes A. Read the expression $y = \overline{A}$ as y equals NOT A. The bar over A means you change or complement the quantity to the alternative digit (same as the 1's complement of Chap. 2). In other words, when $A = 0$,

$$y = \overline{A} = \overline{0} = 1 \qquad \text{because NOT 0 is 1}$$

When $A = 1$,

$$y = \overline{A} = \overline{1} = 0 \qquad \text{because NOT 1 is 0}$$

The NOT operation is also called "negation" or "inversion," and the prime is sometimes used instead of a bar to signify the NOT operation. That is,

$$y = A'$$

may be used instead of $y = \overline{A}$.

Figure 4-16 shows the symbol for a NOT circuit. It should be memorized.

The OR, AND, and NOT operations of Boolean algebra may seem strange. Why make up new operations like these? Because these operations describe OR, AND, and NOT circuits—the building blocks of complex digital systems. With Boolean algebra you can analyze and design digital systems more easily. This will become clearer in the remainder of the book.

Let us summarize the OR, AND, and NOT rules:

OR	AND	NOT
$0 + 0 = 0$	$0 \cdot 0 = 0$	$\overline{0} = 1$
$0 + 1 = 1$	$0 \cdot 1 = 0$	$\overline{1} = 0$
$1 + 0 = 1$	$1 \cdot 0 = 0$	
$1 + 1 = 1$	$1 \cdot 1 = 1$	

Example 4-4

Write the Boolean equation for the output of Fig. 4-17.

Solution

The OR gate has two inputs. One of them is $\overline{A}$, and the other is B. Therefore, the output is

$$y = \overline{A} + B$$

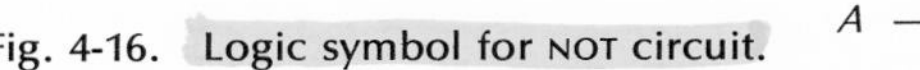

Fig. 4-16. Logic symbol for NOT circuit.

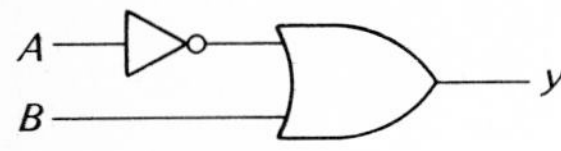

Fig. 4-17. Logic circuit.

Example 4-5

Solve the Boolean equation of the preceding example for all possible input conditions.

Solution

1. $A = 0$ and $B = 0$:

$$y = \overline{A} + B = \overline{0} + 0 = 1 + 0 = 1$$

2. $A = 0$ and $B = 1$:

$$y = \overline{A} + B = \overline{0} + 1 = 1 + 1 = 1$$

3. $A = 1$ and $B = 0$:

$$y = \overline{A} + B = \overline{1} + 0 = 0 + 0 = 0$$

4. $A = 1$ and $B = 1$:

$$y = \overline{A} + B = \overline{1} + 1 = 0 + 1 = 1$$

Example 4-6

Write the Boolean expression for the output of Fig. 4-18*a*.

Solution

One of the inputs to the OR gate is $\overline{A}$; the other input is $\overline{B}$. Therefore, the final output must be

$$y = \overline{A} + \overline{B}$$

Fig. 4-18. Two more logic circuits.

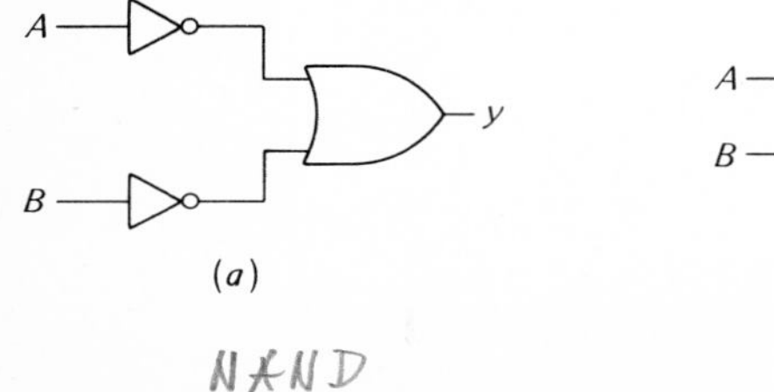

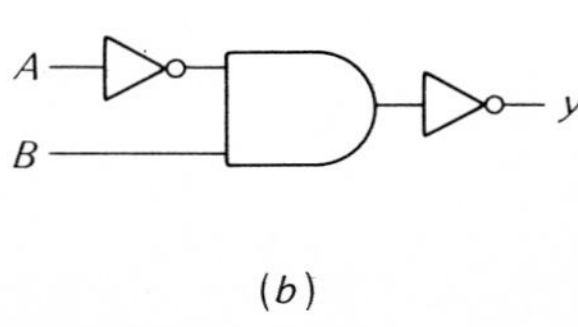

Example 4-7

Evaluate the Boolean expression of the preceding example for the four possible input combinations.

Solution

When $A = 0$ and $B = 0$,

$$y = \overline{A} + \overline{B} = \overline{0} + \overline{0} = 1 + 1 = 1$$

When $A = 0$ and $B = 1$,

$$y = \overline{A} + \overline{B} = \overline{0} + \overline{1} = 1 + 0 = 1$$

When $A = 1$ and $B = 0$,

$$y = \overline{A} + \overline{B} = \overline{1} + \overline{0} = 0 + 1 = 1$$

Finally, when $A = 1$ and $B = 1$,

$$y = \overline{A} + \overline{B} = \overline{1} + \overline{1} = 0 + 0 = 0$$

Example 4-8

(*a*) Find the Boolean expression for the output of Fig. 4-18*b*.
(*b*) Evaluate this Boolean expression for all possible input combinations.

Solution

(*a*) The inputs to the AND gate are A and B. Therefore, the input to the NOT circuit is $\overline{A}B$. The final output is

$$y = \overline{\overline{A}B}$$

(*b*) When $A = 0$ and $B = 0$,

$$y = \overline{\overline{A}B} = \overline{\overline{0}\cdot 0} = \overline{1\cdot 0} = \overline{0} = 1$$

When $A = 0$ and $B = 1$,

$$y = \overline{\overline{A}B} = \overline{\overline{0}\cdot 1} = \overline{1\cdot 1} = \overline{1} = 0$$

When $A = 1$ and $B = 0$,

$$y = \overline{\overline{A}B} = \overline{\overline{1}\cdot 0} = \overline{0\cdot 0} = \overline{0} = 1$$

When $A = 1$ and $B = 1$,

$$y = \overline{\overline{A}B} = \overline{\overline{1}\cdot 1} = \overline{0\cdot 1} = \overline{0} = 1$$

As a general rule, when evaluating expressions like the foregoing:

1. Take the NOT of all individual terms first.
2. When a NOT is applied to more than one term (like $\overline{0 \cdot 1}$), work out the AND or OR operation first, and then take the NOT of the result.

4-8 DE MORGAN'S THEOREMS

De Morgan was a great logician and mathematician, as well as a friend of Boole. Among De Morgan's important contributions to logic are these two theorems:

$$\overline{A+B} = \overline{A} \cdot \overline{B} \tag{4-1}$$

$$\overline{A \cdot B} = \overline{A} + \overline{B} \tag{4-2}$$

The first equation says *the complement of a sum equals the product of the complements*. The second equation says *the complement of a product equals the sum of the complements*.

These two theorems can easily be proved. To prove

$$\overline{A+B} = \overline{A} \cdot \overline{B}$$

we need to show that the left side equals the right side for all possible values of A and B. Here are the four possible cases:

Case 1 $A = 0$ and $B = 0$.

Left $\qquad \overline{A+B} = \overline{0+0} = \overline{0} = 1$

Right $\qquad \overline{A} \cdot \overline{B} = \overline{0} \cdot \overline{0} = 1 \cdot 1 = 1$

Case 2 $A = 0$ and $B = 1$.

$$\overline{A+B} = \overline{0+1} = \overline{1} = 0$$

$$\overline{A} \cdot \overline{B} = \overline{0} \cdot \overline{1} = 1 \cdot 0 = 0$$

Table 4-11

A	B	$\overline{A+B}$
0	0	1
0	1	0
1	0	0
1	1	0

Table 4-12

A	B	$\overline{A} \cdot \overline{B}$
0	0	1
0	1	0
1	0	0
1	1	0

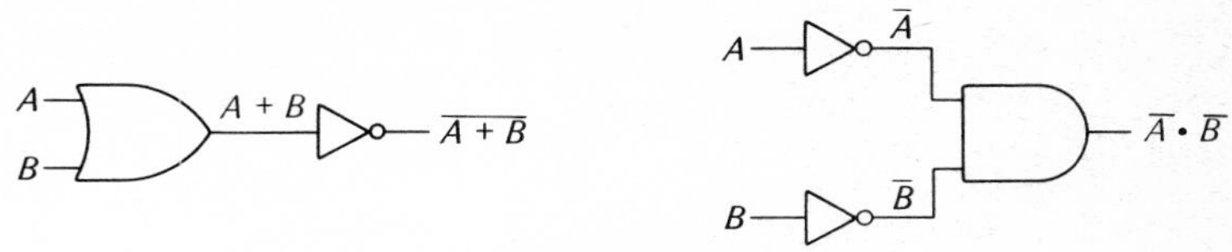

Fig. 4-19. De Morgan's first theorem. NOR gate

Case 3 $A = 1$ and $B = 0$.

$$\overline{A + B} = \overline{1 + 0} = \bar{1} = 0$$
$$\bar{A} \cdot \bar{B} = \bar{1} \cdot \bar{0} = 0 \cdot 1 = 0$$

Case 4 $A = 1$ and $B = 1$.

$$\overline{A + B} = \overline{1 + 1} = \bar{1} = 0$$
$$\bar{A} \cdot \bar{B} = \bar{1} \cdot \bar{1} = 0 \cdot 0 = 0$$

Since no other combinations of A and B exist, we have proved De Morgan's first theorem: $\overline{A + B} = \bar{A} \cdot \bar{B}$. To summarize the proof we can show the truth table for each expression (Tables 4-11 and 4-12). These tables show that $\overline{A + B}$ does equal $\bar{A} \cdot \bar{B}$ for each case; therefore, the expressions are identical.

The circuit meaning of the first theorem is important. $\overline{A + B}$ represents a logic system in which a NOT circuit *follows* an OR gate (Fig. 4-19*a*). Also, $\bar{A} \cdot \bar{B}$ describes a logic system in which the outputs of two NOT circuits are used as the inputs to an AND gate (Fig. 4-19*b*). De Morgan's theorem tells us these two systems are interchangeable. Incidentally, in Fig. 4-19*a* a NOT follows an OR gate; we call this particular combination a NOT-OR, or simply a NOR gate.

The second De Morgan theorem is

$$\overline{A \cdot B} = \bar{A} + \bar{B}$$

(*The complement of a product equals the sum of the complements.*) This theorem is easily proved (see Example 4-9). Tables 4-13 and 4-14 give the truth tables for each expression. Note that the truth tables are identical.

Table 4-13

A	B	$\overline{A \cdot B}$
0	0	1
0	1	1
1	0	1
1	1	0

Table 4-14

A	B	$\bar{A} + \bar{B}$
0	0	1
0	1	1
1	0	1
1	1	0

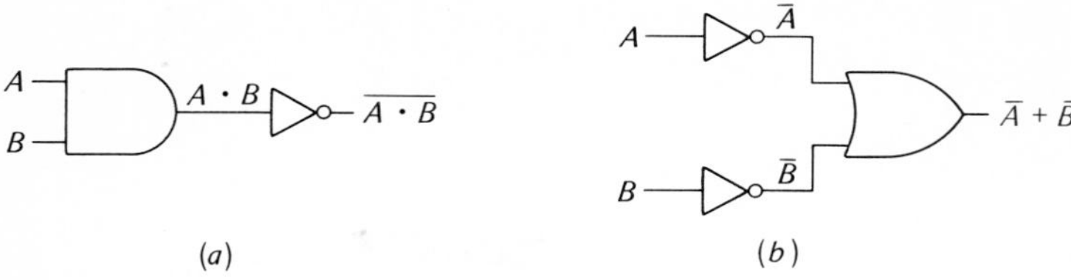

Fig. 4-20. De Morgan's second theorem.

Therefore, the expressions are equivalent and the logic circuits represented by $\overline{A \cdot B}$ and $\overline{A} + \overline{B}$ are interchangable. Figure 4-20 shows these circuits.

In Fig. 4-20*a* a NOT follows an AND gate; this particular combination is called a NOT-AND gate, or simply a NAND gate. This book uses the NAND gate a great deal. Because of this, we shall abbreviate the NAND-gate symbol as shown in Fig. 4-21. Note that the NAND-gate symbol looks like a *D* connected to a small circle.

De Morgan's theorems are important and should be memorized. We repeat them again in words:

> The complement of a sum equals the product of the complements.
> The complement of a product equals the sum of the complements.

These theorems are useful in changing Boolean expressions to equivalent forms.

As already indicated, to apply the De Morgan theorems, change plus signs to multiplication signs or vice versa, and take the complement of the individual terms rather than the entire expression. For instance, for $\overline{A + B}$

1. Change the + sign to a · sign to get $A \cdot B$.
2. Take the complement of each term to get $\overline{A} \cdot \overline{B}$.

The *A* and *B* can represent complicated expressions; we can still apply De Morgan's theorem. As an example, suppose

$$y = \overline{(C + DE)(CE + DF)}$$

This is the complement of a product, $(C + DE)$ times $(CE + DF)$. Therefore, we can use the second De Morgan theorem and rewrite the expression as

$$y = \overline{(C + DE)} + \overline{(CE + DF)}$$

Fig. 4-21. Logic symbol for NAND gate.

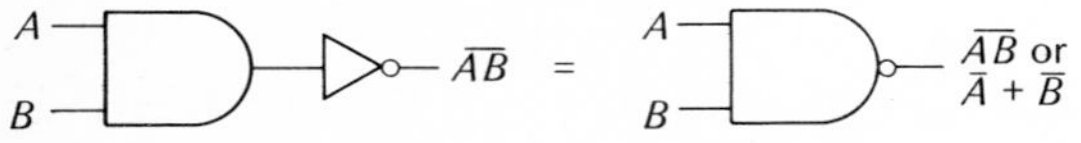

(Geometrically, this is like cutting the original expression down the middle, and replacing the multiplication by addition.)

Example 4-9

Prove the second De Morgan theorem: $\overline{A \cdot B} = \overline{A} + \overline{B}$.

Solution

When $A = 0$ and $B = 0$,

$$\overline{A \cdot B} = \overline{0 \cdot 0} = \overline{0} = 1$$

$$\overline{A} + \overline{B} = \overline{0} + \overline{0} = 1 + 1 = 1$$

When $A = 0$ and $B = 1$,

$$\overline{A \cdot B} = \overline{0 \cdot 1} = \overline{0} = 1$$

$$\overline{A} + \overline{B} = \overline{0} + \overline{1} = 1 + 0 = 1$$

When $A = 1$ and $B = 0$,

$$\overline{A \cdot B} = \overline{1 \cdot 0} = \overline{0} = 1$$

$$\overline{A} + \overline{B} = \overline{1} + \overline{0} = 0 + 1 = 1$$

When $A = 1$ and $B = 1$,

$$\overline{A \cdot B} = \overline{1 \cdot 1} = \overline{1} = 0$$

$$\overline{A} + \overline{B} = \overline{1} + \overline{1} = 0 + 0 = 0$$

Since there are no other input combinations, we have proved the second De Morgan theorem. Tables 4-13 and 4-14 show the truth tables for these expressions.

Example 4-10

Prove the following identity:

$$\overline{\overline{A} \cdot B} = A + \overline{B}$$

Solution

Use the second De Morgan theorem: the complement of a product equals the sum of the complements. First,

$$\overline{\overline{A} \cdot B}$$

means the complement of $\overline{A} \cdot B$. In turn $\overline{A} \cdot B$ is the product of $\overline{A}$ and B. Rewrite

$\overline{\overline{A} \cdot B}$ as the sum of the complements of $\overline{A}$ and B:

$$\overline{\overline{A} \cdot B} = \overline{\overline{A}} + \overline{B}$$

(We changed the · sign to a + sign, and took the complements of $\overline{A}$ and B individually. From the geometrical viewpoint, we cut the expression down the middle and changed the · sign to a + sign.)

The final step is to recognize that the double complement of a variable is equal to the variable itself: $\overline{\overline{A}} = A$. If $A = 0$, $\overline{\overline{0}} = \overline{1} = 0$. On the other hand, if $A = 1$, $\overline{\overline{1}} = \overline{0} = 1$. Therefore,

$$\overline{\overline{A} \cdot B} = \overline{\overline{A}} + \overline{B} = A + \overline{B}$$

This proves what we set out to prove.

Example 4-11

Prove the three-input NAND gate of Fig. 4-22*a* is interchangeable with the logic circuit of Fig. 4-22*b*.

Solution

The output of the NAND gate of Fig. 4-22*a* is the NOT-AND of the inputs. The AND of the inputs is ABC. The NOT of this is $\overline{ABC}$. So the output of the NAND gate is $\overline{ABC}$.

The output of Fig. 4-22*b* is easily found. The OR gate has inputs of $\overline{A}$, $\overline{B}$, and $\overline{C}$. Therefore, the final output is $\overline{A} + \overline{B} + \overline{C}$.

To show the two systems are equivalent or interchangeable, we must prove

$$\overline{ABC} = \overline{A} + \overline{B} + \overline{C}$$

We can do this with the second De Morgan theorem as follows: Think of ABC as the product of AB and C. Now, apply the second De Morgan theorem to get

$$\overline{ABC} = \overline{AB \cdot C} = \overline{AB} + \overline{C}$$

Applying De Morgan's theorem again to $\overline{AB}$ gives

$$\overline{AB} + \overline{C} = \overline{A} + \overline{B} + \overline{C}$$

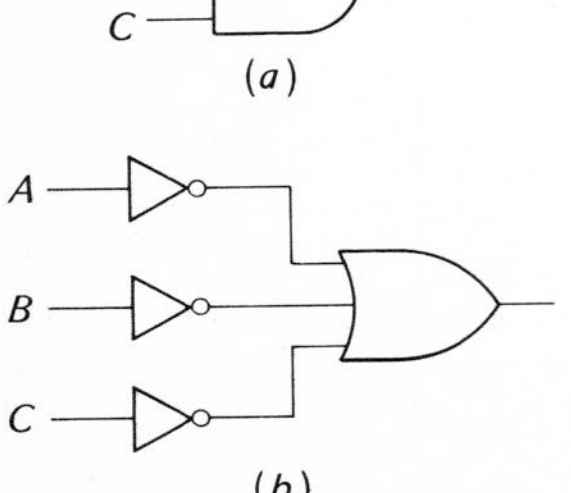

Fig. 4-22. Example 4-11.

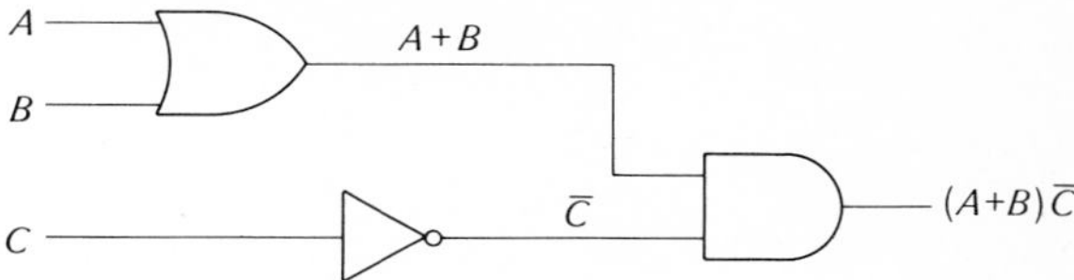

Fig. 4-23. Producing a given Boolean expression.

So you see, the two circuits of Fig. 4-22 are equivalent and interchangeable. Another way to prove equivalence is to show the truth table for $\overline{ABC}$ is identical to the one for $\bar{A} + \bar{B} + \bar{C}$ (see Prob. 4-19).

4-9 THE UNIVERSAL BUILDING BLOCK

Given any Boolean expression, we can build a logic circuit for it. Conversely, given a logic circuit, we can write a Boolean expression. As an example, the expression $(A + B)\bar{C}$ suggests a logic circuit in which A and B are first ORed, and then ANDed with the complement of C. Figure 4-23 shows the logic circuit.

To build the logic circuit associated with *any* Boolean expression, we can use OR gates for the + signs, AND gates for · signs, and NOT circuits for the overhead bars. In other words, OR, AND, and NOT circuits are the basic building blocks of all logic circuits.

The NAND gate has an interesting property: it can be used to build an OR gate or an AND gate or a NOT circuit. Because of this, NAND gates are all we need to build any logic circuit. Therefore, the NAND gate is a universal building block. (A similar statement can be made for the NOR gate.)

A NOT circuit can be made out of a NAND gate by connecting all inputs together as shown in Fig. 4-24*a*. If A is 0, the output of the NAND gate is

$$\overline{A \cdot A} = \overline{0 \cdot 0} = \bar{0} = 1 = \bar{A}$$

And if A is 1, the output is

$$\overline{A \cdot A} = \overline{1 \cdot 1} = \bar{1} = 0 = \bar{A}$$

From now on, when using a NAND gate as a NOT circuit, we shall show the symbol of Fig. 4-24*b*—it will be understood that all inputs are connected together.

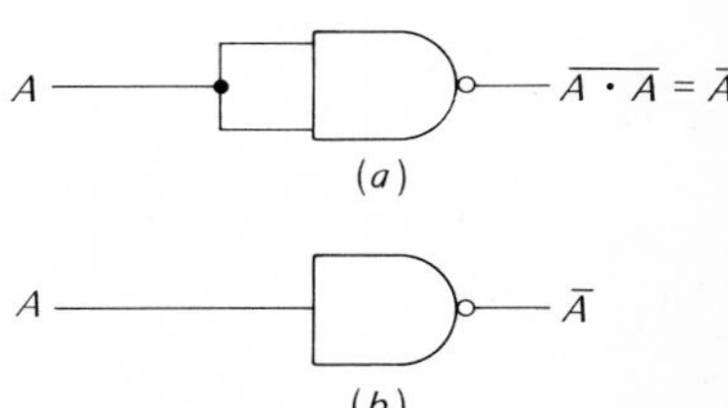

Fig. 4-24. Using a NAND gate as an inverter.

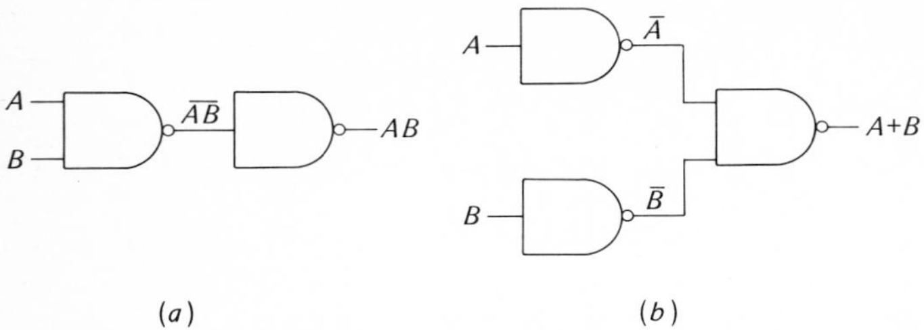

Fig. 4-25. (a) NAND gates produce ANDing. (b) NAND gates produce ORing.

We can use NAND gates to make an AND gate. Figure 4-25a shows how. The output of the first NAND gate is $\overline{AB}$. The second NAND gate then complements this, producing $\overline{\overline{AB}}$. Since the double complement of a quantity is the quantity itself, $\overline{\overline{AB}} = AB$. Therefore, NAND gates can be connected to perform the AND function.

We can also make an OR gate using NAND gates. Figure 4-25b shows how to connect the NAND gates. The first two NAND gates invert A and B to produce $\overline{A}$ and $\overline{B}$. The two-input NAND gate then produces an output of $\overline{\overline{A} \cdot \overline{B}}$. With De Morgan's second theorem,

$$\overline{\overline{A} \cdot \overline{B}} = \overline{\overline{A}} + \overline{\overline{B}} = A + B$$

(Cut down the middle and change the · sign to a + sign.) Since the final output is A OR B, we have shown that NAND gates can be used to get the OR function.

The point is that we can use NAND gates to build OR, AND, and NOT circuits. This, in turn, means we can build any logic circuit using only NAND gates. (A similar statement applies to the NOR gate.)

4-10 LAWS AND THEOREMS OF BOOLEAN ALGEBRA

When we simplify a complicated Boolean expression, we are changing a complicated digital circuit into a simpler one. For instance, to build the circuit for

$$y = ABC + A\overline{B}C + AB\overline{C}$$

we need the following:

A three-input OR gate to add ABC, $A\overline{B}C$, and $AB\overline{C}$
Three three-input AND gates to produce, ABC, $A\overline{B}C$, and $AB\overline{C}$
Two NOT circuits to produce $\overline{B}$ and $\overline{C}$

Figure 4-26a shows the proper connection of these elements. By using Boolean laws and theorems, we can simplify the original expression (this is done in Example

4-14), and get

$$y = A(B + C)$$

This leads to a simpler logic circuit. Only one two-input AND gate and one two-input OR gate are needed, as shown in Fig. 4-26*b*.

The logic circuits of Fig. 4-26 are identical as far as input-output operation is concerned; in other words, they are interchangeable. The circuit of Fig. 4-26*b* is preferred since it uses less hardware and is easier to construct.

You should be familiar enough with Boolean algebra to make obvious simplifications. Here are some basic laws and theorems of Boolean algebra.

The first group of laws in Boolean algebra is the same as in ordinary algebra.

Commutative laws $$A + B = B + A \tag{4-3}$$

$$A \cdot B = B \cdot A \tag{4-4}$$

Associative laws $$A + (B + C) = (A + B) + C \tag{4-5}$$

$$A(BC) = (AB)C \tag{4-6}$$

Distributive law $$A(B + C) = AB + AC \tag{4-7}$$

The commutative laws indicate that the order of adding or multiplying is unimportant. In other words, we get the same answer by adding *A* to *B* as we do by ad-

Fig. 4-26. Equivalent logic circuits.

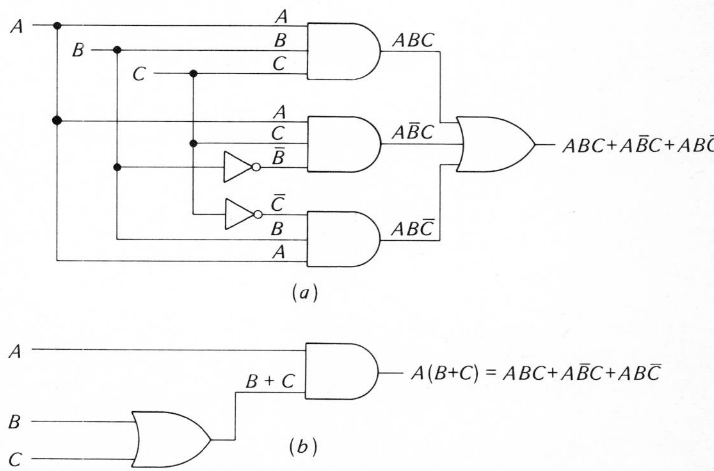

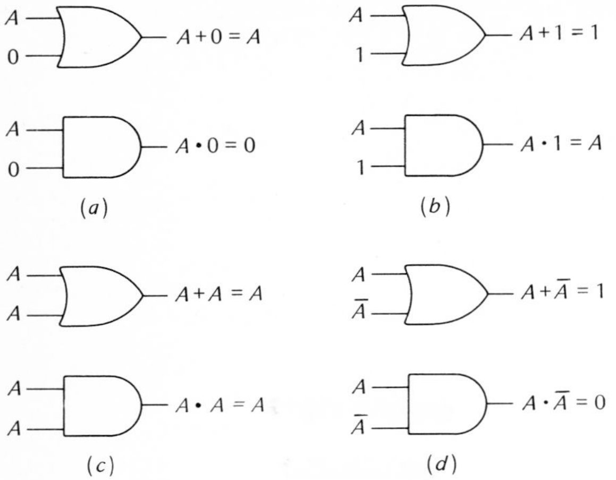

A, 0 — $A + 0 = A$ A, 1 — $A + 1 = 1$

A, 0 — $A \cdot 0 = 0$ A, 1 — $A \cdot 1 = A$

(*a*) (*b*)

A, A — $A + A = A$ A, $\overline{A}$ — $A + \overline{A} = 1$

A, A — $A \cdot A = A$ A, $\overline{A}$ — $A \cdot \overline{A} = 0$

(*c*) (*d*)

Fig. 4-27. Fundamental Boolean relations.

ding *B* to *A*. Likewise, the same answer results whether we multiply *A* by *B* or *B* by *A*.

The associative laws say you can group any two terms of a sum or any two factors of a product. In other words, given $A + B + C$, you can first add *B* and *C*, and then add the result to *A*; or, you can first add *A* and *B*, and then add the result to *C*. A similar statement applies to multiplication.

The distributive law, Eq. (4-7), indicates that we can expand expressions by multiplying term by term as in ordinary algebra. This law also implies that we can factor expressions. In other words, given a sum of two terms, each containing a common variable, we can factor this common variable. [This is like $AB + AC$. Each term contains *A*, so we can factor the *A* to get $A(B + C)$.]

These first five laws present no difficulties. They are identical to those of ordinary algebra and therefore are already ingrained in our minds.

The next group of laws is the backbone of Boolean algebra. First,

$$A + 0 = A \tag{4-8}$$

$$A \cdot 0 = 0 \tag{4-9}$$

These should be thought of in terms of OR and AND gates. $A + 0 = A$ should remind us of an OR gate with inputs of *A* and 0 as shown in Fig. 4-27*a*. The output of the OR gate depends on *A* only; if *A* is 0, the output must be 0; if *A* is 1, the output must be 1. Similarly, $A \cdot 0 = 0$ suggests the AND gate of Fig. 4-27*a*. Since the AND gate is an *all-or-nothing* gate, the output must be 0. The remaining identities in the group are

$$A + 1 = 1 \qquad \textit{new} \tag{4-10}$$

$$A \cdot 1 = A \tag{4-11}$$

$$A + A = A \qquad \textit{new} \tag{4-12}$$

$$A \cdot A = A \qquad \textit{new} \tag{4-13}$$

$$A + \overline{A} = 1 \qquad \textit{new} \tag{4-14}$$

$$A \cdot \overline{A} = 0 \qquad \textit{new} \tag{4-15}$$

Again, visualize these formulas as logic circuits. Figure 4-27 shows the circuits for each. If there is any doubt, you can prove each formula by substituting the two possible values of A: 0 and 1. In each case, the left side of the formula equals the right side. These formulas are very basic and should be memorized.

The next three formulas are already familiar: the double-complement theorem and De Morgan's theorems.

$$\overline{\overline{A}} = A \qquad \textit{new} \tag{4-16}$$

$$\overline{A + B} = \overline{A} \cdot \overline{B} \qquad \textit{new} \tag{4-17}$$

$$\overline{A \cdot B} = \overline{A} + \overline{B} \qquad \textit{new} \tag{4-18}$$

Finally, there is a group of miscellaneous theorems, all of which can be derived from the identities already listed. Some of these theorems are

$$A + AB = A \tag{4-19}$$

$$A(A + B) = A \tag{4-20}$$

$$(A + B)(A + C) = A + BC \tag{4-21}$$

$$A + \overline{A}B = A + B \tag{4-22}$$

$$A(\overline{A} + B) = AB \tag{4-23}$$

$$(A + B)(\overline{A} + C) = AC + \overline{A}B \tag{4-24}$$

$$AB + \overline{A}C = (A + C)(\overline{A} + B) \tag{4-25}$$

For this book, you should know Eqs. (4-3) through (4-18). Many of these are the same as in ordinary algebra. The ones that are different are marked *new*. Almost all of these are easily remembered if you visualize the corresponding logic circuits (Fig. 4-27).

Often, you can simplify a complicated Boolean expression by using the foregoing laws and theorems. This is important; it means you can build a simpler circuit instead of a more complicated one. For instance, suppose a digital circuit has an output given by

$$y = AC + ABC$$

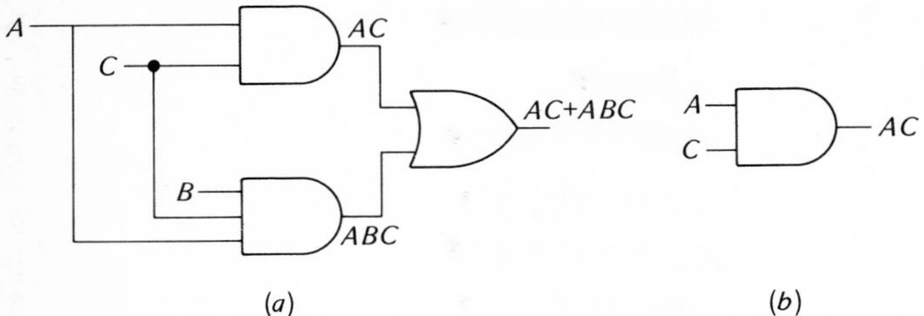

Fig. 4-28. Simplifying a logic circuit.

This equation says the output y is obtained by

1. ANDing A and C
2. ANDing A, B, and C
3. ORing AC and ABC

Figure 4-28*a* shows the digital circuit for $y = AC + ABC$; it uses two AND gates and an OR gate. This circuit is more complicated than it has to be because the original expression simplifies as follows:

$$\begin{aligned} y &= AC + ABC \\ &= AC(1 + B) \\ &= AC \end{aligned}$$

Figure 4-28*b* illustrates the logic circuit for $y = AC$; it uses only one AND gate. The point is clear. Whenever you can simplify a Boolean expression, do so; this results in less hardware when the circuit is built.

One more point. Positive and negative logic give rise to a basic *duality* in all the identities. When changing from one logic system to another, 0 becomes 1, and vice versa. Furthermore, AND gates become OR gates, and OR gates become AND gates. This means that, given any Boolean identity, we can produce a dual identity by

Changing + signs to · signs, and vice versa
Complementing all 0s and 1s

For instance, given Eq. (4-8),

$$A + 0 = A$$

complement the + sign and the 0 to get the dual identity, Eq. (4-11):

$$A \cdot 1 = A$$

As another example, Eq. (4-14) is

$$A + \overline{A} = 1$$

To get the dual identity, complement the + sign and the 1 to get

$$A \cdot \overline{A} = 0$$

This property of duality is helpful. The Boolean identities are listed at the end of the chapter in dual pairs.

Example 4-12

Prove the following:

(*a*) $A + A = A$
(*b*) $A \cdot A = A$

Solution

(*a*) Prove this by substituting the possible values of *A*, which are 0 and 1. When $A = 0$,

$$\begin{aligned} A + A &= A \\ 0 + 0 &= 0 \\ 0 &= 0 \quad \text{check} \end{aligned}$$

When $A = 1$,

$$\begin{aligned} A + A &= A \\ 1 + 1 &= 1 \\ 1 &= 1 \quad \text{check} \end{aligned}$$

(*b*) Prove that $A \cdot A = A$ in the same way. When $A = 0$,

$$\begin{aligned} A \cdot A &= A \\ 0 \cdot 0 &= 0 \\ 0 &= 0 \quad \text{check} \end{aligned}$$

When $A = 1$,

$$\begin{aligned} A \cdot A &= A \\ 1 \cdot 1 &= 1 \\ 1 &= 1 \quad \text{check} \end{aligned}$$

Example 4-13

Prove Eq. (4-19), which says $A + AB = A$.

Solution

One way to prove this is to check the four possible cases.

Case 1 When $A = 0$ and $B = 0$,

$$
\begin{aligned}
A + AB &= A \\
0 + 0 \cdot 0 &= 0 \\
0 + 0 &= 0 \\
0 &= 0 \qquad \text{check}
\end{aligned}
$$

Case 2 When $A = 0$ and $B = 1$,

$$
\begin{aligned}
A + AB &= A \\
0 + 0 \cdot 1 &= 0 \\
0 + 0 &= 0 \\
0 &= 0 \qquad \text{check}
\end{aligned}
$$

Case 3 When $A = 1$ and $B = 0$,

$$
\begin{aligned}
A + AB &= A \\
1 + 1 \cdot 0 &= 1 \\
1 + 0 &= 1 \\
1 &= 1 \qquad \text{check}
\end{aligned}
$$

Case 4 When $A = 1$ and $B = 1$,

$$
\begin{aligned}
A + AB &= A \\
1 + 1 \cdot 1 &= 1 \\
1 + 1 &= 1 \\
1 &= 1 \qquad \text{check}
\end{aligned}
$$

A different way to prove $A + AB = A$ is the following:

$$A + AB = A(1 + B) = A(1) = A$$

This approach uses Eqs. (4-7) and (4-10).

Example 4-14

Show that $y = ABC + A\bar{B}C + AB\bar{C}$ can be simplified to $y = A(B + C)$.

Solution

$$
\begin{aligned}
y &= ABC + A\bar{B}C + AB\bar{C} \\
&= AC(B + \bar{B}) + AB\bar{C} && \text{by Eq. (4-7)} \\
&= AC \qquad\qquad + ABC && \text{by Eq. (4-14)} \\
&= A(C + B\bar{C}) && \text{by Eq. (4-7)} \\
&= A(C + B) && \text{by Eq. (4-22)} \\
&= A(B + C) && \text{by Eq. (4-3)}
\end{aligned}
$$

4-11 TTL NAND GATES

NAND gates are the least expensive circuits in the 7400 series. Because of this, they are the most widely used TTL gates. Figure 4-29 shows the schematic diagram for a two-input NAND gate. Notice the *multiple-emitter* input transistor (Q_1) and the *totem-pole* output transistors (Q_3 and Q_4).

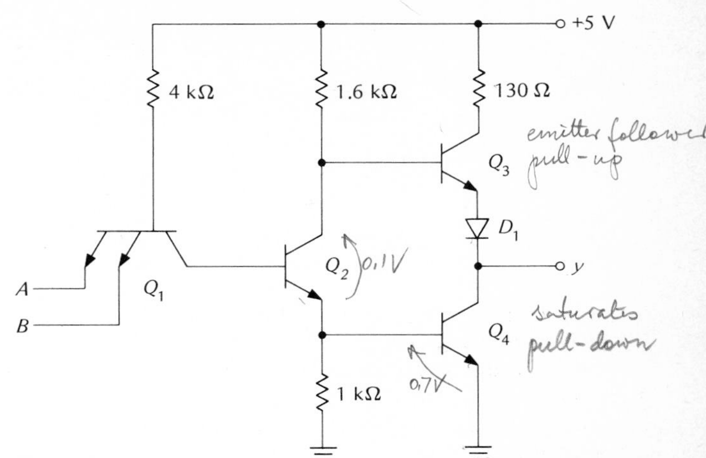

Fig. 4-29. Two-input TTL NAND gate.

When either A or B is low, Q_1 saturates, cutting off Q_2 and Q_4. Then Q_3 acts like an emitter follower and couples a high voltage to the output. This is why the first three entries of Table 4-15 show 1 outputs.

When both A and B are high, forward current through the Q_1 collector diode forces Q_2 and Q_4 into saturation. Because Q_4 is saturated, the output voltage is low. This explains the fourth output entry in Table 4-15. As you see, the truth table confirms that Fig. 4-29 is a NAND gate.

Incidentally, diode D_1 prevents Q_3 from turning on when Q_4 is saturated. Here's why. The V_{BE} of Q_4 is approximately 0.7 V, and the V_{CE} of Q_2 is around 0.1 V. This means a total of 0.8 V is applied to the base of Q_3. Without D_1 in the circuit, this 0.8 V would be sufficient to turn on Q_3. But D_1 is in the circuit and reduces the V_{BE} of Q_3 below 0.7 V, the amount needed to turn on Q_3. This is why Q_3 is cut off when Q_4 is saturated.

Why are totem-pole output transistors used? Mainly for their low output impedance. When Q_3 is on, it acts like an emitter follower with a Thevenin resistance of approximately 70 Ω. On the other hand, when Q_4 is on, it is saturated and has a Thevenin resistance of about 12 Ω. In either case, the output impedance is low enough to prevent excessive loading when other TTL gates are connected.

Table 4-15

A	B	y
0	0	1
0	1	1
1	0	1
1	1	0

Table 4-16

TTL NAND GATES

Number	Description
7400	Quad two-input NAND gate
7410	Triple three-input NAND gate
7420	Dual four-input NAND gate
7430	Single eight-input NAND gate

Low output impedance also helps the switching speed. As you know, output voltage cannot change until all stray and load capacitance across the output has been charged or discharged. With a low impedance in either output state, the *RC* time constant is short and the output voltage can change rapidly.

By adding more emitters to the multiple-emitter input transistor, we can get three-input and four-input NAND gates. Table 4-16 shows some of the more popular NAND gates available in the 7400 series. The 7400 is a quadruple two-input NAND gate; this means it is four separate two-input NAND gates in a single package like Fig. 4-30.

The remaining entries of Table 4-16 are clear. The 7410 contains three separate NAND gates, each with three inputs. The 7420 has two separate NAND gates, each with four inputs. Finally, the 7430 is one NAND gate with eight inputs.

Chapter 6 will discuss a neat way of connecting NAND gates to get the logic circuit for any Boolean expression.

1-27-77

4-12 TTL SPECIFICATIONS

The 7400 series operates normally over an ambient temperature range of 0 to 70°C. This means each TTL device in the 7400 series satisfies the values specified on its data sheet for surrounding air temperatures as low as 0°C and as high as 70°C.

Fig. 4-30. Quad two-input NAND gate.

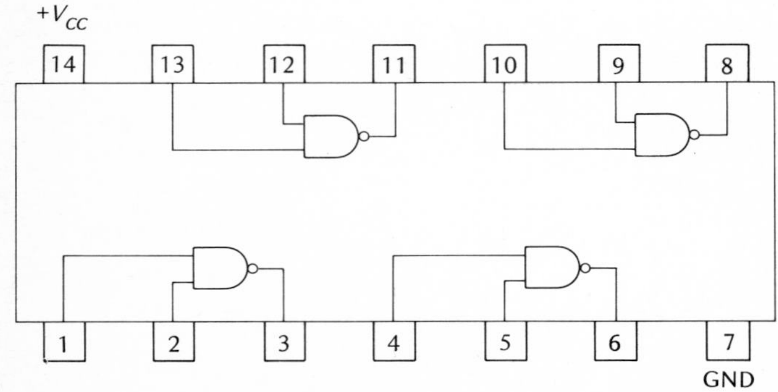

why is NOR gate wired in reverse?

Ideally, the low and high voltages for 7400 devices are 0 V and 5 V. But the voltages across the transistors cause deviations from these ideal values. For instance, an input voltage from 0 to 0.8 V qualifies as *low*. The low-state input voltage is designated on the data sheet for each 7400 device as V_{IL}, where the subscripts stand for *input low*. Data sheets give the worst-case values, and this is why V_{IL} is listed as

$$V_{IL} = 0.8 \text{ V}$$

Similarly, the ideal high value of input voltage is 5 V. In actuality, any input voltage from 2 to 5 V qualifies as a high input because it can cause the output to change states. Again, 7400 data sheets will list the worst-case value as follows:

$$V_{IH} = 2 \text{ V}$$

This is the smallest input voltage that qualifies as a *high* input.

Ideally, the output voltage across the saturated transistor should be 0 V. In reality, the output voltage may be as large as 0.4 V in the low state. This is why data sheets list a worst-case value of

$$V_{OL} = 0.4 \text{ V}$$

The subscripts on V_{OL} stand for *output low.*

Finally, the ideal value of high output voltage is 5 V. But reliable operation occurs when the high output is from 2.4 to 5 V. In other words, an output voltage as small as 2.4 V is still considered a high output because it can adequately drive other TTL circuits. On a data sheet the worst-case output high voltage is listed as

$$V_{OH} = 2.4 \text{ V}$$

Table 4-17 summarizes all the worst-case values.

Fan-out (also called "loading factor") is another thing to know about. Fan-out is defined as the number of TTL loads that a TTL device can drive reliably. The 7400 series has a fan-out of 10; this means you can connect the output of a TTL gate to 10 other TTL gates. If you exceed this number of loads, the worst-case values on the data sheet are no longer guaranteed.

Finally, there's *propagation delay time* t_p. This is the amount of time that elapses between a change in input state and the resulting change in output state. If $t_p =$ 10 ns, it will take 10 ns for the output to change states after the input has changed states. If several TTLs are cascaded, the total propagation delay time

Table 4-17

Quantity	Qualifying condition
V_{IL}	Less than 0.8 V
V_{IH}	More than 2 V
V_{OL}	Less than 0.4 V
V_{OH}	More than 2.4 V

equals the sum of the individual propagation delay times. For instance, if three TTL gates are cascaded and each has a t_p of 15 ns, then the total propagation delay time equals 45 ns.

4-13 NOISE IMMUNITY

Noise immunity is defined as the maximum induced noise voltage a TTL device can withstand without a false change in output state. For instance, Fig. 4-31*a* shows one TTL device driving another. In the low state the maximum or worst-case output voltage of the first device is 0.4 V, as shown in Fig. 4-31*b*. If no noise voltage is induced on the connecting line, the input voltage to the second device is 0.4 V.

In a noisy environment, stray electric and magnetic fields can induce unwanted noise voltages on the connecting line between the two TTL devices. It's possible, therefore, to have 0.4 V of induced noise voltage with the polarity shown in Fig. 4-31*c*. In this case, the noise raises the input voltage of the second TTL device to 0.8 V. Look at Table 4-17 and notice that V_{IL} must be less than 0.8 V to qualify as a low-state input voltage. Because of this, the second TTL device of Fig. 4-31*c* is on the verge of falsely changing output states. If the noise voltage were more than 0.4 V, the second TTL device would undergo a false change in output voltage under the worst-case conditions.

Figure 4-32 shows the other possibility, false triggering in the high state. In the high state the minimum or worst-case output voltage of the first TTL device is 2.4 V. If 0.4 V of noise is induced on the connecting line with the polarity shown, then the input voltage to the second TTL device equals 2 V. Table 4-17 shows V_{IH} must be more than 2 V to qualify as a high-state input. For this reason, the second TTL of Fig. 4-32 is on the verge of a false change in output state. If the noise voltage were more than 0.4 V, the second TTL could falsely trigger to the opposite state under the worst-case conditions.

In either state, therefore, the noise immunity of any 7400 device is 0.4 V in the worst case. This means the manufacturer guarantees that cascaded 7400 devices can withstand a maximum of 0.4 V of induced noise without a false change in output state.

Fig. 4-31. Noise produces false triggering to high state.

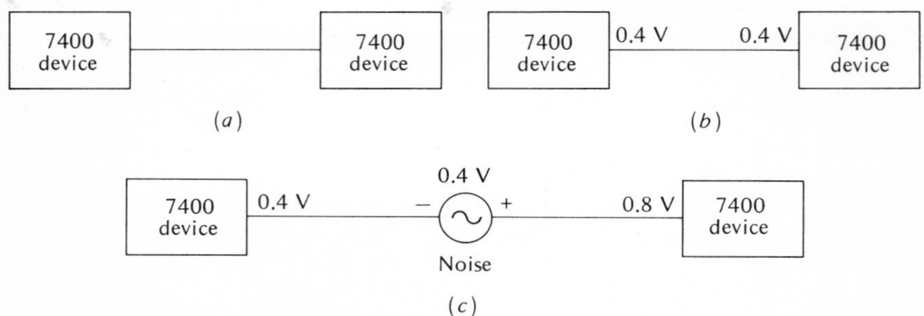

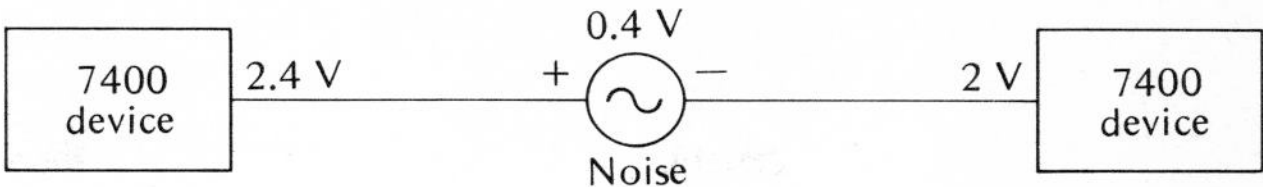

Fig. 4-32. Noise produces false triggering to low state.

4-14 OPEN-COLLECTOR TTL AND WIRE-AND

As mentioned earlier, totem-pole output transistors improve switching speed. The lower transistor provides *active pull-down;* when it saturates, it pulls the output voltage down to a low value. On the other hand, the upper transistor provides *active pull-up;* when it turns on and acts as an emitter follower, it pulls the output voltage up to a high value. In either case, an active device (a transistor) charges or discharges any capacitance across the output; this improves switching speed.

Instead of a totem-pole output, some TTL devices have an *open-collector output*. This means they use only the lower transistor of a totem-pole pair. For instance, Fig. 4-33*a* shows a two-input NAND gate with an open-collector output. Because the collector of Q_4 is open, a gate like this won't work properly until you connect an external *pull-up resistor,* shown in Fig. 4-33*b*.

Q_4 still provides active pull-down when it saturates. But when it cuts off, there's no upper transistor to pull up the output voltage. Because of this, any stray capacitance or other load capacitance across the output has to charge through the pull-up resistor. Since the charging is through a passive component, switching from low to high output is called "passive pull-up."

For 7400 TTL devices, the pull-up resistor has to be hundreds to thousands of ohms. For this reason, passive pull-up is much slower than active pull-up. This is

Fig. 4-33. (*a*) Open-collector TTL gate. (*b*) Pull-up resistor.

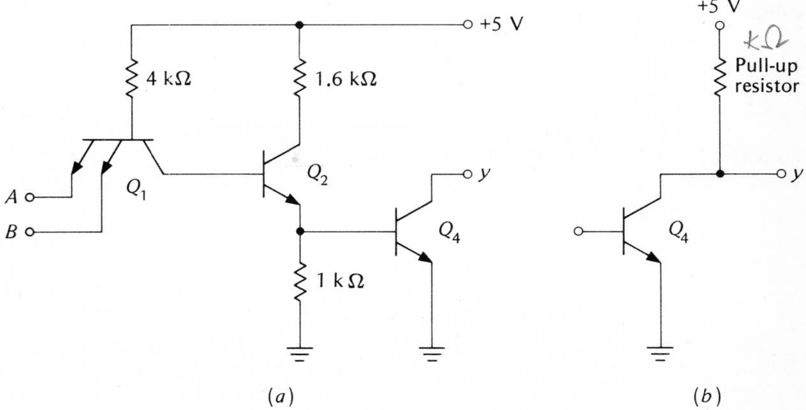

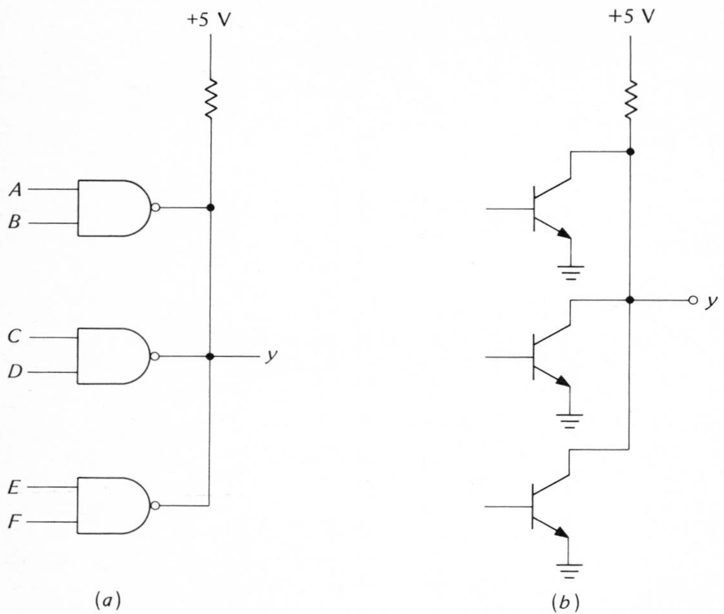

Fig. 4-34. One way to AND the outputs of NAND gates.

the main disadvantage of open-collector devices as compared with totem-pole devices.

So why bother with open-collector devices? Because their outputs can be wired together and connected to a common pull-up resistor; this direct connection produces ANDing action and eliminates the need for an AND gate.

For instance, Fig. 4-34*a* shows the outputs of three NAND gates connected to a common pull-up resistor. If these are open-collector devices, each output transistor connects to the pull-up resistor of Fig. 4-34*b*. When any or all transistors are saturated (low state), the output voltage is pulled down to a low value. The only way to get a high output is for all transistors to be cut off; then the pull-up resistor pulls the output voltage up to a high value.

In other words, wiring the outputs of open-collector devices to a common pull-up resistor automatically produces ANDing action. Because the ANDing is the result of a direct wire connection, it's called *wire*-AND. This is why the output of Fig. 4-34a is

$$y = \overline{AB} \cdot \overline{CD} \cdot \overline{EF}$$

Notice that each NAND output is ANDed with the others.

Figure 4-35 shows how we shall indicate the wire-AND connection on a *logic diagram* (one with devices and signal paths, but no supply voltages or grounds). When you see the wire-AND symbol of Fig. 4-35, remember it means open-collector devices are connected to a common pull-up resistor (not shown). The output

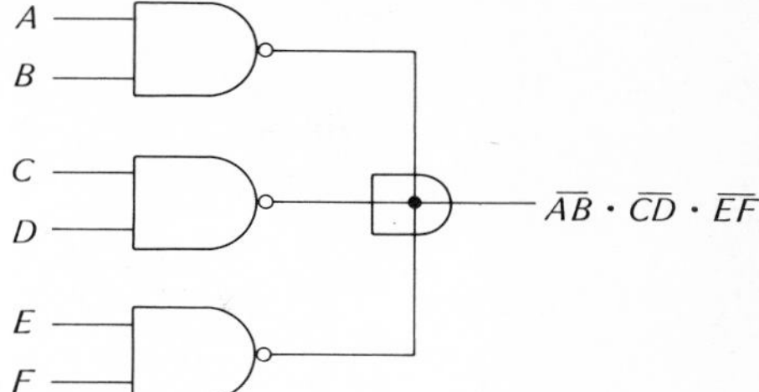

Fig. 4-35. Another way to AND the outputs of NAND gates.

of the wire-AND symbol is the ANDing of its inputs; this is why the final output in Fig. 4-35 is

$$y = \overline{AB} \cdot \overline{CD} \cdot \overline{EF}$$

Wire-AND is tremendously important in applications where the outputs of many TTL devices have to be ANDed. For instance, suppose it's necessary to AND the outputs of 25 TTL devices. If totem-pole devices are used, then a 25-input AND gate is needed. But if open-collector devices are used, we can wire-AND the outputs and eliminate the 25-input AND gate.

One more point. Wire-AND won't work with standard totem-pole devices, because excessive current destroys internal transistors for certain output conditions. In particular, trouble arises when the pull-up transistor of a totem-pole device drives a saturated pull-down transistor in another totem-pole device. This is what led to *tristate* TTL, a new variety of totem-pole TTL devices introduced in the early 1970s. With tristate TTL, it is possible to wire-AND totem-pole outputs and avoid the loss of switching speed that occurs with open-collector outputs.[1]

4-15 WIRE-OR

The wire-AND connection of Fig. 4-35 results in an output of

$$y = \overline{AB} \cdot \overline{CD} \cdot \overline{EF} \tag{4-26}$$

With De Morgan's first theorem, this is equivalent to

$$y = \overline{AB + CD + EF} \tag{4-27}$$

Figure 4-36 shows the original circuit with this equivalent output. The inputs are ANDed, then ORed, and finally inverted. Because of the ORing action, the term *wire-*OR is widely used as a synonym for wire-AND.

[1] For an in-depth paper, see John Sheets, "Three-state Switching Brings Wire-OR to TTL," *Electronics,* p. 78, Sept. 14, 1970.

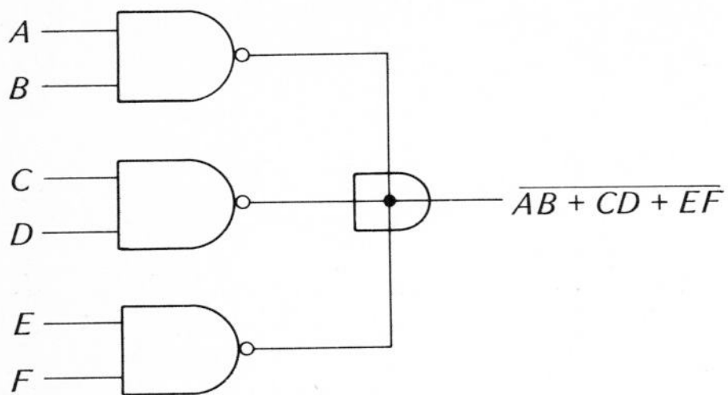

Fig. 4-36. Logic symbol for the wire-AND connection.

Both terms mean the same thing physically. They mean the outputs of NAND gates are connected to a common pull-up resistor. If you use Eq. (4-26), you should refer to the direct connection of outputs as wire-AND. On the other hand, when using Eq. (4-27), refer to the connection as wire-OR.

STUDY AIDS

Summary

The OR gate is an *any-or-all* gate; it has a 1 output when any or all of the inputs are 1. The AND gate is an *all-or-nothing* gate; it has 1 output only when all inputs are 1. The NOT circuit is an inverter or complementer; it changes a 0 to a 1, and vice versa.

Boolean algebra is the algebra of logic circuits. The + sign stands for OR addition, the way an OR gate combines its inputs to produce its output. The · sign denotes AND multiplication, and the — sign denotes the operation of a NOT circuit.

De Morgan's theorems are useful in changing Boolean expressions to equivalent forms. These two theorems suggest the NOT-OR (NOR) gate and the NOT-AND (NAND) gate. The NAND gate has a 0 output only when all inputs are 1. The NAND gate is a universal building block and can be used to build any logic circuit.

We can list the laws and theorems of Boolean algebra in dual pairs as follows:

1.	$A + B = B + A$	$AB = BA$
2.	$A + (B + C) = (A + B) + C$	$A(BC) = (AB)C$
3.	$A(B + C) = AB + AC$	$A + BC = (A + B)(A + C)$
4.	$A + 0 = A$	$A \cdot 1 = A$
5.	$A + 1 = 1$	$A \cdot 0 = 0$
6.	$A + A = A$	$A \cdot A = A$
7.	$A + \overline{A} = 1$	$A \cdot \overline{A} = 0$
8.	$\overline{\overline{A}} = A$	$\overline{\overline{A}} = A$
9.	$\overline{A + B} = \overline{A} \cdot \overline{B}$	$\overline{A \cdot B} = \overline{A} + \overline{B}$
10.	$A + AB = A$	$A(A + B) = A$
11.	$A + \overline{A}B = A + B$	$A(\overline{A} + B) = AB$

Noise immunity is a measure of how much noise a digital IC can withstand without a false change in output state. The fan-out or loading factor of a gate is the number of gates that can be driven. Propagation delay time is the time that elapses between an input change and the resulting output change.

TTL gates are available with totem-pole outputs, the main advantage being low output impedance in either state. Alternatively, TTL gates come with open-collector outputs; this allows their use in wire-AND connections.

Glossary

AND *gate* An all-or-nothing gate. The output is a 1 only when all the inputs are 1s.
complement This refers to the NOT operation. The complement of 0 is 1, and the complement of 1 is 0.
gate A device with one output and two or more inputs, designed so that there is an output only for certain combinations of input signals.
inverter A NOT circuit; also called a "complementing" circuit or a "negating" circuit.
NAND *gate* An AND gate followed by a NOT circuit; sometimes called a NOT-AND gate.
negative logic This means 1 stands for the more negative of the two voltage levels.
noise immunity A measure of how much noise a digital IC can withstand without a false change in output state.
NOR *gate* An OR gate followed by a NOT circuit; sometimes called a NOT-OR gate.
NOT *circuit* A circuit that inverts or complements the input signal; it makes 1s out of 0s, and vice-versa.
OR *gate* An any-or-all gate. It has a 1 output if any or all of its inputs are 1s.
positive logic This means 1 stands for the more positive of the two voltage levels.
totem-pole output A pair of output transistors often used in the 7400 series ICs to provide a low output impedance in either state.
truth table A list of all the input-output possibilities of a logic circuit.
*wire-*AND Connecting the outputs of open-collector TTL devices to a common pull-up resistor. (The term may also be used with tristate TTL devices.)
*wire-*OR This means a direct connection of open-collector NAND gates to a common pull-up resistor.

Review Questions

1. When does an OR gate have a 1 output?
2. Describe the truth table for a two-input OR gate. For a three-input OR gate.
3. How many horizontal rows are there in a truth table if there are n inputs?
4. When does an AND gate have a 1 output? What does the truth table of an AND gate look like?
5. Define a positive logic system. A negative logic system.
6. In Boolean algebra why does $1 + 1 = 1$?
7. What logic circuits do the + sign, · sign, and — sign represent?
8. What are the two De Morgan theorems (in words)?
9. What is a NOR gate? A NAND gate? Under what input conditions is the output of a NOR gate equal to 1? For what input conditions is the NAND-gate output equal to 0?

10. Why is the NAND gate called a universal building block?

11. How do you find the dual identity for any given identity?

12. What is the main advantage of a totem-pole output?

13. Define propagation delay time.

14. What is the fan-out of a gate?

Problems

4-1. In Fig. 4-1*a* show the truth table for the OR gate if the batteries can have values of 3 or 12 V. Use actual voltages in the table and ideal diodes.

4-2. In Fig. 4-3 suppose that the *A, B, C* inputs are at either 2 or 10 V, and that the diodes are ideal. Show the truth table using actual voltages.

4-3. Show the truth table for Fig. 4-6 using actual voltages. Treat the diodes as ideal, and use voltage levels of 1 and 10 V.

4-4. In Fig. 4-7 the two voltage levels are 2 and 10 V. Construct the truth table assuming a negative logic system.

4-5. A system has three inputs: *A, B,* and *C*. Show the truth table if the output *y* of this system is

$$y = (A + B)(A + C)$$

4-6. A logic circuit has inputs of *A, B,* and *C*. Show its truth table if its output *y* is

$$y = A + BC$$

Also, draw the logic circuit that has this output.

4-7. Write the Boolean expression for the output *y* of Fig. 4-37*a*. Work out the value of *y* for all possible input conditions, and show the truth table.

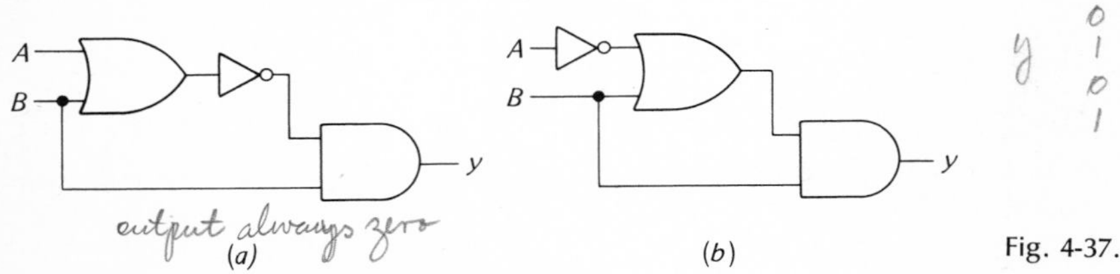

(*a*) (*b*)

Fig. 4-37.

4-8. In Fig. 4-37*b* write the expression for v. What is the truth table for this system?

4-9. Work out the value of $y = A\overline{B} + \overline{A}B$ for the four possible input conditions.

4-10. Draw the logic circuit for $y = A\overline{B} + \overline{A}B$.

4-11. Prove that $AB + B = B$ by showing that the left side equals the right side for all four input conditions.

4-12. Given that $y = A + \overline{B}C(A + \overline{C}) + B$, what is the value of y when
(a) $A = 0,\ B = 1,\ C = 0.$
(b) $A = 1,\ B = 0,\ C = 1.$

4-13. Show that $\overline{ABCD} = \overline{A} + \overline{B} + \overline{C} + \overline{D}$.

4-14. Prove that $\overline{A + B + C + D} = \overline{A} \cdot \overline{B} \cdot \overline{C} \cdot \overline{D}$.

4-15. Rewrite $\overline{\overline{A} + B}$ using the first De Morgan theorem.

4-16. Use De Morgan's second theorem to rewrite

$$y = \overline{(AB + CD)(ABC + D)}$$

4-17. Show how NAND gates can be used to build the logic circuit for $y = A + B\overline{C}$.

4-18. Show how NAND gates can be used to build the logic circuit for $y = AB + CD$.

4-19. Use truth tables to show that $\overline{ABC} = \overline{A} + \overline{B} + \overline{C}$.

4-20. Prove Eq. (4-22) by constructing the truth table of each side of the equation.

4-21. Prove Eq. (4-25) by showing the truth table of the left side is identical to that of the right side.

4-22. Prove Eqs. (4-7) and (4-10) using the truth-table approach.

4-23. Reduce $y = (A + B)(A + \overline{B})(\overline{A} + B)$ by using the laws and theorems of Boolean algebra.

4-24. Simplify $y = AB + ABC + \overline{A}B + A\overline{B}C$.

4-25. Three TTL gates are cascaded. Two of them have propagation delay times of 15 ns, and the third has a propagation delay time of 10 ns. What is the overall propagation delay time?

4-26. Write the Boolean equation for the output of Fig. 4-38*a*.

4-27. In Fig. 4-38*b*, $A = 1,\ B = 0,\ C = 1,\ D = 1,\ E = 0,\ F = 0,\ G = 1$, and $H = 0$. What does y equal?

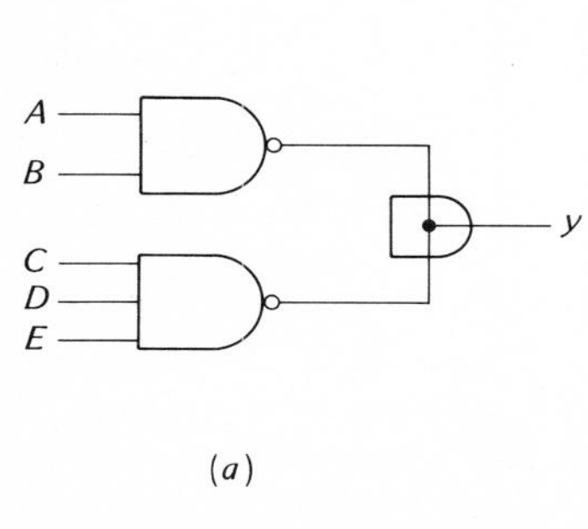

(*a*)

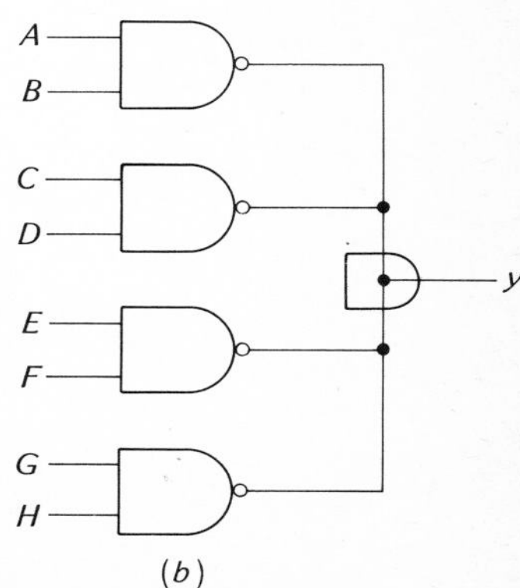

(*b*)

Fig. 4-38.

Arithmetic Circuits

Digital electronics is exciting; it enables us to build circuits that duplicate some of the processes of our minds—the best known being that of computing. By combining AND, OR, and NOT circuits in the right way, we can build circuits that add and subtract. Since these circuits are electronic, they are fast. Typically, an addition problem is done in microseconds.

In this chapter we shall discuss some basic arithmetic circuits like the exclusive-OR gate, the half- and full-adders, the half- and full-subtractors, 8421 adders, and excess-3 adders. Besides giving us a first idea of how a computer works, these circuits lay the base upon which to build our later discussions of digital systems.

After you've had a chance to study this chapter, you should be able to

1. Construct truth tables for the exclusive-OR gate, the half-adder, and the full-adder.
2. Describe how binary numbers are added in a parallel binary adder like Fig. 5-11.
3. Define small-scale and medium-scale integration.

5-1 THE EXCLUSIVE-OR GATE

Figure 5-1 shows an *exclusive-*OR *gate*. It has two inputs and one output. Each input goes into an inverter; the outputs of the inverters are $\bar{A}$ and $\bar{B}$. As shown in Fig. 5-1, $\bar{A}$ and B go into the upper AND gate, so its output is $\bar{A}B$. Likewise, $A\bar{B}$ comes out of the lower AND gate. The OR gate has inputs of $A\bar{B}$ and $\bar{A}B$, so the final output is

$$\boxed{y = A\bar{B} + \bar{A}B} \qquad (5\text{-}1)$$

Why is the circuit of Fig. 5-1 called an exclusive-OR gate? To answer this, we shall find the value of y for the four input conditions.

1. When $A = 0$ and $B = 0$,

$$y = 0 \cdot \bar{0} + \bar{0} \cdot 0 = 0 \cdot 1 + 1 \cdot 0 = 0 + 0 = 0$$

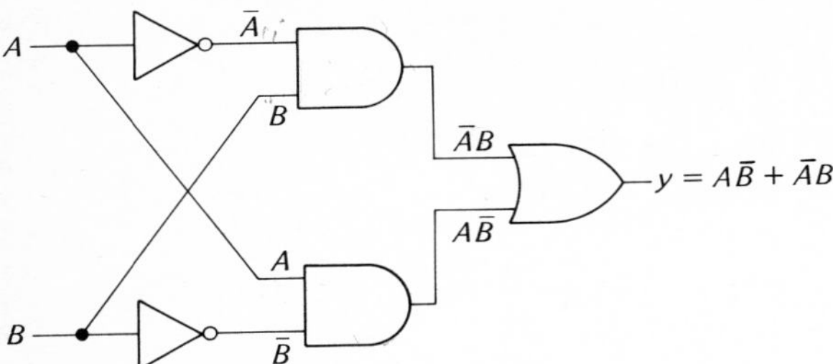

Fig. 5-1. Exclusive-OR gate.

2. When $A = 0$ and $B = 1$,

$$y = 0 \cdot \bar{1} + \bar{0} \cdot 1 = 0 \cdot 0 + 1 \cdot 1 = 0 + 1 = 1$$

3. When $A = 1$ and $B = 0$,

$$y = 1 \cdot \bar{0} + \bar{1} \cdot 0 = 1 \cdot 1 + 0 \cdot 0 = 1 + 0 = 1$$

4. When $A = 1$ and $B = 1$,

$$y = 1 \cdot \bar{1} + \bar{1} \cdot 1 = 1 \cdot 0 + 0 \cdot 1 = 0 + 0 = 0$$

These results are summarized in the truth table of Table 5-1.

Table 5-1
EXCLUSIVE-OR TRUTH TABLE

A	B	y
0	0	0
0	1	1
1	0	1
1	1	0

The reason for the name *exclusive*-OR is this: a 1 output occurs when A or B is 1, *but not both*. Stated another way, the exclusive-OR gate has a 1 output only when the inputs are different; the output is 0 when the inputs are the same.

The exclusive-OR gate gives us a new kind of function to work with. We will use the symbol ⊕ to stand for this function. That is, to describe an exclusive-OR gate, we write

$$y = A \oplus B \tag{5-2}$$

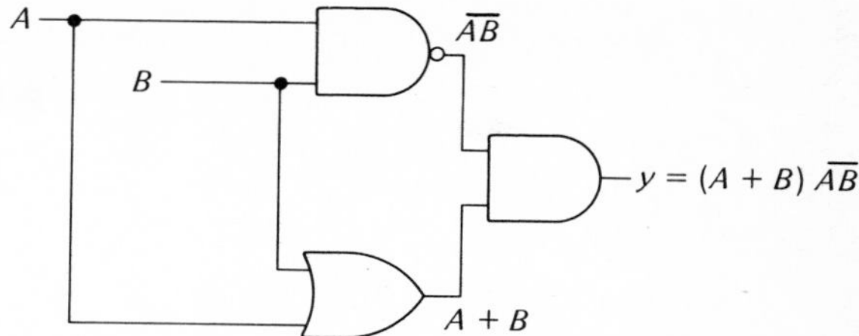

Fig. 5-2. Another way to build an exclusive-OR gate.

Whenever we see $y = A \oplus B$, we shall know the output is given by the truth table of Table 5-1.

The exclusive-OR operation is sometimes called "mod-2 addition." Here are the rules for this addition:

$$
\begin{aligned}
0 \oplus 0 &= 0 \\
0 \oplus 1 &= 1 \\
1 \oplus 0 &= 1 \\
1 \oplus 1 &= 0
\end{aligned}
$$

Mod-2 addition is the same as binary addition, provided we disregard carries.

The circuit of Fig. 5-1 is only one of many ways to build an exclusive-OR gate. Figure 5-2 shows another way. $\overline{AB}$ comes out of the NAND gate and the final output is

$$y = (A + B)\overline{AB} \tag{5-3}$$

At first, this may not seem equal to $A\overline{B} + \overline{A}B$ (the output of an exclusive-OR gate). However, Eqs. (5-1) and (5-3) are equal; by De Morgan's second theorem,

$$y = (A + B)\overline{AB} = (A + B)(\overline{A} + \overline{B})$$

Fig. 5-3. Exclusive-OR gates.

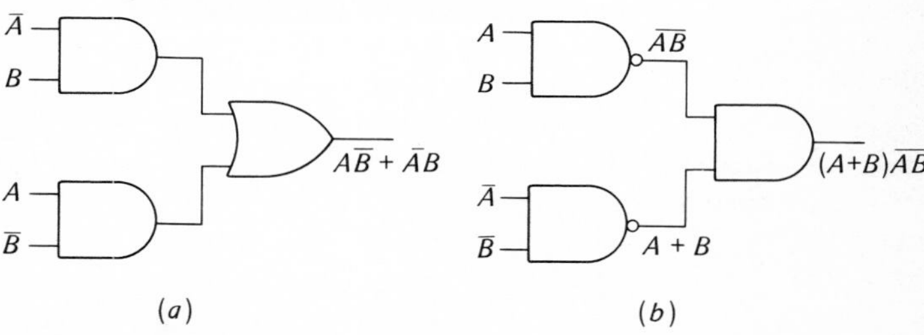

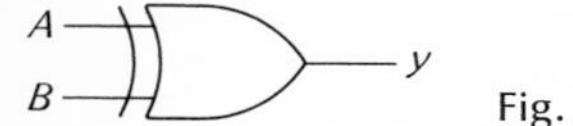

Fig. 5-4. Standard symbol for exclusive-OR gate.

Next, multiply the binomials to get

$$y = A\bar{A} + A\bar{B} + \bar{A}B + B\bar{B}$$

The AND of a variable and its complement must always be 0 because either the variable or its complement is 0. Therefore, $A\bar{A} = 0$ and $B\bar{B} = 0$, so the expression becomes

$$y = A\bar{B} + \bar{A}B$$

which is the exclusive-OR function. Therefore, the circuits of Figs. 5-1 and 5-2 are equivalent.

Sometimes both a variable and its complement are stored in a digital system. We can then use the simpler circuits of Fig. 5-3 for mod-2 addition.

To represent the exclusive-OR gate, we use the standard symbol of Fig. 5-4. Whenever you see this logic symbol, remember the output y is a 1 when A or B is 1, but not both.

Example 5-1

Show one way of building a parity checker for a four-bit word.

Solution

Figure 5-5*a* shows a way to check the parity of the word *ABCD*. The circuit adds the bits of *ABCD;* a final sum of 0 implies even parity; a sum of 1 means odd parity. For instance, suppose $ABCD = 1001$. Figure 5-5*b* shows the circuit for this condition. Note that the final sum is 0, which means even parity.

As another example, the circuit for $ABCD = 1110$ is shown in Fig. 5-5*c*. The final output is 1, denoting odd parity.

Example 5-2

Show a way of building a six-bit binary-to-Gray converter.

Solution

Figure 5-6*a* shows one way to do it. The binary number is *ABCDEF*. The circuit adds the adjacent bits using mod-2 addition and produces the Gray equivalent.

As a numerical example, suppose $ABCDEF = 100110$. Figure 5-6*b* shows the circuit for this input. The circuit adds the bits, two at a time, to get the Gray equivalent, 110101.

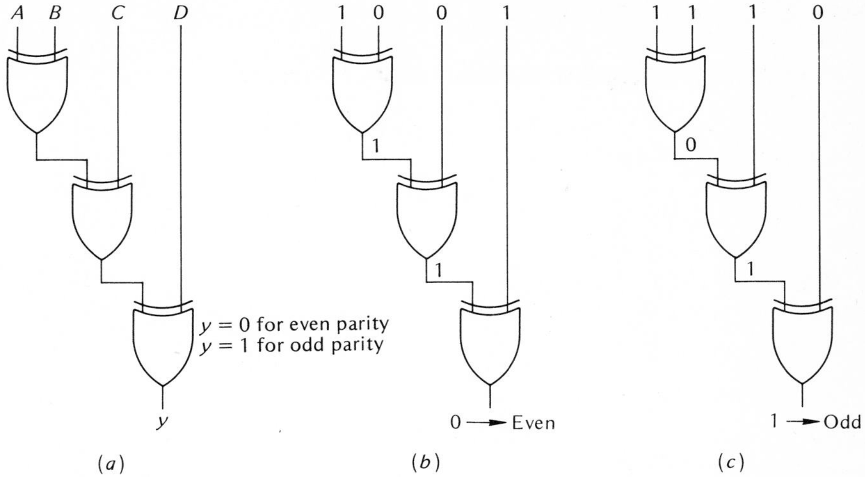

Fig. 5-5. Checking parity with exclusive-OR gates.

Fig. 5-6. Converting from binary to Gray code.

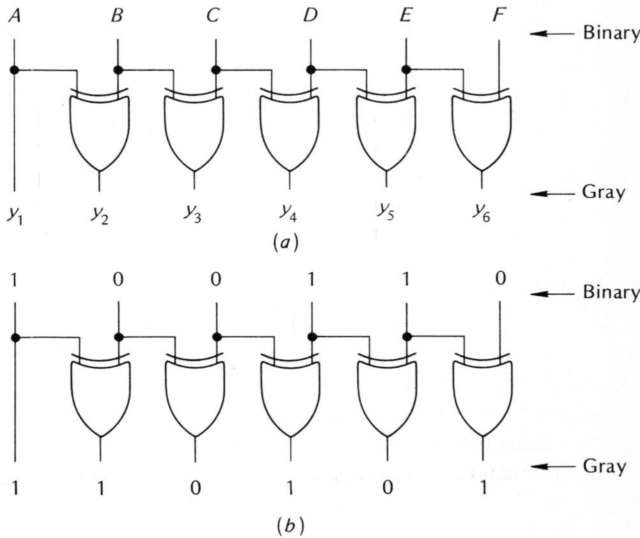

5-2 THE HALF-ADDER

The *half-adder* adds two binary digits at a time. Figure 5-7 shows how to make a half-adder. The output of the exclusive-OR gate is the sum, and the output of the AND gate is the carry. As with any two-input circuit, there are four distinct cases. These are:

1. When $A = 0$ and $B = 0$,

$$\text{Sum} = A \oplus B = 0 \oplus 0 = 0$$
$$\text{Carry} = AB = 0 \cdot 0 = 0$$

2. When $A = 0$ and $B = 1$,

$$\text{Sum} = 0 \oplus 1 = 1$$
$$\text{Carry} = 0 \cdot 1 = 0$$

3. When $A = 1$ and $B = 0$,

$$\text{Sum} = 1 \oplus 0 = 1$$
$$\text{Carry} = 1 \cdot 0 = 0$$

4. When $A = 1$ and $B = 1$,

$$\text{Sum} = 1 \oplus 1 = 0$$
$$\text{Carry} = 1 \cdot 1 = 1$$

Table 5-2 summarizes these results.

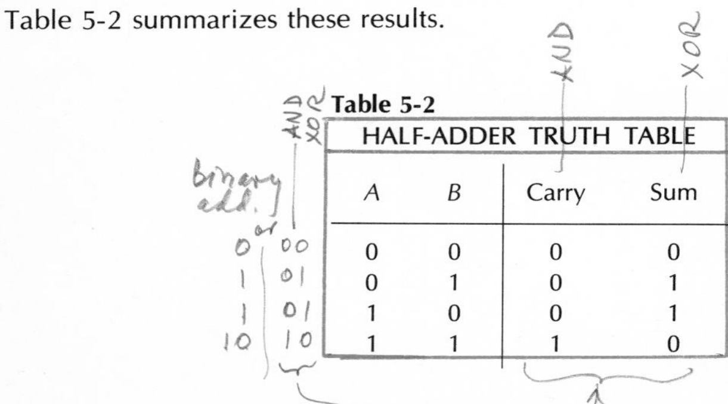

Table 5-2

HALF-ADDER TRUTH TABLE

A	B	Carry	Sum
0	0	0	0
0	1	0	1
1	0	0	1
1	1	1	0

The half-adder of Fig. 5-7 performs binary addition. It does what we do mentally when we add two binary digits. For instance, 0 combined with 0 gives 0 carry a 0. 1 combined with 1 gives 0 carry a 1.

Figure 5-8*a* shows another way to build a half-adder. The upper AND gate produces AB, the carry output, and the NAND gate produces $\overline{AB}$. The sum output is

$$\text{Sum} = (A + B)\overline{AB} = A\overline{B} + \overline{A}B = A \oplus B$$

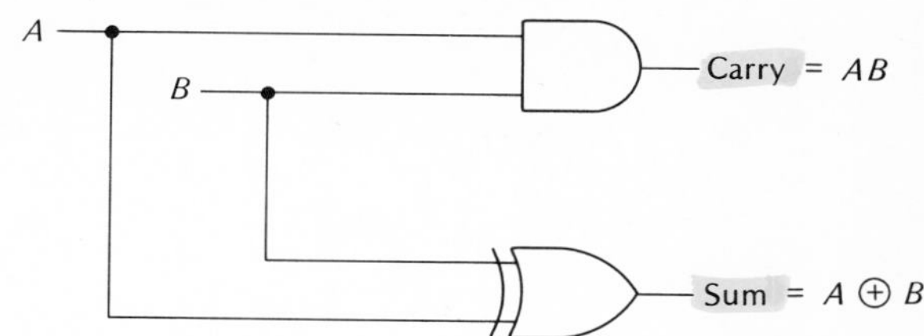

Fig. 5-7. Half-adder.

and the carry is

$$\text{Carry} = AB$$

Digital systems often contain a variable and its complement. This leads to simpler forms of the half-adder. For instance, Fig. 5-8*b* shows a half-adder that can be used when the inputs are A and B, along with the complements $\bar{A}$ and $\bar{B}$.

Figure 5-8*c* shows another way to make a half-adder. This time only NAND gates are used.

There are many ways to build half-adders. The important thing to remember is the half-adder adds two binary digits. It is an elementary circuit, but it does take the first step toward circuits capable of more difficult arithmetic.

5-3 THE FULL-ADDER

When adding two binary numbers, you may have a carry from one column to the next. For instance,

$$\begin{array}{r} 111 \\ +101 \\ \hline 1100 \end{array}$$

In the least significant column,

$$1 + 1 = 0 \qquad \text{carry a 1}$$

In the next column, you must add three digits because of the carry.

$$1 + 0 + 1 = 0 \qquad \text{carry a 1}$$

In the last column, you again must add three digits because of the carry.

$$1 + 1 + 1 = 1 \qquad \text{carry a 1}$$

To add binary numbers electronically, we need a circuit that can handle three digits at a time. By connecting two half-adders and an OR gate, we get a *full-adder,*

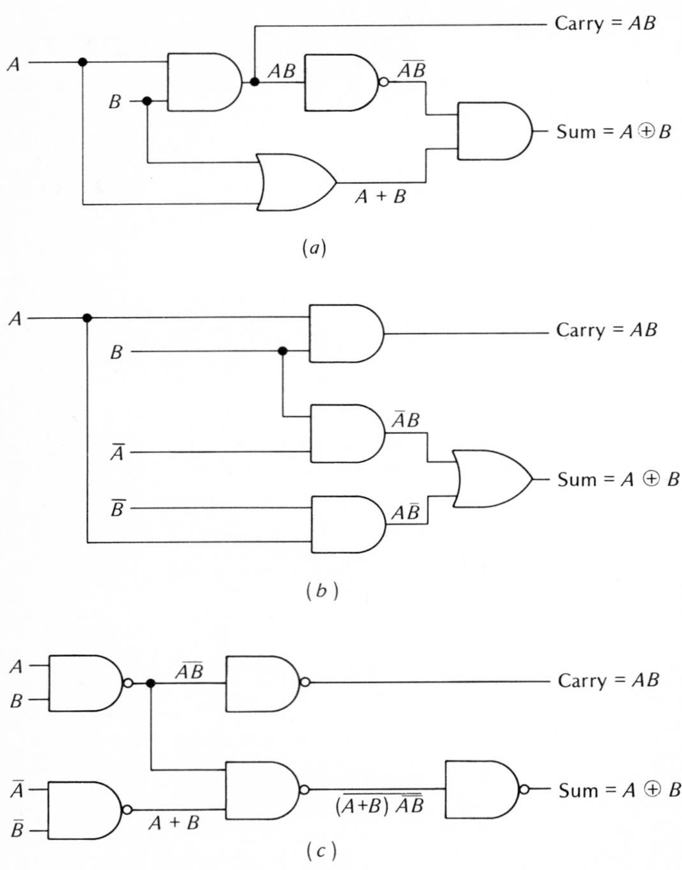

Fig. 5-8. Other half-adder designs.

a circuit that can add three digits at a time. Figure 5-9 shows a full-adder. The boxes labeled *HA* are half-adders. Since we know how a half-adder and an OR gate work, we can easily determine the final output.

For instance, suppose $A = 1$, $B = 1$, and $C = 0$. Figure 5-10*a* shows the full-adder with these inputs. The first half-adder has a sum of 0 with a carry of 1. The

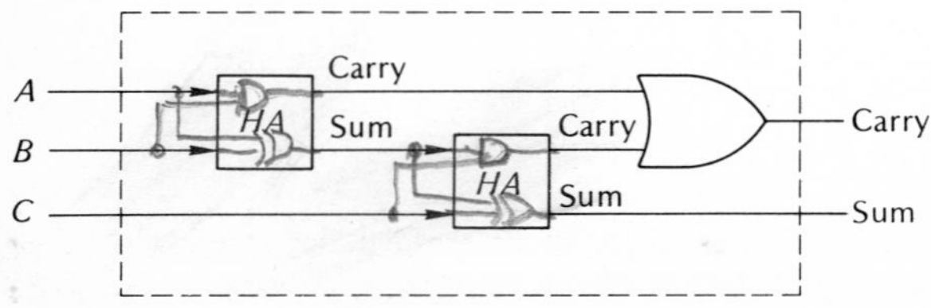

Fig. 5-9. Full-adder.

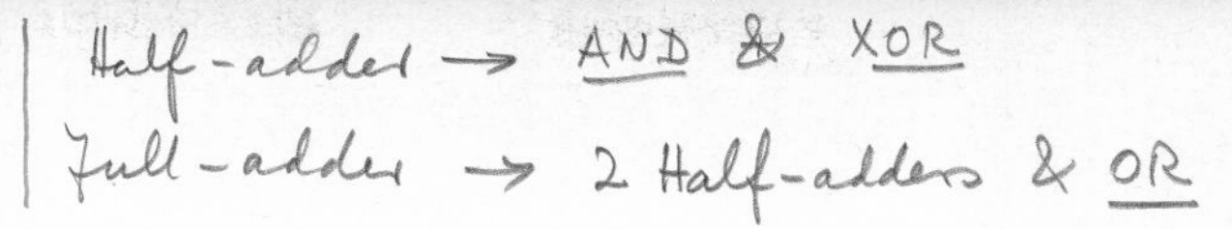

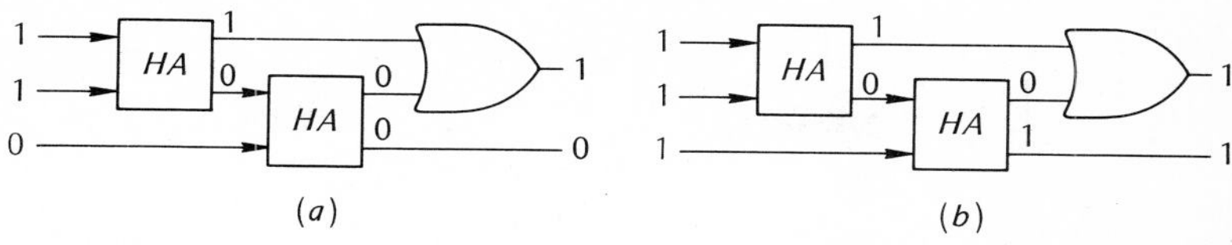

Fig. 5-10. Examples of how a full-adder works.

second half-adder has a sum of 0 with a carry of 0. Therefore, the final output is a sum of 0 with a carry of 1.

If the inputs are $A = 1$, $B = 1$, and $C = 1$, we get a sum of 1 with a carry of 1. Figure 5-10*b* shows the half-adder with these inputs. Each half-adder has the outputs shown, so the final output is a sum of 1 with a carry of 1.

If we work out the sum and carry outputs in the same way for the other input combinations, we get the truth table of Table 5-3. (You should work out enough of these to satisfy yourself that the full-adder does work.)

Table 5-3

FULL-ADDER TRUTH TABLE

A	B	C	Carry	Sum
0	0	0	0	0
0	0	1	0	1
0	1	0	0	1
0	1	1	1	0
1	0	0	0	1
1	0	1	1	0
1	1	0	1	0
1	1	1	1	1

Remember the key idea: the full-adder adds 3 binary digits at a time. The next section shows how to use full-adders to get the sum of binary numbers with more than one bit.

5-4 A PARALLEL BINARY ADDER

We can connect adders as shown in Fig. 5-11 to add two binary numbers. (The boxes labeled *FA* are full-adders.) The binary numbers being added are $A_4A_3A_2A_1$ and $B_4B_3B_2B_1$. The answer is

$$\begin{array}{r} A_4A_3A_2A_1 \\ +B_4B_3B_2B_1 \\ \hline S_5S_4S_3S_2S_1 \end{array}$$

The first column requires only a half-adder. For any column above the first, there

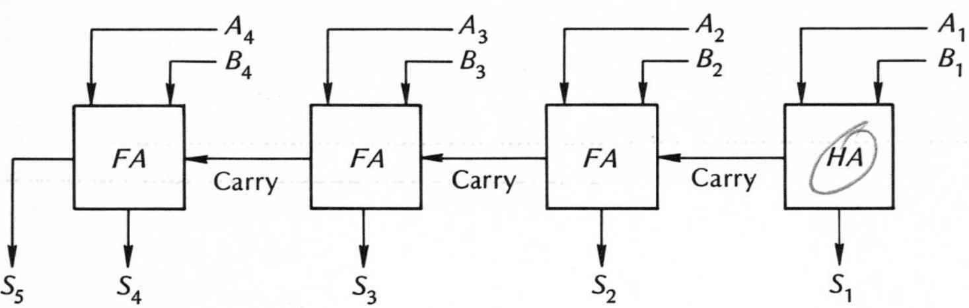

Fig. 5-11. Parallel four-bit binary adder.

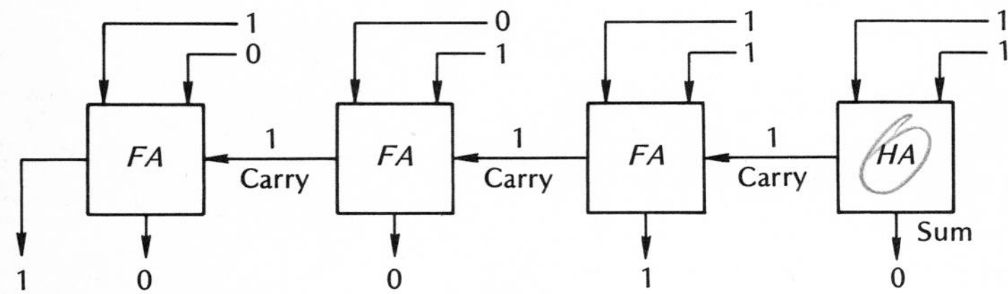

Fig. 5-12. Example of adding four-bit numbers.

may be a carry from the preceding column; therefore, we must use a full-adder for each column above the first.

As an example of how the adder of Fig. 5-11 works, suppose we want to add decimal numbers 11 and 7. The binary equivalent of decimal 11 is 1011, and the binary equivalent of decimal 7 is 0111. Figure 5-12 shows the binary adder with these inputs. Starting with the half-adder, we know the sum must be 0 with a carry of 1 as shown. The carry goes into the first full-adder, which adds 1 + 1 + 1 to get a sum of 1 with a carry of 1. This carry goes into the next full-adder, which adds 0 + 1 + 1 to get a sum of 0 with a carry of 1. The last full-adder adds 1 + 0 + 1 to get a sum of 0 with a carry of 1. The final output of the system is 10010. The decimal equivalent of this is

$$\begin{array}{ccccc} 1 & 0 & 0 & 1 & 0 \\ 16 & \not{8} & \not{4} & 2 & \not{1} = 18 \end{array}$$

which is the correct decimal sum of 11 and 7. Therefore, the parallel binary adder of Fig. 5-11 gives the binary sum of two four-bit numbers. (You should work out a few other cases to convince yourself that the system does add binary numbers.)

The adder of Fig. 5-11 has limited capacity. The largest binary numbers that can be added are 1111 and 1111. So, the maximum capacity is

$$\begin{array}{r} 15 \\ +15 \\ \hline 30 \end{array} \qquad \begin{array}{r} 1111 \\ +1111 \\ \hline 11110 \end{array}$$

To increase the capacity, more full-adders can be connected to the left end of the system. For instance, to add six-bit numbers, connect two more full-adders.

How about subtraction? Figure 5-13 shows a system that subtracts $B_4B_3B_2B_1$ from $A_4A_3A_2A_1$. Here is what it does. First, the four inverters complement each B bit to get $\overline{B}_4\overline{B}_3\overline{B}_2\overline{B}_1$, the 1's complement of $B_4B_3B_2B_1$. The full-adders now add $A_4A_3A_2A_1$, $\overline{B}_4\overline{B}_3\overline{B}_2\overline{B}_1$, and the end-around carry. In other words, the system does what we do in our minds when we use the 1's complement to subtract. (Sometimes the inverters are unnecessary because the 1's complement of the number is already stored in the system.) see p. 25

At last we are beginning to remove some of the mystery about digital computers. We have just seen how logic circuits can be connected to perform binary addition and subtraction. There are other aspects of computer operation such as storing numbers, bringing the numbers out of memory into the arithmetic section, and putting the answer back into memory. These processes are covered in later chapters.

5-5 AN 8421 ADDER

Some digital systems use binary numbers throughout. But there are many machines that use BCD numbers. As mentioned earlier, the 8421 code is the most popular BCD code; it is often called "the BCD code."

In the 8421 code each decimal digit is converted to its four-bit equivalent. For instance, decimal 539 becomes

$$\begin{array}{cccc} 5 & 3 & 9 & \text{decimal} \\ \downarrow & \downarrow & \downarrow & \downarrow \\ 0101 & 0011 & 1001 & \text{8421 equivalent} \end{array}$$

In the 8421 code there is no four-bit group above 1001; combinations like 1010, 1011, 1100, 1101, 1110, and 1111 do not exist. Because of this, we run into trouble when trying to add 8421 numbers whose sum exceeds 9. For instance, if we add 8 and 5 using binary addition, we get

$$\begin{array}{rr} 8 & 1000 \\ +5 & +0101 \\ \hline 13 & 1101 \end{array}$$

Fig. 5-13. Parallel four-bit binary subtractor.

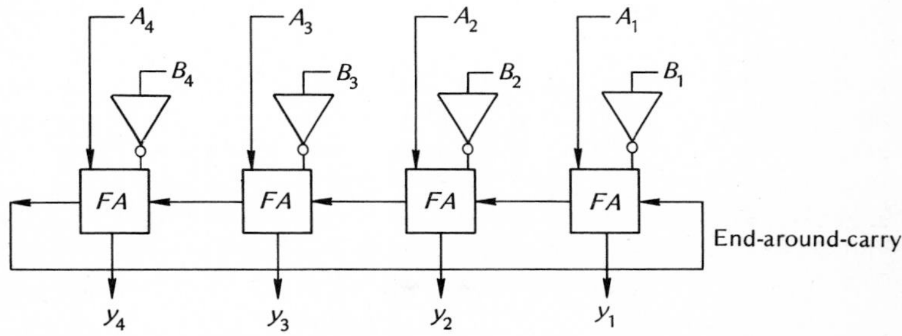

The answer 1101 is fine in straight binary code, but it is meaningless in the 8421 code. The 8421 answer should be 0001 0011.

The problem arises because the 8421 code uses only 10 of 16 possible four-bit groups. When the decimal sum exceeds 9, we must somehow skip the six forbidden groups to restore the answer to 8421 form. To skip these six forbidden combinations, we can add 6 (0110) to the sum. For instance, if we add 8 and 5 using 8421 numbers, we get

1000	
+0101	
1101	binary equivalent of 13
+0110	add 6 to return to 8421 form
0001 0011	8421 equivalent of 13

The final answer is the 8421 equivalent of 13. As another example, we add 9 and 3.

9	1001	
+3	+0011	
12	1100	binary answer
	+0110	add 6 to return to 8421
	0001 0010	8421 answer

We can summarize the rules for 8421 addition as follows:

> Add 8421 numbers using the rules of binary addition.
> If the sum of the four-bit groups is greater than decimal 9, add 0110 to that sum to return to 8421 form.
> If the sum is 9 or less, leave it alone because it is in 8421 form.

Figure 5-14 shows an 8421 adder. This system adds 8421 digits using binary addition, and adds the correction of 0110 when the sum exceeds 9.

As an example of how the 8421 adder of Fig. 5-14 works, let us trace the addition for 8 and 5. At the top of the figure the two input BCD digits are $A_4A_3A_2A_1$ (1000) and $B_4B_3B_2B_1$ (0101). The upper row of adders combines these two inputs to get

$A_4A_3A_2A_1$	1000
$+B_4B_3B_2B_1$	+ 0101
$S_5S_4S_3S_2S_1$	01101

The sum is greater than 9, detected by feeding S_4 and S_3 into the upper AND gate. Both the inputs are 1s, so the AND gate produces a 1 output; this 1 output is fed into the OR gate, so a 1 comes out of the OR gate.

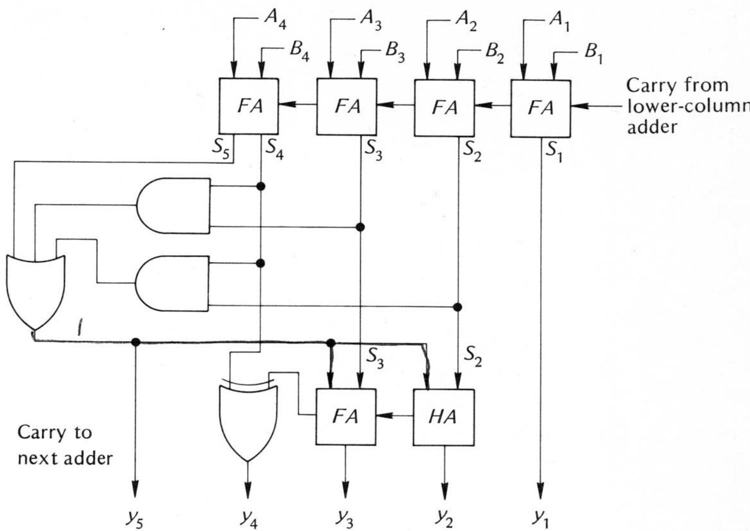

Fig. 5-14. Four-bit 8421 adder.

Next, to restore $S_4S_3S_2S_1$ (1101) to 8421 form, 0110 is added to it.

$$\begin{array}{r} S_4S_3S_2S_1 \\ +0\ 1\ 1\ 0 \\ \hline y_5y_4y_3y_2y_1 \end{array}$$

In the least significant column y_1 must equal S_1, so S_1 comes down to the final output as shown in Fig. 5-14. The bottom row of adders adds the 1 out of the OR gate to S_3 and S_2. This restores the answer to 8421 form.

The system of Fig. 5-14 will correctly add any two 8421 BCD digits. It detects when the decimal sum is greater than 9—this is the purpose of the two AND gates and the OR gate—and it adds 0110 to the answer. When the sum is 9 or less, 0000 is added.

Remember: The system of Fig. 5-14 takes care of only one column of decimal digits. To add decimal numbers with several digits, use several 8421 adders, one for each decimal column. For instance, to add 531 and 326, three 8421 adders like that of Fig. 5-14 are needed, one adder for each decimal column. By cascading 8421 adders, you can add numbers of any size.

5-6 AN EXCESS-3 ADDER

We discussed the excess-3 code in Chap. 3. Recall that a decimal number is expressed as an excess-3 number by adding 3 to each decimal digit before converting to its four-bit equivalent. Also recall the rules for excess-3 addition. When-

see p. 42, etc.

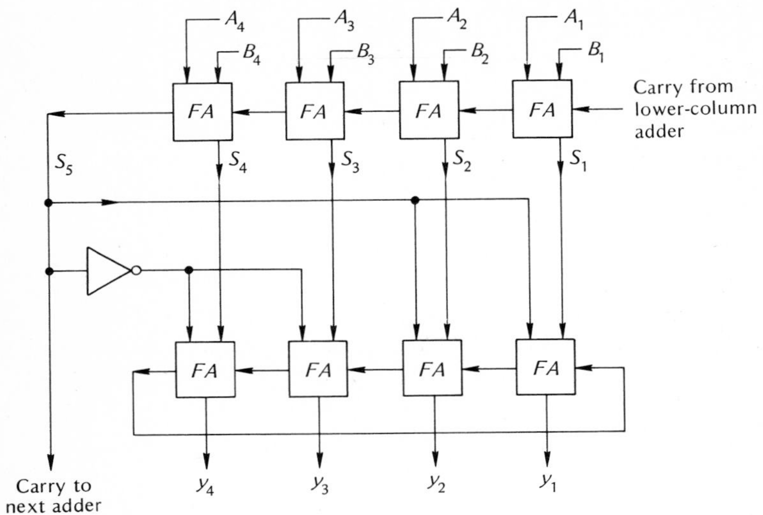

Fig. 5-15. Four-bit excess-3 adder.

ever a four-bit group does not produce a carry, subtract 0011 (3) from it to return to excess-3 form. But when a group does produce a carry, add 3 to it. (See Sec. 3-2.)

Figure 5-15 shows how to build an excess-3 adder. The two excess-3 inputs go into the upper row of adders. The uncorrected sum is $S_5S_4S_3S_2S_1$. Then 0011 is added or subtracted depending upon whether there is a carry or not. S_5 is a 1 when there is a carry, and 0 when there is no carry. Note that $\overline{S}_5\overline{S}_5S_5S_5$ is added to the uncorrected sum $S_4S_3S_2S_1$. When there is a carry, S_5 is 1, and $\overline{S}_5\overline{S}_5S_5S_5$ is 0011. In other words, we are adding 3 to return the answer to excess-3 form. When there is no carry, we subtract 3 from the answer by using the 1's complement. Therefore, when $S_5 = 0$, $\overline{S}_5\overline{S}_5S_5S_5 = 1100$ (the 1's complement of 0011). The full-adder on the left produces the end-around carry, so the final answer is the correct excess-3 sum.

As a numerical example, we trace the addition of 5 and 2. In the upper adders,

$A_4A_3A_2A_1$	1000	excess-3 for 5
$+B_4B_3B_2B_1$	+ 0101	excess-3 for 2
$S_5S_4S_3S_2S_1$	1101	excess-6 for 7

Since the decimal sum is less than 10, there is no final carry, and S_5 is 0. The second row of adders subtracts 0011 by the 1's-complement approach.

1101	uncorrected sum
+1100	1's complement of 0011
1 1001	
→ 1	end-around carry
1010	excess-3 for 7

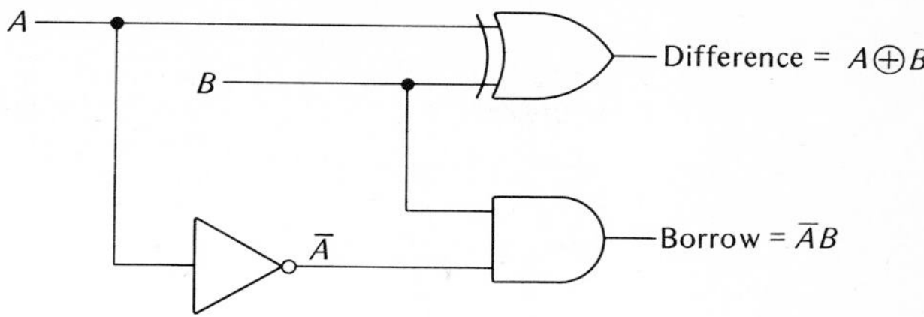

Fig. 5-16. Half-subtractor.

The excess-3 adder of Fig. 5-15 handles one column of decimal digits. By cascading adders of this type, you can add larger excess-3 numbers. For instance, to add 3946 and 2679 use four excess-3 adders, one for each decimal column.

5-7 HALF- AND FULL-SUBTRACTORS

Instead of using complements to subtract, circuits can subtract binary numbers directly. Recall the rules for binary subtraction:

$$
\begin{aligned}
0 - 0 &= 0 \quad \text{with a borrow of } 0 \\
0 - 1 &= 1 \quad \text{with a borrow of } 1 \\
1 - 0 &= 1 \quad \text{with a borrow of } 0 \\
1 - 1 &= 0 \quad \text{with a borrow of } 0
\end{aligned}
$$

Table 5-4 summarizes these results by listing the subtraction rules for $A - B$.

What kind of logic circuit has a truth table like Table 5-4? First, the difference output is 0 whenever the inputs A and B are the same; the difference output is 1 whenever A and B are different. So, we can use an exclusive-OR gate to produce the difference output. Second, the borrow output is 1 only when A is 0 and B is 1. We can get this borrow output by ANDing $\overline{A}$ and B.

Figure 5-16 shows one way to build a *half-subtractor,* a circuit that subtracts one binary digit from another. The circuit of Fig. 5-16 has a truth table identical to Table 5-4. You can see there will be a borrow only when $A = 0$ and $B = 1$. Further, the difference output will be correct for each of the four possible $A - B$ combinations.

Table 5-4
HALF- AND FULL-SUBTRACTORS

A	B	Borrow	Difference
0	0	0	0
0	1	1	1
1	0	0	1
1	1	0	0

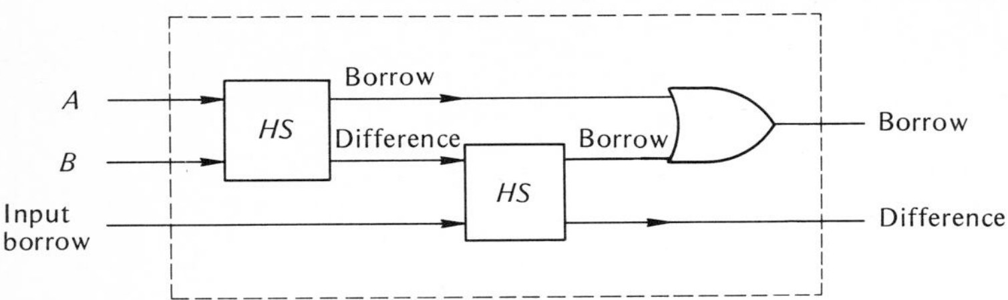

Fig. 5-17. Full-subtractor.

The half-subtractor handles only two bits at a time and can be used for the least significant column of a subtraction problem. To take care of a higher-order column, we need a *full-subtractor*. Figure 5-17 shows a full-subtractor; it uses two half-subtractors and an OR gate.

Half- and full-subtractors are analogous to half- and full-adders; by cascading half- and full-subtractors as shown in Fig. 5-18, we have a system that directly subtracts $B_4B_3B_2B_1$ from $A_4A_3A_2A_1$.

The adders and subtractors give us the basic circuits needed for binary arithmetic; multiplication and division can be done by repeated additions and subtractions (discussed in later chapters, after we have studied registers).

5-8 TTL MSI

Small-scale integration (SSI) means the complexity of an IC is comparable to a logic circuit with fewer than 12 gates. Examples of SSI are NAND gates (7400, 7410, 7420, 7430), AND gates (7408, 7411, 7421), OR gates (7432), and hex inverters (7404 through 7407; these are six separate NOT circuits in one package).

Depending on the technology used (bipolar or MOS) and the chip size, manufacturers can produce ICs much more complicated than SSI. *Medium-scale integration* (MSI) means the complexity of an IC is comparable to a logic circuit with 12 to 100 gates. Examples of MSI include the 74141 and the 7447 decoder/drivers dis-

Fig. 5-18. Parallel four-bit binary subtractor.

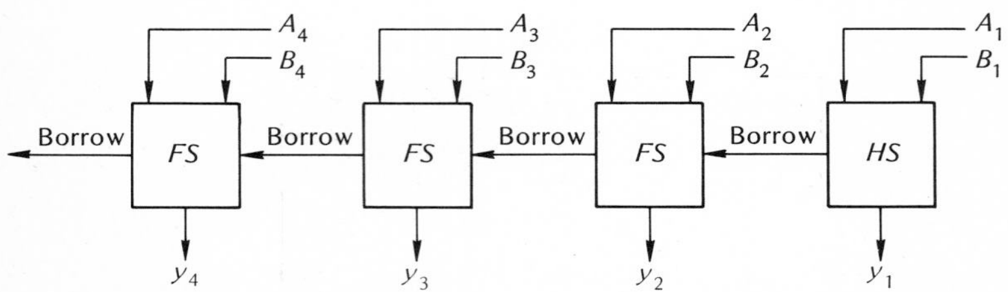

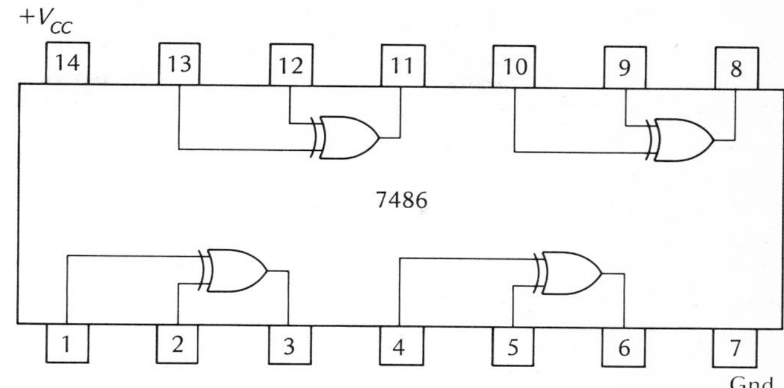

Fig. 5-19. Quad two-input exclusive-OR gate.

cussed earlier, exclusive-OR gates, parity checkers, full adders, and Gray-to-binary converters.

The 7486 is an MSI device that contains four two-input exclusive-OR gates with totem-pole outputs. This means four separate exclusive-OR gates in a single IC package as shown in Fig. 5-19. Besides the uses discussed earlier for exclusive-OR gates, we can build a *controlled four-bit inverter* (see Fig. 5-20). The idea is simple. When the voltage on the control line is low, the output equals the four-bit input. On the other hand, when the control voltage is high, the output equals the 1's complement of the input.

So a circuit like Fig. 5-20 can transfer either the input number or the 1's complement of the input number. When the input word has more than four bits, it can be processed in four-bit *bytes* (subdivisions of a word). For instance, a 20-bit word can be visualized as five bytes, each with four bits. In this case, five 7486s can produce the 1's complement of the 20-bit number; each 7486 handles one four-bit byte as shown in Fig. 5-21.

The 74180 is an example of a parity checker. This MSI circuit can be set up to check for even or odd parity of a nine-bit input (one parity bit and eight data bits).

Fig. 5-20. Controlled four-bit inverter.

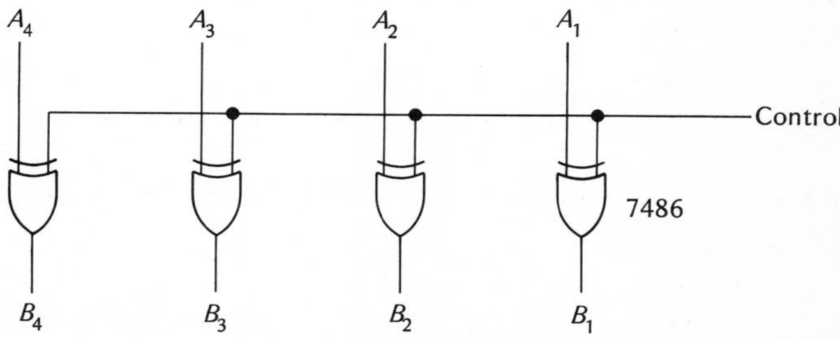

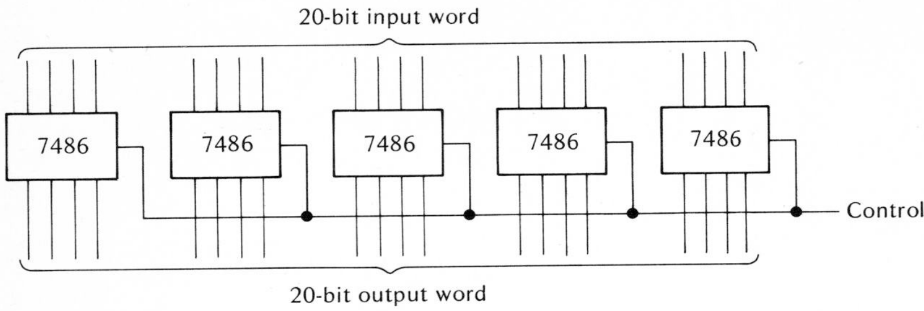

Fig. 5-21. Inverting a 20-bit word in four-bit bytes.

Figure 5-22 shows a 74180 testing for even parity. Here's the idea. When a high voltage is applied to pin 3 (even input), the 74180 performs exclusive-OR addition on the nine input bits. If the number of input 1s is even, a high voltage comes out of pin 5 (Σ even output); if the number of input 1s is odd, a low voltage comes out.

One-bit, two-bit, and four-bit adders are available in the 7400 series. The 7483 is an MSI device that performs binary addition of two four-bit numbers. Figure 5-23 illustrates the idea. $A_4A_3A_2A_1$ is one of the input binary numbers; $B_4B_3B_2B_1$ is the other. The 7483 adds these binary numbers and delivers an output of $S_4S_3S_2S_1$ plus a carry of C_5.

For longer binary numbers, the bits can be processed in four-bit bytes. To add 20-bit numbers, five 7483s can operate in parallel, each 7483 handling a four-bit byte. The C_5 carry of one 7483 becomes the C_0 input to the next higher 7483.

Table 5-5 shows a few of the widely used 7400-series MSI devices. All except the 7490 and the 7493 have already been discussed. The next chapter includes a discussion of the 7490 and the 7493. The 7490 is a counting device with a base of 10; the 7493 is a counting device with a base of 16.

Fig. 5-22. Integrated parity checker.

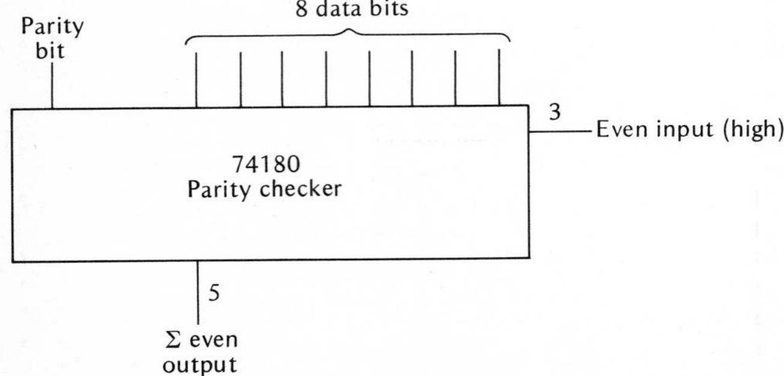

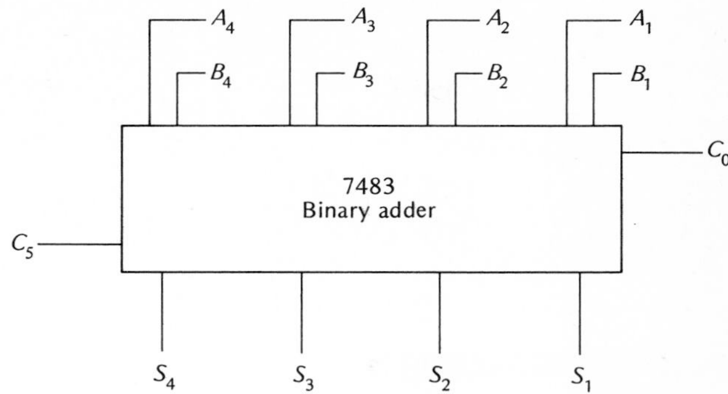

Fig. 5-23. Four-bit binary adder.

STUDY AIDS

Summary

The exclusive-OR gate has two inputs and one output. The output is a 1 when either input, but not both, is a 1. Stated another way, the output is a 1 when the inputs are different.

The exclusive-OR gate does mod-2 addition. This kind of addition is the same as binary addition, provided we disregard the carries. The symbol for mod-2 addition is $\oplus$.

The half-adder has two inputs and two outputs; it adds two bits at a time. The full-adder has three inputs and two outputs; it adds three bits at a time.

By connecting half-adders and full-adders, we can make a parallel binary adder capable of adding numbers with many bits. With a few modifications this binary adder can subtract using the 1's complement.

To make an 8421 adder we must add 0110 (6) whenever the decimal sum of the four-bit groups exceeds 9. Cascaded 8421 adders can add numbers of any size.

An excess-3 adder is also easy to make. First, use a straight binary adder to get the uncorrected sum. Next, add or subtract 0011 (3), depending on whether the un-

Table 5-5

TTL MSI DEVICES

Number	Description
7447	BCD-to-seven-segment decoder/driver
7483	Binary adder (two four-bit inputs)
7486	Quad two-input exclusive-OR gate
7490	Decade counter
7493	Four-bit binary counter
74141	BCD-to-decimal decoder/driver

corrected sum is greater or less than 9. Cascaded excess-3 adders can handle numbers of any size.

The use of complements to subtract numbers is not necessary. Half- and full-subtractors give us a direct method. By cascading half- and full-subtractors, we get a parallel binary subtractor.

Small-scale integration (SSI) means the complexity of an IC is comparable to a logic circuit with fewer than 12 gates. Medium-scale integration (MSI) means the complexity is comparable to a logic circuit with 12 to 100 gates.

Glossary

byte A subdivision of a word.

full-adder A logic circuit with three inputs and two outputs. This circuit adds three bits at time, giving a sum and a carry output.

full-subtractor A three-input two-output circuit. It is used to subtract one binary digit from another, as well as to subtract a borrow produced by a lower-order column.

half-adder A logic circuit with two inputs and two outputs. It adds two bits at a time, producing a sum and a carry output.

half-subtractor A two-input two-output logic circuit. It subtracts one binary digit from another.

mod-2 addition A binary-type addition whose rules are the same as ordinary binary addition, except that carries are disregarded. It is also called exclusive-OR addition.

MSI Abbreviation for medium-scale integration. Refers to ICs whose complexity is comparable to logic circuits with 12 to 100 gates.

SSI Abbreviation for small-scale integration. Refers to ICs whose complexity is comparable to logic circuits with fewer than 12 gates.

Review Questions

1. What is the output of an exclusive-OR gate for the four-input cases?

2. What are the rules for mod-2 addition? What symbol do we use for this addition?

3. Describe the truth table of a half-adder. What kind of logic circuit can we use to produce the carry output? The sum output?

4. What does a full-adder do? How many inputs and outputs does it have?

5. To make a parallel binary adder that can handle 10-bit inputs, how many half-adders and how many full-adders do we need?

6. In an 8421 adder, how much is added when the decimal sum is less than 10? What do we add if the decimal sum is greater than 9?

7. In an excess-3 adder, it is necessary to correct the answer to return it to excess-3 form. What number is used to correct the answer, and when is it added or subtracted?

8. What is the truth table for a half-subtractor? How is the borrow output produced?

9. What is a full-subtractor? How many half-subtractors and OR gates make a full-subtractor?

10. Define SSI and MSI.

11. What is a byte?

Problems

5-1. If we let 1 stand for a minus sign, and 0 for a plus sign, numbers like −110 and +101 would be written as 1 110 and 0 101. In multiplying binary numbers we may get plus or minus answers. Show how an exclusive-OR gate can be used to produce the correct sign in a multiplication problem.

5-2. To check the parity of a 10-bit word using exclusive-OR gates, how many of these gates do we need?

5-3. Draw a logic circuit that converts 10-bit binary numbers into Gray-code numbers. Use exclusive-OR gates.

5-4. Show a logic system that converts 10-bit Gray numbers into binary numbers. Use exclusive-OR gates.

5-5. Instead of using De Morgan's second theorem to show that

$$A\overline{B} + \overline{A}B = (A + B)\overline{AB}$$

use the truth-table method of proof. In other words, show that the truth table of the left member equals the truth table of the right member.

5-6. Work out the truth table of the circuit shown in Fig. 5-24. What kind of circuit is this?

Fig. 5-24.

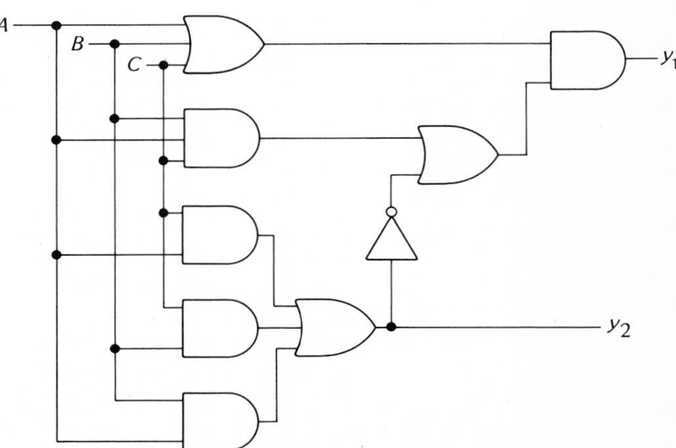

5-7. Redraw the 8421-adder of Fig. 5-14, showing the inputs and outputs of all logic circuits for input numbers of $A_4A_3A_2A_1 = 0110$, and $B_4B_3B_2B_1 = 1001$.

5-8. Redraw the excess-3 adder of Fig. 5-15, showing the inputs and outputs of all logic circuits for decimal-number inputs of 4 and 7.

5-9. Show a logic system that converts a four-bit 8421 number into a four-bit excess-3 number.

5-10. Draw a logic circuit that converts a four-bit excess-3 number into the equivalent four-bit 8421 number.

5-11. To subtract 35-bit numbers, how many half-subtractors and how many full-subtractors are needed?

5-12. A controlled inverter (similar to Fig. 5-21) is needed to process a 48-bit word. How many 7486s are required? To produce the 1's complement of the 48-bit word, should a low or high voltage be applied to the control line?

Simplifying Logic Circuits

Given a truth table, what is the corresponding Boolean equation? Given a Boolean equation, what is the simplest logic circuit for it? These are a few of the important questions answered in this chapter.

When finished with this chapter, you should be able to

1. Draw the AND-OR network corresponding to a truth table.
2. Convert a truth table to a Karnaugh map.
3. Use a Karnaugh map to simplify a logic circuit.

6-1 FUNDAMENTAL PRODUCTS

In digital computers and most other digital systems, signals are available in either complemented or uncomplemented form. For instance, if two variables A and B exist, their complements $\overline{A}$ and $\overline{B}$ normally are also present.

Figure 6-1 shows the four possible ways to AND two input signals that are in complemented or uncomplemented form. Look at Fig. 6-1*a* and note the inputs are $\overline{A}$ and $\overline{B}$. Therefore, the output is

$$y = \overline{A}\overline{B}$$

It's clear the output can equal a 1 only when $\overline{A} = 1$ and $\overline{B} = 1$. Equivalently, the output is a 1 only when

$$A = 0 \qquad \text{and} \qquad B = 0$$

From now on, we are going to abbreviate the foregoing input condition by

$$A\ B = 0\ 0$$

(Notice the space between A and B to avoid confusion with the product AB.)

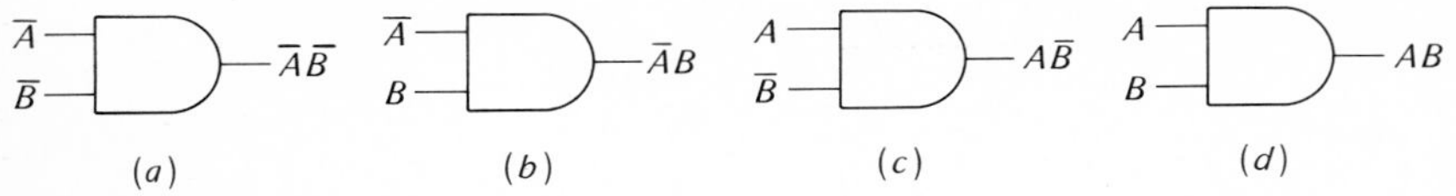

Fig. 6-1. ANDing two variables and complements.

Figure 6-1*b* shows another possibility. Here the inputs are $\overline{A}$ and B. So the output is

$$y = \overline{A}B$$

This output can equal a 1 only when

$$A\ B = 0\ 1$$

Any other input condition produces an output of 0.

In Fig. 6-1*c*, the inputs are A and $\overline{B}$. The output

$$y = A\overline{B}$$

equals a 1 only for the input condition

$$A\ B = 1\ 0$$

Finally, in Fig. 6-1*d*, the inputs are A and B. The output

$$y = AB$$

can equal a 1 only when

$$A\ B = 1\ 1$$

Table 6-1 summarizes the four possible ways of ANDing two signals in complemented or uncomplemented form. The four logical products are $\overline{A}\overline{B}$, $\overline{A}B$, $A\overline{B}$, and AB. These products are called *fundamental products* because they represent the basic ways to combine two signals with an AND gate.

A similar idea applies to three signals in complemented or uncomplemented

Table 6-1

FUNDAMENTAL PRODUCTS

A	B	$\overline{A}\overline{B}$	$\overline{A}B$	$A\overline{B}$	AB
0	0	1	0	0	0
0	1	0	1	0	0
1	0	0	0	1	0
1	1	0	0	0	1

Table 6-2
FUNDAMENTAL PRODUCTS OF THREE VARIABLES

A	B	C	Fundamental product
0	0	0	$\overline{A}\overline{B}\overline{C}$
0	0	1	$\overline{A}\overline{B}C$
0	1	0	$\overline{A}B\overline{C}$
0	1	1	$\overline{A}BC$
1	0	0	$A\overline{B}\overline{C}$
1	0	1	$A\overline{B}C$
1	1	0	$AB\overline{C}$
1	1	1	ABC

form. Let $\overline{A}$, $\overline{B}$, and $\overline{C}$ represent the three signals in complemented form, and let A, B, and C stand for the three signals in uncomplemented form. Then there are eight ways to combine these three signals with an AND gate: $\overline{A}\overline{B}\overline{C}$, $\overline{A}\overline{B}C$, $\overline{A}B\overline{C}$, $\overline{A}BC$, $A\overline{B}\overline{C}$, $A\overline{B}C$, $AB\overline{C}$, and ABC. Again, we refer to these basic possibilities as fundamental products.

Table 6-2 lists the fundamental products for three input signals. For convenience, each fundamental product appears adjacent to the input condition resulting in an output of 1. For instance, $\overline{A}\overline{B}\overline{C}$ equals 1 only when

$$A\ B\ C = 0\ 0\ 0$$

The fundamental product $\overline{A}\overline{B}C$ is a 1 only for an input condition of

$$A\ B\ C = 0\ 0\ 1$$

Likewise, $\overline{A}B\overline{C}$ equals 1 only for

$$A\ B\ C = 0\ 1\ 0$$

and so forth.

When there are four input variables ($A\ B\ C\ D$), there are 16 possible input conditions: 0 0 0 0 through 1 1 1 1. The corresponding fundamental products are from $\overline{A}\overline{B}\overline{C}\overline{D}$ through $ABCD$. A quick method for finding the fundamental product corresponding to any input condition is this: Whenever an input variable is a 0, the same variable is complemented in the fundamental product. For instance, if the input condition is

$$A\ B\ C\ D = 0\ 1\ 1\ 0$$

the corresponding fundamental product is $\overline{A}BC\overline{D}$; this is the product that results in a 1 output for the given input condition. If you want the fundamental product for

$$A\ B\ C\ D = 0\ 1\ 0\ 0$$

you have to complement *A*, *C*, and *D* to get $\bar{A}B\bar{C}\bar{D}$.

Example 6-1

Given five variables *A* through *E*, what is the fundamental product for each of these input conditions:

(*a*) $A\ B\ C\ D\ E = 0\ 0\ 1\ 1\ 0$
(*b*) $A\ B\ C\ D\ E = 0\ 1\ 1\ 0\ 1$
(*c*) $A\ B\ C\ D\ E = 1\ 0\ 0\ 0\ 1$

Solution

When an input variable is a 0, the same variable is complemented in the fundamental product. Therefore, the fundamental products are

(*a*) $\bar{A}\bar{B}CD\bar{E}$
(*b*) $\bar{A}BC\bar{D}E$
(*c*) $A\bar{B}\bar{C}\bar{D}E$

6-2 SUM OF PRODUCTS

Given a truth table, you can find the Boolean equation for the output by ORing the fundamental products that produce 1 outputs. The following examples show you how.

Suppose we're given a truth table like Table 6-3 and we want to find the Boolean equation corresponding to this table. To get a 1 output for the input condition

$$A\ B = 1\ 0$$

we need a fundamental product of $A\bar{B}$. Similarly, to get a 1 output for an input of

$$A\ B = 1\ 1$$

we need a fundamental product of *AB*. ORing these products gives

$$y = A\bar{B} + AB \qquad (6\text{-}1)$$

This is the Boolean equation for *y* because it results in a 1 for the input conditions $A\ B = 1\ 0$ and $A\ B = 1\ 1$; the equation produces a 0 for the other input conditions.

As another example of getting a Boolean equation from a truth table, look at Table 6-4. In this case, the fundamental products that produce the 1 outputs are $\bar{A}B$

Table 6-3

A	B	y
0	0	0
0	1	0
1	0	$1 \rightarrow A\bar{B}$
1	1	$1 \rightarrow AB$

Table 6-4

A	B	y
0	0	0
0	1	$1 \rightarrow \bar{A}B$
1	0	$1 \rightarrow A\bar{B}$
1	1	0

and $A\bar{B}$. Therefore, the Boolean equation is

$$y = \bar{A}B + A\bar{B} \tag{6-2}$$

As a check, you can substitute the four input conditions of Table 6-4; if you do, you will find y equals a 1 only for these input conditions:

$$A\ B = 0\ 1$$
$$A\ B = 1\ 0$$

When three input variables are involved, you again OR the appropriate fundamental products. For instance, suppose we're given a truth table like Table 6-5. The fundamental products that result in 1 outputs are listed. By ORing these products, we get the Boolean equation of the output:

$$y = \bar{A}BC + A\bar{B}C + AB\bar{C} + ABC \tag{6-3}$$

This is the equation that results in a 1 output only for the input conditions:

$$A\ B\ C = 0\ 1\ 1$$
$$A\ B\ C = 1\ 0\ 1$$
$$A\ B\ C = 1\ 1\ 0$$
$$A\ B\ C = 1\ 1\ 1$$

Table 6-5

A	B	C	y
0	0	0	0
0	0	1	0
0	1	0	0
0	1	1	$1 \rightarrow \bar{A}BC$
1	0	0	0
1	0	1	$1 \rightarrow A\bar{B}C$
1	1	0	$1 \rightarrow AB\bar{C}$
1	1	1	$1 \rightarrow ABC$

In each of the foregoing examples, the same method is used:

1. Find the fundamental product corresponding to each 1 output in the truth table.
2. OR the foregoing fundamental products.

These two steps always result in a Boolean equation that is a logical sum of fundamental products. For this reason, this method of getting the Boolean equation is called the "sum-of-products" method.

Example 6-2

A truth table has four input variables. A 1 output occurs only for these input conditions:

$$A\ B\ C\ D = 0\ 0\ 1\ 0$$
$$A\ B\ C\ D = 1\ 0\ 0\ 1$$
$$A\ B\ C\ D = 1\ 1\ 0\ 1$$

What is the Boolean equation for the output?

Solution

The fundamental products for the given inputs are $\bar{A}\bar{B}C\bar{D}$, $A\bar{B}\bar{C}D$, and $AB\bar{C}D$. ORing these products gives

$$y = \bar{A}\bar{B}C\bar{D} + A\bar{B}\bar{C}D + AB\bar{C}D$$

This equation results in a 1 output only for the given input conditions.

6-3 AND-OR NETWORKS

The sum-of-products method results in a Boolean equation where the inputs are first ANDed, then ORed. Because of this, the logic circuit corresponding to a sum-of-products equation is a group of AND gates working into an OR gate.

Fig. 6-2.

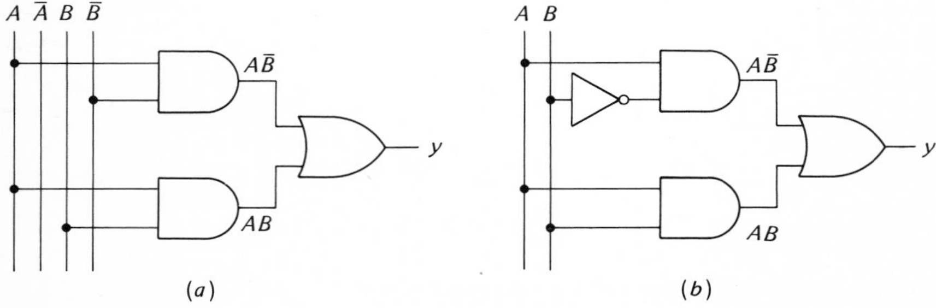

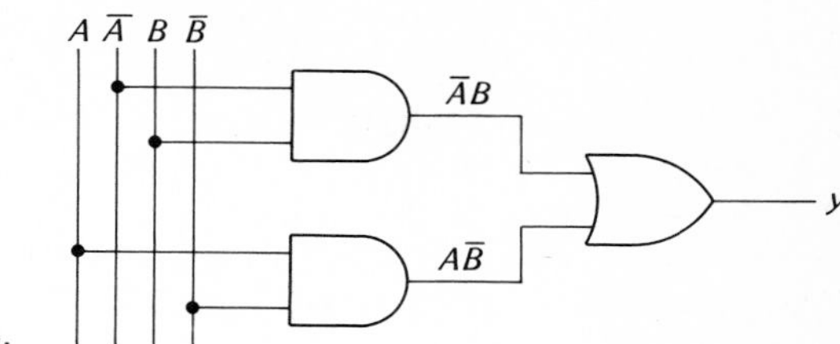

Fig. 6-3.

For instance, Table 6-3 led to the Boolean equation

$$y = A\overline{B} + AB$$

This implies that A is ANDed with $\overline{B}$, A is ANDed with B, and $A\overline{B}$ is ORed with AB. Figure 6-2*a* shows the logic circuit to use when variables and their complements are available from other circuits. If complements are not available, you have to add an inverter as shown in Fig. 6-2*b*. In most digital circuits discussed in this book, variables and their complements are available from other digital circuits.

Here's another example of *implementing* (finding the circuit for) a Boolean equation. Table 6-4 led to this Boolean equation:

$$y = \overline{A}B + A\overline{B}$$

Figure 6-3 shows the corresponding logic circuit. It produces a 1 output only for

$$A\ B = 0\ 1$$
$$A\ B = 1\ 0$$

Another example: Applying the sum-of-products method to Table 6-5 gave

$$y = \overline{A}BC + A\overline{B}C + AB\overline{C} + ABC$$

To implement this, we need four three-input AND gates working into a four-input OR gate as shown in Fig. 6-4.

In summary, given a truth table, we can find its Boolean equation by ORing the fundamental products producing 1 outputs. Because this equation is always a sum of products, the corresponding logic circuit consists of a group of AND gates driving an OR gate. From now on, we shall refer to logic circuits like Fig. 6-4 as AND-OR networks.

Example 6-3

Suppose a truth table leads to this Boolean equation:

$$y = A\overline{B}\overline{C}D + ABCD$$

What is the corresponding AND-OR network using TTL gates?

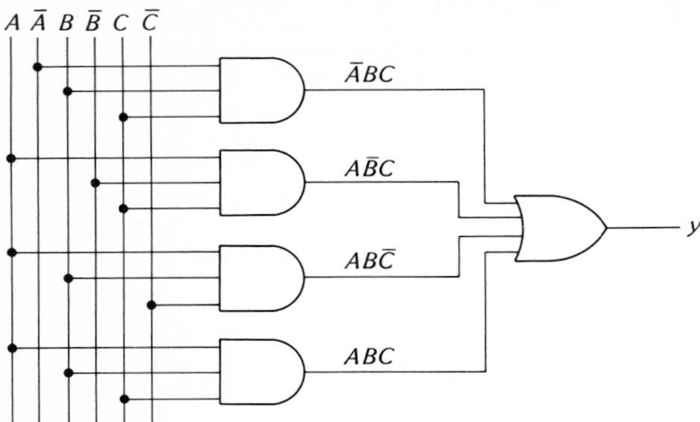

Fig. 6-4. AND-OR network.

Solution

Figure 6-5 shows the AND-OR network. The 7421 is a dual four-input AND gate with totem-pole outputs. The 7432 is a quad two-input OR gate with totem-pole outputs. Note that we need to use only one-quarter of the 7432.

6-4 ALGEBRAIC SIMPLIFICATION

The sum-of-products method always works. Given a truth table, you can always find a Boolean equation by ORing the appropriate fundamental products; then you can implement the equation with an AND-OR network. On small production runs, an AND-OR network like Fig. 6-4 may be all right to use as is. But on large production runs, the cost of each gate contributes significantly to the overall cost of production. In this case, it pays to look for a simpler logic circuit.

A preliminary guide for comparing one logic circuit with another is to count the number of input gate leads; the circuit with fewer input gate leads is usually less expensive. For instance, the AND-OR network of Fig. 6-5 has a total of 10 input gate

Fig. 6-5. Example 6-3.

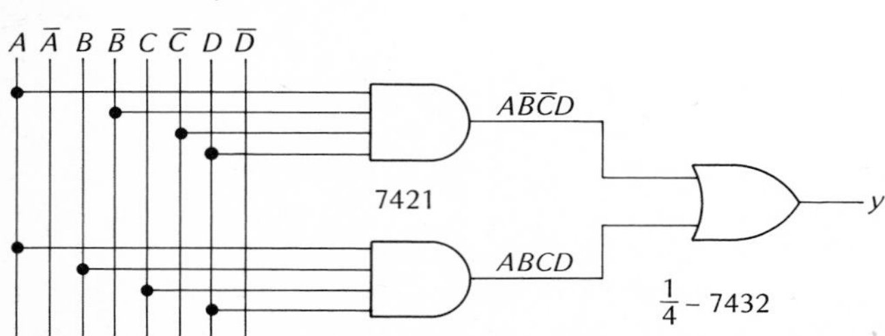

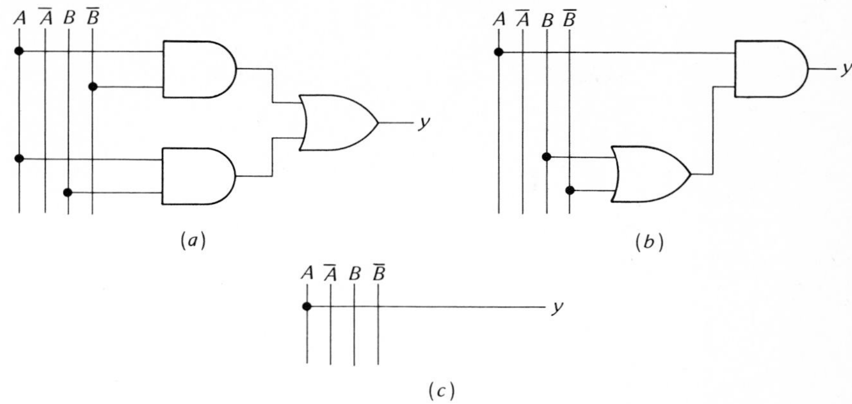

Fig. 6-6. Equivalent logic circuits.

leads (four on each AND gate and two on the OR gate). The AND-OR network of Fig. 6-4, on the other hand, has a total of 16 input gate leads (three on each AND gate and four on the OR gate). The AND-OR network of Fig. 6-5 will cost less to build than the AND-OR network of Fig. 6-4.

One way to reduce the number of input gate leads is to factor the Boolean equation, if possible. For instance, Table 6-3 led to the Boolean equation

$$y = A\overline{B} + AB \tag{6-4}$$

Figure 6-6*a* shows the AND-OR network; it has six input gate leads. By factoring Eq. (6-4), we obtain

$$y = A(\overline{B} + B)$$

The logic circuit for this is shown in Fig. 6-6*b*; only four input gate leads are required, so this circuit would cost less than the AND-OR network of Fig. 6-6*a*.

As a matter of fact, recall that a variable ORed with its complement always equals 1; therefore,

$$y = A(\overline{B} + B) = A \cdot 1 = A$$

This equation says the output equals the value of A. This being the case, all we need is a connecting wire from input to output as shown in Fig. 6-6*c*. In other words, we don't need any gates at all to implement Eq. (6-4).

Here's another example of using algebra to simplify a Boolean equation and the corresponding logic circuit. Suppose a truth table has a 1 output for these input

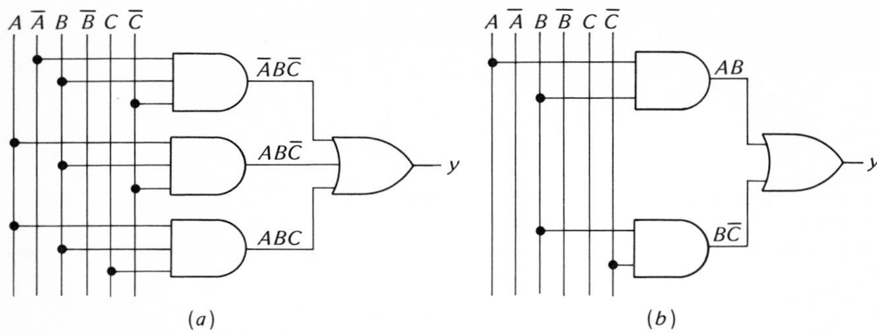

Fig. 6-7. Simplifying an AND-OR network.

conditions:

$$ABC = 010$$
$$ABC = 110$$
$$ABC = 111$$

The sum-of-products equation is

$$y = \overline{A}B\overline{C} + AB\overline{C} + ABC \tag{6-5}$$

Figure 6-7*a* shows the AND-OR network; it has 12 input gate leads. Is this the least expensive logic circuit?

Recall that a variable ORed with itself equals the variable; that is,

$$A + A = A$$

In a similar way, it's valid to write

$$AB\overline{C} + AB\overline{C} = AB\overline{C}$$

which is equivalent to

$$AB\overline{C} = AB\overline{C} + AB\overline{C}$$

Substitute this into Eq. (6-5) to get

$$y = \overline{A}B\overline{C} + AB\overline{C} + AB\overline{C} + ABC$$

Next, factor to get

$$y = (\overline{A} + A)B\overline{C} + AB(\overline{C} + C)$$

But a variable ORed with its complement always equals 1; therefore, the foregoing equation reduces to

$$y = B\overline{C} + AB \tag{6-6}$$

Figure 6-7*b* is the AND-OR network for this simplified equation; it has only six input gate leads.

The two circuits of Fig. 6-7 are equivalent, meaning they produce the same output for each input condition. In other words, both circuits have the same truth table. The circuit on the left has 12 input gate leads, while the one on the right has only six input gate leads. There's no question about which circuit is simpler.

This example brings out some important ideas. First, factoring a Boolean equation may lead to a simpler equation and a less expensive logic circuit. Second, there appear to be algebraic tricks (like adding the extra $AB\overline{C}$) that lead to simpler equations and logic circuits. It is possible to become an expert at manipulating equations using the algebra rules in Chap. 4. But most engineers and technicians don't simplify equations with algebra. Instead, they use a much simpler geometric method called the *Karnaugh map,* discussed in the remainder of this chapter.

Example 6-4

What is the AND-OR network for

$$y = AB + AC + BD$$

If this equation is factored, what is the corresponding logic circuit?

Solution

Figure 6-8*a* shows the AND-OR network; it has nine input gate leads. Factoring the given equation gives

$$y = A(B + C) + BD$$

The corresponding logic circuit is shown in Fig. 6-8*b*; this has only eight input gate leads.

Even though Fig. 6-8*b* has one fewer input gate lead than Fig. 6-8*a*, it may be better to use Fig. 6-8*a*. Here's one reason. In the worst case, the overall propagation delay time of Fig. 6-8*a* is the sum of only two gate delays, whereas the overall propagation delay time of Fig. 6-8*b* may be the sum of as many as three gate delays. For instance, if each gate has a t_p of 10 ns, the overall propagation delay time is 20 ns for Fig. 6-8*a* and 30 ns for Fig. 6-8*b* in the worst case. If speed is important, Fig. 6-8*a* is preferred in spite of its extra input gate lead. (Section 6-10 discusses another reason why the AND-OR network is preferred.)

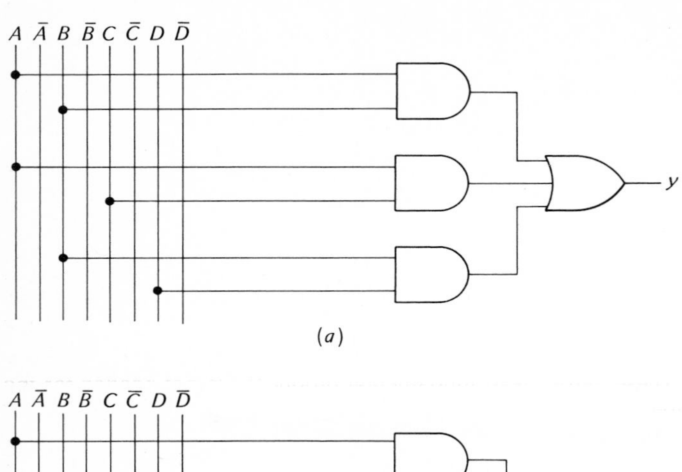

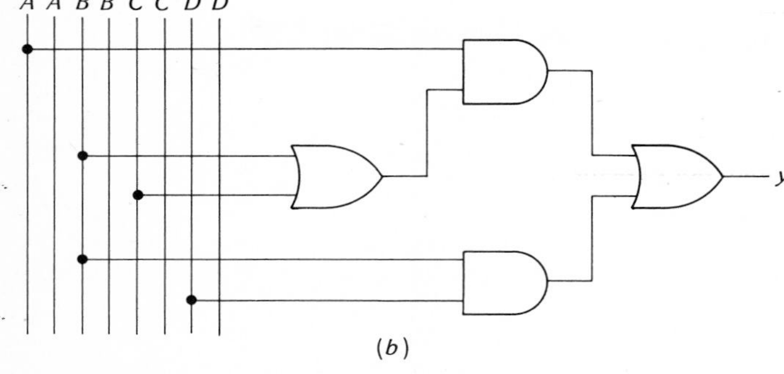

Fig. 6-8. Factoring to reduce input gate leads.

6-5 TRUTH TABLE TO KARNAUGH MAP

Suppose you're given a truth table like Table 6-6. Here's how to construct the *Karnaugh map* for this truth table. Begin by drawing Fig. 6-9*a*. Note the order of the variables and their complements: the vertical column has $\overline{A}$ followed by A, and the horizontal row has $\overline{B}$ followed by B.

Next, look for output 1s in Table 6-6. The first output 1 to appear is for an input of

$$A\ B = 1\ 0$$

The fundamental product for this is $A\overline{B}$. Now, enter a 1 on the Karnaugh map as shown in Fig. 6-9*b*. This 1 represents the product $A\overline{B}$ because the 1 is in the A row and $\overline{B}$ column.

Similarly, Table 6-6 has an output 1 appearing for an input of

$$A\ B = 1\ 1$$

Table 6-6

A	B	y
0	0	0
0	1	0
1	0	1
1	1	1

The fundamental product for this is AB. When you enter a 1 on the Karnaugh map to represent AB, you get the map of Fig. 6-9*c*.

The final step in the construction of the Karnaugh map is to enter 0s in the remaining spaces. These zeros mean fundamental products are not needed for the corresponding inputs

$$A\ B = 0\ 0$$
$$A\ B = 0\ 1$$

Figure 6-9*d* shows how the Karnaugh map looks in final form.

How does a Karnaugh map differ from a truth table? A truth table shows the output for each input condition. A Karnaugh map, on the other hand, shows the fundamental products needed to produce the output 1s for corresponding input conditions. Later sections will tell you why this is important and how to use a Karnaugh map to simplify the logic circuit.

Here's another example of constructing a two-variable Karnaugh map. In the truth table of Table 6-7, notice that output 1s appear for inputs of

$$A\ B = 0\ 1$$
$$A\ B = 1\ 0$$

The fundamental products for these inputs are $\overline{A}B$ and $A\overline{B}$. When 1s are entered on the Karnaugh map for these products and 0s in the remaining spaces, the completed map looks like Fig. 6-10.

Here's how to construct a Karnaugh map when there are three input variables: A, B, C. Suppose the truth table is that of Table 6-8. First, draw a blank Karnaugh map like Fig. 6-11*a*. It is especially important to notice the order of the variables and

Fig. 6-9. Constructing a Karnaugh map.

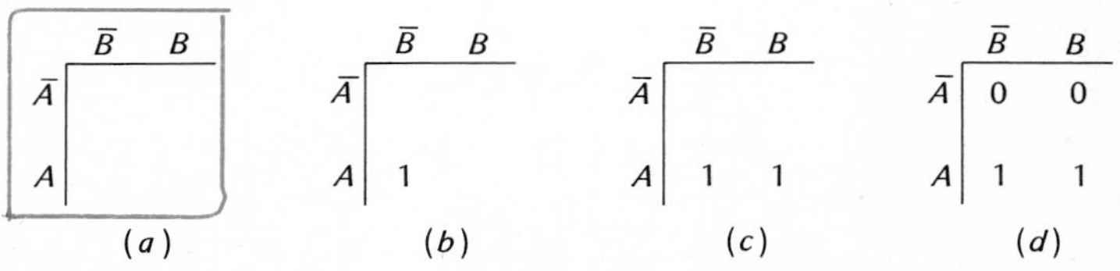

(*a*)

	$\overline{B}$	B
$\overline{A}$		
A		

(*b*)

	$\overline{B}$	B
$\overline{A}$		
A	1	

(*c*)

	$\overline{B}$	B
$\overline{A}$		
A	1	1

(*d*)

	$\overline{B}$	B
$\overline{A}$	0	0
A	1	1

Table 6-7

A	B	y
0	0	0
0	1	1
1	0	1
1	1	0

complements. The vertical column is labeled $\overline{A}\overline{B}$, $\overline{A}B$, AB, and $A\overline{B}$. This order is not a binary progession; rather, it follows the Gray-code progression of 00, 01, 11, and 10. The reason for this is explained in the derivation of the Karnaugh-map method; briefly, it's done so that only one variable will change from complemented to uncomplemented form (or vice versa). Whenever you construct a Karnaugh map, you must use the Gray-code order of $\overline{A}\overline{B}$, $\overline{A}B$, AB, and $A\overline{B}$.

Next, look for output 1s in Table 6-8. Output 1s appear for inputs of

$$A\ B\ C = 0\ 1\ 0$$
$$A\ B\ C = 1\ 1\ 0$$
$$A\ B\ C = 1\ 1\ 1$$

The fundamental products for these inputs are $\overline{A}B\overline{C}$, $AB\overline{C}$, and ABC.

Enter 1s for these products on the Karnaugh map (Fig. 6-11*b*). Each entry should be clear. The 1 for $\overline{A}B\overline{C}$ appears to the right of $\overline{A}B$ and below $\overline{C}$; the 1 for $AB\overline{C}$ is right of AB and under $\overline{C}$; and the 1 for ABC is right of AB and below C.

The final step is to enter 0s in the remaining spaces (Fig. 6-11c). This completed Karnaugh map is useful because it shows at a glance which fundamental products are needed to implement the Boolean equation corresponding to the truth table.

6-6 FOUR-VARIABLE KARNAUGH MAPS

As already mentioned, many digital computers and systems process words in bytes of 4 bits each. For this reason, many logic circuits are designed to handle four input variables (or their complements). This is why the four-variable Karnaugh map is the most important one to know about. From here on, we concentrate on truth tables and Karnaugh maps involving four input variables.

Here's an example of constructing a four-variable Karnaugh map. Suppose you're given a truth table like Table 6-9. The first step in constructing the Karnaugh map is

	$\overline{B}$	B
$\overline{A}$	0	1
A	1	0

Fig. 6-10. Another example of a two-variable Karnaugh map.

Table 6-8

A	B	C	y
0	0	0	0
0	0	1	0
0	1	0	1
0	1	1	0
1	0	0	0
1	0	1	0
1	1	0	1
1	1	1	1

to draw the blank map of Fig. 6-12*a*. Again, the order is crucial; the vertical column is labeled $\bar{A}\bar{B}$, $\bar{A}B$, AB, and $A\bar{B}$; the horizontal row is labeled $\bar{C}\bar{D}$, $\bar{C}D$, CD, and $C\bar{D}$. You must always follow this Gray-code progression when constructing Karnaugh maps.

In Table 6-9, output 1s appear for these inputs:

$$ABCD = 0001$$
$$ABCD = 0110$$
$$ABCD = 0111$$
$$ABCD = 1110$$

The fundamental products corresponding to these inputs are $\bar{A}\bar{B}\bar{C}D$, $\bar{A}BC\bar{D}$, $\bar{A}BCD$, and $ABC\bar{D}$. After entering 1s on the Karnaugh map, we have Fig. 6-12*b*. The final step of filling in 0s results in the completed map of Fig. 6-12*c*.

6-7 PAIRS, QUADS, AND OCTETS

Look at Fig. 6-13*a*. The map contains a pair of 1s that are horizontally adjacent (next to each other). The first 1 represents the product $ABCD$; the second 1 stands for the product $ABC\bar{D}$. As we move from the first 1 to the second 1, only one variable goes from uncomplemented to complemented form (D to $\bar{D}$); the other vari-

Fig. 6-11. Three-variable Karnaugh map.

	$\bar{C}$	C
$\bar{A}\bar{B}$		
$\bar{A}B$		
AB		
$A\bar{B}$		

(*a*)

	$\bar{C}$	C
$\bar{A}\bar{B}$		
$\bar{A}B$	1	
AB	1	1
$A\bar{B}$		

(*b*)

	$\bar{C}$	C
$\bar{A}\bar{B}$	0	0
$\bar{A}B$	1	0
AB	1	1
$A\bar{B}$	0	0

(*c*)

Table 6-9

A	B	C	D	y
0	0	0	0	0
0	0	0	1	1
0	0	1	0	0
0	0	1	1	0
0	1	0	0	0
0	1	0	1	0
0	1	1	0	1
0	1	1	1	1
1	0	0	0	0
1	0	0	1	0
1	0	1	0	0
1	0	1	1	0
1	1	0	0	0
1	1	0	1	0
1	1	1	0	1
1	1	1	1	0

ables don't change form (A, B, and C remain uncomplemented). Whenever this happens, you can *eliminate the variable that changes form.*

Here's the proof. The sum-of-products equation corresponding to Fig. 6-13*a* is

$$y = ABCD + ABC\bar{D}$$

which factors into

$$y = ABC(D + \bar{D})$$

Since D is ORed with $\bar{D}$, the equation reduces to

$$y = ABC$$

Fig. 6-12. Constructing a four-variable Karnaugh map.

(*a*)

	$\bar{C}\bar{D}$	$\bar{C}D$	CD	$C\bar{D}$
$\bar{A}\bar{B}$				
$\bar{A}B$				
AB				
$A\bar{B}$				

(*b*)

	$\bar{C}\bar{D}$	$\bar{C}D$	CD	$C\bar{D}$
$\bar{A}\bar{B}$		1		
$\bar{A}B$			1	1
AB				1
$A\bar{B}$				

(*c*)

	$\bar{C}\bar{D}$	$\bar{C}D$	CD	$C\bar{D}$
$\bar{A}\bar{B}$	0	1	0	0
$\bar{A}B$	0	0	1	1
AB	0	0	0	1
$A\bar{B}$	0	0	0	0

	$\bar{C}\bar{D}$	$\bar{C}D$	CD	$C\bar{D}$
$\bar{A}\bar{B}$	0	0	0	0
$\bar{A}B$	0	0	0	0
AB	0	0	1	1
$A\bar{B}$	0	0	0	0

(*a*)

	$\bar{C}\bar{D}$	$\bar{C}D$	CD	$C\bar{D}$
$\bar{A}\bar{B}$	0	0	0	0
$\bar{A}B$	0	0	0	0
AB	0	0	1	1
$A\bar{B}$	0	0	0	0

(*b*)

Fig. 6-13. Horizontally adjacent 1s.

A pair of horizontally adjacent 1s like those of Fig. 6-13*a* always means the sum-of-products equation will have a factorable variable and a complement that drop out as shown above.

For easy identification, it is customary to encircle a pair of adjacent 1s as shown in Fig. 6-13*b*. Then, when you look at the map, you can tell at a glance that one variable and its complement will drop out of the Boolean equation. In other words, an encircled pair of 1s like those of Fig. 6-13*b* no longer stand for the ORing of two separate products, $ABCD$ and $ABC\bar{D}$: rather, the encircled pair should be visualized as representing a single reduced product ABC.

Here's another example of a reduced product. Figure 6-14*a* shows a pair of 1s that are vertically adjacent. These 1s correspond to products of $ABC\bar{D}$ and $A\bar{B}C\bar{D}$. Notice that only one variable changes from uncomplemented to complemented form (B to $\bar{B}$); all other variables retain their original form. Therefore, we can predict that B and $\bar{B}$ can be factored and eliminated algebraically.

Here's the proof. The sum-of-products equation is

$$y = ABC\bar{D} + A\bar{B}C\bar{D}$$

which factors into

$$y = AC\bar{D}(B + \bar{B})$$

This reduces to

$$y = AC\bar{D}$$

Therefore, the encircled pair of 1s in Fig. 6-14*a* represents the simplified product $AC\bar{D}$.

From now on, we don't have to bother with algebra. Whenever we see a pair of horizontally or vertically adjacent 1s, we can *eliminate the variable that goes from complemented to uncomplemented form* (or vice versa). The remaining variables or their complements will be the only ones appearing in the single product term corresponding to the pair of 1s. For instance, a glance at Fig. 6-14*b* indicates that B

goes from complemented to uncomplemented form when we move from the upper to the lower 1; the other variables ($\bar{A}$, C, and D) remain the same. Therefore, the encircled pair of 1s in Fig. 6-14*b* represents the product $\bar{A}CD$.

Likewise, given the pair of 1s in Fig. 6-14*c*, the only change is from $\bar{D}$ to D. So the encircled pair of 1s stands for the product $A\bar{B}\bar{C}$.

If more than one pair exists on a Karnaugh map, you can OR the simplified products to get the Boolean equation. For instance, the lower pair of Fig. 6-14*d* represents the simplified product $A\bar{C}\bar{D}$; the upper pair stands for $\bar{A}BD$. The corresponding Boolean equation for this map is

$$y = A\bar{C}\bar{D} + \bar{A}BD$$

From now on, a *pair* means two 1s that are horizontally or vertically adjacent. Because one variable changes form, a pair always represents a simpler product. (Diagonally adjacent 1s are worthless; they lead to no simplifications.)

A *quad* is a group of four 1s that are horizontally or vertically adjacent. The 1s may be end-to-end as shown in Fig. 6-15*a* or in the form of a square as in Fig. 6-15*b*. When you see a quad, always encircle it because it leads to a simpler product. In fact, a quad means *two variables and their complements drop out of the Boolean equation.*

Here's why a quad eliminates two variables and their complements. Visualize the four 1s of Fig. 6-15*a* as two pairs (see Fig. 6-15*c*). The first pair represents $AB\bar{C}$;

Fig. 6-14. Examples of pairs.

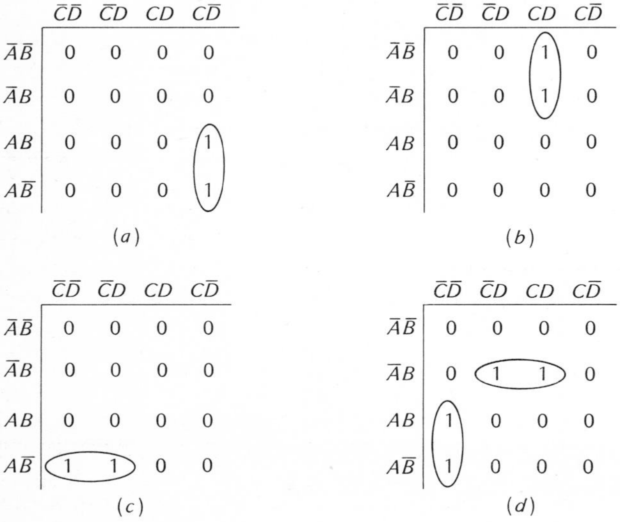

(*a*)

	$\bar{C}\bar{D}$	$\bar{C}D$	CD	$C\bar{D}$
$\bar{A}\bar{B}$	0	0	0	0
$\bar{A}B$	0	0	0	0
AB	0	0	0	1
$A\bar{B}$	0	0	0	1

(*b*)

	$\bar{C}\bar{D}$	$\bar{C}D$	CD	$C\bar{D}$
$\bar{A}\bar{B}$	0	0	1	0
$\bar{A}B$	0	0	1	0
AB	0	0	0	0
$A\bar{B}$	0	0	0	0

(*c*)

	$\bar{C}\bar{D}$	$\bar{C}D$	CD	$C\bar{D}$
$\bar{A}\bar{B}$	0	0	0	0
$\bar{A}B$	0	0	0	0
AB	0	0	0	0
$A\bar{B}$	1	1	0	0

(*d*)

	$\bar{C}\bar{D}$	$\bar{C}D$	CD	$C\bar{D}$
$\bar{A}\bar{B}$	0	0	0	0
$\bar{A}B$	0	1	1	0
AB	1	0	0	0
$A\bar{B}$	1	0	0	0

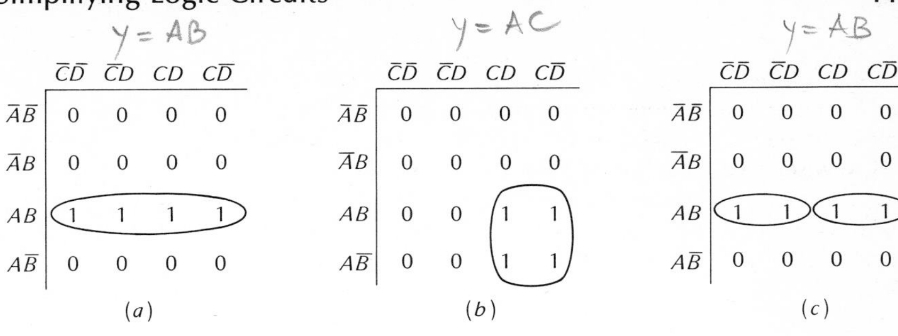

(a)	$\bar{C}\bar{D}$	$\bar{C}D$	CD	$C\bar{D}$
$\bar{A}\bar{B}$	0	0	0	0
$\bar{A}B$	0	0	0	0
AB	1	1	1	1
$A\bar{B}$	0	0	0	0

(b)	$\bar{C}\bar{D}$	$\bar{C}D$	CD	$C\bar{D}$
$\bar{A}\bar{B}$	0	0	0	0
$\bar{A}B$	0	0	0	0
AB	0	0	1	1
$A\bar{B}$	0	0	1	1

(c)	$\bar{C}\bar{D}$	$\bar{C}D$	CD	$C\bar{D}$
$\bar{A}\bar{B}$	0	0	0	0
$\bar{A}B$	0	0	0	0
AB	1	1	1	1
$A\bar{B}$	0	0	0	0

(a) (b) (c)

Fig. 6-15. Examples of quads.

the second pair stands for ABC. The Boolean equation for these two pairs is

$$y = AB\bar{C} + ABC$$

This factors into

$$y = AB(\bar{C} + C)$$

But this reduces to

$$y = AB$$

So, the quad of Fig. 6-15*a* represents a product where two variables and their complements have dropped out.

A similar proof applies to any quad. You can visualize it as two pairs whose Boolean equation leads to a single product involving only two variables or their complements. There's no need to go through the algebra any more. Merely step through the different 1s in the quad and determine which two variables go from complemented to uncomplemented form (or vice versa); these are the variables that drop out.

For instance, look at the quad of Fig. 6-15*b*. Pick any 1 as a starting point. When you move horizontally, D is the variable that changes form; when you move vertically, B changes form. Therefore, the remaining variables (A and C) are the only ones appearing in the simplified product. In other words, the simplified equation for the quad of Fig. 6-15*b* is

$$y = AC$$

Besides pairs and quads, there's one more group of adjacent 1s to look for: the *octet*. This is a group of eight 1s like those of Fig. 6-16*a*. An octet like this *eliminates three variables and their complements*. Here's why. Visualize the octet

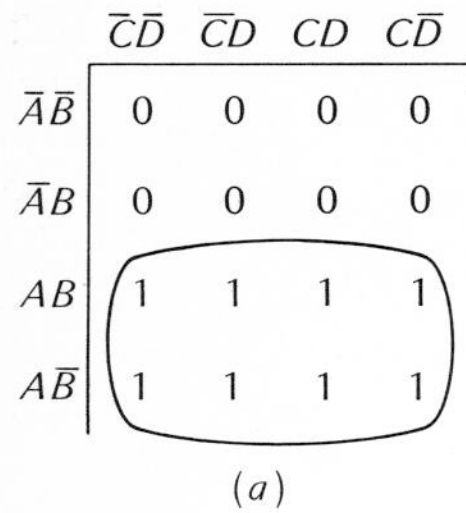

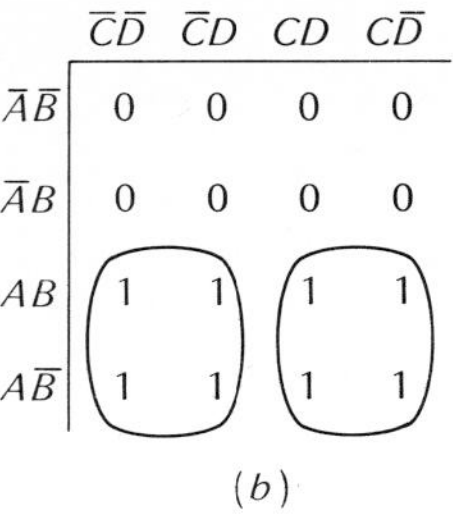

Fig. 6-16. Example of octet.

as two quads (see Fig. 6-16*b*). The equation for these two quads is

$$y = A\bar{C} + AC$$

After factoring,

$$y = A(\bar{C} + C)$$

But this reduces to

$$y = A$$

So the octet of Fig. 6-16*a* means three variables and their complements drop out of the corresponding product.

A similar proof applies to any octet. So, from now on, don't bother with the algebra. Merely step through the 1s of the octet and determine which three variables change form. These are the variables that drop out.

6-8 KARNAUGH SIMPLIFICATIONS

You've just learned that a pair eliminates one variable and its complement, a quad eliminates two variables and their complements, and an octet eliminates three variables and their complements. Because of this, after you construct a Karnaugh map, you should encircle the *octets first,* the *quads second,* and the *pairs last.* In this way, the greatest simplification results.

Here's an example. Suppose you've translated a truth table into the Karnaugh map shown in Fig. 6-17*a*. To find the simplified Boolean equation, look for octets first. There are none in Fig. 6-17*a*. Next, look for quads. When you find them, encircle them. Finally, look for and encircle pairs. If you do this correctly, you arrive at Fig. 6-17*b*.

The pair represents the simplified product $\bar{A}\bar{B}D$, the lower quad stands for $A\bar{C}$, and the quad on the right represents $C\bar{D}$. By ORing these simplified products, we get

Pair eliminates 1 variable & compl.
Quad 2
Octet 3

	$\bar{C}\bar{D}$	$\bar{C}D$	CD	$C\bar{D}$
$\bar{A}\bar{B}$	0	1	1	1
$\bar{A}B$	0	0	0	1
AB	1	1	0	1
$A\bar{B}$	1	1	0	1

(a)

	$\bar{C}\bar{D}$	$\bar{C}D$	CD	$C\bar{D}$
$\bar{A}\bar{B}$	0	1	1	1
$\bar{A}B$	0	0	0	1
AB	1	1	0	1
$A\bar{B}$	1	1	0	1

(b)

Fig. 6-17. Encircling octets, quads, and pairs.

the Boolean equation corresponding to the entire Karnaugh map:

$$y = \bar{A}\bar{B}D + A\bar{C} + C\bar{D} \tag{6-10}$$

When you encircle groups, you are allowed to use the same 1 more than once. In other words, the same 1 can be common to two or more groups. Figure 6-18*a* illustrates this idea. The 1 representing the fundamental product $AB\bar{C}D$ is part of the pair and part of the octet. The simplified equation for the overlapping groups is

$$y = A + B\bar{C}D \tag{6-11}$$

It is valid to encircle the 1s as shown in Fig. 6-18*b*, but then the isolated 1 results in a more complicated equation:

$$y = A + \bar{A}B\bar{C}D$$

This will require a more complicated logic circuit than Eq. (6-11). So always *overlap* groups if possible; that is, use the 1s more than once to get the largest groups you can.

Another thing to know about is *rolling*. Look at Fig. 6-19a. The pairs result in this equation:

$$y = B\bar{C}\bar{D} + BC\bar{D} \tag{6-12}$$

Fig. 6-18. Overlapping groups.

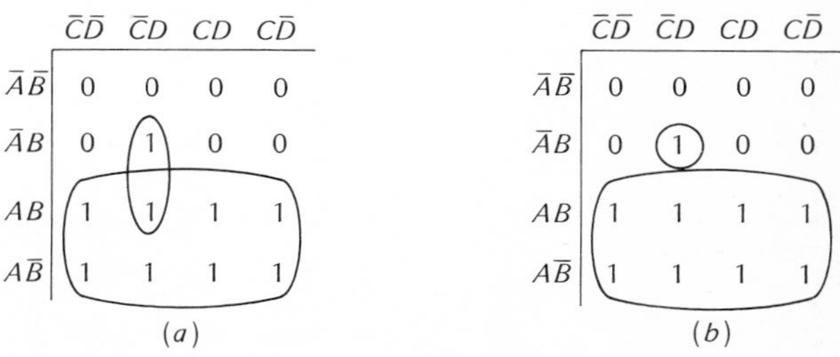

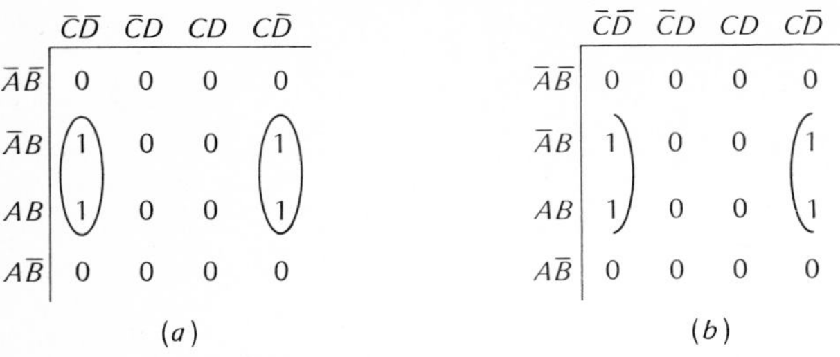

Fig. 6-19. Rolling the Karnaugh map.

Visualize picking up the Karnaugh map and rolling it so that the left side touches the right side. If you're visualizing correctly, you will realize the two pairs actually form a quad. To indicate this, draw half circles around each pair as shown in Fig. 6-19*b*. From this viewpoint, the quad of Fig. 6-19*b* has the equation

$$y = B\bar{D} \tag{6-13}$$

Why is rolling valid? Because Eq. (6-12) can be algebraically simplified to Eq. (6-13). The proof is to start with Eq. (6-12):

$$y = B\bar{C}\bar{D} + BC\bar{D}$$

This factors into

$$y = B\bar{D}(\bar{C} + C)$$

This reduces to

$$y = B\bar{D}$$

But this final equation is the one that represents a rolled quad like Fig. 6-19*b*. Therefore, 1s on the edges of a Karnaugh map can be grouped with 1s on opposite edges.

If possible, roll and overlap to get the largest groups you can find. For instance, Fig. 6-20*a* shows an inefficient way to encircle groups. The octet and pair have a Boolean equation of

$$y = \bar{C} + BC\bar{D}$$

You can do better by rolling and overlapping as shown in Fig. 6-20*b*; the Boolean equation now becomes

$$y = \bar{C} + B\bar{D}$$

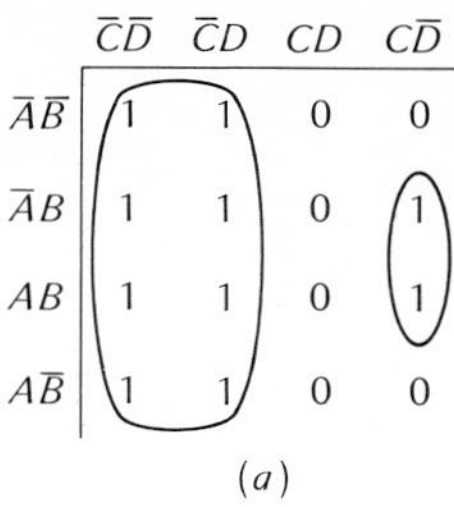

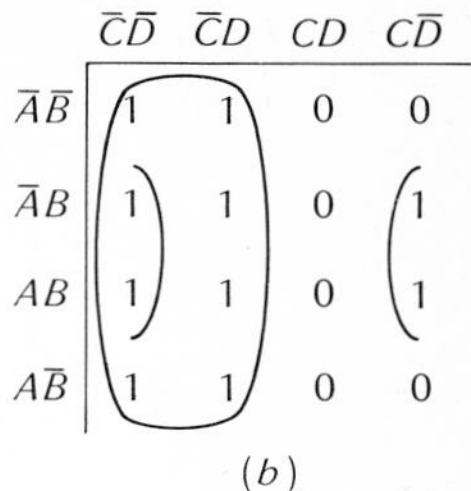

Fig. 6-20. Rolling and overlapping.

Here's another example. Figure 6-21*a* shows an inefficient grouping of 1s; the corresponding equation is

$$y = \overline{C} + \overline{A}C\overline{D} + A\overline{B}C\overline{D}$$

If we roll and overlap as shown in Fig. 6-21*b*, the equation is simpler:

$$y = \overline{C} + \overline{A}\overline{D} + A\overline{B}\overline{D} \qquad (6\text{-}14)$$

It's possible to group the 1s as shown in Fig. 6-21*c*. The equation now becomes

$$y = \overline{C} + \overline{A}\overline{D} + \overline{B}C\overline{D} \qquad (6\text{-}15)$$

Compare this with Eq. (6-14). As you can see, the equations are comparable in simplicity. Because of this, the corresponding logic circuits have the same number of input gate leads. Either grouping (Fig. 6-21*b* or *c*) is valid; therefore, you can use whichever you like.

After you've finished encircling groups, there's one more thing you should do before writing the simplified Boolean equation: eliminate any group whose 1s are completely overlapped by other groups. Here's an example. Given a Karnaugh map like Fig. 6-22*a*, the first thing you can do is encircle the quad in the center of the map to get Fig. 6-22*b*. Next, you can group the remaining 1s into pairs by overlap-

Fig. 6-21. Different ways of encircling groups.

	$\overline{C}\overline{D}$	$\overline{C}D$	CD	$C\overline{D}$
$\overline{A}\overline{B}$	1	1	0	1
$\overline{A}B$	1	1	0	1
AB	1	1	0	0
$A\overline{B}$	1	1	0	1

(*a*)

	$\overline{C}\overline{D}$	$\overline{C}D$	CD	$C\overline{D}$
$\overline{A}\overline{B}$	1	1	0	1
$\overline{A}B$	1	1	0	1
AB	1	1	0	0
$A\overline{B}$	1	1	0	1

(*b*)

	$\overline{C}\overline{D}$	$\overline{C}D$	CD	$C\overline{D}$
$\overline{A}\overline{B}$	1	1	0	1
$\overline{A}B$	1	1	0	1
AB	1	1	0	0
$A\overline{B}$	1	1	0	1

(*c*)

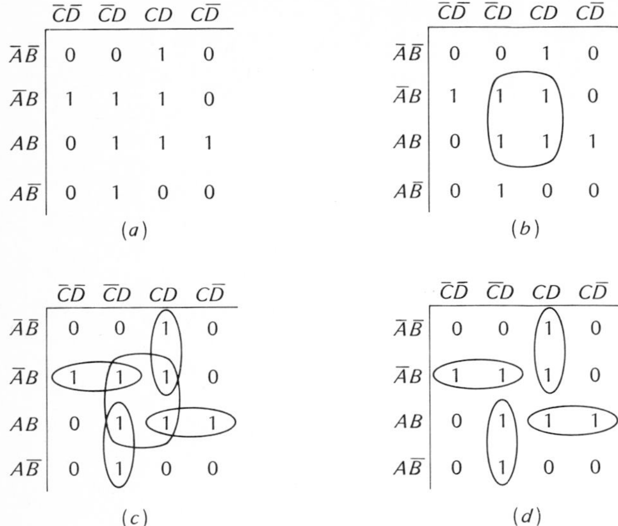

Fig. 6-22. Eliminating an unnecessary group.

ping (Fig. 6-22*c*). The final step is to review the encircled groups to see if any group has all its 1s overlapped by other groups. In Fig. 6-22*c*, each 1 in the quad is overlapped by a pair. Because of this, the quad can be eliminated to get Fig. 6-22*d*. The equation for Fig. 6-22*d* will contain one less product than the equation for Fig. 6-22*c*; therefore, Fig. 6-22*d* is the most efficient way to group the 1s.

Having all 1s in a group overlapped by other groups does not happen often. But when it does, you should take advantage of it because it eliminates one product from the Boolean equation.

Here's a summary of the Karnaugh-map method for simplifying Boolean equations:

1. Enter a 1 on the Karnaugh map for each fundamental product that produces a 1 output in the truth table. Enter 0s elsewhere.
2. Encircle the octets, quads, and pairs. Remember to roll and overlap to get the largest groups possible.
3. If any isolated 1s remain, encircle each.
4. Review the groups, and eliminate any group whose 1s are entirely overlapped by other groups.
5. Write the Boolean equation by ORing the products corresponding to the encircled groups.

Example 6-5

What is the simplified Boolean equation corresponding to the Karnaugh map of Fig. 6-23*a*?

Solution

There are no octets, but there is a quad as shown in Fig. 6-23*b*. By overlapping, we can find two more quads (see Fig. 6-23c). We can encircle the remaining 1 by making it part of an overlapped pair (Fig. 6-23*d*). Finally, in reviewing the map, it's clear all groups are essential because each contains a 1 not included in another group.

The horizontal quad of Fig. 6-23*d* corresponds to a simplified product *AB*. The square quad on the right corresponds to *AC*, while the one on the left stands for *AD*. The pair represents *BCD*. By ORing these products, we get the simplified Boolean equation:

$$y = AB + AC + AD + BCD \qquad \text{13 leads} \qquad (6\text{-}16)$$

$$= A(B + C + D) + BCD \qquad \text{10 leads}$$

Example 6-6

What is the AND-OR network corresponding to the simplified Boolean equation of the preceding example? If the equation is factored, what is the resulting logic circuit?

Solution

Figure 6-24*a* shows the AND-OR network for Eq. (6-16). Notice it has 13 input gate leads.

When Eq. (6-16) is factored, it becomes

$$y = A(B + C + D) + BCD$$

Fig. 6-23. Example 6-5.

	$\bar{C}\bar{D}$	$\bar{C}D$	CD	$C\bar{D}$
$\bar{A}\bar{B}$	0	0	0	0
$\bar{A}B$	0	0	1	0
AB	1	1	1	1
$A\bar{B}$	0	1	1	1

(*a*)

	$\bar{C}\bar{D}$	$\bar{C}D$	CD	$C\bar{D}$
$\bar{A}\bar{B}$	0	0	0	0
$\bar{A}B$	0	0	1	0
AB	1	1	1	1
$A\bar{B}$	0	1	1	1

(*b*)

	$\bar{C}\bar{D}$	$\bar{C}D$	CD	$C\bar{D}$
$\bar{A}\bar{B}$	0	0	0	0
$\bar{A}B$	0	0	1	0
AB	1	1	1	1
$A\bar{B}$	0	1	1	1

(*c*)

	$\bar{C}\bar{D}$	$\bar{C}D$	CD	$C\bar{D}$
$\bar{A}\bar{B}$	0	0	0	0
$\bar{A}B$	0	0	1	0
AB	1	1	1	1
$A\bar{B}$	0	1	1	1

(*d*)

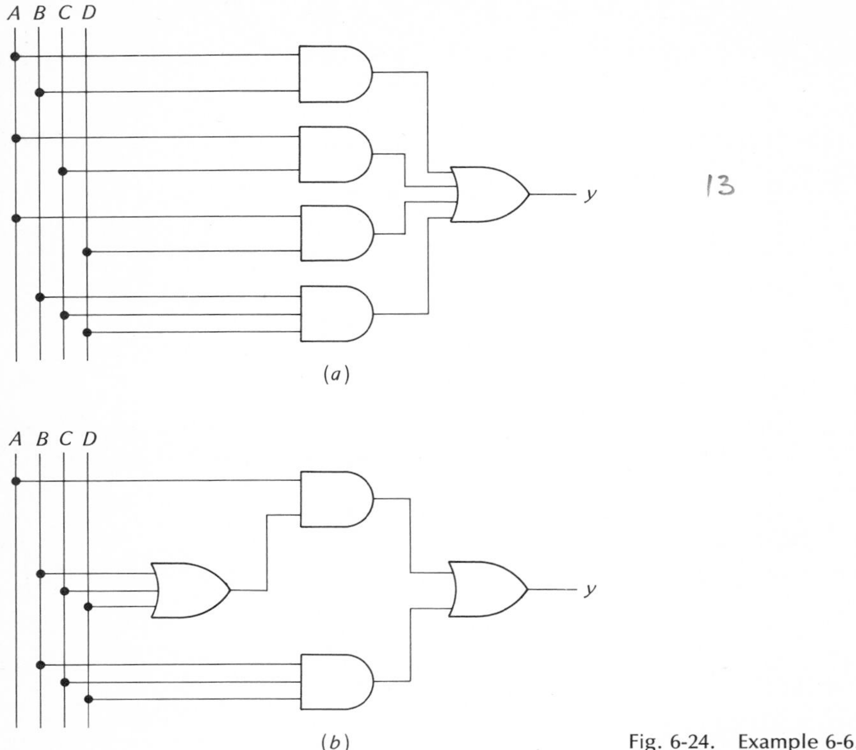

Fig. 6-24. Example 6-6.

Figure 6-24*b* shows the corresponding logic circuit; it has only 10 input gate leads.

Figure 6-24*b* is simpler than Fig. 6-24*a*. Nevertheless, as mentioned earlier, Fig. 6-24*a* might be preferred because it has a smaller propagation delay time (the sum of two gate delays instead of three). Furthermore, as will be discussed in Sec. 6-10, the AND-OR network can be converted into a NAND-NAND network (a group of NAND gates working into a final NAND gate). A NAND gate is the least expensive TTL gate. Because of this, the AND-OR network of Fig. 6-24*a* can be built with fewer NAND gates than Fig. 6-24*b*. More about this later.

The point of the example is this. Even though it may be possible to factor a simplified sum-of-products equation like Eq. (6-16) and reduce the number of input gate leads, this is not normally done because it increases the overall propagation delay time and may cost more to implement the logic circuit with NAND gates.

6-9 DON'T-CARE CONDITIONS

In Fig. 6-25*a*, a BCD input drives a decoder. As indicated, the decoder produces a 1 output only for a BCD input of 1001 (equivalent to decimal 9). As you already know, BCD digits are restricted to the four-bit numbers from 0000 through 1001;

Table 6-10

A	B	C	D	y
0	0	0	0	0
0	0	0	1	0
0	0	1	0	0
0	0	1	1	0
0	1	0	0	0
0	1	0	1	0
0	1	1	0	0
0	1	1	1	0
1	0	0	0	0
1	0	0	1	1

1010 through 1111 cannot occur for normal operation. This is why the truth table for the decoder of Fig. 6-25*a* lists only the inputs from 0000 through 1001 (see Table 6-10).

What is the logic circuit inside the decoder of Fig. 6-25*a*? Let's use the Karnaugh-map method to find the answer. To begin with, Table 6-10 has a 1 output for only one input condition:

$$A\ B\ C\ D = 1\ 0\ 0\ 1$$

Fig. 6-25. *Don't care* conditions.

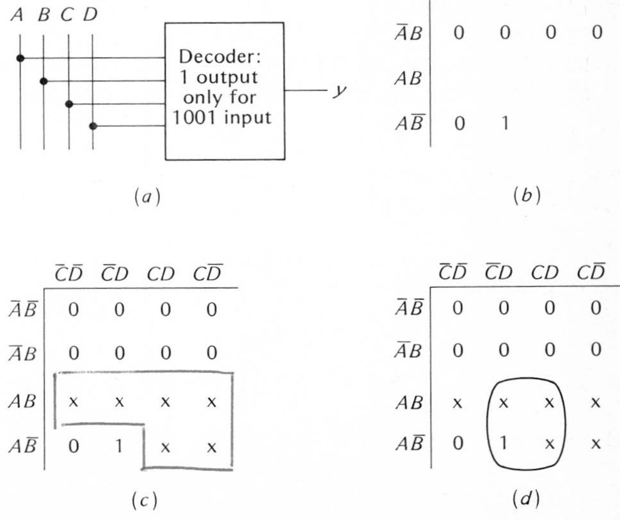

The fundamental product for this is $A\bar{B}\bar{C}D$. Figure 6-25*b* shows the 1 representing this fundamental product. Also shown are the 0s corresponding to the other input conditions of Table 6-10. The empty spaces on the map are for the forbidden BCD inputs that are not listed in the truth table.

Since forbidden BCD inputs don't occur under normal operating conditions, the empty spaces in Fig. 6-25*b* can be treated as 0s or 1s, whichever is more convenient. To indicate this, we mark X's as shown in Fig. 6-25*c*. These X's are called *don't cares* because they can be treated either as 0s or as 1s. (*Don't cares* are like wild cards in poker; you can let them stand for whatever you like.)

Figure 6-25*d* shows the most efficient way to encircle the 1. Notice two crucial ideas. First, the 1 is included in a quad, the largest group you can find if you visualize all X's as 1s. Second, after the 1 has been encircled, all X's outside the quad are visualized as 0s. In this way, the X's are used to the best possible advantage. As already mentioned, you are free to do this because *don't cares* correspond to BCD inputs that cannot occur under normal operating conditions.

The quad of Fig. 6-25*d* results in the Boolean equation

$$y = AD$$

The logic circuit inside the decoder is therefore nothing more than an AND gate (Fig. 6-26). When A and D are both 1s (which only happens for a BCD input of 1001), the output y equals a 1.

In summary, *don't cares* can simplify the logic circuit if you remember these ideas:

1. Enter 1s on the Karnaugh map for the fundamental products that produce output 1s in the truth table. Enter 0s for the other inputs listed in the truth table, and enter X's for the forbidden inputs (those that cannot occur during normal operation).
2. Encircle the actual 1s on the Karnaugh map in the largest groups you can find by treating the *don't cares* as 1s.
3. After the actual 1s have been included in groups, disregard the remaining *don't cares* by visualizing them as 0s.

Example 6-7

What is the simplest logic circuit for a decoder that produces a 1 output when the *BCD* input is 0000?

Solution

The truth table has a 1 output only for the input condition

$$A\ B\ C\ D = 0\ 0\ 0\ 0$$

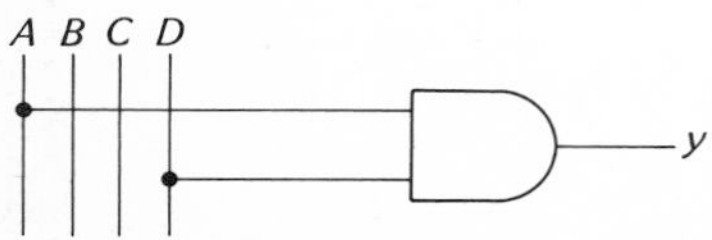

Fig. 6-26. Logic circuit for decoding 1001.

	$\bar{C}\bar{D}$	$\bar{C}D$	CD	$C\bar{D}$
$\bar{A}\bar{B}$	(1)	0	0	0
$\bar{A}B$	0	0	0	0
AB	x	x	x	x
$A\bar{B}$	0	0	x	x

(*a*)

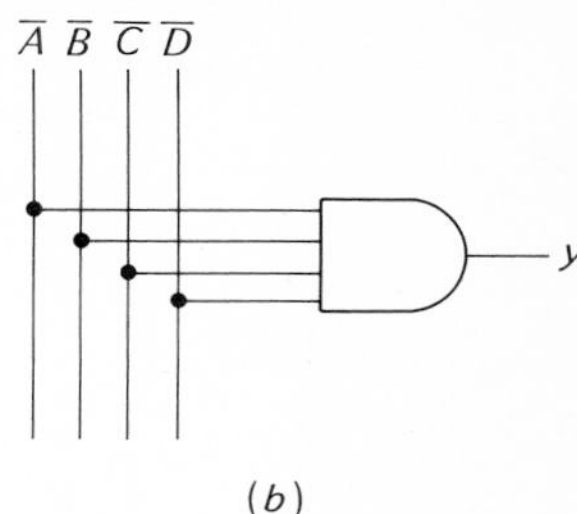

(*b*)

Fig. 6-27. Decoding 0000.

The corresponding fundamental product is $\bar{A}\bar{B}\bar{C}\bar{D}$. Figure 6-27*a* shows the Karnaugh map with a 1 for this product, 0s for the other inputs of the truth table, and X's for the forbidden inputs (1010 through 1111). In this case, the *don't cares* are of no help. The best we can do is encircle the isolated 1, while treating the *don't cares* as 0s. So the Boolean equation is

$$y = \bar{A}\,\bar{B}\,\bar{C}\,\bar{D}$$

Figure 6-27*b* shows the corresponding logic circuit. This four-input AND gate produces a 1 output only for the input condition

$$A\,B\,C\,D = 0\,0\,0\,0$$

Example 6-8

What's the simplest logic circuit for decoding a BCD input of 0111 as a 1 output?

Solution

Figure 6-28*a* is the Karnaugh map. The most efficient encircling is to group the 1 into a pair using the *don't care* as shown. Since this is the largest group possible, all remaining *don't cares* are treated as 0s. The equation for the pair is

$$y = BCD$$

Fig. 6-28. Decoding 0111.

	$\bar{C}\bar{D}$	$\bar{C}D$	CD	$C\bar{D}$
$\bar{A}\bar{B}$	0	0	0	0
$\bar{A}B$	0	0	(1	0
AB	x	x	x)	x
$A\bar{B}$	0	0	x	x

(*a*)

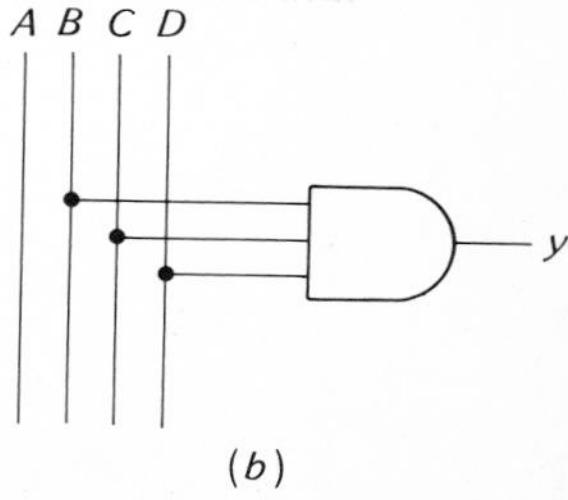

(*b*)

and Fig. 6-28*b* is the logic circuit. This three-input AND gate produces a 1 output only for a BCD input of 0111.

Example 6-9

Show one way of building a BCD-to-decimal decoder that produces a 1 output on one of 10 output lines.

Solution

One way to do this is to use a decoder for each 1 output. In other words, we can use an AND gate like Fig. 6-27*b* to decode a BCD input of 0000, an AND gate like Fig. 6-28*b* for the 0111 input, an AND gate like Fig. 6-26 for the 1001 input, and so forth. To get the AND gates for other inputs, we'd have to construct a Karnaugh map

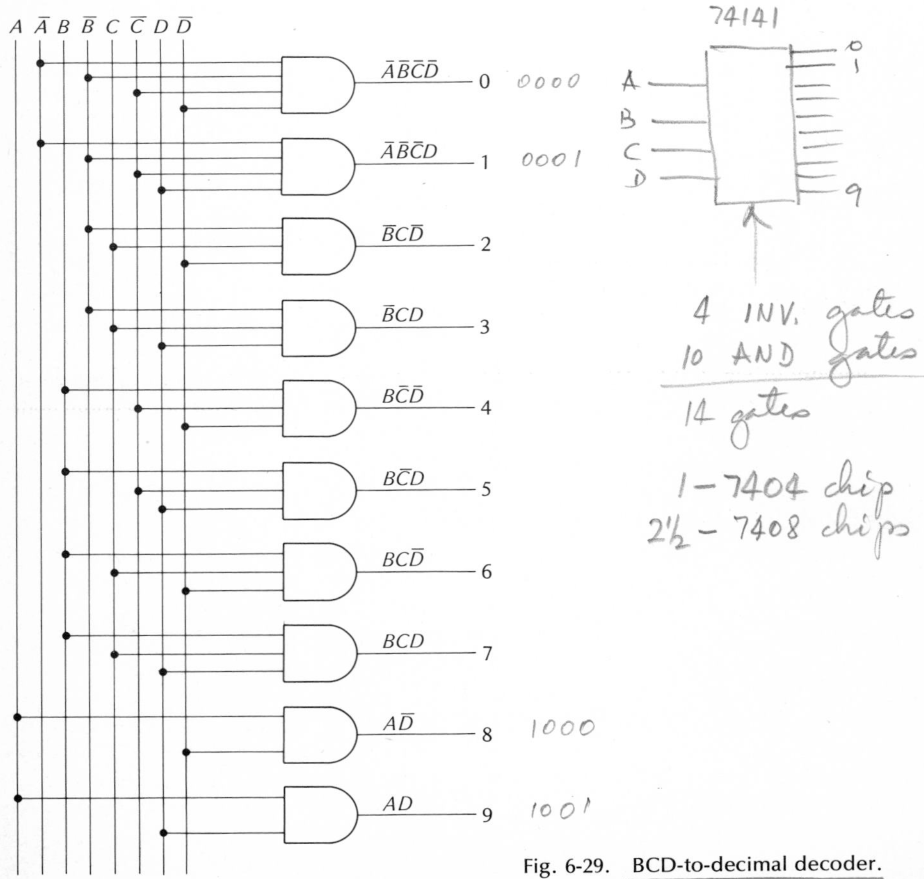

Fig. 6-29. BCD-to-decimal decoder.

for each of the remaining inputs, similar to Examples 6-7 and 6-8. When we finished, there would be one AND gate for each input to be decoded.

Figure 6-29 shows the final result. Given any BCD input from 0000 through 1001, only one AND gate will be on. Therefore, a 1 appears on only one of the 10 output lines.

This example is for instructional purposes only. At one time, BCD-to-decimal decoders were built by connecting SSI gates as shown in Fig. 6-29. But this is no longer done because BCD-to-decimal decoders are now available as MSI circuits (discussed in Secs. 3-9 and 3-10). see p. 59

6-10 NAND-NAND NETWORKS

The NAND gate is the backbone of the 7400 TTL series. Not only is it the least expensive gate in the line, but other circuits in the 7400 series often use the NAND gate as a basic building block. As mentioned earlier, 7400 series NAND gates are available as quad two-input gates, triple three-input gates, dual four-input gates, and single eight-input gates.

Any AND-OR network can be converted into a NAND-NAND network by replacing each AND gate by a NAND gate, and the final OR gate by a NAND gate. For example, if we have an AND-OR network like Fig. 6-30*a*, we can replace it with the NAND-NAND network of Fig. 6-30*b*. Both networks have the same truth table.

Here's the proof. To begin with, we need two ideas discussed earlier. Recall that an AND gate followed by an inverter is equivalent to a NAND gate as shown in Fig. 6-31*a*. Also, if you review the discussion of De Morgan's second theorem, you will agree that an OR gate preceded by inverters is equivalent to a NAND gate as shown in Fig. 6-31*b*.

With the foregoing equivalent circuits we can now prove that any AND-OR network is replaceable by a NAND-NAND network. Consider the simple AND-OR network of Fig. 6-32*a*. We can double complement each signal path as shown in Fig. 6-32*b* without changing the value of the final output. Because of the equivalent circuits given in Fig. 6-31, the circled portions of Fig. 6-32*c* can be replaced by NAND gates as shown in Fig. 6-32*d*.

A similar proof applies to any number of AND gates driving an OR gate. Because of this, any AND-OR network can be replaced by a NAND-NAND network; simply replace each AND gate by a NAND gate, and replace the final OR gate by a NAND gate.

Now we have a way of getting from a truth table to an inexpensive NAND-NAND network. First, the truth table is translated into a Karnaugh map. Second, the 1s are grouped as efficiently as possible. Third, we write a simplified Boolean equation. Fourth, the corresponding AND-OR network is drawn. Fifth, we replace each AND gate by a NAND gate, and the final OR gate by a NAND gate.

Incidentally, since a NAND gate is equivalent to an AND gate followed by an inverter (Fig. 6-31*a*), a wire-AND connection and an inverter can replace the final NAND gate of a NAND-NAND network. For instance, suppose we want to implement

$$y = AB + CD + EF \qquad (6\text{-}17)$$

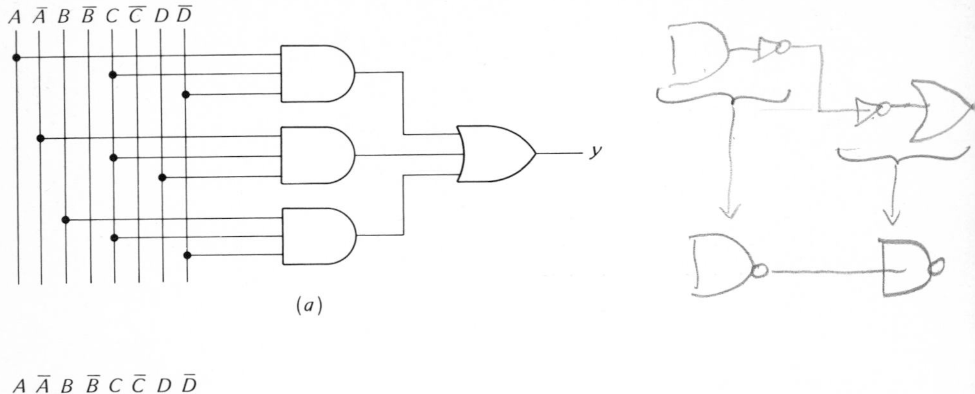

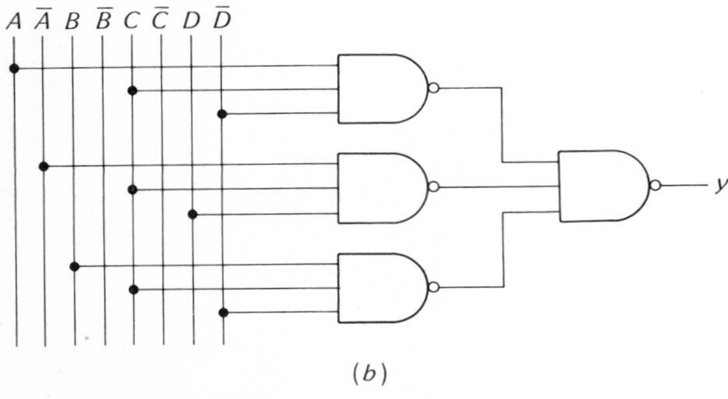

Fig. 6-30. Converting AND-OR to NAND-NAND.

Figure 6-33*a* shows the AND-OR network for this, and Fig. 6-33*b* is the equivalent NAND-NAND network. But the NAND function is the same as the AND function followed by inversion. For this reason, Fig. 6-33*c* is equivalent to Fig. 6-33*b*. Therefore, you can implement Eq. (6-17) with any of the three circuits shown in Fig. 6-33.

Example 6-10

Use De Morgan's first theorem to prove Eq. (6-17) gives the output of Fig. 6-33*c*.

Fig. 6-31. Equivalent logic circuits.

(*a*) (*b*)

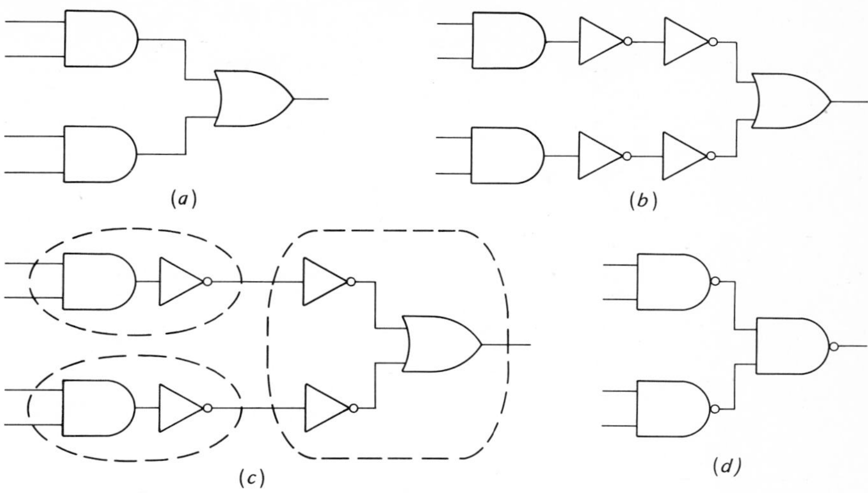

Fig. 6-32. Proving a NAND-NAND is equivalent to an AND-OR.

Fig. 6-33. Wire-AND and inverter replace final NAND gate.

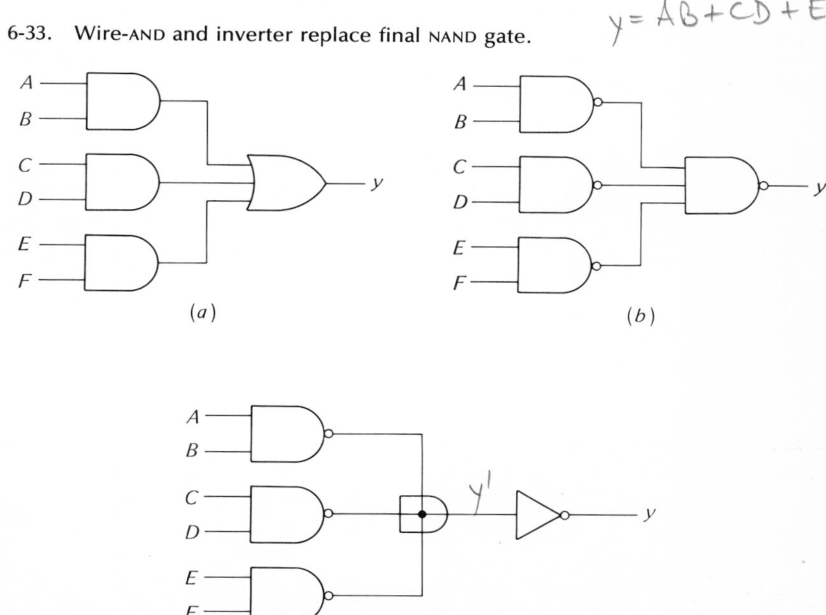

Solution

Let y_1 stand for the input to the inverter. Then,

$$y_1 = \overline{AB} \cdot \overline{CD} \cdot \overline{EF}$$

With De Morgan's first theorem, this becomes

$$y_1 = \overline{AB + CD + EF}$$

The output of the inverter is the complement of y_1; therefore,

$$y = \overline{y}_1 = \overline{\overline{AB + CD + EF}}$$

or

$$y = AB + CD + EF$$

By including an inverter after a wire-AND connection, we wind up with the ORing of logical products. This is why the wire-AND connection is often referred to as a wire-OR connection.

6-11 OTHER METHODS

Besides the sum-of-products simplification method discussed in this chapter, there's another approach based on the product of fundamental sums. This *product-of-sums* method is useful when the final logic circuit is to be a NOR-NOR network (a group of NOR gates working into a NOR gate).

In problems where a logic circuit has more than one output (like Example 6-9), it's sometimes possible to get further simplification. This design problem is called the *multi-output problem*.

Advanced books on digital design cover the product-of-sums method and the multi-output problem.[1]

STUDY AIDS

Summary

A fundamental product is the logical product that results in a 1 output for a given input condition. We can translate a truth table into its Boolean equation by ORing the fundamental products corresponding to the output 1s.

A Karnaugh map displays the fundamental products. A pair is two adjacent 1s, a

[1] If interested, see E. J. McCluskey,: "Introduction to the Theory of Switching Circuits," pp. 127–139, 157–174, McGraw-Hill Book Company, New York, 1965. Another discussion is in R. K. Richards, "Digital Design," pp. 53–68, John Wiley & Sons, Inc., New York, 1971.

quad is four adjacent 1s, and an octet is eight adjacent 1s. Efficient grouping of 1s is equivalent to algebraic simplification of the Boolean equation. By ORing the products for the octets, quads, and pairs, we can arrive at a simplified Boolean equation.

A NAND-NAND network is equivalent to an AND-OR network.

Glossary

adjacent On a Karnaugh map this means the 1s are horizontally or vertically next to one another.

AND-OR *network* A group of AND gates whose outputs drive a single OR gate.

don't cares These are the output conditions for input conditions that cannot exist normally. Because of this, *don't cares* can be treated as 0s or 1s, whichever is more convenient.

fundamental product A logical product of input variables in complemented or uncomplemented form that results in a 1 output for a given input condition.

Karnaugh map A simplified way of displaying the fundamental products corresponding to a truth table.

NAND-NAND *network* A group of NAND gates working into a final NAND gate.

octet A group of eight adjacent 1s on a Karnaugh map.

overlap Groups of 1s may overlap because the same 1s can be used more than once in the forming of octets, quads, and pairs.

pair Two adjacent 1s on a Karnaugh map.

quad Four adjacent 1s on a Karnaugh map.

rolling Refers to visualizing a Karnaugh map rolled so that its opposite edges are adjacent.

Review Questions

1. What is a fundamental product? If there are four input variables, how many factors are there in the fundamental product?

2. Given the fundamental products, how do you get the Boolean equation?

3. The sum-of-products method leads to what kind of network?

4. The sum of a variable and its complement always equals what?

5. In a four-variable Karnaugh map, what progression do the lettered rows and columns follow?

6. How does a Karnaugh map differ from a truth table?

7. Define pairs, quads, and octets.

8. What can you do with a pair of diagonally adjacent 1s?

9. What does overlapping mean? Rolling?

10. How do you treat *don't cares?*

11. What is a NAND-NAND network? What logic circuit can it replace?

Problems

6-1. What are the fundamental products for each of these input conditions:
(a) $A\ B\ C = 0\ 1\ 0$.
(b) $A\ B\ C\ D = 1\ 0\ 0\ 1$.

6-2. A truth table has output 1s for each of these input conditions:
(a) $A\ B\ C\ D = 0\ 0\ 1\ 1$.
(b) $A\ B\ C\ D = 0\ 1\ 1\ 1$.
(c) $A\ B\ C\ D = 1\ 0\ 0\ 0$.
(d) $A\ B\ C\ D = 1\ 1\ 0\ 1$.

What are the corresponding fundamental products?

6-3. A truth table has output 1s for these input conditions:

$$A\ B\ C = 0\ 1\ 1$$
$$A\ B\ C = 1\ 0\ 1$$

What is the Boolean equation corresponding to the truth table?

6-4. Output 1s appear in a truth table for these input conditions:

$$A\ B\ C\ D = 0\ 0\ 0\ 1$$
$$A\ B\ C\ D = 0\ 1\ 0\ 0$$
$$A\ B\ C\ D = 1\ 0\ 1\ 1$$

Write the corresponding Boolean equation.

6-5. Draw the AND-OR network for Prob. 6-3.

6-6. Draw the AND-OR network for Prob. 6-4.

6-7. What is the AND-OR network for

$$y = \overline{A}\overline{B}C + A\overline{B}C + AB\overline{C} + ABC$$

6-8. What is the AND-OR network for

$$y = \overline{A}B\overline{C}\overline{D} + \overline{A}B\overline{C}D + A\overline{B}\overline{C}D$$

6-9. Use algebra to simplify this equation:

$$y = A\overline{B}C + ABC$$

How many input gate leads does the simplified logic circuit have?

6-10. A truth table has output 1s for these inputs:

$$A\ B\ C\ D = 0\ 0\ 1\ 1$$
$$A\ B\ C\ D = 0\ 1\ 1\ 0$$
$$A\ B\ C\ D = 1\ 0\ 0\ 0$$
$$A\ B\ C\ D = 1\ 1\ 0\ 0$$

Draw the Karnaugh map.

6-11. A truth table has four input variables. The first eight outputs are 1s, and the remaining eight outputs are 0s. Draw the Karnaugh map.

6-12. A four-variable truth table has outputs that alternate between 0 and 1 for each input condition. That is, the output column looks like 0, 1, 0, 1, 0, 1, and so forth. Draw the Karnaugh map.

6-13. Using the Karnaugh map, find the simplified Boolean equation for the truth table given in Prob. 6-10.

6-14. Use the Karnaugh map to find the simplified equation for the truth table given in Prob. 6-11.

6-15. With the Karnaugh map, work out the simplified equation corresponding to the truth table of Prob. 6-12.

6-16. A truth table has a 1 output for the input condition

$$A\ B\ C\ D = 0\ 1\ 1\ 0$$

Inputs from 1010 through 1111 are forbidden. All other inputs produce output 0s. What is the simplest logic circuit that implements this truth table?

6-17. We want to build a decoder that produces an output 1 for the input condition

$$A\ B\ C\ D = 1\ 0\ 0\ 0$$

Input conditions from 1010 through 1111 are forbidden. What is the simplest logic circuit that does the job?

6-18. Given the Boolean equation

$$y = A\overline{B} + AC + BD$$

What is the corresponding NAND-NAND network?

6-19. A truth table leads to this Boolean equation:

$$y = AB\overline{C} + A\overline{C}D + AB\overline{D}$$

Show the NAND-NAND network for this equation.

6-20. Refer to Table 5-3, the truth table of a full adder. Write the sum-of-products equation for the *carry* output, and simplify the Boolean equation by the Karnaugh-map method. Finally, show the NAND-NAND network for the simplified Boolean equation of the *carry* output.

6-21. Refer to Table 5-3 and write the Boolean equation for the *sum* output. After Karnaugh simplification, show the NAND-NAND network for the *sum* output.

Multivibrators

7

A *multivibrator* is a regenerative circuit with two active devices, designed so that one device conducts while the other cuts off. Multivibrators can store binary numbers, count pulses, synchronize arithmetic operations, and perform other essential functions in digital systems.

After studying multivibrators, you should be able to

1. Explain why four cascaded flip-flops can count from 0 through 15.
2. Recall the input-output tables of a *D* flip-flop and a *JK* flip-flop.
3. Describe the action of a Schmitt trigger, an astable multivibrator, and a monostable multivibrator.

7-1 THE *RS* FLIP-FLOP

A *flip-flop* is another name for a *bistable* multivibrator, one whose output is either a low or a high voltage, a 0 or a 1. This output stays low or high; to change it, the circuit must be driven by an input called a *trigger*. Until the trigger arrives, the output voltage remains low or high indefinitely.

The first type of bistable multivibrator we discuss is the *RS* flip-flop. Look at Fig. 7-1, and notice the cross-coupling from each collector to the opposite base. The cross-coupling results in positive feedback. Because of this, if Q_1 is saturated, the low Q_1 collector voltage will force Q_2 to cut off.[1] Similarly, if we can once get Q_2 saturated, it will force Q_1 to go into cutoff. So there are two stable operating conditions: Q_1 saturated, Q_2 cut off; or Q_1 cut off, Q_2 saturated.

To ensure adequate saturation and cutoff, β_{dc} must be greater than approximately R_B/R_C, the ratio of base resistance to collector resistance. In Fig. 7-1, R_B/R_C equals 50. With β_{dc} greater than approximately 50, the *on* transistor will be saturated and the *off* transistor will be cut off.

To control the state of a flip-flop, we have to add trigger inputs as shown in Fig. 7-2. If a high voltage is applied to the *S* (set) input, then Q_1 saturates; this forces Q_2 into cutoff. Once Q_1 is saturated and Q_2 is cut off, the trigger at the *S* input can be

[1] Assuming silicon transistors, $V_{CE(sat)}$ is only 0.1 V or thereabouts, not enough to turn on the base, because the required V_{BE} is approximately 0.7 V.

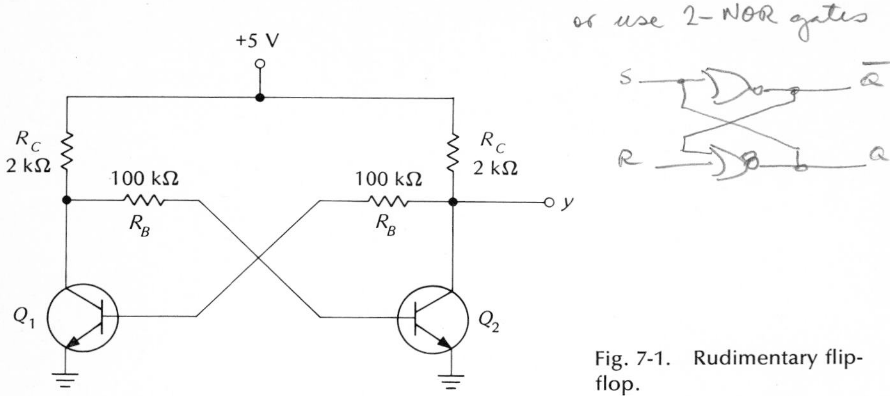

Fig. 7-1. Rudimentary flip-flop.

removed. Likewise, a high voltage can be applied to the *R* (reset) input; this saturates Q_2 and forces Q_1 into cutoff.

Applying a high voltage to the *S* input is called *setting* the flip-flop and results in a binary output of

$$y = 1$$

On the other hand, applying the high voltage to the *R* input is known as *resetting* the flip-flop and results in a binary output of

$$y = 0$$

Incidentally, the flip-flop of Fig. 7-2 also has a complementary output $\overline{y}$. So, a

Fig. 7-2. *RS* flip-flop.

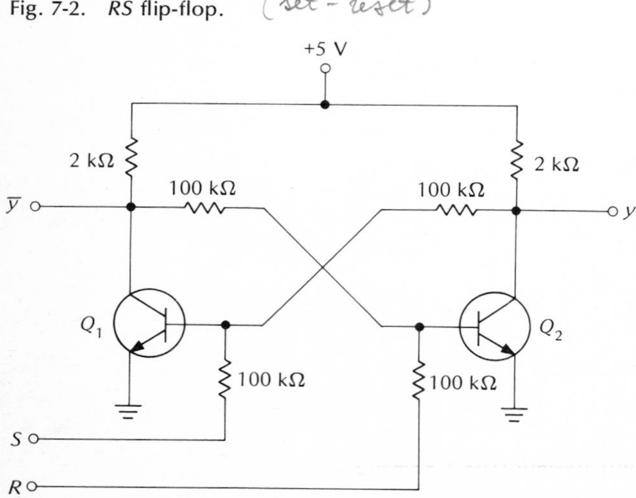

Table 7-1
RS FLIP-FLOP

R	*S*	*y*
0	0	Last value
0	1	1
1	0	0

flip-flop is a natural circuit for generating a variable and its complement. This ties in with our discussion of Karnaugh maps and simplified logic circuits. As you recall, a variable and its complement are usually available for *inputing* to logic circuits.

Table 7-1 summarizes the input-output possibilities for the *RS* flip-flop of Fig. 7-2. The first input condition is

$$RS = 0\,0$$

This means no triggers have been applied. In this case, the *y* output retains the last value it had.

The second input condition

$$RS = 0\,1$$

means a trigger has been applied to the *S* input. As we know, this sets the flip-flop and results in a *y* output of 1.

The third input condition is

$$RS = 1\,0$$

This indicates a trigger has been applied to the *R* input. The resulting *y* output is a 0.

The input condition

$$RS = 1\,1$$

is not listed. Why? Because it's a forbidden input. It means applying a trigger to both the *S* and *R* inputs at the same time. This is contradictory because it implies we're trying to get a *y* output that's simultaneously equal to a 1 and a 0. This doesn't make sense and explains why the forbidden input is deleted from Table 7-1.

Figure 7-2 is an elementary design intended only to show the key ideas behind *RS* flip-flops. Many advanced designs are possible, to improve switching speed, output impedance, etc. In this book, we shall use the symbol of Fig. 7-3 to represent an *RS* flip-flop of any design. (The *y* output is often shown as the *Q* output on data sheets.)

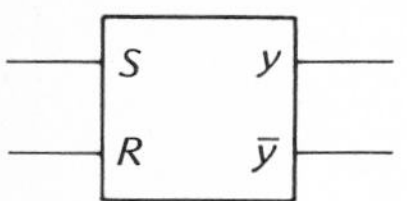

Fig. 7-3. Logic symbol for *RS* flip-flop.

Example 7-1

A simple *RS* flip-flop can be converted into a *clocked RS* flip-flop as shown in Fig. 7-4. Describe the circuit action.

Solution

When the input labeled "clock" is low, both AND gates are *disabled*. This ensures that

$$R\,S = 0\,0$$

which means the y output remains in the last state it was in.

When the clock input goes high, however, both AND gates are *enabled*. This allows the S and R signals to reach the *RS* flip-flop. In this way, the flip-flop either sets or resets, depending on the value of $R\,S$. Therefore, the clocked *RS* flip-flop can't change states until the clock signal occurs.

Clocking a flip-flop as just described is important in large digital systems where hundreds of flip-flops may be interconnected. The clock is applied to all flip-flops simultaneously; this ensures that they all change states in unison. This synchronization is essential in many digital systems.

7-2 THE *T* AND *RST* FLIP-FLOPS

Figure 7-5 is one way to build a *toggle* (*T*) flip-flop. A train of extremely narrow triggers drives the T input. Each time one of these triggers appears, the output of the flip-flop changes state. For instance, suppose y equals 0 just before point A in time. Then the upper AND gate is enabled, and the lower AND gate is disabled. When the trigger comes in at point A in time, it results in a high S input. This sets the y output to 1. (The width of the trigger is less than the propagation delay time; this ensures only one change in the output state.)

When the next trigger appears at point B in time, the lower AND gate is enabled

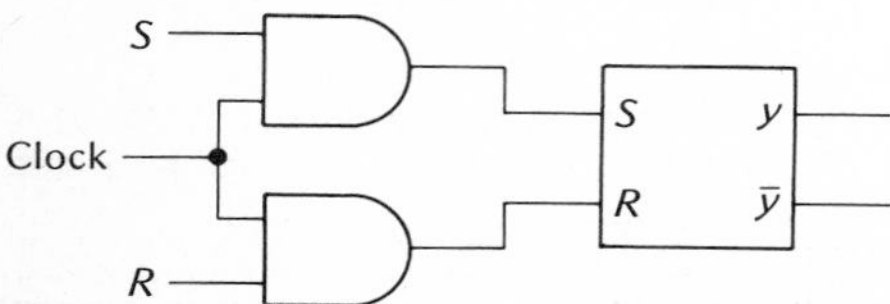

Fig. 7-4. Clocked *RS* flip-flop.

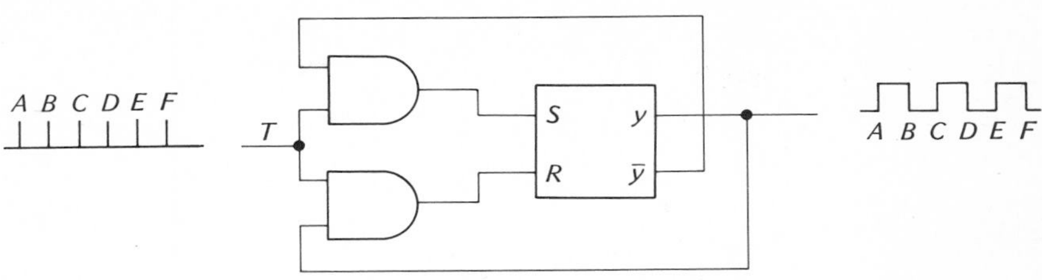

Fig. 7-5. One way to build a *T* flip-flop.

and the trigger passes through to the *R* input. This forces the flip-flop to reset.

Since each incoming trigger is alternately steered into the set and reset inputs, the flip-flop *toggles* (repeatedly sets and resets). Notice it takes two triggers to produce one cycle of the output waveform. This means the output has half the frequency of the input. Stated another way, a *T* flip-flop divides the input frequency by two.

Many designs are possible for a *T* flip-flop. For instance, we can build a *T* flip-flop to respond only to negative triggers. Or we can design a *T* flip-flop to respond to the negative-going edge of a square wave. This latter case is the one most often encountered in practice. In this book, therefore, all *T* flip-flops discussed from here on will respond to the *negative edge of an input square wave.* Figure 7-6 shows the symbol we shall use for a *T* flip-flop. When you see this symbol, remember the flip-flop changes state only on the negative-going edge of the input signal. This is why the *y* output changes at points *B, D, F, H,* and so forth.

An *RST flip-flop* combines an *RS* flip-flop and a *T* flip-flop. It can reset, set, or toggle. Many designs are possible. Especially important is the case of an *RST* flip-flop that responds to positive-going *R* and *S* inputs, and to negative-going *T* inputs. (Section 8-4 explains why this particular choice is important; it's related to the master-slave concept.)

Figure 7-7 shows the symbol used in this book for an *RST* flip-flop. Remember: *it responds only to positive-going R and S inputs, and to negative-going T inputs.*

Example 7-2

The waveform of Fig. 7-8*a* drives the *T* input of Fig. 7-7. Initially, $y = 0$. Sketch the y and $\overline{y}$ outputs.

Fig. 7-6. Logic symbol for *T* flip-flop.

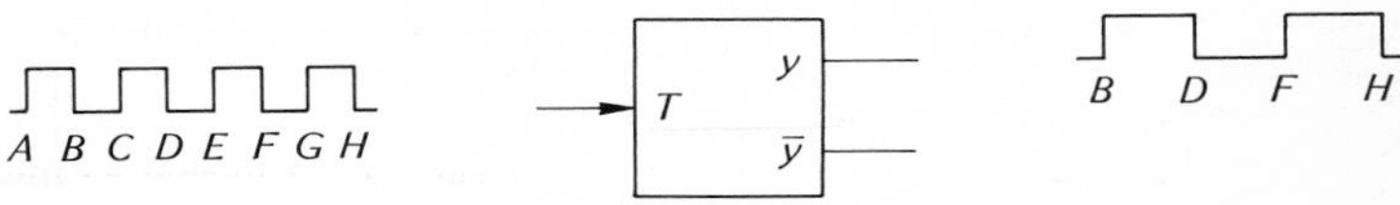

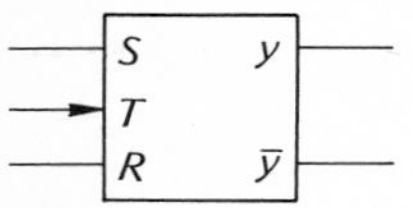

Fig. 7-7. Logic symbol for *RST* flip-flop.

Solution

Each time the input waveform goes from a 1 to a 0, the flip-flop changes states. So, we can draw the y output as shown in Fig. 7-8*b*. Figure 7-8*c* shows the $\overline{y}$ output.

Example 7-3

Suppose we connect three *T* flip-flops as shown in Fig. 7-9. If the input to the first flip-flop is a 1,000-Hz square wave, what comes out of the final flip-flop?

Solution

Each time the input square wave goes from a 1 to a 0, the first flip-flop changes state. Therefore, the output of the first flip-flop is a 500-Hz square wave.

The 500-Hz square wave drives the second flip-flop. On each negative-going transition of this 500-Hz signal, the second flip-flop changes state. So, the output of the second flip-flop is a 250-Hz square wave.

The 250-Hz square wave drives the final flip-flop; therefore, the final output is a 125-Hz square wave.

Figure 7-9 is an example of a *scaler,* a circuit that delivers one cycle of output after a specified number of input cycles. In this case, we have an 8-to-1 scaler.

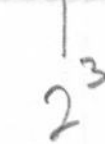

Fig. 7-8. Example 7-2.

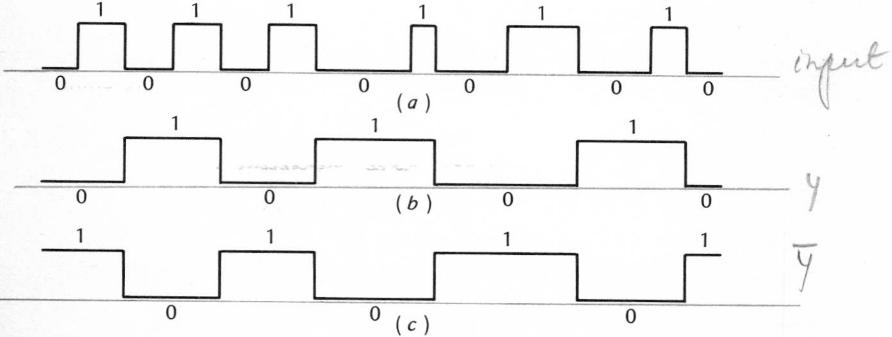

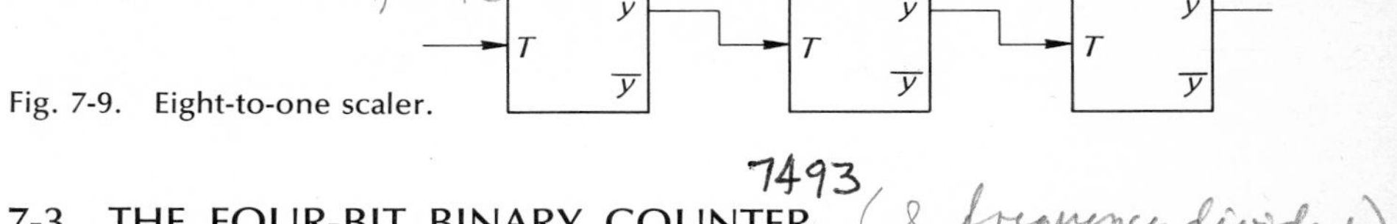

Fig. 7-9. Eight-to-one scaler.

7-3 THE FOUR-BIT BINARY COUNTER

We can connect flip-flops to get an *electronic counter,* a device that counts the number of input triggers. Figure 7-10 shows four *RST* flip-flops in cascade. A square wave drives the *A* flip-flop. From now on, we shall call this square wave the "clock" signal. Note that the output of the *A* flip-flop drives the *B* flip-flop. The *B* flip-flop in turn drives the *C* flip-flop, which then drives the *D* flip-flop.

What happens as each clock pulse comes in? Let's assume all flip-flops are initially reset to produce 0 outputs. Therefore, the output condition is

$$DCBA = 0\,0\,0\,0$$

just before the first clock pulse. (We can produce this condition by simultaneously feeding a positive-going trigger to the *R* inputs of all the flip-flops.) When the first clock pulse comes in, the *A* flip-flop changes state on the negative-going part of the pulse. So, at the end of the first input cycle, the output condition is

$$DCBA = 0\,0\,0\,1$$

Note that the *A* output has gone from a 0 to a 1. This is a positive change. When fed to the *T* input of the *B* flip-flop, this positive change has no effect because the *T* flip-flops respond only to negative-going changes.

When the second clock pulse occurs, the *A* flip-flop again changes state on the negative-going edge of the input square wave. As it changes, *A* goes from a 1 to a 0, a negative change. This negative-going change triggers the *B* flip-flop; therefore, *B* goes from a 0 to a 1. This positive-going change in *B* has no effect on *C*. So, at the end of the second input pulse, the output condition of the four flip-flops is

$$DCBA = 0\,0\,1\,0$$

After the third clock pulse, *A* has changed from a 0 to a 1; this positive-going

Fig. 7-10. Four-bit binary counter.

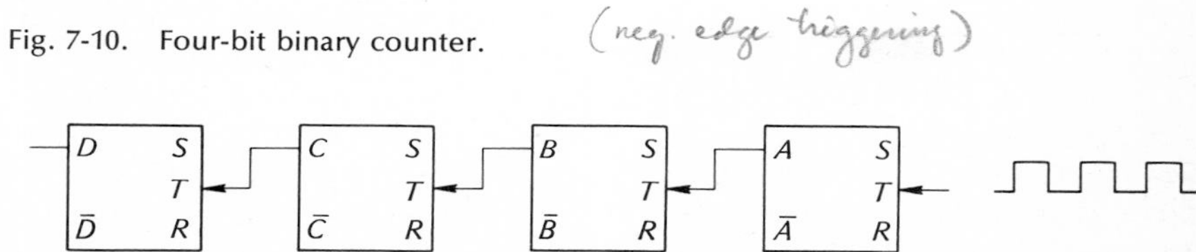

change has no effect on the remaining flip-flops, so the output condition is

$$D\,C\,B\,A = 0\,0\,1\,1$$

On the negative-going edge of the fourth clock pulse, A changes from a 1 to a 0. This causes B to change from a 1 to a 0. In turn, the negative-going change in B forces C to change from a 0 to a 1. The output condition of the four flip-flops then is

$$D\,C\,B\,A = 0\,1\,0\,0$$

Notice that the output condition of the flip-flops is a binary number equivalent to the number of clock pulses received. For the first four clock pulses received, the output condition of the flip-flops has been 0001, 0010, 0011, and 0100 (straight binary progression). If we analyze the action of Fig. 7-10 for ensuing clock pulses, we find the output conditions agree with those given in Table 7-2. (You should analyze Fig. 7-10 from the fifth clock pulse on, to convince yourself of the output conditions listed.)

Because each output condition of Table 7-2 is the binary equivalent of the number of clock pulses, the four cascaded flip-flops of Fig. 7-10 comprise a four-bit binary counter. We can use this counter to count the number of clock pulses up to a maximum of 15.

What happens on the 16th clock pulse? A goes from a 1 to a 0, a negative-going change. This causes B to go from a 1 to a 0, another negative-going change. This forces C to go from a 1 to a 0; this negative-going change causes D to go from 1 to 0. So the entire counter resets, and

$$D\,C\,B\,A = 0\,0\,0\,0$$

Ensuing clock pulses reproduce the output conditions of Table 7-2.

If we cascade n flip-flops, we get 2^n output conditions. For instance, five flip-flops have 2^5 or 32 output conditions (00000 through 11111). Six flip-flops have 2^6 or 64 output states (000000 through 111111). The largest binary number counted by n cascaded flip-flops has a decimal equivalent of $2^n - 1$. For example, four flip-flops reach a maximum decimal value of 15, five flip-flops reach decimal 31, six flip-flops have a maximum count of 63, and so on.

When the output of one flip-flop drives another, we call the counter a *ripple counter*; the A flip-flop has to change states before it can trigger the B flip-flop; B has to change before it can trigger C; and so forth. The triggers move through the flip-flops like a ripple in water. Because of this, the overall propagation delay time is the sum of the individual delays. If each flip-flop in a four-bit binary counter has a t_p of 10 ns, the overall t_p equals 40 ns.

Incidentally, four-bit binary counters are available commercially as MSI circuits. A good example in the TTL 7400 series is the 7493, whose truth table is the same as Table 7-2. The 7493 has a totem-pole output, a worst-case propagation delay of 70 ns, a typical power dissipation of 130 mW, an input for resetting the counter to zero, and other features described on its data sheet.

Table 7-2
FOUR-BIT BINARY COUNTER

Clock pulses	D	C	B	A
0	0	0	0	0
1	0	0	0	1
2	0	0	1	0
3	0	0	1	1
4	0	1	0	0
5	0	1	0	1
6	0	1	1	0
7	0	1	1	1
8	1	0	0	0
9	1	0	0	1
10	1	0	1	0
11	1	0	1	1
12	1	1	0	0
13	1	1	0	1
14	1	1	1	0
15	1	1	1	1

7-4 THE DECADE COUNTER

Because we use the decimal number system so much, we often prefer to have a decimal or decade counter instead of a binary counter. A decade counter uses the base 10; it has 10 distinct output states. The usual way to get 10 distinct output states is to modify a four-bit binary counter so that it skips six of its possible output states.

There are many ways to advance a binary counter to skip some of its natural states. (This chapter discusses one way; the next chapter examines several ways.) Here's a way to make a BCD counter (same as an 8421 counter). Initially, the counter of Fig. 7-11 is set to zero, so that

$$D\ C\ B\ A = 0\ 0\ 0\ 0$$

Because of this, the first clock pulse causes A to change from a 0 to a 1; this positive-going change is coupled through the AND gate to the T input of the B flip-

Fig. 7-11. Decade counter.

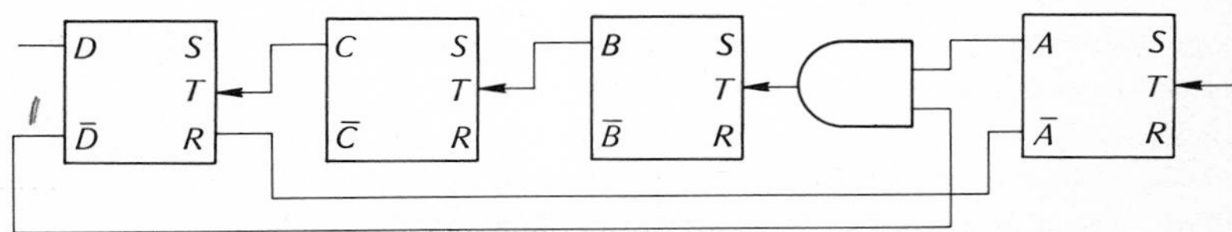

flop. Since the trigger is positive-going, it has no effect on B. At the end of the first clock pulse, therefore, the output condition of the counter is

$$DCBA = 0001$$

The AND gate stays enabled as long as the $\bar{D}$ output is a 1. For this reason, the counter acts like a straight binary counter up to the seventh clock pulse. At the end of the seventh clock pulse, the output condition is

$$DCBA = 0111$$

On the negative-going edge of the eighth clock pulse, A changes from a 1 to a 0. The AND gate is still enabled, and this negative-going change passes through to the B flip-flop, causing B to change from a 1 to a 0. In turn, C changes from a 1 to a 0, and D changes from a 0 to a 1. When D changes from a 0 to a 1, $\bar{D}$ goes from a 1 to a 0. The change in $\bar{D}$ is fed back to the AND gate; this disables the AND gate. At the end of the eighth clock pulse, the output condition of the counter is

$$DCBA = 1000$$

On the negative-going edge of the ninth clock pulse, A goes from a 0 to a 1, and $\bar{A}$ goes from a 1 to a 0. This negative-going change has no effect on the D flip-flop. So, at the end of the ninth pulse,

$$DCBA = 1001$$

When the tenth clock pulse comes in, the counter resets to zero because the following things happen in rapid order:

1. The A output goes to 0, temporarily producing

$$DCBA = 1000$$

2. Since A went from 1 to 0, $\bar{A}$ went from 0 to 1. This positive-going change is fed to the reset input of the D flip-flop, forcing D to go from 1 to 0. Therefore, on the tenth clock pulse the output condition quickly goes to

$$DCBA = 0000$$

Therefore, the counter of Fig. 7-11 acts like a BCD or 8421 counter. (In the foregoing explanation we've assumed triggering occurs only on the edges of waveforms; the changes in voltage cause the triggering.) Table 7-3 summarizes the output states of the decade counter just discussed.

It is also possible to build decade counters using other BCD codes. Most of the time, however, decade counters use the 8421 code. In fact, the industry standard

Table 7-3
8421 DECADE COUNTER

Count	D	C	B	A
0	0	0	0	0
1	0	0	0	1
2	0	0	1	0
3	0	0	1	1
4	0	1	0	0
5	0	1	0	1
6	0	1	1	0
7	0	1	1	1
8	1	0	0	0
9	1	0	0	1

for a decade counter is the 7490, an MSI circuit in the 7400 TTL series. The 7490 of Fig. 7-12 counts in the 8421 code from 0000 through 1001; then it resets to zero. This well-known MSI circuit has a typical power dissipation of 145 mW, a propagation delay time of 50 ns (worst case), a totem-pole output, and other features described on its data sheet.

To count beyond 10, all we have to do is cascade decade counters. For instance, Fig. 7-13 shows three 7490s. Initially, all 7490s are reset to

$$DCBA = 0000$$

(This is done by applying a reset pulse simultaneously to all 7490s.) The clock pulses drive the first 7490. After nine clock pulses, the *units* counter has an output condition of

$$DCBA = 1001$$

All other counters still have

$$DCBA = 0000$$

On the tenth clock pulse, the D output of the units counter goes from a 1 to a 0. This negative-going change triggers the *tens* counter. Because of this, the output condition of all three 7490s is

0000 0001 0000

after the tenth clock pulse.

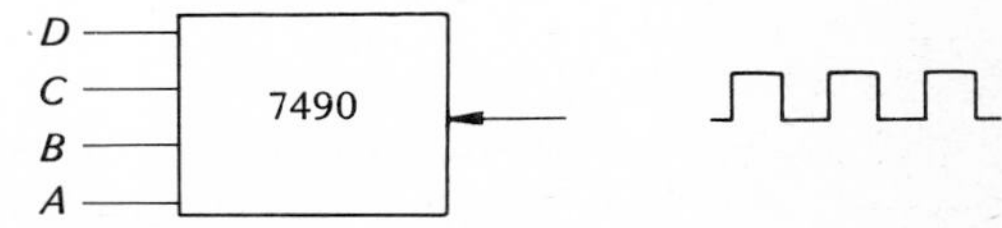

Fig. 7-12. Logic symbol for a 7490.

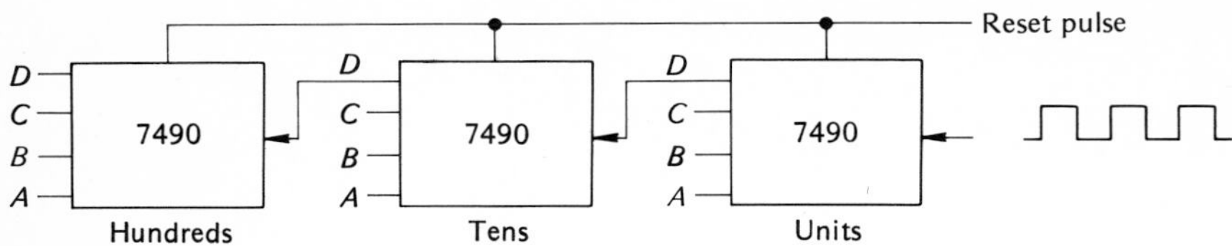

Fig. 7-13. Cascaded 7490s can count up to 999.

As each additional clock pulse arrives, the *units* counter advances one count. Every time the *units* counter resets to 0000, it triggers the *tens* counter, which advances one count. After 99 clock pulses, the BCD output of the three 7490s is

0000 1001 1001

equivalent to decimal 99.

On the 100th clock pulse, the *tens* counter resets to 0000, producing a negative-going trigger for the *hundreds* counter. Therefore, after 100 clock pulses, the BCD output of the 7490s is

0001 0000 0000

equivalent to decimal 100.

You should be able to see the point. The three 7490s count the number of clock pulses, using the BCD code. The maximum number of clock pulses that can be counted by three 7490s is 999. To increase the capacity, we cascade more 7490s. For instance, to count up to 99,999,999, we use eight 7490s.

Cascaded 7490s like those of Fig. 7-13 are commonly used in frequency counters, digital voltmeters, etc.

7-5 GATING A COUNTER

Gating a counter means turning it on for a specified period of time during which it counts the number of pulses that arrive at its input. Figure 7-14 shows a simple way to gate a counter. An AND gate drives a group of cascaded 7490s (similar to Fig. 7-13). A train of clock pulses goes into the upper input of the AND gate, while a rectangular pulse drives the lower input. This rectangular pulse resets the counter to zero at point *A* in time; thereafter, the rectangular pulse enables the AND gate, allowing the clock pulses to reach the counter.

Between points *A* and *B* in time, the counter counts each clock pulse that comes in. At point *B* in time, the AND gate is disabled and the counter stops counting. The final count holds until the next gating pulse. Therefore, with a system like Fig. 7-14, we can count the number of clock pulses that occur in a given time.

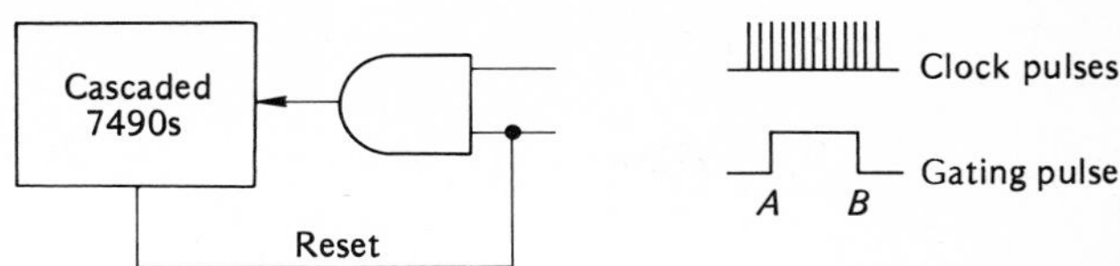

Fig. 7-14. Gated counter.

Since the accuracy of the count depends on the width of the gating pulse, the gating pulse is usually derived from a highly accurate frequency source. Figure 7-15 shows a way to get a very accurate gating pulse. A 5-MHz frequency standard is successively divided by 10. After seven divisions, the final output pulse has a period of 2 s. Therefore, the positive half cycle is a very accurate 1-s wide pulse. This highly accurate 1-s wide pulse can then be used for the gating pulse in Fig. 7-14.

Example 7-4

Suppose the gating pulse of Fig. 7-14 is exactly 10 s wide. If the clock pulses have a frequency of 579 Hz, what does the counter read at the end of the gating pulse? What is the BCD output of the 7490s?

Solution

There are 579 clock pulses during 1 s. Therefore, during 10 s there will be 5,790 clock pulses. The counter reads the number of pulses during a 10-s gating period, so the final count is 5,790.

The corresponding BCD output of the 7490s is

0101 0111 1001 0000

Example 7-5

In Fig. 7-15, suppose we want the positive half cycle to have a width of 10 ms instead of 1 s. How many 7490s would be needed?

Solution

To get a half-cycle time of 10 ms, we need a final cycle time of 20 ms. This represents a frequency of 50 Hz. To divide the 5-MHz source down to 50 Hz, we need five 7490s in cascade.

Fig. 7-15. Deriving an accurate gating pulse.

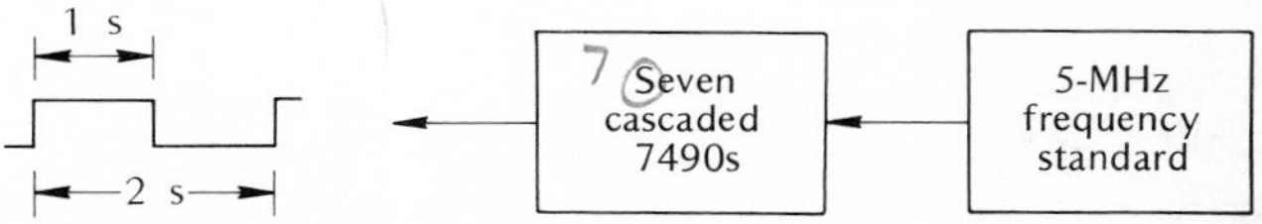

1 s

2 s

Seven cascaded 7490s

5-MHz frequency standard

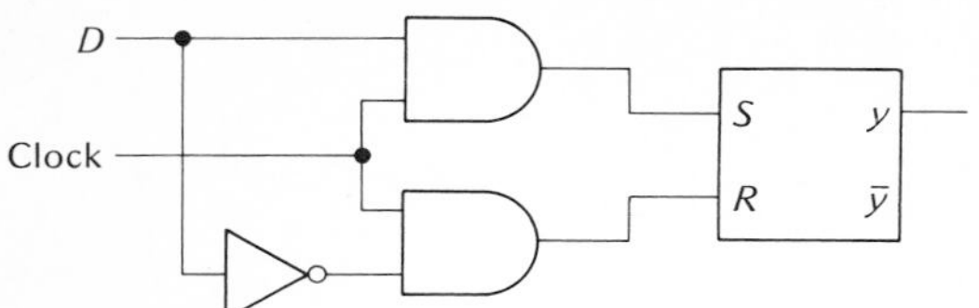

Fig. 7-16. One way to build a D flip-flop.

7-6 THE *D* FLIP-FLOP

Figure 7-16 shows a simple way to build a *delay* (*D*) flip-flop. This kind of flip-flop prevents the value of *D* from reaching the *y* output until a clock pulse occurs. The action of the circuit is straightforward, as follows. When the clock is low, both AND gates are disabled; therefore, *D* can change values without affecting the value of *y*. On the other hand, when the clock is high, both AND gates are enabled. In this case, *y* is forced to equal the value of *D*. When the clock again goes low, *y* retains or stores the last value of *D*.

There are many ways to design *D* flip-flops. In general, a *D* flip-flop is a bistable multivibrator whose *D* input is transferred to the output after a clock pulse is received. Figure 7-17 shows the logic symbol used in this book for any type of *D* flip-flop.

We're especially interested in the *D* flip-flop in which *y* can follow the value of *D* while the clock is high (similar to Fig. 7-16). This type of *D* flip-flop is so important it's usually called a "*D*-type latch." The *D*-type latch is widely used for temporarily storing data in frequency counters, digital voltmeters, etc.

Figure 7-18 illustrates the idea of temporary data storage. Four *D* flip-flops are driven by the same clock pulse. When the clock goes high, input data is transferred to the output. Then when the clock goes low, the output retains the data. For instance, suppose the data input is

$$D_1D_2D_3D_4 = 0\ 1\ 1\ 1$$

When the clock pulse goes high, this data transfers to the output, resulting in

$$y_1y_2y_3y_4 = 0\ 1\ 1\ 1$$

After the clock goes low, the output data is retained or stored. As long as the clock is low, the *D* values can change without affecting the *y* values.

The 7475 is a TTL MSI circuit that contains four *D*-type latches; it's sometimes called a "quad bistable latch." The 7475 is ideal for handling four-bit bytes of data.

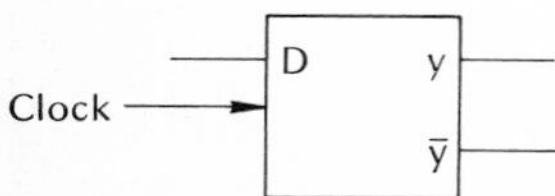

Fig. 7-17. Logic symbol for *D* flip-flop.

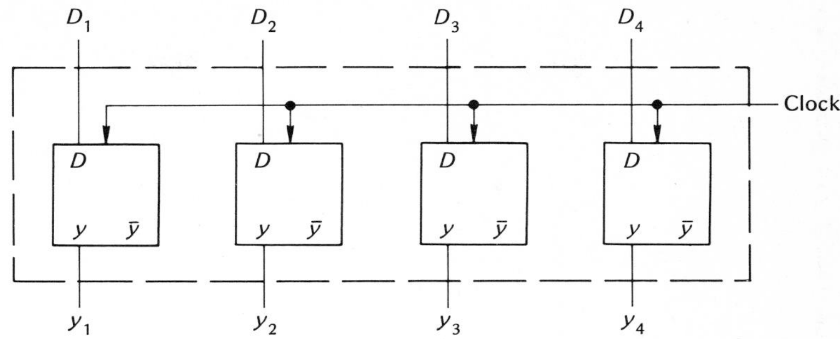

Fig. 7-18. Temporary storage of four-bit data.

From now on, we shall use the symbol of Fig. 7-19 to represent the 7475. Remember the basic action: while the clock is high, input data is transferred to the output; when the clock goes low, the output retains the four-bit data present at the input just before the negative-going clock transition.

Figure 7-20 illustrates one of the standard uses for the 7475. Basically, the 7475s provide temporary storage of the final count in the 7490s. The point of this is to avoid a blinking readout while the 7490s are counting. Briefly, here's how the system works. Between points *A* and *B* in time, the 7490s are counting, and their BCD outputs are continuously changing. These changing BCD outputs are the data inputs to the 7475s. Because of the inverter, however, the 7475s are disabled while the counting is taking place.

At the end of the count, the BCD output of the 7490s is

0111 0101 0010

because the frequency of the narrow clock pulses is 752 Hz. At point *B* in time, the 7475s are enabled; therefore, the BCD output of the 7490s transfers through the 7475s to the 7447s. As you may recall, the 7447 is a BCD decoder/driver suitable for driving seven-segment indicators. Therefore, starting at point *B* in time, the number 752 is displayed.

At point *C* in time, the 7490s are reset to zero and a new count begins. At the

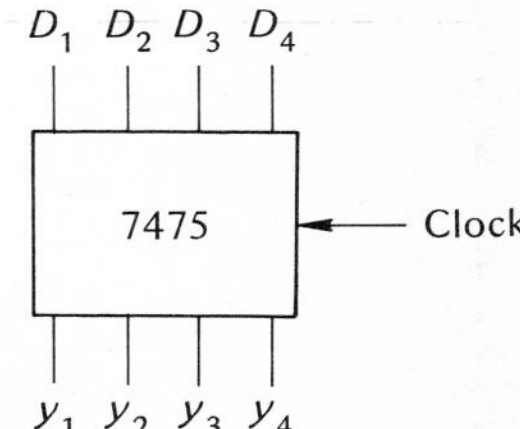

Fig. 7-19. Logic symbol for a 7475.

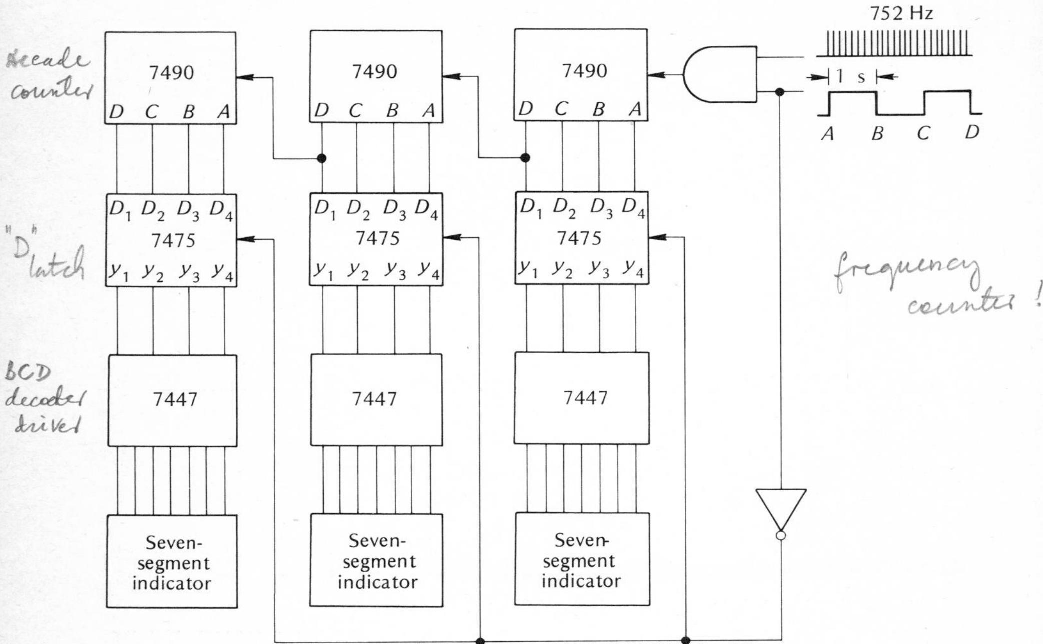

Fig. 7-20. Decade counter with temporary storage, decoding and display.

same instant, the 7475s are disabled and retain the BCD output of

0111 0101 0010

This BCD number is temporarily stored while the 7490s are going through a new count. In this way, the final readout continues displaying decimal 752 while a new count is in progress. Then, at point *D* in time, the display is *updated* (the new count is shown).

Before the 7475 was available, the BCD outputs of the 7490s went directly into the 7447s. Because of this, the final readout displayed a blinking value while the count was in progress. Nowadays, typical frequency counters, digital voltmeters, etc., use 7475s to avoid such blinking displays.

7-7 THE *JK* FLIP-FLOP

Figure 7-21 shows one way to build a *JK* flip-flop. For reasons given in Chap. 8 (under the master-slave idea), we want triggering to occur on the negative-going edge of the clock pulse. This is why an inverter is included on the clock-pulse input.

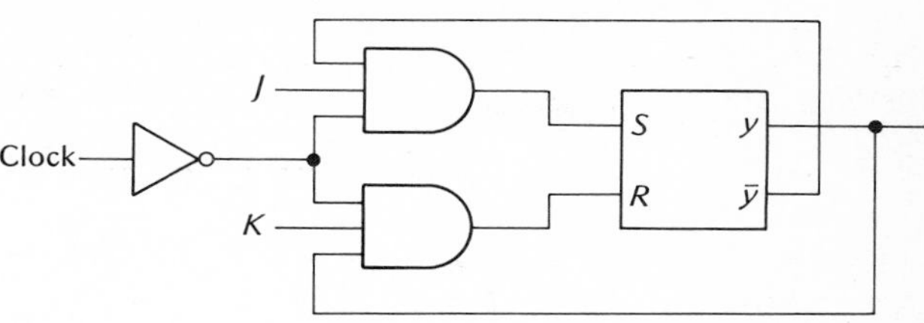

Fig. 7-21. One way to build a *JK* flip-flop.

There are four cases to analyze.

Case 1 $J=0$ and $K=0$. These *J* and *K* inputs disable the AND gates; therefore, clock pulses have no effect on the state of the flip-flop. In other words, *y* retains its last value.

Case 2 $J=0$ and $K=1$. The upper AND gate is disabled. The lower AND gate is enabled if *y* is a 1. Because of this, the next negative-going edge of the clock will reset the flip-flop if it is not already in this state.

Case 3 $J=1$ and $K=0$. The lower AND gate is disabled. The upper AND gate is enabled if $\bar{y}$ is a 1. As a result, the next negative-going edge of the clock will set the flip-flop if it is not already in this state.

Case 4 $J=1$ and $K=1$. If *y* is a 0, the lower AND gate is disabled, but the upper AND gate is enabled. The negative-going edge of the clock pulse will then set the flip-flop. On the other hand, if *y* is a 1, the lower AND gate is enabled; the next negative-going edge of the clock forces the flip-flop to reset. In other words, when *J* and *K* are both high, clock pulses cause the *JK* flip-flop to toggle.

Table 7-4 summarizes the action. Notice that *b* is the value of *y* just before the negative-going edge of the clock pulse. With this in mind, you can see that the output does one of four things: stays the same, resets, sets, or toggles. Memorize this table because it applies to any *JK* flip-flop.

Figure 7-22 shows the symbol we shall use for a *JK* flip-flop. Remember, the

Table 7-4

JK FLIP-FLOP		
J	*K*	*y* output after negative-going edge of clock*
0	0	*b* (same)
0	1	0 (reset)
1	0	1 (set)
1	1	$\bar{b}$ (toggle)

* Note: *b* is the value of the *y* output just before the negative-going edge of the clock.

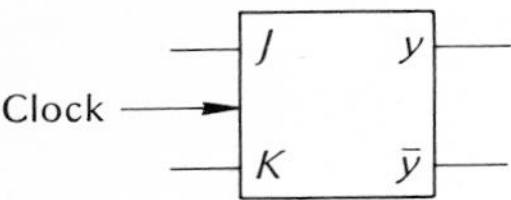

Fig. 7-22. Logic symbol for a *JK* flip-flop.

and *K* inputs are either 0s or 1s, and triggering occurs on the negative-going edge of the clock pulse.

One more point. For convenience, we often show the toggle condition ($J = 1$ and $K = 1$) by leaving the *J* and *K* inputs unconnected. When we do this, remember that even though not shown, 1s are going into the *J* and *K* inputs (discussed further in Sec. 8-3).

Example 7-6

The *JK* flip-flop of Fig. 7-23 has its outputs cross-coupled back to the *J* and *K* inputs. What is the *y* output like?

Solution

When *y* is a 0, *J* is a 1 and *K* is a 0; therefore, the flip-flop will set on the trailing edge of the next input pulse. On the other hand, when *y* becomes a 1, *J* goes to 0 and *K* to 1. In this case, the next trailing edge resets *y* to 0. Therefore, the action of Fig. 7-23 is similar to a *T* flip-flop; the output toggles on the negative-going edge of each clock pulse.

Incidentally, because of the historical relation between Boolean algebra and logic, many people use the terms *true* and *false* as substitutes for 1 and 0. With this in mind, we can describe the action of Fig. 7-23 as follows. When the *y* output is false, the *J* input is true and the *K* input is false. The next trailing edge of the clock will set the *y* output to true. Once this happens, the *J* input becomes false and the *K* input becomes true. The following negative edge of the clock pulse resets the *y* output to false.

7-8 THE SCHMITT TRIGGER

The Schmitt trigger is stable in either of two states and therefore falls into the category of bistable multivibrators. However, instead of coupling from the collector of one transistor to the base of another, it couples by way of a common-emitter resistor.

Fig. 7-23. *JK* flip-flop connected to act like *T* flip-flop.

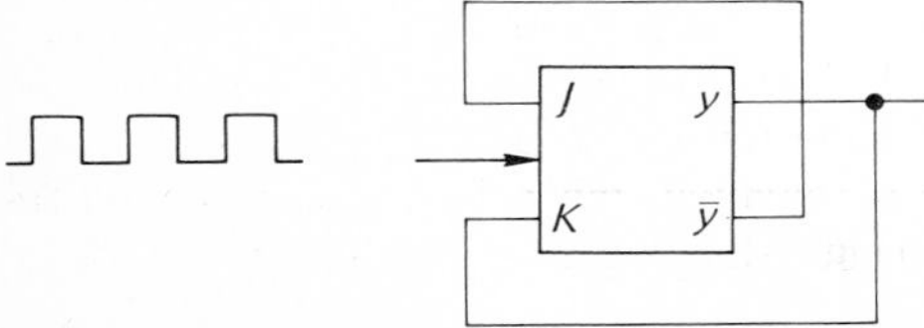

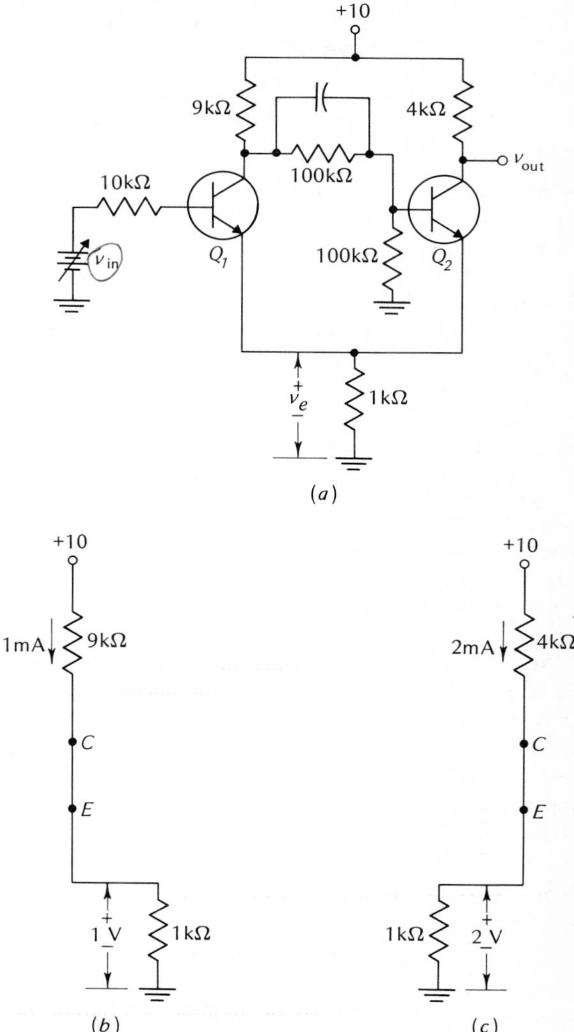

Fig. 7-24. Schmitt trigger.

Figure 7-24a shows one way of building a Schmitt trigger. When Q_1 is on, Q_2 is off, and vice versa. To understand how the circuit works, assume the input voltage v_{in} is set to 0 V. Under this condition no base current flows in Q_1; therefore, there is no collector current in Q_1. The base of Q_2 is connected to a voltage divider formed by the 100-kΩ resistors. The voltage at the top of the divider is enough to turn on the base and saturate Q_2. With Q_2 saturated, the collector and emitter are effectively shorted together, as shown in Fig. 7-24c. As a result, the voltage from the emitter to ground is 2 V.

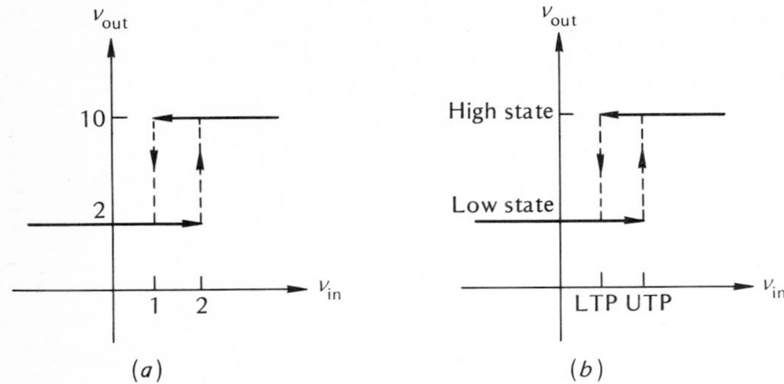

Fig. 7-25. Schmitt-trigger characteristics.

With the input battery at 0 V, we have Q_1 cut off and Q_2 saturated. Under this condition the output voltage of the Schmitt trigger is about 2 V, as in Fig. 7-24c. Since the emitter voltage is 2 V, no base current can flow in Q_1 for any value of v_{in} less than 2 V (the base diode remains back-biased). Therefore, for any input less than 2 V, the output voltage of the Schmitt trigger stays at 2 V because Q_2 remains saturated.

If we raise the input voltage above 2 V, base current flows in Q_1. Once this happens, collector current begins to flow, and the voltage at the Q_1 collector decreases. The speed-up capacitor couples this drop into the base of Q_2, bringing Q_2 out of saturation. The drop in Q_2 current reduces the voltage from the emitter to ground. Since v_e has been reduced, Q_1 is turned on harder. The increase in the Q_1 current drops the collector voltage of Q_1 even further, and we have regeneration. This regeneration continues until Q_1 saturates and Q_2 cuts off. With Q_2 cut off, the output voltage goes to 10 V because there is no current through the 4-kΩ load resistor. The regenerative switching action takes place very quickly, so we can think of v_{out} as suddenly having gone from 2 to 10 V.

Any further increase in the input voltage will increase the Q_1 base current. However, since Q_2 is already cut off, the output voltage must stay at 10 V.

Note that with Q_1 saturated the emitter voltage to ground is only 1 V, as shown in Fig. 7-24*b*.

With the output at 10 V, the only way of bringing the output back to 2 V is by reducing the input voltage driving Q_1. Since the emitter voltage with Q_1 saturated is only 1 V, the input voltage must be reduced to about 1 V to bring Q_1 out of saturation. At this value the collector of Q_1 comes out of saturation. The collector voltage will increase, and this causes Q_2 to conduct. With Q_2 conducting, the emitter voltage increases, causing the current in Q_1 to decrease further. This raises the collector voltage of Q_1, and again we have regenerative action. This regeneration continues until Q_1 cuts off and Q_2 saturates. The circuit is then back in the original state with an output voltage of 2 V.

Figure 7-25*a* summarizes the action of a Schmitt trigger. The output voltage is either 2 or 10 V. When v_{out} is in the low state, it is necessary to raise v_{in} to slightly more than 2 V to cause v_{out} to jump to the high state. Once in the high state, v_{out} stays at 10 V until v_{in} drops to slightly more than 1 V. The output then jumps back to the low state of 2 V. The rapid switching action is indicated by the dashed lines.

Figure 7-25*b* shows the graph for any Schmitt trigger. The value of v_{in} that causes the output to jump from a low to a high state is called the *upper trip point* (UTP); the value of v_{in} that causes the output to jump from the high state to the low state is called the *lower trip point* (LTP).

The Schmitt trigger is an amplitude-sensitive circuit. Once the input voltage exceeds the upper trip point, the output voltage goes from a low value to a high value, from a 0 to a 1. When the input voltage drops below the lower trip point, the output voltage drops from a 1 to a 0. Because of this, we can use the Schmitt trigger to detect when the input voltage crosses certain voltage levels.

In Fig. 7-25*b* the difference between the UTP and the LTP produces a *hysteresis* in the switching action. It is possible to eliminate this hysteresis by making the LTP and UTP equal. But a small amount of hysteresis is preferred because it ensures fast switching from one state to another.

The Schmitt trigger is an example of a category of circuits known as *voltage comparators*. These circuits detect when a voltage has crossed a specified level. Comparators are useful in analog-to digital converters, discussed in Chap. 11.

Example 7-7

A Schmitt trigger has the following characteristics: UTP = 3 V, LTP = 1 V, high state = 20 V, low state = 2 V. Sketch the output voltage for a sine-wave input with a peak value of 6 V.

Solution

First we can sketch the Schmitt-trigger characteristic and the input sinusoid as shown in Fig. 7-26*a*. When the sinusoid exceeds 3 V, the Schmitt trigger goes from the low to the high state. This occurs at point *A* in time. The output stays in the high state until the input sinusoid drops below 1 V. Then the output drops back to the low state at point *B* in time. Therefore, we can sketch the input and output as

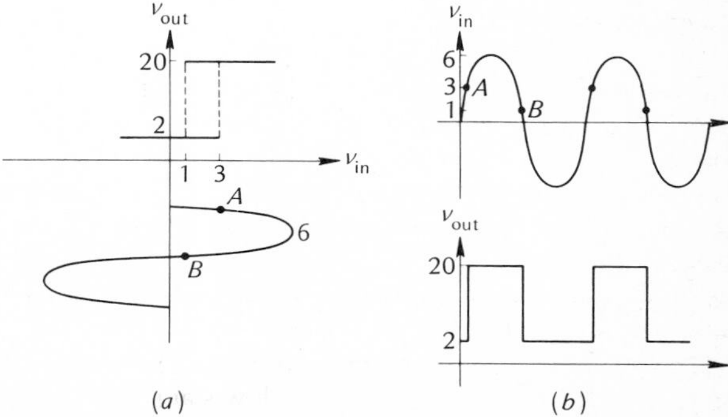

Fig. 7-26. Example 7-7.

shown in Fig. 7-26*b*. This shows one major use of the Schmitt trigger—to change sine waves into rectangular waves.

Example 7-8

The Schmitt trigger of the preceding example is driven by a sawtooth with a peak value of 10 V. Sketch the Schmitt-trigger output.

Solution

The Schmitt-trigger characteristic and the sawtooth are shown in Fig. 7-27*a*. The Schmitt trigger trips as the sawtooth rises above the 3-V level. On the sawtooth flyback, the Schmitt trigger trips at 1 V. Figure 7-27*b* shows the output voltage.

Whenever the Schmitt trigger is driven by a periodic signal whose peak value exceeds the UTP, the output will be a rectangular waveform. Because of this, the Schmitt trigger is sometimes known as a *squaring circuit*.

The trip points of a Schmitt trigger can be moved. Figure 7-28*a* shows a variable emitter resistor. This allows us to adjust the trip points. Figure 7-28*b* shows the emitter resistor returned to a negative supply; this permits us to use negative trip points if we want.

7-9 THE ASTABLE MULTIVIBRATOR (free-running)

The *astable multivibrator* has two states but is stable in neither; it switches back and forth between these two states, producing a square-wave output. In other words, we can think of an astable multivibrator as being a square-wave oscillator. *Free-running* multivibrator is another name for the astable multivibrator.

Often in digital systems a single astable multivibrator keeps all the circuits in step with one another. This multivibrator is like a master clock that synchronizes all parts of the digital system. Because of this, the astable multivibrator is often called a clock. (not very frequency-stable)

Figure 7-29 shows a transistorized astable multivibrator. The transistors alternately saturate and cut off. A complete analysis of the circuit shows the period of the output square wave is

$$\boxed{T \cong 1.4RC} \qquad (7\text{-}1)$$

where T = period of the square wave
R = value of the base resistor
C = value of the coupling capacitor

The frequency of the output square wave is

$$\boxed{f \cong \frac{0.7}{RC}} \qquad (7\text{-}2)$$

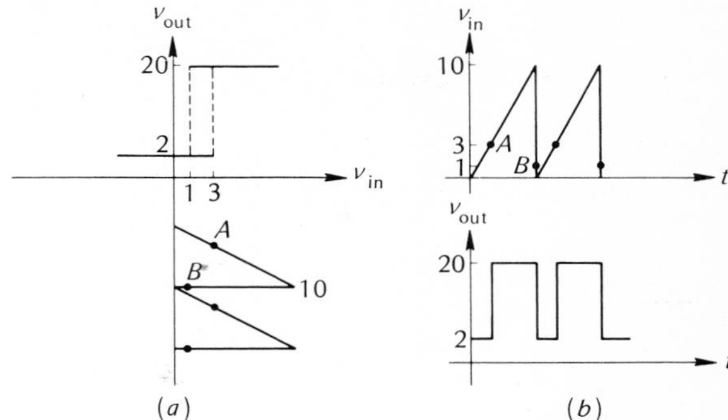

Fig. 7-27. Example 7-8.

To ensure that the transistor saturates, the approximate beta of each transistor must be greater than the ratio of the base to load resistance. That is,

$$\beta > \frac{R}{R_L} \tag{7-3}$$

For instance, if the base resistor R is 400 kΩ, and if the load resistor R_L is 20 kΩ, then the beta of each transistor must be greater than 400/20, or 20.

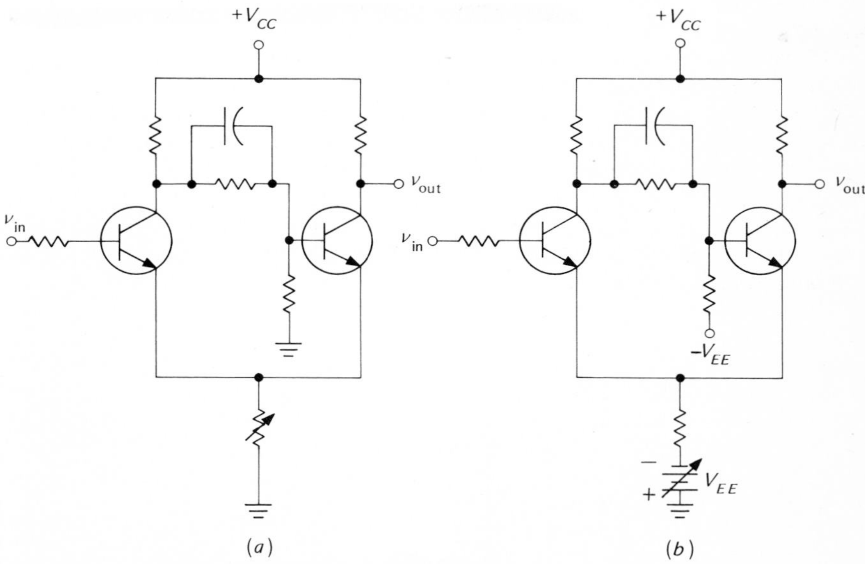

Fig. 7-28. Schmitt triggers with adjustable trip points.

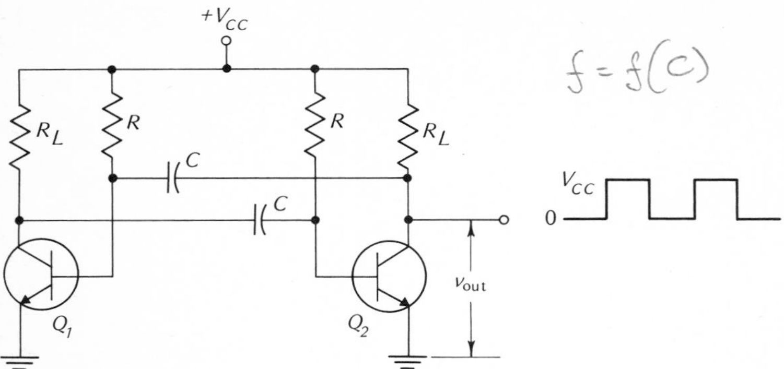

Fig. 7-29. Astable multivibrator.

The circuit of Fig. 7-29 is symmetrical; both base resistors, both load resistors, and both capacitors are equal. Because of this, each transistor stays off for the same amount of time, and the output waveform is square. Occasionally, an unsymmetrical output is needed, and can be obtained by using either two different base resistances or two different capacitor values.

To get a variable-frequency clock, we gang variable capacitors. In this way, the frequency changes when the capacitors are changed.

Figure 7-30 shows another way to control the frequency. A separate supply drives the base resistors. When the base supply voltage changes, the frequency changes. In terms of natural logarithms, the period of the square wave is

$$T \cong 2RC \ln\left(1 + \frac{V_{\mathrm{CC}}}{V_{\mathrm{BB}}}\right) \tag{7-4}$$

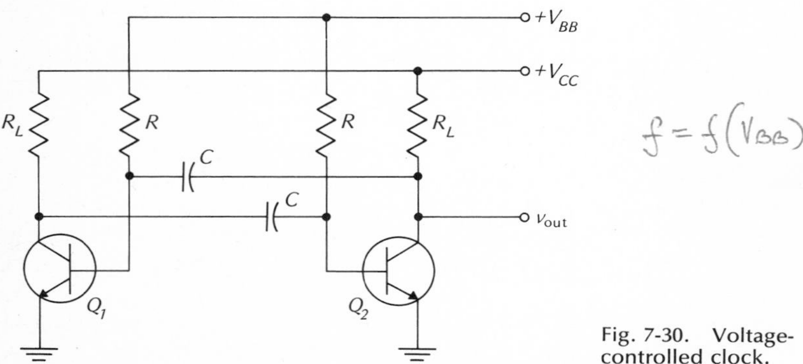

Fig. 7-30. Voltage-controlled clock.

Example 7-9

In the multivibrator of Fig. 7-29, the base resistors have a value of 100 kΩ, and the capacitors have a value of 1,000 pF. Find the approximate frequency of the square wave.

Solution

With Eq. (7-2),

$$f \cong \frac{0.7}{RC} = \frac{0.7}{100(10^3)1{,}000(10^{-12})} = 7 \text{ kHz}$$

The astable multivibrator delivers an output square wave with a frequency of 7 kHz.

Example 7-10

Suppose the betas of the transistors in the preceding example are in the range of 50 to 150. To ensure saturation, what is the smallest permissible value of R_L?

Solution

The worst case is $\beta = 50$. By Eq. (7-3),

$$50 = \frac{100(10^3)}{R_L}$$

or

$$R_L = 2(10^3) = 2 \text{ k}\Omega$$

7-10 THE MONOSTABLE MULTIVIBRATOR

This section discusses the last basic type of multivibrator—the *monostable multivibrator*. This kind of multivibrator is stable in one state but unstable in the other. When triggered, it goes from the stable state to the unstable state. It remains in the unstable state *temporarily* and then returns to the stable state.

Figure 7-31 shows the general idea of a monostable multivibrator. At point A in time a trigger hits the input. This causes the output voltage to go from a low to a high value. The high state is an unstable state, so that after a while the output voltage returns to the low state. The output remains in the low state until the next trigger arrives at point B in time. Again the output jumps to the high state, and after a while returns to the low state, where it stays until the trigger comes in at point C in time.

The basic idea of a monostable multivibrator is now clear. Each time a trigger arrives, the output temporarily changes to the high state but then returns to the low

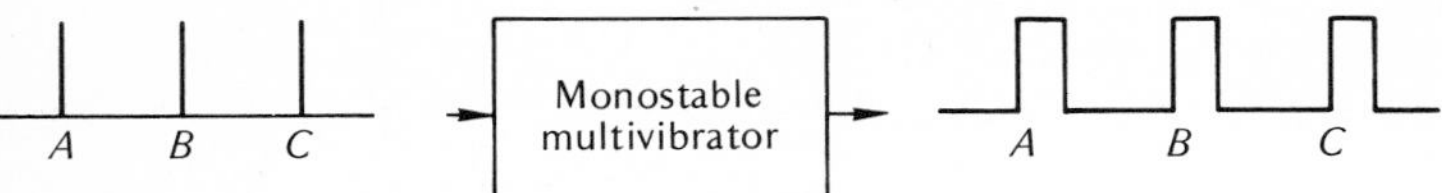

Fig. 7-31. Input-output waveforms of monostable multivibrator.

state. For each input trigger there is one rectangular output pulse. Because of this the monostable multivibrator is often called a *one-shot* multivibrator, or simply a *one-shot*.

There are many ways to build a one-shot. Figure 7-32 shows one of these. The low state has Q_1 off and Q_2 on. When a positive trigger arrives, it turns Q_1 on, which drops the collector voltage of Q_1. This drop couples into the base of Q_2, shutting this transistor off. But this condition, Q_1 on and Q_2 off, is only temporary, because as the charge on the capacitor changes, the back bias on the base of Q_2 disappears. After an amount of time determined by the RC time constant of the base circuit, Q_2 again turns on and Q_1 turns off. Each time a positive trigger hits the base of Q_1, the output voltage y goes from low to high temporarily, but then returns to a low voltage. There is one rectangular output pulse for each input trigger.

A square wave can drive a one-shot multivibrator, provided we differentiate the square wave with a small input capacitor.

The one-shot multivibrator is useful for reshaping ragged pulses. It also can generate gating signals, introduce time delays, and change the width of a pulse.

STUDY AIDS

Summary

A multivibrator is a regenerative circuit with two active devices, designed so that one device conducts while the other cuts off. The three basic types of multivibrators are the bistable, the astable, and the monostable.

Fig. 7-32. One way to build a monostable multivibrator.

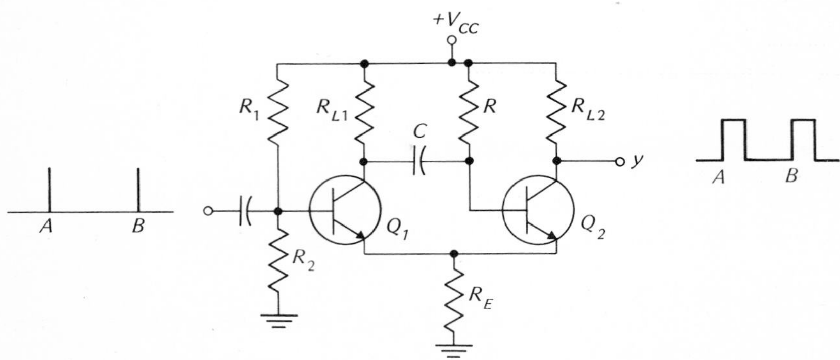

A trigger to the set input of an *RS* flip-flop forces the *y* output to become a 1 if it is not already a 1. On the other hand, a trigger to the reset input forces the *y* output to become a 0 if it is not already in this state.

An *RS* flip-flop may be modified to a *T* flip-flop or an *RST* flip-flop. The *T* input of a flip-flop makes the *y* output toggle. In this book the *T* input must be negative-going to produce toggling; the *R* and *S* inputs must be positive-going to cause a change in output state.

The four-bit binary counter counts from 0000 through 1111. A decade counter counts from 0000 through 1001. Cascading decade counters allows us to count beyond decimal 9. The best known TTL decade counter is the 7490.

A *D*-type latch is widely used for temporary storage of digital data in frequency counters, digital voltmeters, etc. The 7475 is a TTL device containing four *D*-type latches (also called "bistable latches").

The values of *J* and *K* determine what a *JK* flip-flop does on the next clock trigger. If *J K* = 0 0, the flip-flop retains its last state. When *J K* = 0 1, the flip-flop resets. If *J K* = 1 0, the flip-flop sets. When *J K* = 1 1, the flip-flop toggles.

The Schmitt trigger is a bistable multivibrator. It changes states when the input voltage crosses certain levels called the "trip points."

The astable multivibrator free-runs, producing a rectangular output waveform. Because it often synchronizes the circuits in digital systems, the astable multivibrator is sometimes known as a *clock*.

The monostable multivibrator, also called a "one-shot," delivers one rectangular output pulse for each input trigger.

Glossary

astable Without a stable state. In a multivibrator it means the circuit switches back and forth between two unstable states.

bistable Having two stable states. A bistable multivibrator can remain indefinitely in either state.

clock An oscillator that synchronizes all the circuits in a digital system.

monostable One stable state. When a trigger hits a monostable multivibrator, the output goes into the unstable state temporarily but then returns to the stable state.

ripple counter The kind of counter in which the output of one flip-flop triggers the next flip-flop.

scaler A circuit that delivers one output pulse after a specified number of input pulses.

toggle To trigger to the opposite state.

Review Questions

1. Describe the input-output action of an *RS* flip-flop.
2. How is the output frequency of a *T* flip-flop related to the input frequency?
3. What is a scaler?
4. How many flip-flops are there in a four-bit binary counter? How many flip-flops are needed to reach a maximum count of 11111?

5. How many decade counters are needed to count up to 99,999?
6. What does "gating a counter" mean?
7. What does a D flip-flop do? What is a D-type latch?
8. How can we prevent a digital readout from blinking while measuring frequency?
9. Describe the input-output action of a JK flip-flop.
10. What does a Schmitt trigger do? What does UTP mean? LTP?
11. What kind of triggers are used with an astable multivibrator?
12. Describe the action of a monostable multivibrator.

Problems

7-1. A 500-kHz square wave drives a T flip-flop. What is the frequency of the output?

7-2. A square wave with a period of 10 μs drives a T flip-flop. What is the period of the output signal?

7-3. Suppose we cascade seven T flip-flops. If the input frequency is 512 kHz, what is the final output frequency?

7-4. We want to build a scaler that delivers one output pulse for every 1,024 input pulses. How many T flip-flops are needed?

7-5. What are the output states of a binary ripple counter that has five T flip-flops in cascade.

7-6. How many distinct states does a nine-flip-flop ripple counter have?

7-7. Six 7490s are cascaded. What is the highest decimal count possible? If the 7490s are reset and then 4793 pulses are received, what will be the BCD output of the 7490s?

7-8. A counter is gated for 0.2 s. If the frequency of the signal being counted is 250 kHz, what is the BCD output of the counter at the end of the gating pulse?

7-9. An electronic wristwatch has a clock with a frequency of 32 kHz. To divide this frequency down to a 1 Hz, what scaling factor do we need? Given T flip-flops, four-bit binary counters, and decade counters, suggest one way of getting that scaling factor.

7-10. In Fig. 7-33, narrow clock pulses drive the circuit as shown. Sketch the y output for each of these initial conditions:

(a) $y_1 = 1$ and $y_2 = 0$.
(b) $y_1 = 0$ and $y_2 = 1$.
(c) $y_1 = 0$ and $y_2 = 0$.
(d) $y_1 = 1$ and $y_2 = 1$.

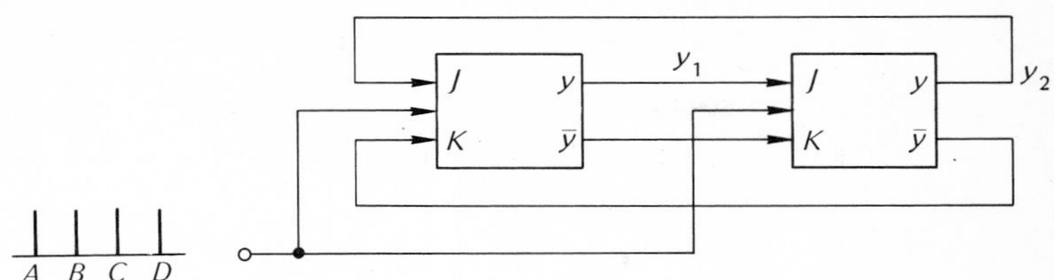

Fig. 7-33.

7-11. In Fig. 7-34, list the pairs of y_1 y_2 values for each of the following initial conditions:

(*a*) $y_1 = 1$ and $y_2 = 1$.
(*b*) $y_1 = 0$ and $y_2 = 0$.

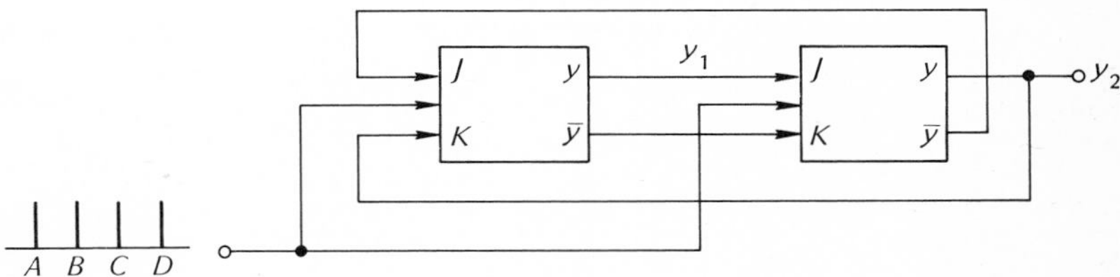

Fig. 7-34.

7-12. A Schmitt trigger has the following characteristics: UTP = 5 V, LTP = 4 V, high state = 12 V, and low state = 3 V. A 10-V-peak sine wave drives the Schmitt trigger. Sketch the output waveform.

7-13. In Fig. 7-29, R is 50 kΩ, and C is 0.005 μF. What is the approximate frequency of the output waveform?

7-14. If R equals 20 kΩ in Fig. 7-29, and if the capacitors are adjustable from 50 to 500 pF, what are the minimum and maximum frequency of the free-running multivibrator?

7-15. The triggers of Fig. 7-35 drive a monostable multivibrator. The one-shot delivers a 3-μs pulse for each trigger. Sketch the output waveform.

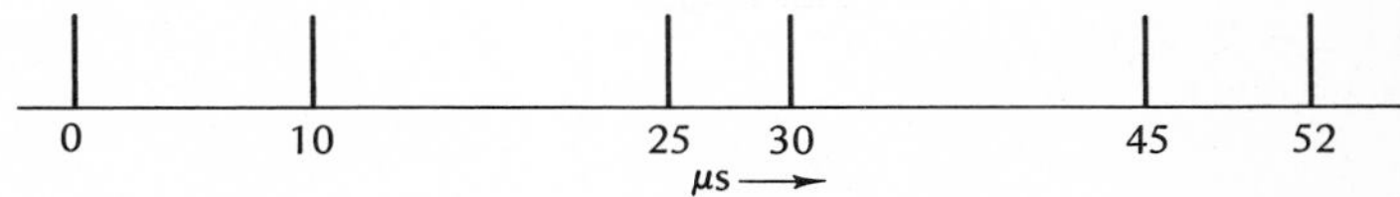

Fig. 7-35.

Counter Techniques

A counter is probably one of the most useful and versatile subsystems in a digital system. As implied by its name, this unit has the ability to count. In the previous chapter, we saw that a counter driven by a clock can be used to count the number of clock cycles. Since the clock pulses occur at known intervals, the counter can be used as an instrument for measuring time (and therefore period or frequency).

The ripple counter previously discussed is the simplest and most straightforward counter. This unit is very simple in operation and construction and usually requires a minimum of hardware. It does, however, have a speed limitation. Each flip-flop is triggered by the previous flip-flop, and thus the counter has a cumulative settling time. An increase in speed of operation can be achieved by use of a parallel counter. Unfortunately, the increase in speed is usually accompanied by an increase in hardware. Series and parallel counters are used in combination to compromise between speed of operation and hardware count.

In this chapter we shall discuss in detail the operation of serial, parallel, and serial-parallel combination counters. We shall investigate the implementation of counters having any desired number of binary counts as well as one form of a BCD counter.

After studying this chapter you should be able to

1. Determine the modulus of a serial or parallel counter.
2. Explain how *master/slave* flip-flops are used to construct counters.
3. Demonstrate how logic gates are used to decode the binary states of a counter.

8-1 BINARY RIPPLE COUNTER

As shown previously, a binary ripple counter can be constructed using *RST* flip-flops. The total number of counts or discrete states through which the counter can progress is given by 2^n, where n is the total number of flip-flops used. It is said to have a *natural count* of 2^n. For example, a basic binary counter consisting of three flip-flops counts through eight discrete states and is said to have a natural count of eight. A binary counter consisting of four flip-flops counts through 16 discrete states, and so on. Thus, we can construct counters which count through 2, 4, 8, 16, 32, etc., states by using the proper number of flip-flops. A three-flip-flop counter is often referred to as a "modulus-8" (or mod-8) counter since it has eight states. Similarly, a four-flip-flop counter is a mod-16 counter. The *modulus* of a counter is the total number of states through which the counter can progress.

It is often desirable to construct counters which have moduli other than two, four, eight, etc. For example, you might like to construct a counter having a modulus of three or five or seven. A smaller-modulus counter can always be constructed from a larger-modulus counter by skipping states. Such counters are said to have a "modified count." It is first necessary to determine the number of flip-flops required. The correct number of flip-flops is determined by choosing the lowest natural count which is greater than the desired modified count. For example, a mod-7 counter requires three flip-flops, since eight is the lowest natural count greater than the desired modified count of seven.

Example 8-1

How many flip-flops are required to construct each of the following counters?

(*a*) mod-3
(*b*) mod-6
(*c*) mod-9

Solution

(*a*) The lowest natural count greater than three is four. Two flip-flops provide a natural count of four. Therefore, it requires a minimum of two flip-flops to construct a mod-3 counter.

(*b*) To construct a mod-6 counter requires at least three flip-flops, since eight is the lowest natural count greater than six.

(*c*) A mod-9 counter requires at least four flip-flops, since 16 is the lowest natural count greater than nine.

The three-flip-flop ripple counter shown in Fig. 8-1*a* can count through the eight states shown in Fig. 8-1*b*. In order to construct a mod-7 counter, we must omit, or skip, one of these states. There is a choice to be made, since any one of the eight states may be omitted. If count 0 is omitted, the counter sequence is 1-2-3-4-5-6-7. On the other hand, count 7 could be eliminated, and the resulting count sequence would be 0-1-2-3-4-5-6. To construct a mod-6 counter, it is necessary to skip two

Fig. 8-1. Three-stage binary ripple counter. (*a*) Logic diagram. (*b*) Truth table showing natural count.

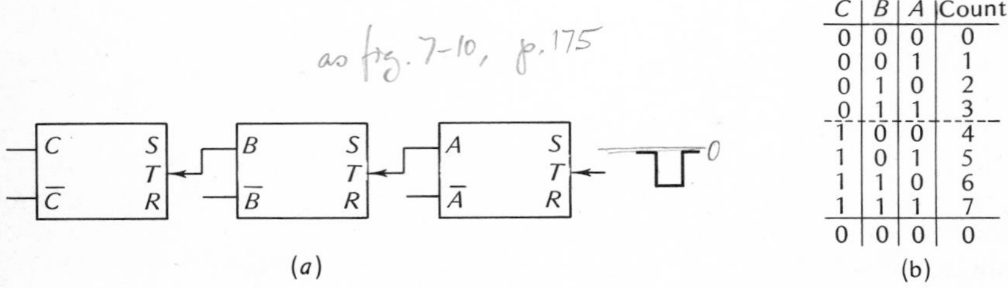

C	*B*	*A*	Count
0	0	0	0
0	0	1	1
0	1	0	2
0	1	1	3
1	0	0	4
1	0	1	5
1	1	0	6
1	1	1	7
0	0	0	0

of the counts in Fig. 8-1*b*. Again there is a choice, since any two states may be eliminated. For example, skipping 0 and 1 would give the count sequence 2-3-4-5-6-7.

Example 8-2

If three flip-flops are used to build a mod-7 counter which counts in sequence, how many possible count sequences are there?

Solution

To construct a mod-7 counter using three flip-flops, one count out of eight must be skipped. Since the skipped count may be any one of the eight, eight possible count sequences are available. They are:

Skipping count 0, 1-2-3-4-5-6-7
Skipping count 1, 0-2-3-4-5-6-7
Skipping count 2, 0-1-3-4-5-6-7
Skipping count 3, 0-1-2-4-5-6-7
Skipping count 4, 0-1-2-3-5-6-7
Skipping count 5, 0-1-2-3-4-6-7
Skipping count 6, 0-1-2-3-4-5-7
Skipping count 7, 0-1-2-3-4-5-6

Example 8-3

If three flip-flops are used to build a mod-6 counter which counts in sequence, how many possible count sequences are there?

Solution

A mod-6 counter using three flip-flops requires two counts out of eight to be skipped. The possible count sequences can be determined by using the solution to Example 8-2. If count 0 is skipped, any one of the remaining seven counts may be skipped in order to form a mod-6 counter. This then provides seven possible count sequences. If count 1 is skipped, any one of the remaining six states (2-3-4-5-6-7) may be omitted. This then provides six more possible sequences. 0 is omitted since it has already been accounted for. Skipping count 2 and any one of the remaining five counts (3-4-5-6-7) provides five more possibilities. Skipping count 3 and any one of the four remaining states (4-5-6-7) provides four more sequences. Skipping count 4 and any one of the remaining three counts (5-6-7) provides three sequences. Skipping count 5 and either 6 or 7 accounts for two more possibilities. Finally, skipping 6 and 7 accounts for the last possible sequence. Thus a total of $7 + 6 + 5 + 4 + 3 + 2 + 1 = 28$ count sequences are possible for a mod-6 counter using three flip-flops.

The count sequence for any particular counter is usually determined by the logical method of implementing the count (i.e., the method used to cause the counter to skip counts), since it is usually desirable to minimize the amount of hardware required.

8-2 MODIFIED COUNTER USING FEEDBACK

One method used to cause a counter to skip counts is to feed back a signal from some flip-flop to some previous flip-flop. Consider the counter shown in Fig. 8-1, and suppose we want to implement a mod-7 counter. Since only one count is to be skipped, it would be convenient to discover some event which occurs only once during the natural count. The truth table in Fig. 8-1 shows that flip-flop *C* changes from a 0 to a 1 only once during the natural eight counts. This represents a single unique occurrence. If *C* is fed back to the set input of flip-flop *A*, as shown in Fig. 8-2, a mod-7 counter is implemented. This is true since each time *C* goes positive (this occurs when the counter advances to state 4), flip-flop *A* is set to the 1 condition, and thus the counter advances itself to state 5 immediately. Since the counter remains in state 4 only momentarily, it essentially skips this state and is therefore a mod-7 counter. The truth table and waveforms for this counter are shown in Fig. 8-2.

This type of counter is sometimes called a "permuting counter" since its natural count has been permuted (changed). Notice that the counter could just as easily be implemented by feeding back a signal from $\bar{C}$. It is possible to use this type of feedback since flip-flops *A* and *B* must have completed their transitions before the feedback signal is generated. Note that it is generally not desirable to feed back a signal which would cause flip-flop *A* to be reset. This causes a negative pulse at the input of flip-flop *B* which causes it to change state, and this may cause flip-flop *C* to change state, etc.

Example 8-4

Design a mod-5 and a mod-10 ripple counter using the feedback method just developed.

Fig. 8-2. Mod-7 counter. (*a*) Logic diagram. (*b*) Truth table. (*c*) Waveforms.

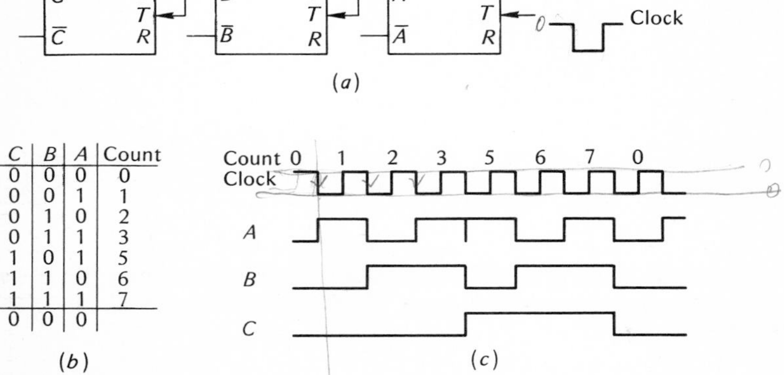

C	B	A	Count
0	0	0	0
0	0	1	1
0	1	0	2
0	1	1	3
1	0	1	5
1	1	0	6
1	1	1	7
0	0	0	

(*b*)

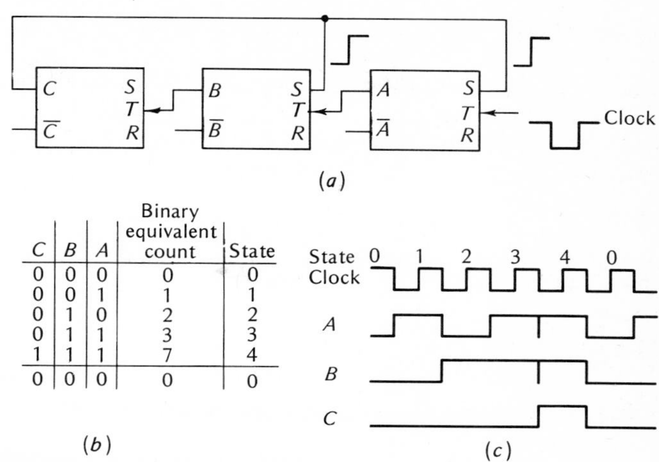

C	B	A	Binary equivalent count	State
0	0	0	0	0
0	0	1	1	1
0	1	0	2	2
0	1	1	3	3
1	1	1	7	4
0	0	0	0	0

(b)

Fig. 8-3. Mod-5 counter. (*a*) Logic diagram. (*b*) Truth table. (*c*) Waveforms.

Solution

To design a mod-5 counter requires three flip-flops. In deciding which count sequence to use, it is convenient to use the feedback scheme used for the mod-7 counter. In this instance it is necessary to skip three counts. If the feedback were directed to the set input of both flip-flops *A* and *B*, a total of three counts would be

Fig. 8-4. Mod-10 counter; decade counter. (*a*) Logic diagram. (*b*) Truth table. (*c*) Waveforms.

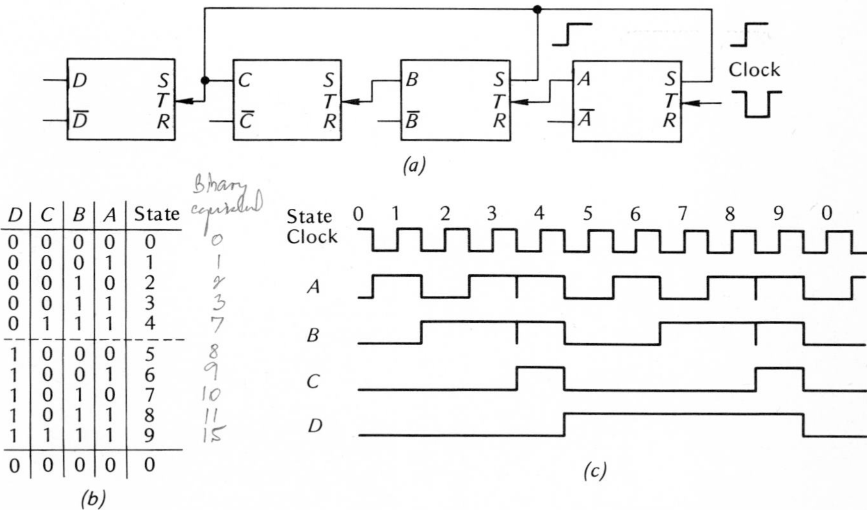

D	C	B	A	State
0	0	0	0	0
0	0	0	1	1
0	0	1	0	2
0	0	1	1	3
0	1	1	1	4
1	0	0	0	5
1	0	0	1	6
1	0	1	0	7
1	0	1	1	8
1	1	1	1	9
0	0	0	0	0

(b)

skipped. This is true since the positive feedback signal developed at the beginning of count 4 would set both flip-flops *A* and *B*, and the counter would immediately advance to count 7. Thus, counts 4, 5, and 6 are eliminated, and we have a mod-5 counter. The complete counter is shown in Fig. 8-3. A mod-10 counter is implemented by adding an extra flip-flop (*D*) which is triggered by *C* as shown in Fig. 8-4. Notice that the output of flip-flop *D* is symmetrical. This is sometimes very convenient, as, for example, when the counter is used as a frequency divider.

8-3 PARALLEL COUNTER

The ripple counter is the simplest to build, but there is a limit to its highest operating frequency. As previously discussed, each flip-flop has a delay time. In a ripple counter these delay times are additive, and the total *settling time* for the counter is approximately the delay time times the total number of flip-flops. This speed limitation can be overcome by the use of a *synchronous* or *parallel* counter. The difference here is that every flip-flop is triggered by the clock. Thus, they all make their transitions simultaneously. The construction of a parallel binary counter is shown in Fig. 8-5, along with the truth table and the waveforms for the natural count sequence. The clocked *JK* flip-flop used responds to a negative clock pulse at the *T* input and toggles when both the *J* and *K* inputs are true. Having no connection to the *J* or *K* input is equivalent to having a positive voltage, or true level, at these inputs. Thus flip-flop *A* in Fig. 8-5 changes state each time a negative pulse appears at the *T* input. The output of NAND gate *X* goes low each time a clock pulse occurs and *A* is high. Thus flip-flop *B* changes state with every other clock pulse. The output of NAND gate *Y* goes low each time a clock pulse occurs and both *A* and *B* are high. Thus, flip-flop *C* changes state with every fourth clock pulse. This represents a mod-8 parallel binary counter.

Fig. 8-5. Mod-8 parallel binary counter. (*a*) Logic diagram. (*b*) Truth table. (*c*) Waveforms.

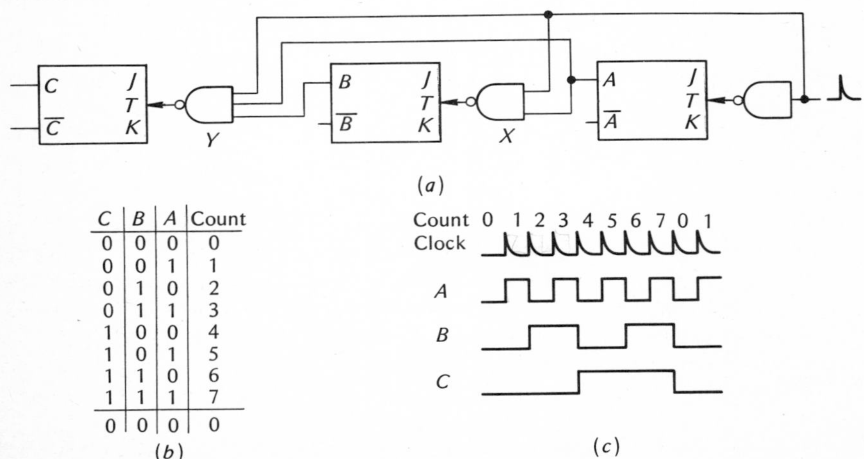

C	B	A	Count
0	0	0	0
0	0	1	1
0	1	0	2
0	1	1	3
1	0	0	4
1	0	1	5
1	1	0	6
1	1	1	7
0	0	0	0

(*b*)

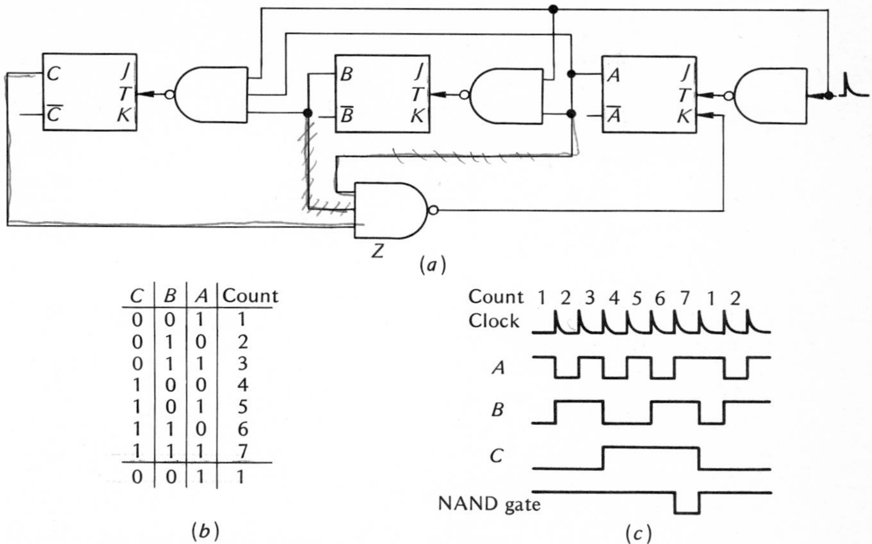

C	B	A	Count
0	0	1	1
0	1	0	2
0	1	1	3
1	0	0	4
1	0	1	5
1	1	0	6
1	1	1	7
0	0	1	1

(b)

Fig. 8-6. Mod-7 parallel counter. (a) Logic diagram. (b) Truth table. (c) Waveforms.

The parallel counter shown in Fig. 8-5 can be used as a basis for building counters of other moduli. Suppose we want to construct a mod-7 parallel binary counter. It is necessary to find some means of eliminating one state from the natural count sequence. The truth table in Fig. 8-5*b* shows that all flip-flops are high during count 7. In changing from count 7 to count 0, all flip-flops change to the 0 state. This is similar to resetting all flip-flops. Now, if some means were found to prevent flip-flop *A* from being reset during the transition from count 7 to count 0, without affecting the operation of flip-flops *B* and *C*, the counter would progress from count 7 to count 1. Thus, count 0 would be skipped, and a mod-7 counter would be formed.

If *A* is high, the next clock pulse causes flip-flop *A* to change state (i.e., the *A* side goes low), and this is equivalent to being reset. This occurs because the *J* and *K* inputs have no connections and are therefore considered true. On the other hand, if the *K* input were held low (false), the flip-flop would be set by the next clock pulse. However, the flip-flop is already set (the *A* output is high), and therefore no change will occur.

Thus, to construct a mod-7 counter from the basic counter shown in Fig. 8-5, it is only necessary to ensure that the *K* input to flip-flop *A* is held low (false) during count 7. This is easily accomplished by using the true outputs of all flip-flops as the inputs to a NAND gate, as shown in Fig. 8-6. The output of NAND gate *Z* will be low only when all three flip-flop outputs are high. Thus the *K* input to flip-flop *A* will be held low only during count 7 and will be high at all other times. The truth table and waveforms for the parallel mod-7 counter are shown in Fig. 8-6*a* and *b* (see Prob. 8-3).

The configuration in Fig. 8-6 can be easily modified to form a mod-6 counter. If

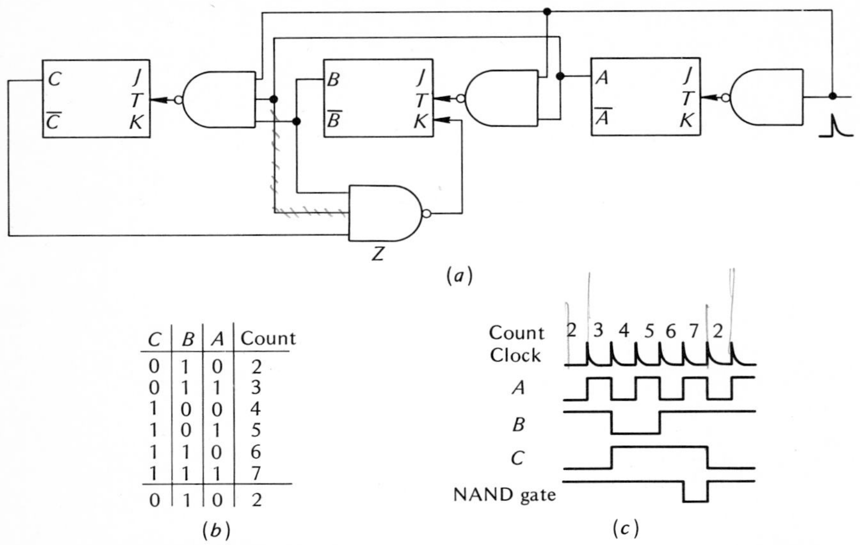

C	B	A	Count
0	1	0	2
0	1	1	3
1	0	0	4
1	0	1	5
1	1	0	6
1	1	1	7
0	1	0	2

Fig. 8-7. Mod-6 parallel counter. (*a*) Logic diagram. (*b*) Truth table. (*c*) Waveforms.

the output of NAND gate *Z* is removed from flip-flop *A* and connected to the *K* input of flip-flop *B*, a mod-6 counter is formed. The proper connection for this mod-6 counter, along with the appropriate truth table and waveforms is shown in Fig. 8-7. Flip-flops *A* and *C* are unaffected and function as before. However, since the output of NAND gate *Z* is low during count 7, flip-flop *B* is prevented from being reset as it would normally be during the transition from count 7 to count 0. Thus the counter progresses from count 7 to count 2, and states 0 and 1 are omitted. (In fact, the counter will work exactly as described even if the *A* input is left off NAND gate *Z*.)

Another method for forming a mod-6 counter from the configuration shown in Fig. 8-6 is simply to remove the input to NAND gate *Z* which comes from *B*. This causes the output of NAND gate *Z* to be low during counts 5 and 7, and the counter thus skips states 6 and 0.

Example 8-5

Design a mod-5 parallel binary counter using the techniques described above.

Solution

The mod-5 counter can be implemented by combining the methods used to form the mod-7 and mod-6 counters. If the output of NAND gate *Z* is connected to the *K* inputs of both flip-flops *A* and *B*, a mod-5 counter is formed. This is true since the NAND gate output is low during count 7, which means that neither flip-flop *A* nor flip-flop *B* is allowed to reset during the natural transition from count 7 to count 0. Flip-flop C is allowed to reset, however, and thus the counter advances from state 7 to state 3. In this way, states 0, 1, and 2 are eliminated, and a mod-5 counter is

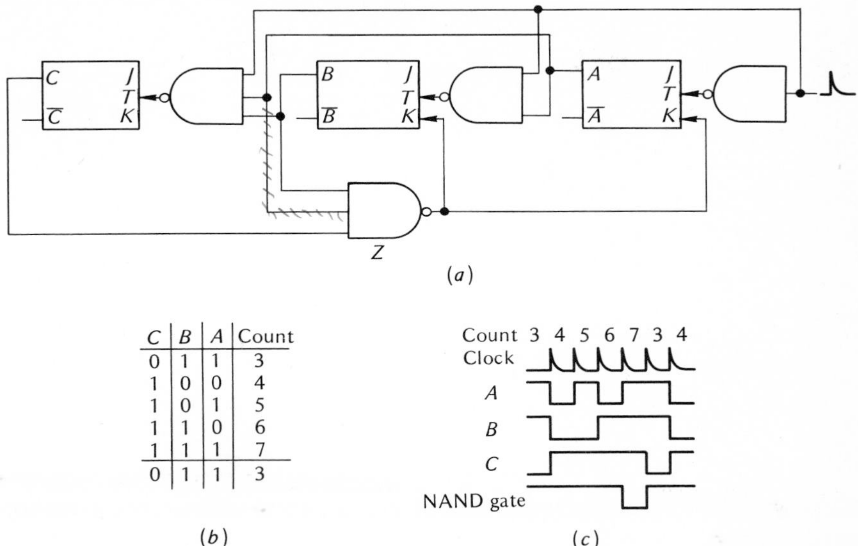

C	B	A	Count
0	1	1	3
1	0	0	4
1	0	1	5
1	1	0	6
1	1	1	7
0	1	1	3

Fig. 8-8. Mod-5 parallel counter. (*a*) Logic diagram. (*b*) Truth table. (*c*) Waveforms.

the result. The counter, along with its truth table and waveforms, is shown in Fig. 8-8. The *A* input to NAND gate *Z* could again be omitted.

There is a problem which occurs in all the parallel counters so far discussed. This is the so-called race problem, and this problem along with its solution is the topic of the next section.

8-4 THE RACE PROBLEM

Generally speaking, a *race problem* may occur whenever two inputs to a gate are undergoing transitions at the same time. Consider the two-input NAND gate shown in Fig. 8-9*a*. When input *A* is low and input *B* is high, the output *C* is obviously

Fig. 8-9. (*a*) The racing problem at the input of a NAND gate. (*b*) Expanded waveforms.

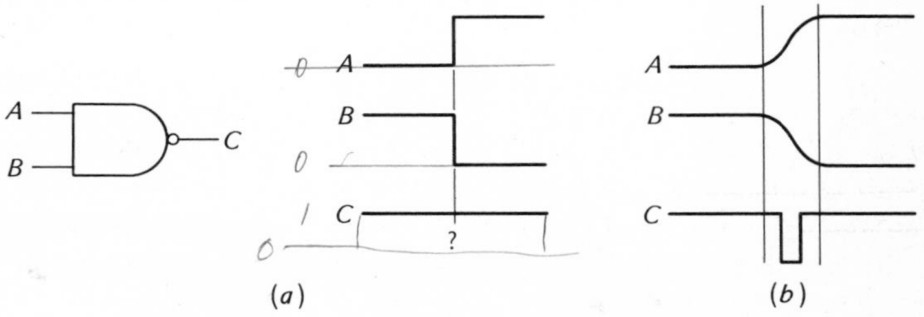

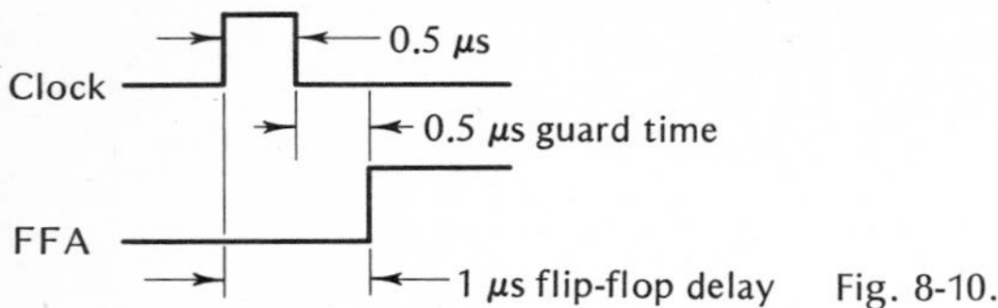

Fig. 8-10.

high. Similarly, when *A* is high and *B* is low, *C* is high. But what happens during the transition time when *A* and *B* are both changing? The answer is, during the transition time the output *C* is indeterminate. The input levels to *A* and *B* must have some rise and fall times, respectively. This is shown in Fig. 8-9*b*. If the signals begin and end their transitions at the same time and have the same rise and fall times, a pulse may appear at *C*. But if *B* is slightly delayed from *A*, there is a definite overlap of positive levels, and a full-amplitude pulse appears at the output of the gate. On the other hand, if *A* is slightly delayed with respect to *B*, no overlap of the input signals occurs, and no output pulse appears at *C*. There are, of course, a number of possibilities between these two extremes. They result in output pulses of various widths. Since the results of this situation are unpredictable, it should be avoided in digital systems. For example, if the output of this gate were connected to the *T* input of a flip-flop, the triggering behavior of the flip-flop would be unpredictable.

As a situation in which the race problem might occur, consider the parallel counter shown in Fig. 8-5. Notice that the clock used for this counter is a series of positive pulses, and *not* a square wave. There is a reason for this. There must be a definite delay between the time when a positive clock pulse appears at the input to a flip-flop, and the time when the flip-flop output changes state. This is known as the flip-flop delay time. This situation is shown in Fig. 8-10 for a flip-flop having a delay time of 1 μs and 0.5 μs clock pulses. The key point here is that the clock returns to its low level before the flip-flop output changes state. In this case, the clock is back down 0.5 μs before the flip-flop output goes high—it is said to have a *guard time* of 0.5 μs.

Now, suppose the clock pulse width is increased. If the clock pulse width is increased to 1 μs, the guard time is 0 μs, and we are approaching a racing condition as depicted in Fig. 8-9. If the clock pulse width increases beyond 1 μs, there is a definite overlap of positive pulse levels at NAND gate *X* in Fig. 8-5 (also NAND gate *Y*). Then the counter does not function properly. That is, racing occurs, as shown in Fig. 8-11. Here's what happens. Examine the levels into and out of the *X* NAND gate as the counter is progressing from count 0 to count 1. The waveforms are shown in Fig. 8-11. Some finite delay time after the clock pulse has gone positive, *A* also goes positive. This overlap of pulses causes a pulse to appear at the output of NAND gate *X*. This pulse goes directly to the *T* input of flip-flop *B*, which is therefore triggered. This is most undesirable since the counter is supposed to progress to count 1; instead, it is progressing to count 3. Thus the race problem at this NAND gate causes the counter to malfunction. As might be expected, a similar situation exists at NAND gate *Y*. Of course, the race problem will not occur if the clock pulse width is kept smaller than the flip-flop delay time—i.e., if some guard time is provided.

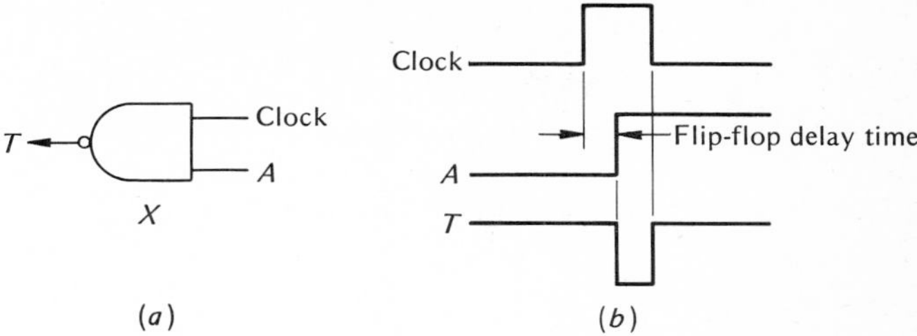

Fig. 8-11. (*a*) NAND gate *X* in Figure 8-5. (*b*) Waveforms going from count 0 to count 1.

Another solution to the race problem makes use of additional circuitry. This method, commonly referred to as "trailing-edge logic," is shown in Fig. 8-12. Examination of the waveforms in this figure shows that AND gate *X* has output pulses every time the clock goes positive—racing. The *T* input to the next flip-flop should have a trigger pulse for every other clock pulse, and so one of the two existing pulses must be eliminated.

The differentiated output of the AND gate which appears at the *T* input of the next flip-flop is shown in the figure; this waveform is achieved by choosing the proper values of resistor and capacitor. Since it requires a negative pulse to trigger the flip-flops, the positive pulses at the *T* input have no effect. The first negative pulse will, however, trigger the flip-flop since it crosses the input threshold. The second negative pulse is too small to cross the threshold of the flip-flop and therefore has no effect. Thus the second pulse out of the AND gate has been eliminated, and all the counters will work properly with this logic. The only difference in counter operation is that the flip-flops will change state on the falling edge of the clock pulse instead of the rising edge. This is the origin of the term *trailing-edge logic*. Think of the operation of the counter as follows: When the clock pulse goes positive, the capacitor is armed (charged). If the positive pulse is too short, as with the racing pulse or with noise pulses, the capacitor is not charged. Then, when the clock falls, the capacitor is fired (discharged) and a trigger pulse is produced at the *T* input of the next flip-flop. The parallel binary counter of Fig. 8-5 has been redrawn in Fig.

Fig. 8-12. Trailing-edge logic.

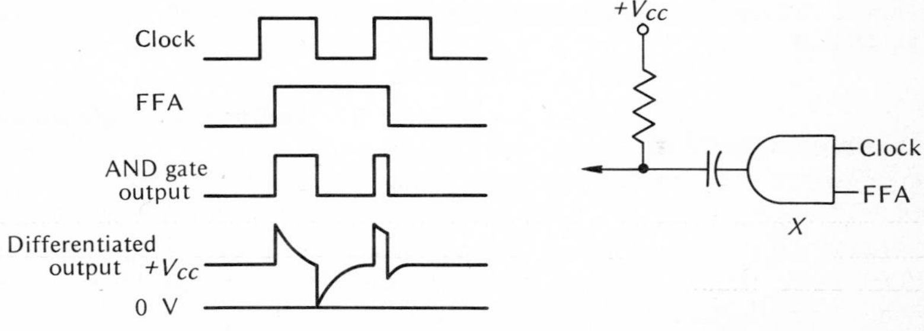

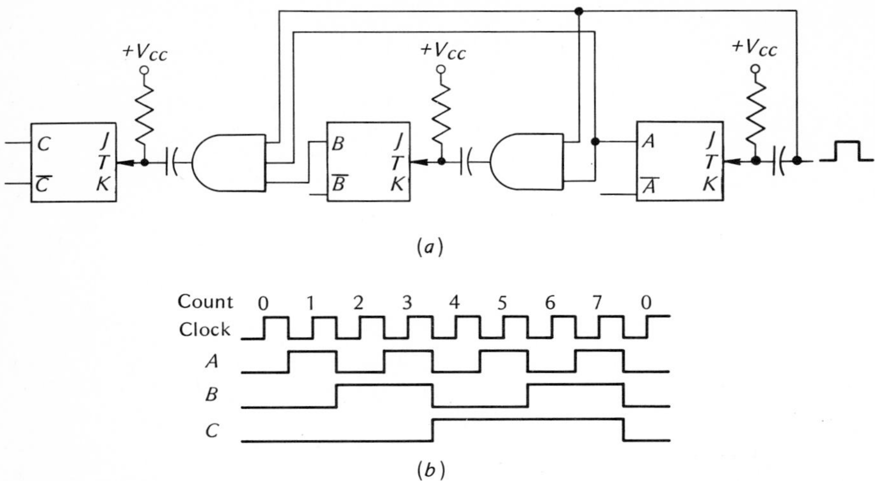

Fig. 8-13. Mod-8 parallel binary counter. (*a*) Logic diagram using trailing-edge logic. (*b*) Waveforms.

8-13 with trailing-edge logic. The only disadvantage of trailing-edge logic is the additional circuitry required.

A third method for overcoming the race problem uses the *master/slave* flip-flop shown in Fig. 8-14. The clock is connected directly to the *R* and *S* inputs of the master flip-flop. Thus, the master is disabled and cannot be changed while the clock is low. The clock level is inverted and applied to the two AND gates at the *R* and *S* inputs to the slave flip-flop. These two AND gates are then enabled, and the slave flip-flop is set or reset while the clock is low according to the state of the master. That is, if the *A* output of the master is high, then the *S* input to the slave must be high, and thus the slave is set. On the other hand, if the $\bar{A}$ side of the master is high, the *R* input to the slave is high, and the slave is reset. When the clock goes high, the inverted clock which appears at the two slave input AND gates goes low. The two AND gates are then disabled, and the slave is thus disabled and cannot change state. The high-level clock, however, enables the *R* and *S* inputs to the master. Thus, the master changes state according to the levels at the *R* and *S* inputs. The master flip-flop is either set or reset when the clock goes high. When the clock goes low again, the master is disabled and cannot change state (since its *R* and *S* inputs are disabled), but the inverted clock which appears at the two slave AND gates is now high. Thus, the AND gates are enabled and the content of the master is now shifted into the slave.

You might now ask how all this helps solve the race problem. Recall that the race problem occurred when the flip-flop output became high while the clock was still high. The output of the *master/slave* flip-flop can change state only after the clock has gone low. Thus, the overlap of positive pulses is avoided, and the race problem for any of the counters discussed here is solved.

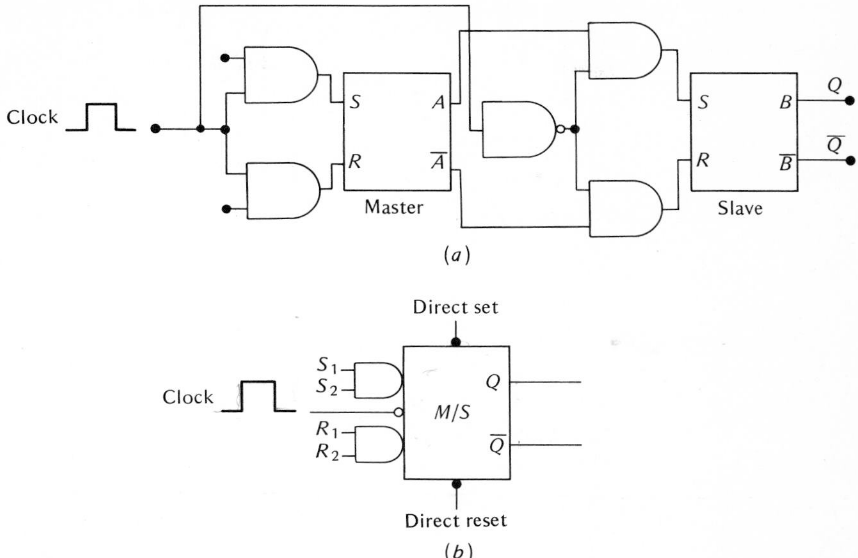

Fig. 8-14. Clocked *master/slave* flip-flop. (*a*) Logic. (*b*) Symbol.

The symbol for the *master/slave* flip-flop is shown in Fig. 8-14*b*. This is a clocked *RS* flip-flop which has two AND-gated *set* inputs and two AND-gated *reset* inputs. This configuration gives greater flexibility in design. There are also a *direct set* and a *direct reset* which go directly to the slave. Thus the flip-flop can be preset to either state without regard for the clock. A *JK* flip-flop can be formed from this unit by simply connecting the *Q* output to either the R_1 or R_2 input, and connecting the $\overline{Q}$ output to either the S_1 or S_2 input. This connection is shown in Fig. 8-15. The flip-flop shown in this figure can be used directly to implement any of the counters discussed so far in this chapter. For example, the counter and waveforms in Fig. 8-5 will appear exactly as shown when the flip-flops used are of the *master/slave* variety, and the NAND gates are changed to AND gates.

This type of *master/slave* flip-flop is so popular in industry, and represents such a great savings in components and cost, that it will be used almost exclusively in the

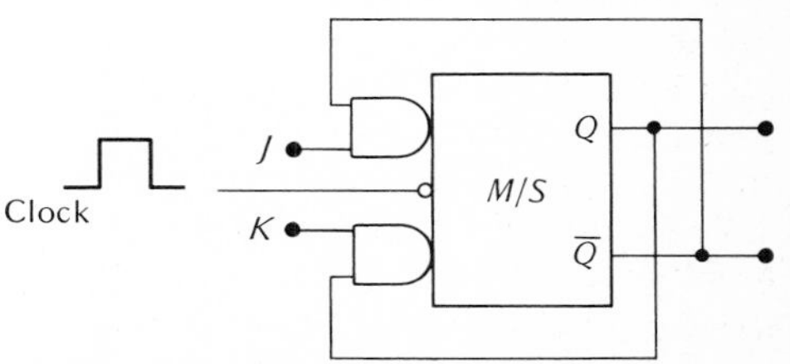

Fig. 8-15. *Master/slave* flip-flop cross-coupled to form a clocked *JK* flip-flop.

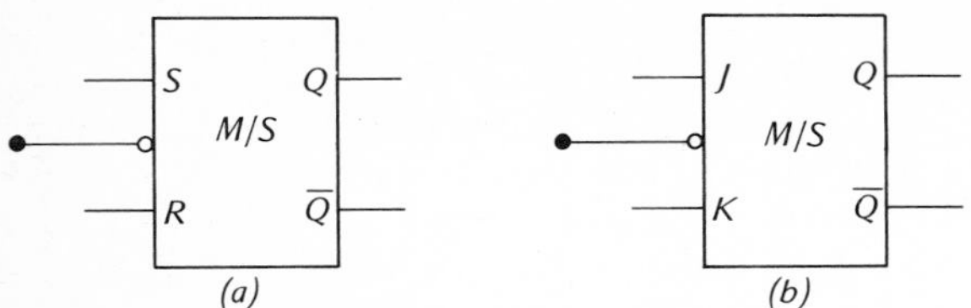

Fig. 8-16. (*a*) Symbol designating a clocked *master/slave* flip-flop used in the *RS* mode. (*b*) Symbol designating a clocked *master/slave* flip-flop which has been cross-coupled to operate in the *JK* mode.

remainder of this book. As a simplification, the symbols shown in Fig. 8-16 will be used to distinguish the clocked *master/slave* flip-flop.

Finally, you should be warned that the race problem is not exclusive to parallel counters and can appear throughout a digital system. Unfortunately, the problem occurs when least expected and can cause a great deal of inconvenience. Therefore, you should at all times be on guard against racing.

8-5 SERIES-PARALLEL COMBINATION COUNTERS

Frequently, a specific counter can be more easily implemented, at a savings in components, by constructing it as a series-parallel combination. Consider the mod-5 counter shown in Fig. 8-17. This counter will progress through a natural binary count as shown in the truth table in Fig. 8-17*a*. The waveforms shown in the

Fig. 8-17. Mod-5 binary counter. (*a*) Truth table. (*b*) Waveforms. (*c*) Logic diagram.

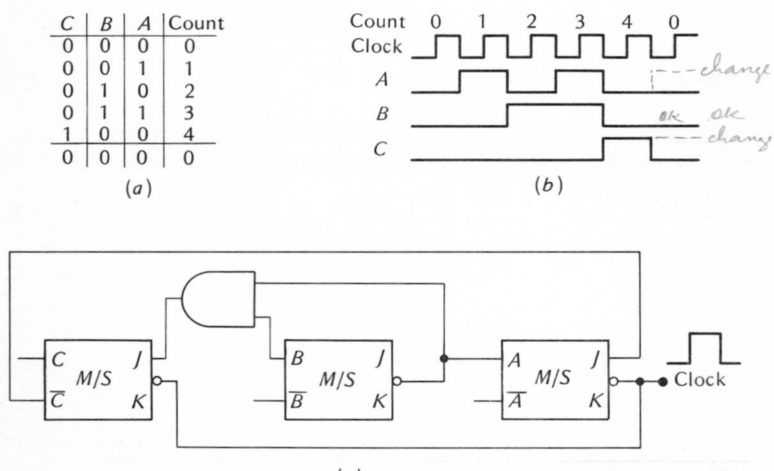

C	*B*	*A*	Count
0	0	0	0
0	0	1	1
0	1	0	2
0	1	1	3
1	0	0	4
0	0	0	0

(*a*)

figure correspond to the desired count sequence. They provide the basis for determining the proper connections to implement the counter.

The waveforms show that flip-flop A changes state each time the clock goes negative, except during the transition from count 4 to count 0. Thus, flip-flop A should be triggered by the clock and must have an inhibit during count 4. That is to say, some signal must be provided during the transition from count 4 to count 0. Notice that $\bar{C}$ is high (true) during all counts except count 4. Thus, if $\bar{C}$ is connected to the J input of flip-flop A, we have the desired inhibit signal. This is true since the J and K inputs to flip-flop A are both true for all counts except count 4; thus the flip-flop triggers each time the clock goes negative. However, during count 4, the J side is low (false), and the next time the clock goes negative the flip-flop will be prevented from being set. The connections which cause flip-flop A to progress through the desired sequence are shown in Fig. 8-17*c*.

Now, the desired waveforms in Fig. 8-17*b* show that flip-flop B must change state each time A goes negative. Thus, the trigger input of flip-flop B will be driven by A as shown in Fig. 8-17*c*. If flip-flop C is triggered by the clock while the J input is held low (false) and the K input is high (no connection), every clock pulse will reset it. Now, if the J input is high only during count 3, C will be high during count 4 and low during all other counts. The necessary levels for the J input can be obtained by ANDing the true outputs of flip-flops A and B. Since A and B are both high only during count 3, the J input to flip-flop C is high only during count 3. Thus, when the clock goes negative during the transition from count 3 to count 4, flip-flop C will be set. At all other times, the J input to flip-flop C is low and it is held in the reset state. The complete mod-5 counter is shown in Fig. 8-17*c*.

In constructing a counter of this type, it is always necessary to examine the omitted states to make sure that the counter will not malfunction. The counter in Fig. 8-17 omits states 5, 6, and 7 during its normal operating sequence. There is, however, a very real possibility that the counter may set up in one of these omitted (illegal) states when power is first applied to the system. It is necessary to check the operation of the counter when starting from each of the three illegal states to ensure that it progresses into the normal count sequence and does not become inoperative.

Begin by assuming that the counter is in state 5 ($A = 1$, $B = 0$, and $C = 1$). When the next clock pulse goes low, the following events occur:

1. Since $\bar{C}$ is low, flip-flop A is reset. Thus A changes from a 1 to a 0.
2. When A changes from 1 to 0, flip-flop B triggers, and B changes from 0 to 1.
3. Since the J input to flip-flop C is low, flip-flop C is reset, and C changes from 1 to 0.

Thus the counter progresses from state 5 to state 2 ($A = 0$, $B = 1$, and $C = 0$) after one clock pulse.

Now, assume that the counter starts in state 6 ($A = 0$, $B = 1$, and $C = 1$). When the next clock pulse goes low, the following events occur:

1. Since $\bar{C}$ is low, flip-flop A is reset. But flip-flop A is already in the reset condition, and therefore A remains at 0.

2. Since A does not change, flip-flop B does not change, and B remains at 1.
3. Since the J input to flip-flop C is low, flip-flop C is reset, and C changes from 1 to 0.

Thus the counter progresses from state 6 to state 2 after one clock pulse.

Finally, assume that the counter begins in state 7 ($A = 1$, $B = 1$, and $C = 1$). When the next clock pulse goes low, the following events occur:

1. Since $\bar{C}$ is low, flip-flop A is reset, and A changes from a 1 to a 0.
2. Since A changes from a 1 to a 0, flip-flop B triggers and B changes from a 1 to a 0.
3. The J input to flip-flop C is true and therefore flip-flop C triggers. (Thus C changes from a 1 to a 0.)

Thus the counter progresses from state 7 to state 0 ($A = 0$, $B = 0$, and $C = 0$) after one clock pulse. Therefore, this counter configuration will automatically work itself out of any of the three illegal states after only one clock pulse.

A straight binary mod-7 counter can be easily constructed using the same techniques. The desired count sequence and waveforms are shown in Fig. 8-18, along with the complete logic diagram. The proper connections for the flip-flops can be determined by examining the waveforms. Notice that flip-flop C must change state each time B goes negative. Therefore, B can be used as the trigger input to flip-flop C.

Flip-flop B must change state each time the clock goes negative and A is high. It must also change state during the transition from state 6 to state 0. State 6 can be

Fig. 8-18. Mod-7 binary counter. (*a*) Truth table. (*b*) Waveforms. (*c*) Logic diagram.

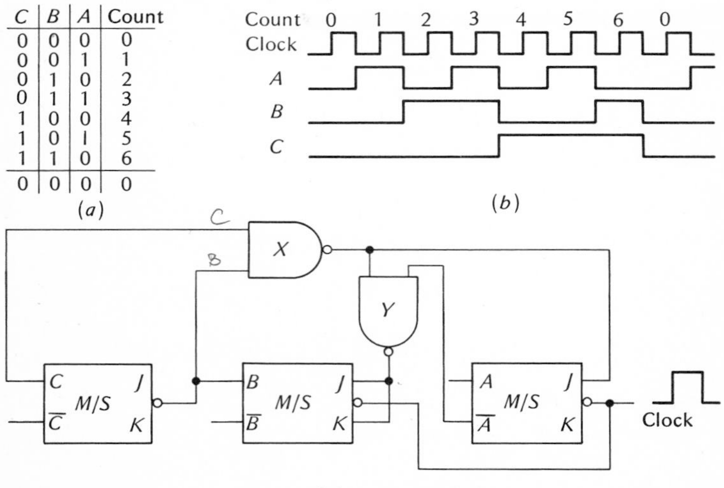

C	B	A	Count
0	0	0	0
0	0	1	1
0	1	0	2
0	1	1	3
1	0	0	4
1	0	1	5
1	1	0	6
0	0	0	0

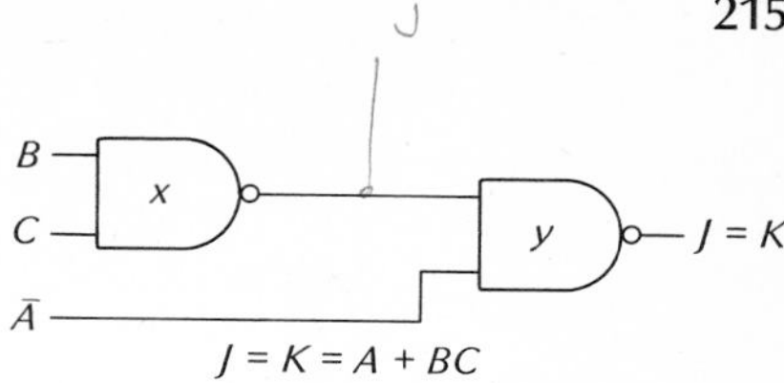

Fig. 8-19. Logic to implement the mod-7 counter in Fig. 8-18.

recognized as the time when both B and C are high. Now, if flip-flop B is triggered by the clock and the J and K inputs are tied together, the action of B is determined by the level on the J and K input lines when the clock goes negative. (When J and K are high, flip-flop B changes state; when the inputs are low, B does not change state.)

The proper action for B can be achieved by using the logic circuit shown in Fig. 8-19. Flip-flop B must change state any time A is high or any time both B and C are high. The logic equation is $K = J = A + BC$. Examination of Fig. 8-19 reveals that the output of NAND gate X will be low only when both B and C are high. Any time the output of NAND gate X is low, the output of NAND gate Y (J or K) must be high and thus B will change state. Furthermore, any time $\bar{A}$ is low (A is high) the output of NAND gate Y (J or K) must be high and thus B will change state. Therefore, any time the condition $A + BC$ is true, flip-flop B will change state.

Finally, notice that flip-flop A must change state every time the clock goes negative, except during the transition from count 6 to count 0. Thus, the clock should be used as the trigger input to flip-flop A. A can be prevented from going positive during the transition from count 6 to count 0 by holding the J input to flip-flop A low during count 6. The output of NAND gate X is low only during count 6; if it is connected to the J input of flip-flop A, A will progress through the desired sequence of levels.

The only illegal state to be considered is count 7 ($A = 1$, $B = 1$, and $C = 1$). If the counter is in this state, the following events occur the first time the clock goes low:

1. Since B and C are both high, the output of NAND gate X is low, and thus flip-flop A is reset. Therefore, A changes from a 1 to a 0.
2. Since the output of NAND gate X is low, the output of NAND gate Y is high, and flip-flop B changes state. Therefore, B goes from a 1 to a 0.
3. Since B goes low, flip-flop C changes state and C goes from a 1 to a 0.

Therefore, the counter progresses automatically from the illegal state 7 to the legal state 0 ($A = 0$, $B = 0$, and $C = 0$) the first time the clock makes a positive-to-negative transition.

Example 8-5

Construct a mod-10 binary counter using the clocked *master/slave* flip-flops.

Solution

We construct a mod-10 counter using the mod-5 counter shown in Fig. 8-17. It is only necessary to add an additional flip-flop (D) which is triggered by C. The complete counter with truth table and waveforms are shown in Fig. 8-20.

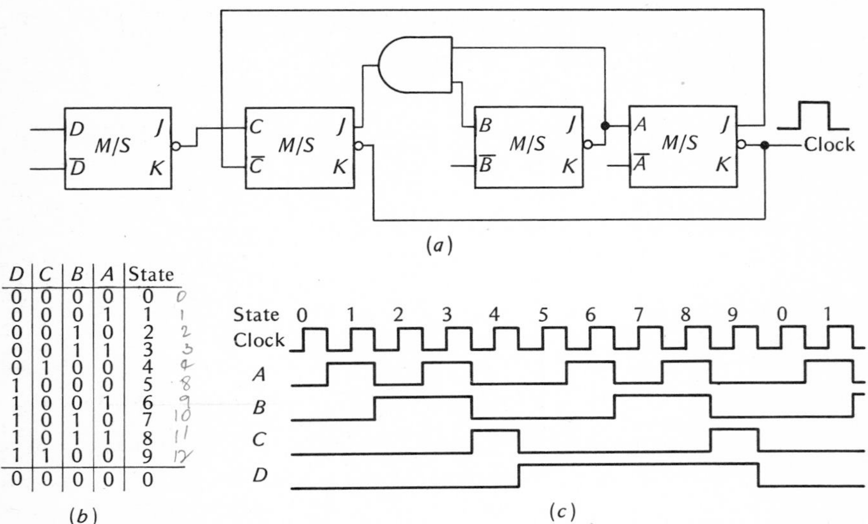

D	C	B	A	State
0	0	0	0	0
0	0	0	1	1
0	0	1	0	2
0	0	1	1	3
0	1	0	0	4
1	0	0	0	5
1	0	0	1	6
1	0	1	0	7
1	0	1	1	8
1	1	0	0	9
0	0	0	0	0

(b)

Fig. 8-20. Decade counter. (*a*) Logic diagram. (*b*) Truth table. (*c*) Waveforms.

8-6 BINARY DECADE COUNTER WITH DECODING GATES

The binary decade counter is a very useful form of counter. It provides a means of changing a count in the less familiar binary mode into an equivalent count in the more familiar decimal mode. If the 10 discrete states of the decade counter are decoded, the series of waveforms shown in Fig. 8-21 are obtained. These waveforms can be used for a number of purposes.

One of the most useful methods of utilizing these 10 output signals is to cause them to control 10 lamps which represent the 10 decimal numbers. Alternatively, they could be used to control the grid voltages on the popular nixie counter tubes. (A nixie counter tube is a neon-filled tube which has 10 grids in it; each grid is in the shape of a decimal number. When the proper voltage is applied to one of the grids, the neon fires and illuminates that grid. Thus any of the decimal numbers from 0 through 9 can be displayed by simply controlling the grid voltages.)

In order to produce the 10 waveforms shown in Fig. 8-21, it is necessary to decode the four flip-flops in the counter. Each of the 10 states represents a unique condition of the four flip-flops in Fig. 8-20 and requires 10 fourlegged AND gates to produce these 10 output waveforms. Count 0 can be recognized as the unique time when A and B and C and D are all low. Alternatively, it is the time when $\bar{A}$ and $\bar{B}$ and $\bar{C}$ and $\bar{D}$ are all high. Thus count 0 can be decoded by means of the AND gate shown in Fig. 8-22. Count 1 can be decoded by recognizing the state when A and $\bar{B}$ and $\bar{C}$ and $\bar{D}$ are high; the four-input AND gate shown in Fig. 8-22 provides the 1 output waveform. The remaining eight counts can be similarly decoded. The

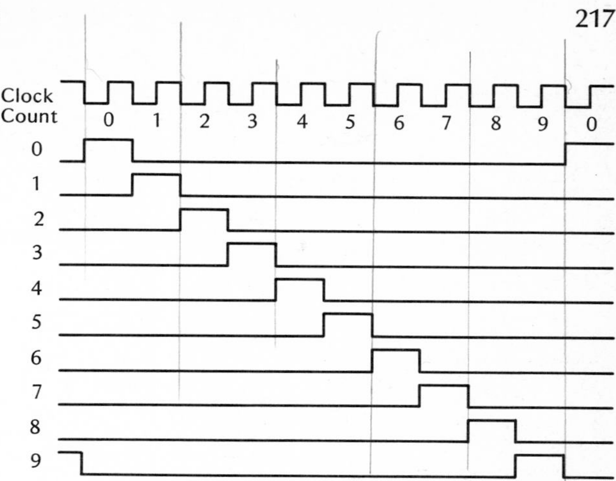

Fig. 8-21. Decade-counter outputs.

complete system of decoding gates is shown in Fig. 8-22. The 10 outputs of these gates produce the 10 decoded waveforms shown in Fig. 8-21.

A decade counter could be formed just as easily by using the mod-5 counter in Fig. 8-17 in conjunction with another flip-flop, but connected as shown in Fig. 8-23. The truth table for this configuration, along with the resulting waveforms, is shown in the figure. This is still a mod-10 (decade) counter since it still has 10 discrete states. If the flip-flops are decoded as in the previous counter, the 10 output waveforms still appear exactly as shown in Fig. 8-21. However, since the

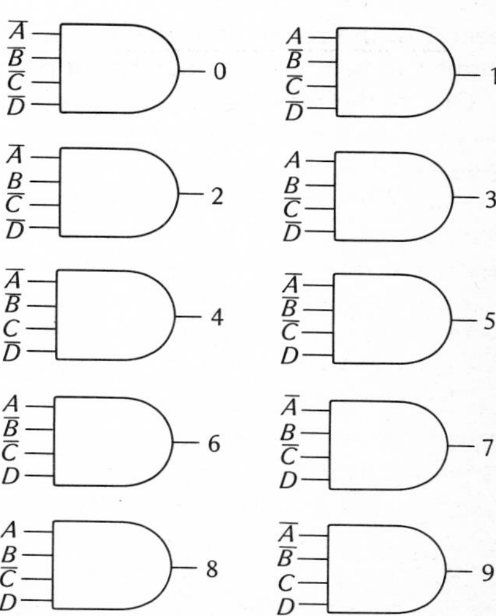

Fig. 8-22. Decoding gates for Fig. 8-20.

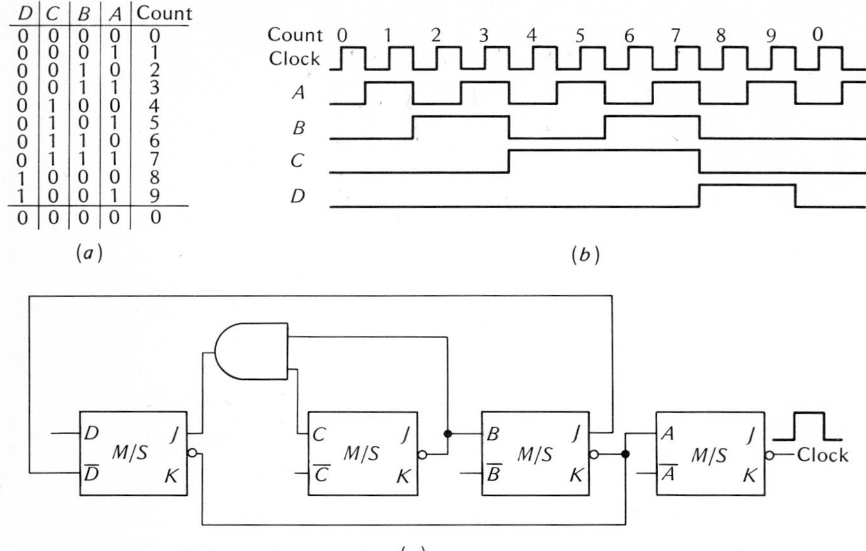

D	C	B	A	Count
0	0	0	0	0
0	0	0	1	1
0	0	1	0	2
0	0	1	1	3
0	1	0	0	4
0	1	0	1	5
0	1	1	0	6
0	1	1	1	7
1	0	0	0	8
1	0	0	1	9
0	0	0	0	0

Fig. 8-23. Decade counter. (*a*) Truth table. (*b*) Waveforms (*c*) Logic diagram.

counter waveforms are different, the decoding gates have to be different. We still proceed in the same manner, however, and state 0 can be recognized as the time when $\overline{A}$ and $\overline{B}$ and $\overline{C}$ and $\overline{D}$ are all high. Thus, the gate to decode state 0 and the other nine decoding gates are shown in Fig. 8-24. Note that the counter in Fig. 8-23 counts in a true binary sequence, while the counter in Fig. 8-20 does not.

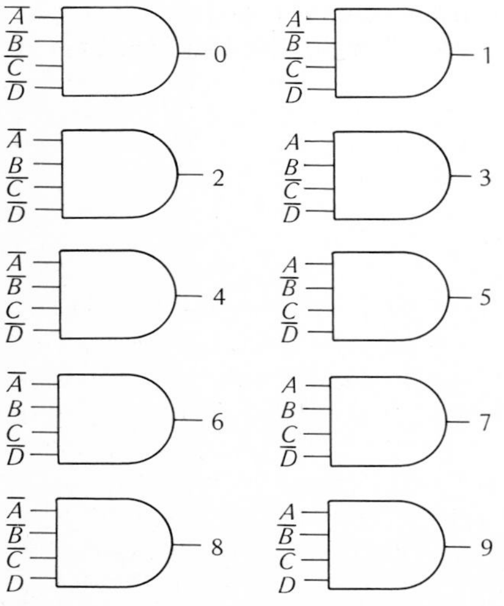

Fig. 8-24. Decoding gates for Fig. 8-23.

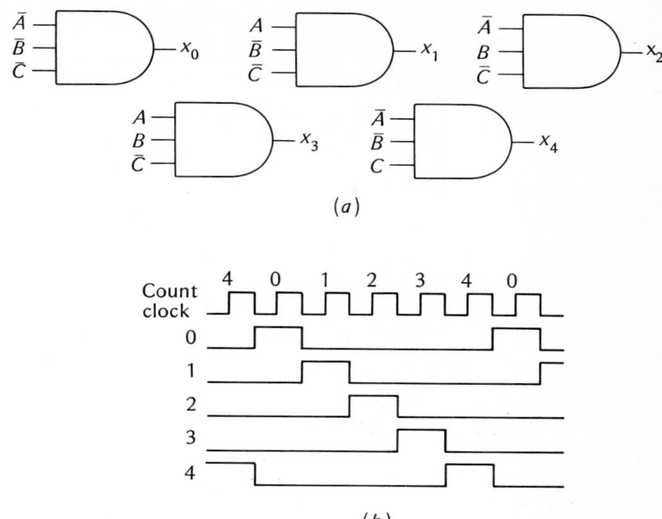

Fig. 8-25. Decoding for the mod-5 counter in Fig. 8-17. (*a*) Gates. (*b*) Decoded output waveforms.

Example 8-6

Determine the gates necessary to decode the mod-5 counter in Fig. 8-17. Draw the waveforms showing the decoded outputs.

Solution

Since the mod-5 counter has five discrete states determined by three flip-flops, it is necessary to use five three-input AND gates to decode the five states. State 0 is uniquely determined when $\bar{A}$ and $\bar{B}$ and $\bar{C}$ are all high. The logic equation is $X_0 = \bar{A}\bar{B}\bar{C}$, and the AND gate is shown in Fig. 8-25. The logic equations for the remaining four states are $X_1 = A\bar{B}\bar{C}$, $X_2 = \bar{A}B\bar{C}$, $X_3 = AB\bar{C}$, and $X_4 = \bar{A}\bar{B}C$. The correct gates and the decoded waveforms are shown in Fig. 8-25.

8-7 HIGHER-MODULUS COUNTERS

In the previous section, a mod-10 counter was constructed by combining a mod-5 counter and a mod-2 counter. Moreover, it was shown that the order in which the two counters are connected affects only the individual waveforms. That is, a mod-10 counter is the result of combining the two counters, regardless of how they are connected. This basic idea makes it possible to construct many different counters by simply using combinations of the counters thus far developed.

As an example, consider the construction of a mod-6 counter by combining a mod-3 counter and a mod-2 counter (single flip-flop). The logic diagram, truth table, and waveforms for the mod-3 counter are shown in Fig. 8-26. The mod-3

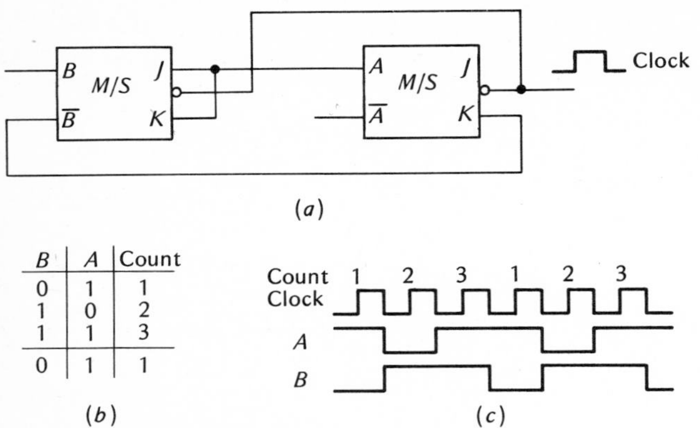

B	A	Count
0	1	1
1	0	2
1	1	3
0	1	1

(b)

Fig. 8-26. (a) Mod-3 counter. (b) Truth table. (c) Waveforms.

counter operates as follows: (1) A will change state every time the clock goes negative, except during the transition from count 3 to count 1. It is only necessary to ensure that flip-flop A cannot be reset during this transition. $\overline{B}$ is low during counts 2 and 3, if it is connected to the K input of flip-flop A, A is prevented from going low during the transition from count 3 to count 1. (2) B changes state every time A is high and the clock goes low. If a flip-flop is now connected in series with this mod-3 counter, a mod-6 counter results. If the flip-flop precedes the mod-3 counter, this is said to be a 2 × 3 mod-6 counter. This counter is shown in Fig. 8-27 along with the corresponding truth table and waveforms. If the mod-3 counter is used to drive the added flip-flop, a 3 × 2 mod-6 counter is formed. This counter and the appropriate truth table and waveforms are shown in Fig. 8-28.

This method of implementing higher-modulus counters seems quite reasonable if we consider a basic flip-flop as a mod-2 counter. Then, a mod-4 counter (two flip-flops in series) is simply two mod-2 counters in series, which forms a 2 × 2 mod-4 counter. Similarly, a mod-8 counter is simply a 2 × 2 × 2 connection, and so on. Thus, a great number of higher-modulus counters can be formed by using the product of any number of lower-modulus counters.

Example 8-7

How could a mod-12 counter be formed using the product method just described?

Solution

A mod-12 counter is divisible by 2 and 2 and 3; that is, 12 = 2 × 2 × 3. Therefore, a mod-12 counter could be implemented by using the following combinations of lower-modulus counters: (1) 2 × 2 × 3, (2) 4 × 3, (3) 2 × 6.

8-8 A BCD COUNTER

It is sometimes desirable to construct counters which count in other than a straight binary code. One other very useful form would be a decade counter which counts

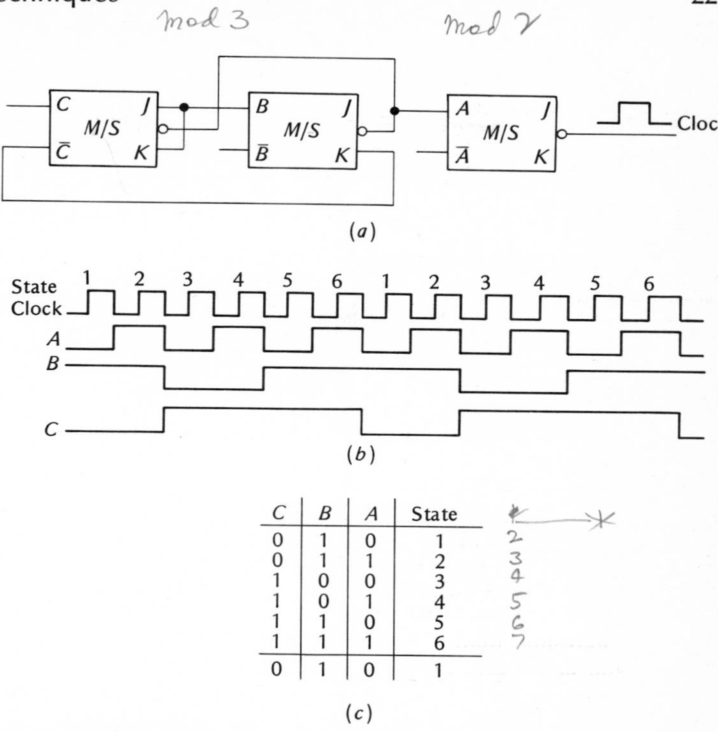

C	B	A	State
0	1	0	1
0	1	1	2
1	0	0	3
1	0	1	4
1	1	0	5
1	1	1	6
0	1	0	1

(c)

Fig. 8-27. (a) 2 × 3 mod-6 counter. (b) Waveforms. (c) Truth table.

Fig. 8-28. (a) 3 × 2 mod-6 counter. (b) Waveforms. (c) Truth table.

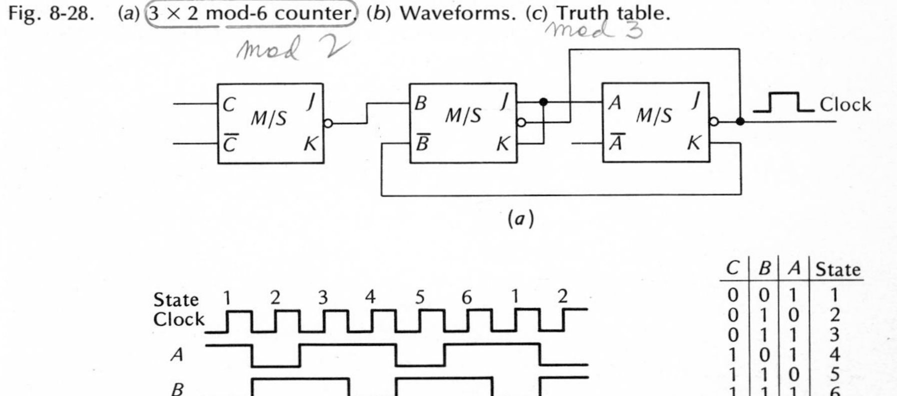

C	B	A	State
0	0	1	1
0	1	0	2
0	1	1	3
1	0	1	4
1	1	0	5
1	1	1	6
0	0	1	1

(c)

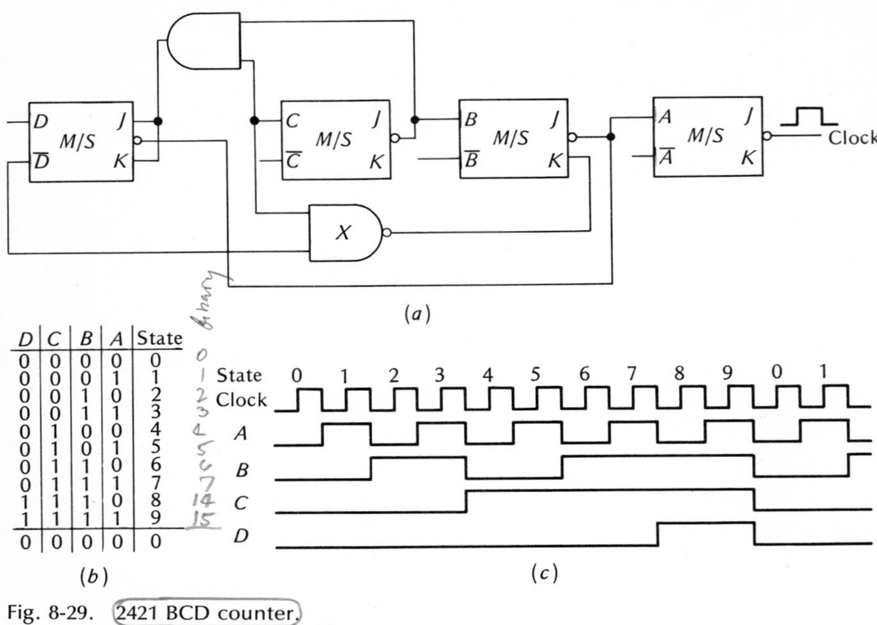

D	C	B	A	State
0	0	0	0	0
0	0	0	1	1
0	0	1	0	2
0	0	1	1	3
0	1	0	0	4
0	1	0	1	5
0	1	1	0	6
0	1	1	1	7
1	1	1	0	8
1	1	1	1	9
0	0	0	0	0

(b)

Fig. 8-29. 2421 BCD counter.

in a modified 2421 binary-coded decimal. This basic counter was discussed in Chap. 7; because of its importance, we shall consider a second means of construction using *JK* flip-flops. Such a counter is shown in Fig. 8-29, along with the appropriate truth table and waveforms. The operation of the counter can be described as follows:

1. *A* must change state each time the clock falls, and therefore the clock input to flip-flop *A* should be the clock.
2. *B* must change state each time *A* goes low except during the transition from count 7 to count 8. The output of NAND gate *X* is low whenever *C* and $\overline{D}$ are both high. This prevents *B* from going low during the transition from count 7 to count 8, since the output of NAND gate *X* holds the *K* input to flip-flop *B* low during counts 4, 5, 6, and 7. Notice that the *J* input to flip-flop *B* is always high (true), and therefore *B* is able to go high during the transition from count 5 to count 6.
3. The clock input to flip-flop *C* is driven by *B*, since *C* must change state each time *B* goes low.
4. *D* must change state each time both *B* and *C* are high and *A* goes low. If the *J* and *K* inputs to flip-flop *D* are conditioned by the output of an AND gate whose inputs are *B* and *C*, it is possible for *D* to change state whenever both *B* and *C* are high. If the clock input to flip-flop *D* is now driven by *A*, *D* will change states during the transition from 7 to 8 and from 9 to 0.

This is a decade counter, and therefore it has only 10 states. Since four flip-flops are used, there are six illegal (omitted) states which must be examined. The six omitted states are 1000, 1001, 1010, 1011, 1100, and 1101.

Let's see what happens if the counter comes up in state 1000 when the power is first applied. The first clock pulse causes *A* to change state, but *B, C,* and *D* remain unchanged. Thus, the counter advances one count to 1001. This is the second illegal state.

The second clock pulse again causes *A* to change state, and since *A* is going low, *B* changes state. *C* and *D* again remain unchanged. Thus the counter again advances one count to 1010, which is the third illegal state.

The third clock pulse causes *A* to go high, but *B, C,* and *D* remain unchanged. Thus the counter advances one count to state 1011, which is the fourth illegal state.

The fourth clock pulse causes *A* to go low, which causes *B* to go low, which causes *C* to change state. *D* remains unchanged, and the counter advances to 1100, which is the fifth illegal state.

The fifth clock pulse causes *A* to change state while *B, C,* and *D* remain unchanged, and the counter advances to the sixth illegal state, 1101.

The sixth clock pulse causes *A* to go low, which causes *B* to change state while *C* and *D* remain unchanged. Thus the counter advances to state 1110, which is count 8 in the desired count sequence.

Therefore, this counter will not lock up in any of the six illegal states and will, in the worst case, require six clock pulses to count itself out of an illegal state into the proper count sequence.

The BCD counter shown in Fig. 8-29 represents one method of implementing this counter. You should realize that there are a number of other ways of accomplishing this task.

STUDY AIDS

Summary

A counter has a natural count of 2^n, where n is the number of flip-flops in the counter. Counters of any modulus can be constructed by incorporating logic which causes certain states to be skipped over or omitted. In serial counters, the technique for skipping counts is to derive a feedback signal from certain flip-flops to reset previous flip-flops. This method takes advantage of the inherent delay in each flip-flop. In parallel counters, the technique is to precondition the logic inputs to each flip-flop in order to omit certain states. This is sometimes called "look-ahead logic." The race problem is encountered in parallel counters. Possible solutions to the race problem are:

1. Delaying the outputs of the flip-flops
2. Trailing-edge logic
3. The *master/slave* flip-flop

Higher-modulus counters can be easily constructed using combinations of lower-modulus counters. The series-parallel counter configuration is an example of this.

These configurations also represent a compromise between speed and hardware count.

Feedback around a basic binary counter can also be used to construct various BCD counters.

It should be mentioned that counters can be constructed using any binary element. This includes all types of flip-flops as well as such other elements as magnetic cores. Moreover, they can be constructed to count in any desired sequence. This, however, is a topic covered in texts on logical design.

Glossary

master/slave flip-flop A clocked *RS* (or *JK*) flip-flop composed of two individual flip-flops wired in such a way as to avoid racing.

modulus Defines the number of states through which a counter can progress.

natural count The maximum number of states through which a counter can progress. Given by 2^n, where n is the number of flip-flops in the counter.

parallel counter A counter in which all flip-flops change state simultaneously since all clock inputs are driven by the clock.

race problem The inability to determine the output of a gate when the inputs to the gate are making simultaneous transitions.

ripple counter A counter in which each flip-flop is triggered by the output of the previous flip-flop.

Review Questions

1. How many possible states are there in a counter composed of the following number of flip-flops?
   (*a*) 3.
   (*b*) 5.
   (*c*) 6.
   (*d*) 7.
   (*e*) 10.

2. What is meant by a count sequence?

3. Explain why a parallel counter is capable of faster operation than a ripple counter.

4. Define the race problem.

5. How does a delay in the output of a flip-flop cure the race problem?

6. Of what use is trailing-edge logic?

7. Describe how the *master/slave* flip-flop is used to cure the race problem.

8. In how many different ways can a mod-9 counter be implemented using lower-modulus counters?

9. What are the ways in which a mod-24 counter can be implemented using lower-modulus counters?

Problems

8-1. How many flip-flops would be required to build the following counters?
(*a*) mod-6.
(*b*) mod-11.
(*c*) mod-15.
(*d*) mod-19.
(*e*) mod-31.

8-2. Draw the logic diagram, truth table, and waveforms for the mod-7 counter shown in Fig. 8-2 if the feedback signal is taken from $\overline{C}$ instead of *C*.

8-3. What modulus counter is formed if the *A* input is removed from NAND gate *Z* of Fig. 8-6?

8-4. Draw the truth table and waveforms for the mod-6 counter formed by removing the *B* input from NAND gate *Z* in the counter in Fig. 8-6.

8-5. Construct a mod-3 parallel binary counter beginning with flip-flops *A* and *B* shown in Fig. 8-5.

8-6. Assume the counter in Fig. 8-5 is constructed using flip-flops whose outputs are delayed by 0.5 μs. If the clock for this counter is a 1-MHz square wave, will there be a race problem?

8-7. In Prob. 8-6, if the clock is modified to a series of 0.2-μs positive pulses occurring at a 1-MHz rate, is the race problem cured? Draw the waveforms for this situation.

8-8. Redraw the mod-5 counter in Fig. 8-8 using trailing-edge logic, and draw the resulting waveforms.

8-9. Draw the gates necessary to decode the seven outputs of the mod-7 counter in Fig. 8-18.

8-10. Draw the gates necessary to decode the three-stage ripple counter in Fig. 8-1.

8-11. Assume the clock for the ripple counter in Fig. 8-1 is a 1-MHz square wave, and assume that each flip-flop has a 0.25-μs delay. Carefully draw waveforms for the clock, each flip-flop, and the output decoded signals. Do you foresee any sources of difficulty?

8-12. Draw the logic diagram, truth table, and waveforms for a mod-9 counter using two mod-3 counters connected in series.

8-13. In how many different ways can a mod-18 counter be implemented using lower-modulus counters? Draw the logic diagram, truth table, and waveforms for one of these configurations.

Special Counters and Registers

A register is a very important logical block in most digital systems. Registers are often used to store (momentarily) binary information which appears at the output of an encoding matrix. Similarly, they are used to store (momentarily) binary data which are being decoded. Thus, registers form a very important link between the main digital system and the input-output channels.

A binary register also forms the basis for some very important arithmetic operations. For example, the operations of complementation, multiplication, and division are frequently implemented by means of a register.

A shift register can be quite easily modified to form a number of different types of counters. These counters offer some very distinct advantages, and we shall therefore investigate them in detail.

After studying this chapter, you should be able to

1. Describe the proper operation of a shift register.
2. Determine the modulus of a ring counter or a shift counter.
3. Draw the waveform at the output of each flip-flop and each decoding gate of a binary counter.
4. Demonstrate how counters can be used as subunits in a more complicated digital system.

9-1 A SERIAL SHIFT REGISTER

A register is nothing more than a group of flip-flops which can be used to store a binary number.[1] There must be one flip-flop for each bit in the binary number. Naturally the flip-flops must be wired so that the binary number can be entered (shifted) into the register and possibly shifted out. A group of flip-flops wired to provide either or both of these functions is called a *shift register.*

There are two methods for shifting binary information into a register. The first in-

[1] Registers can be constructed using other binary elements, for example, the magnetic core, as we shall see in Chap. 12.

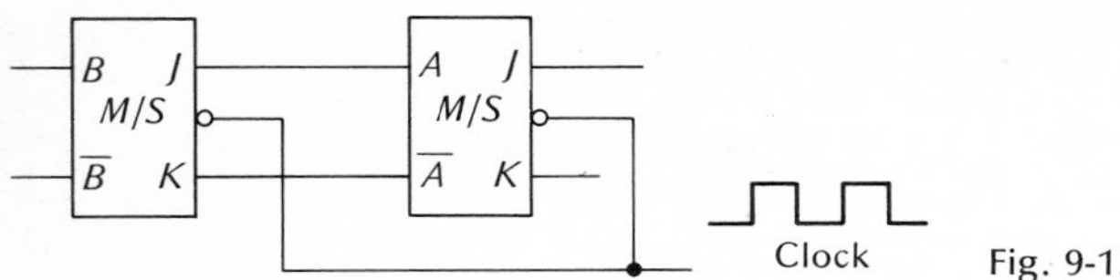

Fig. 9-1.

volves shifting the information into the register one bit at a time in a series fashion and leads to the development of a *serial shift register.* The second method involves shifting all the bits into the register at the same time and leads to the development of a *parallel shift register*. The serial shift register is discussed here, and the parallel register is discussed in a later section.

To see how a serial shift register can be implemented, consider the flip-flops shown in Fig. 9-1. Assume *A* and *B* are both low. Now, with the *J* and *K* inputs to flip-flop *A* held low, allow the clock to go through one cycle. During this period, *A* will not change state since the *J* and *K* inputs to flip-flop *A* are both low. Since *A* holds the *J* input to flip-flop *B* low and $\overline{A}$ holds the *K* input high, *B* will not change state during this period either. Therefore, both *A* and *B* will remain as before.

Now, maintain the *K* input to flip-flop *A* low but allow the *J* input to go high. If the clock is now allowed to go through one cycle, *A* will be set high but *B* will still remain unchanged. Thus during the second cycle of the clock, a 1 is placed in flip-flop *A* (since *J* was high and *K* was low), while *B* remains unchanged (since its *J* input was low).

Now, hold the *J* input to flip-flop *A* low, hold the *K* input high, and allow the clock to advance through one more cycle. *A* will be reset low since its *J* input is low and its *K* input is high. At the same time, *B* will change state (since the *J* input to *B* is high). Thus, during the third cycle of the clock, the 1 in flip-flop *A* is effectively shifted into flip-flop *B*.

If the clock is now allowed to progress through one more cycle, *A* will remain in the low state and *B* will go low (since *A* and therefore the *J* input to *B* is low). Thus, during the fourth cycle of the clock, the 1 in flip-flop *B* is effectively shifted out and both flip-flops are again in their reset states.

The important points to be noted are these:

1. To set a 1 in flip-flop *A*, hold the *J* input to *A* high and the *K* input low and allow the clock to progress through one cycle.
2. To set a 0 in flip-flop *A*, hold the *J* input low and the *K* input high and allow the clock to progress through one cycle.
3. Any time a 1 exists in flip-flop *A* (*A* is high), that 1 will be shifted into flip-flop *B* during the next cycle of the clock (that is, *B* will go high).
4. Any time a 0 exists in flip-flop *A* (*A* is low), *B* will be reset low during the next clock cycle.

Consider the application of these basic ideas to form the six-bit serial shift register shown in Fig. 9-2. It is called a six-bit register since it has six flip-flops and is therefore capable of storing six bits.

Suppose you want to shift the binary number 101110 into this register in a serial fashion. First, notice that it takes one clock time to shift the first bit into flip-flop *A*. It takes a second cycle of the clock to shift the second bit into *A*, and at the same time the first bit is shifted from flip-flop *A* into flip-flop *B*. A third clock pulse shifts the third bit into *A*, and at the same time the first bit is shifted from *B* to *C* and the second bit is shifted from *A* to *B*. This process must be repeated three more times, and during the sixth clock pulse the sixth bit is shifted into *A*, the fifth into *B*, the fourth into *C*, the third into *D*, the second into *E*, and the first bit is in *F*. Thus after six clock pulses, the six bits have been shifted serially through the six flip-flops, and the binary number is now in the register. At this time, the clock must be stopped or a portion of the information will be lost.

It is of course necessary to control the levels of the *J* and *K* inputs to flip-flop *A* during the shifting operation to ensure that the proper information is entered into the register. It is also necessary to define the order in which the number is to be shifted into the register. That is, will the least significant bit (LSB) be entered first, or last? In this case, assume that the most significant bit (MSB) is entered first, and thus after the shifting operation the most significant bit will be in flip-flop *F* and the least significant bit will be in flip-flop *A*. The clock, along with the proper *J* and *K* input levels to flip-flop *A*, is shown in the waveforms in Fig. 9-2*b*. The clock is allowed to start just prior to count 1 and is stopped at the end of count 6. During count 1, *J* is high and *K* is low and a 1 is set in flip-flop *A*. During count 2, the 1 in flip-flop *A* is

Fig. 9-2. Six-bit serial shift register. (*a*) Logic diagram. (*b*) The six cycles required to shift the number 101110 into the register, MSB first.

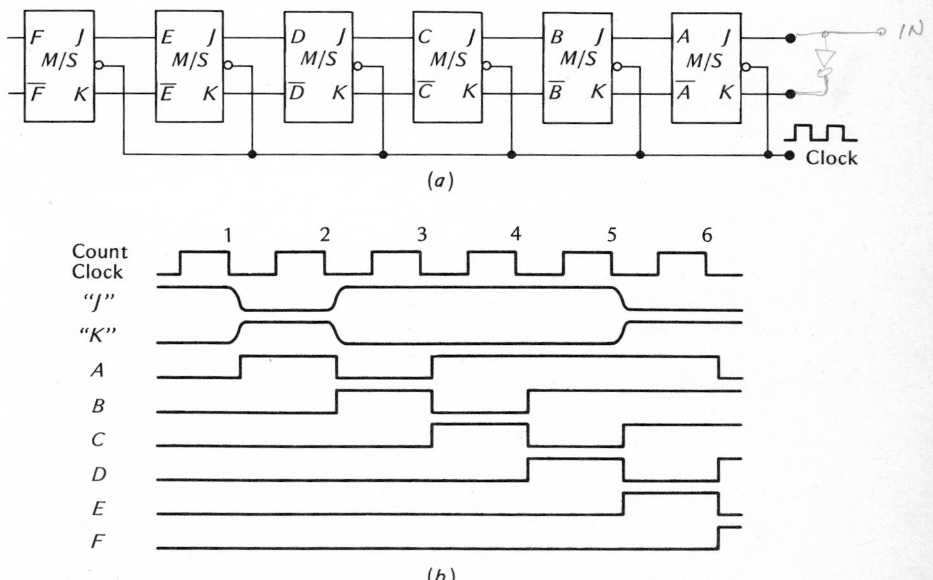

shifted into flip-flop B and a 0 is set in flip-flop A (since J is low and K is high). The remaining four bits are shifted into the register in a similar manner, and the detailed operation can be seen by carefully examining the waveform. Examination of the waveforms also shows that after the shift operation is complete (just after count 6), the desired binary number is indeed stored in the six flip-flops with the MSB in F.

Two things should be stressed. First, the size of a register is determined by the size of the number to be stored. That is, there must be one flip-flop for each bit. Second, in a serial shift register it requires n clock pulses to shift an n-bit number into the register.

9-2 A RING COUNTER

The serial shift register discussed in the previous section must have additional circuitry to control the J and K inputs to flip-flop A in order to ensure proper operation during the shift operation. The logic circuitry that provides these control waveforms would ordinarily be derived from the control section of the system. The control section is the source of the clock and whatever other control signals are necessary. These topics will be discussed in Chap. 14. For now, let's see if any of the feedback techniques discussed in the previous chapter can be applied to this basic shift register with any usable results.

One of the most logical applications of feedback might be to connect F to the input of flip-flop A and at the same time connect $\overline{F}$ to the K input of flip-flop A. This is shown in Fig. 9-3. Now, suppose that all flip-flops are in the reset state and the clock is allowed to run. What will happen? The answer is, nothing will happen since all J inputs are low and all K inputs are high. Therefore, every time the clock goes low, every flip-flop will be reset. But all flip-flops are already reset and thus nothing will happen.

In an effort to get some action, suppose A is high and all other flip-flops are low, and then allow the clock to run. The very first time the clock goes low, the 1 in A will be shifted into B and A will be reset, since F is low and $\overline{F}$ is high. All other flip-flops will remain low. The second clock pulse will shift the 1 into C, and B will reset while all other flip-flops remain low. The third clock pulse will shift the 1 into D, and so on. Thus the 1 will shift down the register, traveling from one flip-flop to the next each time the clock goes low.

The crucial point occurs when the 1 is in flip-flop F. The very next clock pulse will simply shift it into A because of the feedback connection. From that point, the process will be repeated; the 1 will simply circulate around the register as long as the clock is running. For this reason, this configuration is sometimes called a "circulating register" or "ring counter." The waveforms present in this ring counter are shown in Fig. 9-3b.

You will notice that there is a remarkable similarity between these waveforms and the decoded output waveforms of the counters in the last chapter. If waveforms of this type are needed, the ring counter might in some cases be a more logical choice than a binary counter with decoding gates. Ring counters such as this find a great many uses in developing the control waveforms previously mentioned.

There is, however, a problem with such ring counters. In order to produce the

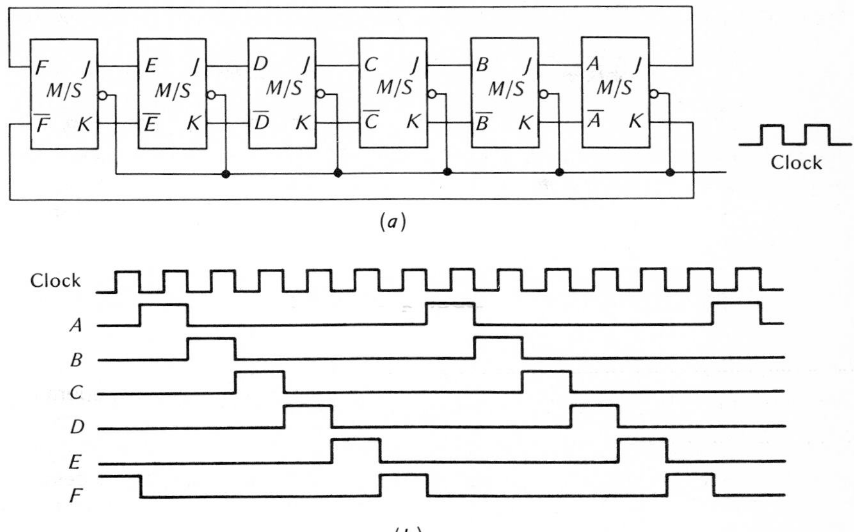

Fig. 9-3. Six-stage ring counter. (*a*) Logic diagram. (*b*) Waveforms with one 1 in the counter.

waveforms shown in Fig. 9-3*b*, it is necessary that the counter have one and only one 1 in it. The chances of this occuring naturally when power is first applied are very remote indeed. If the flip-flops should all happen to be in the reset state when power is first applied, it will not work at all, as we saw previously. On the other hand, if some of the flip-flops come up in the set state while the remainder come up in the reset state, a series of complex waveforms of some kind will be the result. Therefore, it is necessary to preset the counter to the desired state before it can be used.

One method of presetting the counter in Fig. 9-3 would be to tie the *direct reset* inputs of flip-flops *B, C, D, E,* and *F* to the *direct set* of flip-flop *A* and apply a negative pulse to this line just prior to operation of the counter. This will set a 1 in *A* and reset the remaining flip-flops.

Since the ring counter in Fig. 9-3 will function with more than one 1 in it, it might be desirable to operate it in this fashion at some time or other. For example, suppose the waveform in Fig. 9-4 were desired. This could be achieved by constructing a binary counter and decoding it as in the previous chapter. On the other hand, it could be quite easily accomplished by simply presetting the counter in Fig. 9-3 with a 1 in *A* and a 1 in *C* with all the other flip-flops reset. You will notice that

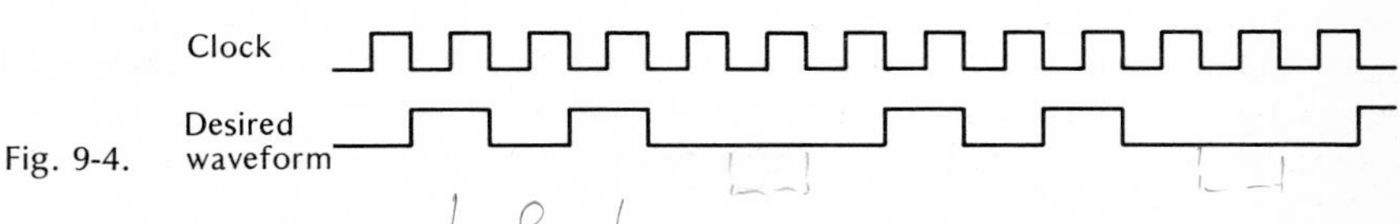

Fig. 9-4.

it is immaterial where the 1s are set initially. It is only necessary to ensure that they are spaced one flip-flop apart.

Example 9-1

How would you preset the ring counter in Fig. 9-3 to obtain a square-wave output which is one-half the frequency of the clock?

Solution

It is only necessary to preset a 1 in every other flip-flop while the remaining flip-flops are reset.

Example 9-2

Show how to use a five-stage ring counter in conjunction with a single flip-flop to form a decade counter.

Solution

The circuit diagram shown in Fig. 9-5*a* will operate as a decade counter if the counter is preset with a 1 in flip-flop *A* and the remaining flip-flops are reset. The waveforms and the decoding gates are shown in Fig. 9-5*b* and *c*. Notice that it requires only 10 two-input AND gates to decode the 10 waveforms.

9-3 A SHIFT COUNTER

neg. feedback

Since the feedback method applied to the shift register in the previous section was quite successful, it might be worthwhile to explore one more possibility. Suppose the true output of the last flip-flop is returned to the *K* input of the first flip-flop, and the false output is returned to the *J* input. This connection applied to a three-flip-flop shift register is shown in Fig. 9-6. Notice that the outputs of the shift register are first crossed and then returned to the inputs of the register. This configuration is sometimes referred to as *inverse* feedback.

Now assume that all flip-flops are in a reset condition and the clock is allowed to run. Since $\bar{C}$ is high and C is low, a 1 is set in flip-flop A during the first cycle of the clock. At the same time, B and C remain low since their J inputs are low and their K inputs are high.

During the second cycle of the clock, A remains high since $\bar{C}$ is still high and C is still low. At the same time, B is set high since A is now high and $\bar{A}$ is low. C remains unchanged since B is low during this period.

During the third clock period, A and B remain high and C is set high since B is now high. Thus after three cycles of the clock, all three flip-flops have been changed from the low state to the high state.

During the fourth clock period, $\bar{C}$ is low and C is high and A therefore is reset to the low state. B and C remain high.

During the fifth cycle of the clock, A remains low, B is reset low (since A is now low and $\bar{A}$ is high), and C remains high.

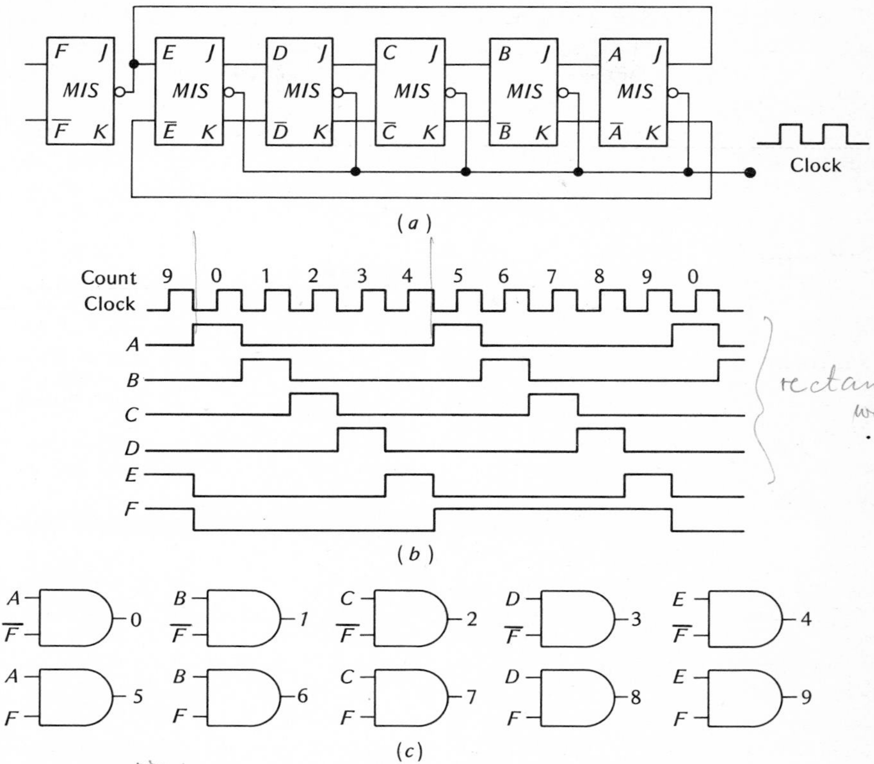

(a)

(b)

(c)

Fig. 9-5. Decade counter. (*a*) Logic diagram. (*b*) Waveforms. (*c*) Decoding gates.

Fig. 9-6. Three-stage shift register using inverse feedback to form a shift counter. (*a*) Logic diagram. (*b*) Waveforms.

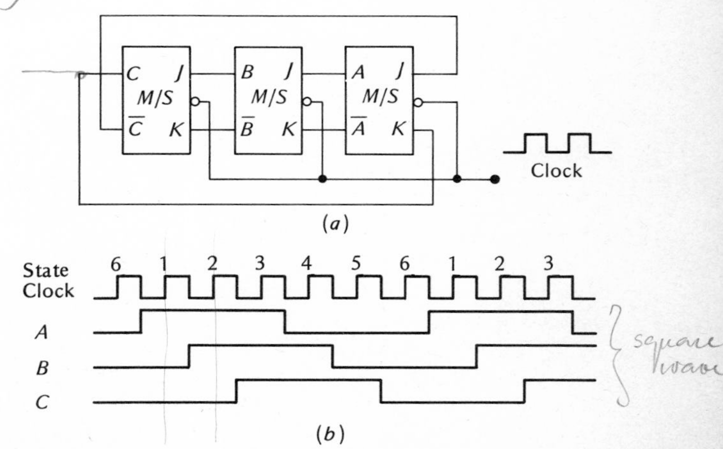

The sixth clock period returns the counter to the initial starting point since *C* is reset low while *A* and *B* both remain low. Thus this shift register with inverse feedback has progressed through a complete cycle of counts in six clock periods.

Examination of Fig. 9-6*b* shows that the waveform of each flip-flop is a square wave which is one-sixth the frequency of the clock. Moreover, all three flip-flop outputs are identical except that they are shifted with respect to one another by one clock period. This square wave apparently shifts through the flip-flops, advancing one flip-flop each time the clock goes low. Since the operation is cyclic and the waveforms shift through the flip-flops, this configuration is commonly called a "shift counter" (it is also called a "Johnson counter").

In order to investigate this counter more carefully, let us make a truth table showing the states through which the counter progresses. This truth table, made with the aid of the counter waveforms, is shown in Fig. 9-7. For easy comparison, a straight binary truth table for three flip-flops is also shown. Notice that the three-flip-flop shift counter counts through six discrete states. The six ordered states through which the shift counter progresses correspond to the binary counts 1-3-7-6-4-0. Thus the six states of the shift counter are indeed discrete and can be decoded.

Note, however, that the shift counter omits binary counts 2 (010) and 5 (101). Therefore, the shift counter must be examined to see whether or not it will work its way out of either of these two states, since it is possible that one of them may occur when power is first applied to the system. In fact, one of these two illegal states could occur during normal operation because of noise or some other malfunction.

First, suppose that the counter is in binary count 2 (010). The next time the clock goes negative, *A* goes high, *B* goes low, and *C* goes high. Thus the counter will advance to binary count 5 (101), which is the second illegal state.

During the second clock cycle, *A* goes low, *B* goes high, and *C* goes low. Thus the second clock period will advance the counter right back into the first illegal state, binary count 2 (010). Therefore, the counter will simply oscillate between the two illegal states and will not function properly.

To avoid this situation, some means is necessary to ensure that the counter cannot remain in one of the illegal states. One method of accomplishing this is to

Fig. 9-7. (*a*) Three-flip-flop shift-counter truth table. (*b*) Truth table for three flip-flops counting in a straight binary sequence.

C	*B*	*A*	State	Equivalent binary count
0	0	1	1	1
0	1	1	2	3
1	1	1	3	7
1	1	0	4	6
1	0	0	5	4
0	0	0	6	0
0	0	1	1	1

(*a*)

C	*B*	*A*	Count
0	0	0	0
0	0	1	1
0	1	0	2
0	1	1	3
1	0	0	4
1	0	1	5
1	1	0	6
1	1	1	7
0	0	0	0

(*b*)

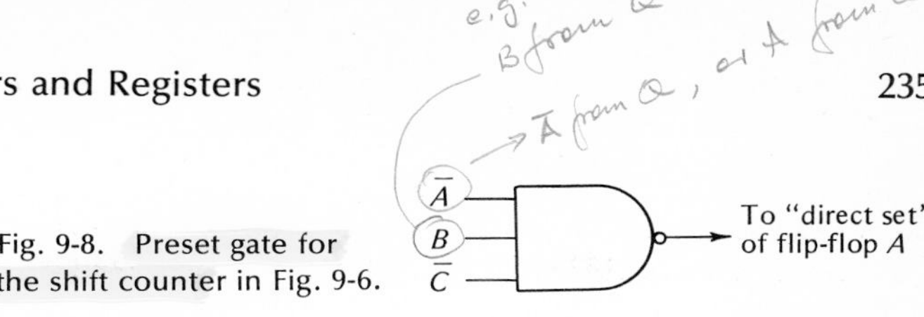

Fig. 9-8. Preset gate for the shift counter in Fig. 9-6.

use the NAND gate shown in Fig. 9-8. When $\bar{A}$, B, and $\bar{C}$ are all high (corresponding to the illegal state 010), the output of the NAND gate is low. If the NAND-gate output is applied to the *direct set* of flip-flop A, then A will be set high any time this condition occurs. Thus the counter will immediately advance from binary count 2 (010) to binary count 3 (011). This is the second state in the normal counting sequence, and the counter will therefore operate as desired (see Prob. 9-7).

Example 9-3

Draw the diagram for a four-flip-flop shift counter. Draw the waveforms for this counter. Using these waveforms, make a truth table showing the desired count

Fig. 9-9. Four-stage shift counter. (*a*) Logic diagram. (*b*) Waveforms. (*c*) Truth table.

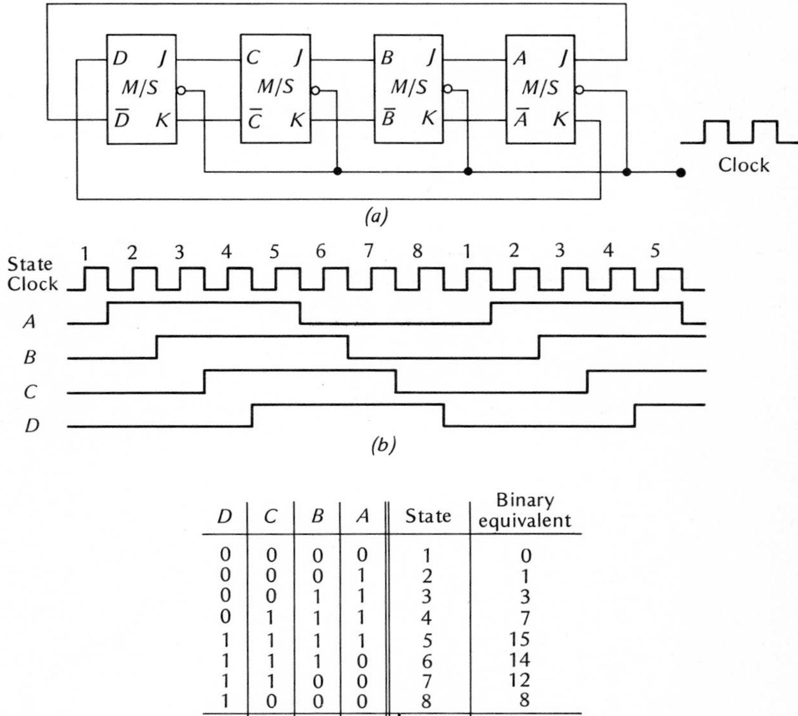

D	C	B	A	State	Binary equivalent
0	0	0	0	1	0
0	0	0	1	2	1
0	0	1	1	3	3
0	1	1	1	4	7
1	1	1	1	5	15
1	1	1	0	6	14
1	1	0	0	7	12
1	0	0	0	8	8
0	0	0	0	1	0

(*c*)

sequence. List the illegal states, and examine them to determine if the counter will get into an undesired mode of operation. If the counter does malfunction because of the illegal states, show a method for curing this problem.

Solution

The desired counter and waveforms are shown in Fig. 9-9*a* and *b*, respectively. The truth table, derived from the desired waveforms, is shown in Fig. 9-9*c*. Now, since this counter is constructed of four flip-flops, there are 16 possible states. In the desired mode of operation, the counter sequences through only eight states, and therefore eight illegal states must be examined. The eight illegal states correspond to the binary counts of 2, 4, 5, 6, 9, 10, 11, and 13, and they are shown in the table in Fig. 9-10. The two columns to the right of the double vertical line in the table show the new state, and the new binary equivalent count, to which the counter will progress after one cycle of the clock. For example, after one clock cycle, the counter will advance from binary count 2 (0010) to binary count 5 (0101). The table shows that if the counter comes up in any one of the eight illegal states, it will count through the binary sequence 2-5-11-6-13-10-4-9-2. Thus the counter will get into an undesired mode of operation and remain there. Notice that the counter still divides by 8, which was part of the original intent. However, to decode the counter it is necessary to have some means of forcing the counter back into the desired count, since the output waveforms in this secondary mode are quite different from

Fig. 9-10. (*a*) The eight illegal states for the counter shown in Fig. 9-9 and the states to which the counter will advance after one clock period. (*b*) Illegal-counting-sequence waveforms.

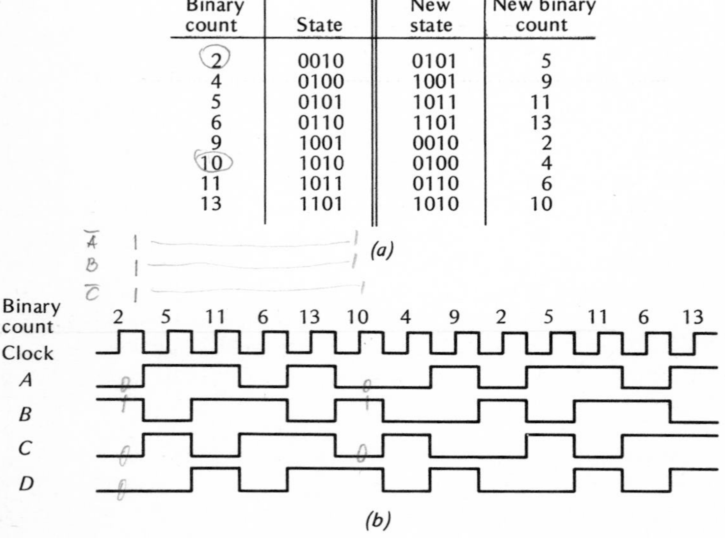

Binary count	State	New state	New binary count
2	0010	0101	5
4	0100	1001	9
5	0101	1011	11
6	0110	1101	13
9	1001	0010	2
10	1010	0100	4
11	1011	0110	6
13	1101	1010	10

(*a*)

(*b*)

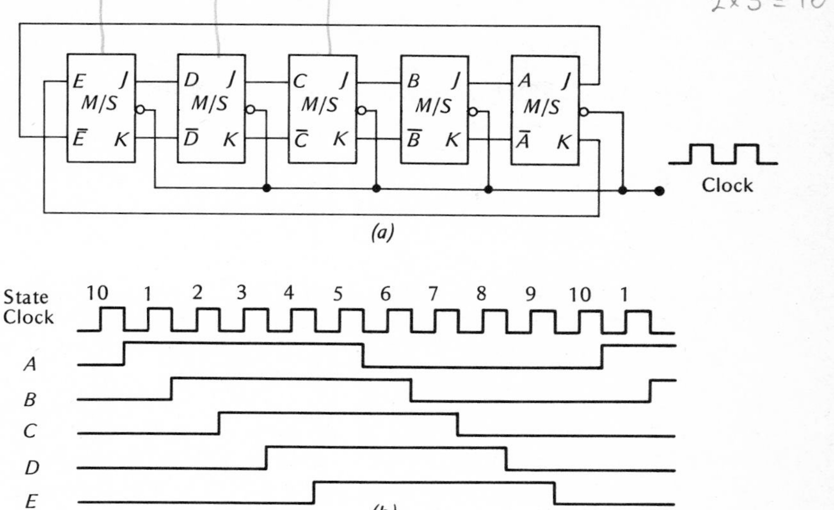

E	D	C	B	A	State	Binary equivalent
0	0	0	0	1	1	1
0	0	0	1	1	2	3
0	0	1	1	1	3	7
0	1	1	1	1	4	15
1	1	1	1	1	5	31
1	1	1	1	0	6	30
1	1	1	0	0	7	28
1	1	0	0	0	8	24
1	0	0	0	0	9	16
0	0	0	0	0	10	0
0	0	0	0	1	1	

(c)

Fig. 9-11. Five-flip-flop shift counter, decade counter. (a) Logic diagram. (b) Decade-counter waveforms. (c) Truth table showing equivalent binary states.

the desired mode. The waveforms for the illegal count sequence are shown in Fig. 9-10*b*. One method of forcing the counter into the desired mode can be found by observing that the condition $\bar{A}B\bar{C}$ is true twice during the undesired count sequence (count 2 and count 10). This condition is never true during the desired count sequence, so the NAND gate shown in Fig. 9-8 can be used to cure the problem. It is necessary, however, to connect the output of the NAND gate to *direct set* on flip-flops C and D. This forces the counter to state 6 whenever the output of the gate goes low (that is, when $\bar{A}B\bar{C}$ is high), and the counter will then be back into the desired count sequence.

A five-flip-flop shift counter can be constructed as shown in Fig. 9-11. This is a very useful form since it divides by 10 and can thus be used as a decade counter. The desired waveforms and corresponding truth table are shown in the figure. Since

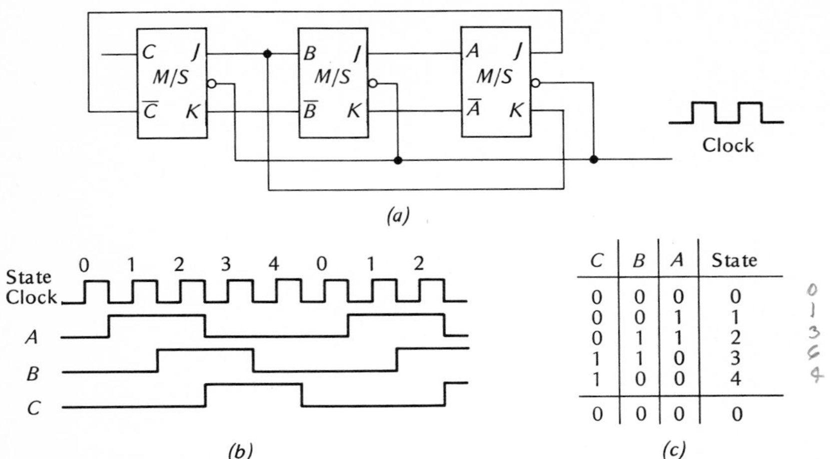

C	B	A	State
0	0	0	0
0	0	1	1
0	1	1	2
1	1	0	3
1	0	0	4
0	0	0	0

Fig. 9-12. Divide-by-5 shift counter. (*a*) Logic diagram. (*b*) Waveforms. (*c*) Truth table.

there are five flip-flops, there are a possible 32 states in which the counter can exist. The desired count sequence uses only 10 of these possible states, and there are therefore 22 illegal states.

It can be shown that (see Prob. 9-10) if the counter comes up in any one of these illegal states it will do one of two things. First, it will continue to divide the clock by 10, but it will advance through one of the following illegal count sequences: 2-5-11-23-14-29-26-20-8-17-2 or 4-9-19-6-13-27-22-12-25-18-4. Second, if the counter comes up in either count 10 or count 21, it will simply alternate back and forth between these two counts. Again, the cure to ensure that the counter operates in the proper sequence is the NAND gate shown in Fig. 9-8. The output of the gate can be connected to the *direct set* of flip-flops *C, D,* and *E* (see Prob. 9-11).

Note that *n* flip-flops can be used to divide the clock by 2*n*. That is, two flip-flops divide the clock by 4, three flip-flops divide the clock by 6, and so on. Alternatively, we can say that *n* flip-flops connected in the shift-counter configuration provide 2*n* discrete states through which the counter will progress. Thus it is possible to construct a counter of any even modulus by simply choosing the proper number of flip-flops and connecting them in the shift-counter configuration.

To construct odd-modulus counters from the basic shift-counter configuration is amazingly simple. The waveforms in Fig. 9-6 show that if *A* is reset low one clock period earlier, *B* is also reset one clock period earlier, as is *C*. This means that *A* and *B* and *C* are all high for only two clock periods but will still be low for three periods. Therefore, the counter is now dividing by 5, instead of 6, and there are five discrete states.

The method of accomplishing this is shown in Fig. 9-12, along with the appropriate waveforms and truth table. Since it is only necessary to cause *A* to go low one clock period earlier than before, this means that the *K* input to flip-flop *A* must go low one period earlier. Therefore, it is only necessary to obtain the *K* input to flip-flop A from B instead of C. Thus, any odd-modulus counter of count n can be made from an even counter of modulus $n + 1$ by simply removing the K input of the first flip-flop from the true output of the last flip-flop and connecting it to the true output of the next-to-last flip-flop.

This counter still has the two illegal counts 2 (010) and 5 (101), and the illegal count 7 (111). The additional count 7 will not cause any problem since the counter will advance naturally from 7 (111) to 6 (110), and this is state 4 in the desired counting sequence. Furthermore it can be seen that the counter will advance from 2 (010) to 5 (101) and then to 3 (011), which is state 2 in the desired counting sequence. Thus this counter has no permanent illegal modes (see Prob. 9-14).

9-4 A MOD-10 SHIFT COUNTER WITH DECODING

The decade counter in Fig. 9-11 has 10 discrete states; therefore, the counter can be decoded to form the 10 waveforms as was done in the last chapter (see Fig. 8-21). In decoding the binary decade counter in the last chapter, we needed 10 four-input AND gates, since there were four flip-flops in the counter. Then we would expect to use 10 five-input AND gates to decode the shift counter in Fig. 9-11, since there are five flip-flops. However, examination of the waveforms in this figure shows that a much simpler arrangement is possible.

Notice that state 1 can be uniquely determined as the time when A is high and B is low. Thus, it requires only a two-input AND gate to decode state 1. The inputs to this AND gate are A and $\overline{B}$, and the appropriate logic equation is $X_1 = A\overline{B}$.

Similarly, state 2 is the only time when B is high and C is low. The inputs to the gate to decode 2 are B and $\overline{C}$, and the appropriate logic equation is $X_2 = B\overline{C}$. The logic equations for the remaining states are found in a similar manner. These equations, along with appropriate decoding gates, are given in Fig. 9-13.

Thus, another of the advantages of using a shift counter is the fact that it requires only two-input AND gates to decode any of the individual counter states.

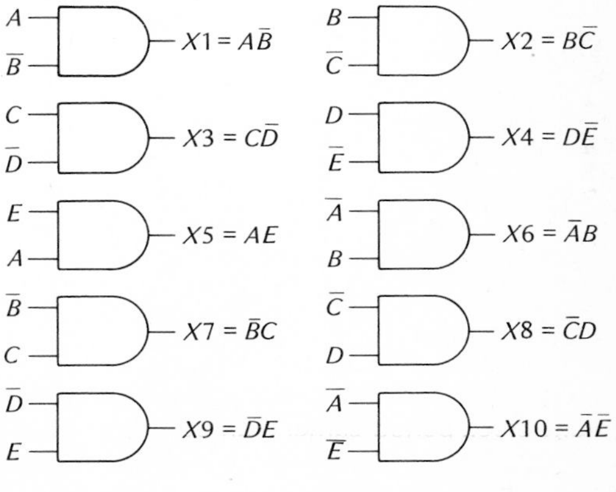

$X1 = A\overline{B}$ $X2 = B\overline{C}$
$X3 = C\overline{D}$ $X4 = D\overline{E}$
$X5 = AE$ $X6 = \overline{A}B$
$X7 = \overline{B}C$ $X8 = \overline{C}D$
$X9 = \overline{D}E$ $X10 = \overline{A}\overline{E}$

Fig. 9-13. Decoding gates for the counter in Fig. 9-11.

Example 9-4

The decade counter in Fig. 9-11 has been constructed and its outputs are available. It is necessary to develop the waveform shown in Fig. 9-14*a* to be used as a control signal. Show the logic circuitry necessary to provide this control signal.

Solution

The desired control signal corresponds to a high level during states 1, 3, and 5, and low at all other times. The most obvious solution is to use the three AND gates shown in Fig. 9-13 for decoding these states. If the outputs of these three AND gates are used as the inputs to an OR gate, as shown in Fig. 9-14*b*, the control signal *X* will be produced at the output of the OR gate.

If the control signal *X* is required to drive a number of circuits, the output circuit which produces *X* must have power available to drive these circuits. A slightly different method for developing the desired waveform *X* is shown in Fig. 9-14*c*. The outputs of NAND gates 1, 3, and 5 will be low only during states 1, 3, and 5, respectively. Any time 1 or 3 or 5 is low, the output *X* must be high, and thus the desired control signal is developed. The output NAND gate which produces *X* can then be chosen such that it has the power capability to drive the desired number of circuits.

9-5 A DIGITAL CLOCK

A very interesting application of counters and decoding arises in the design of a digital clock. Suppose we want to construct an ordinary clock which will display hours, minutes, and seconds. The power supply for this system is the usual 120 volts ac, 60-Hz commercial power. Since the 60-Hz frequency of most power

Fig. 9-14. (*a*) Clock and desired control signal. (*b*) Method for developing control signal *X* using AND gates and one OR gate. (*c*) Method for developing control signal *X* using NAND/NOR gates.

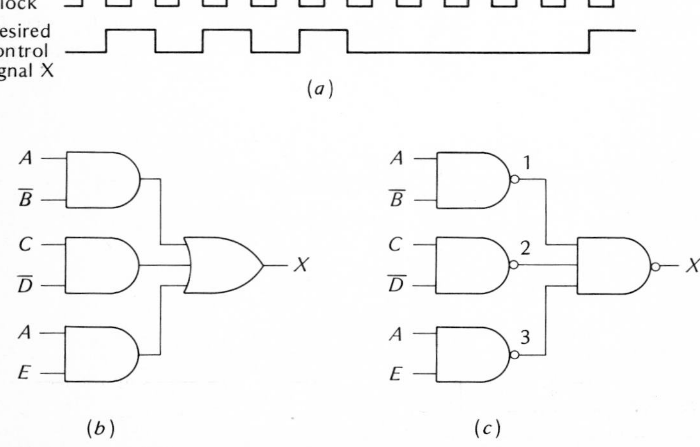

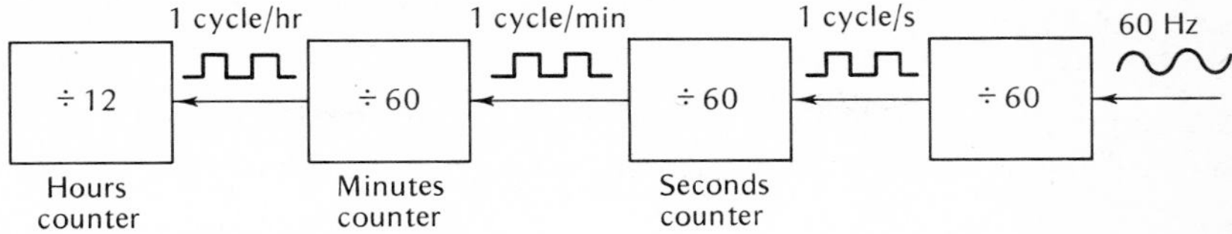

Fig. 9-15. Digital-clock block diagram.

systems is very closely controlled, it is possible to use this signal as the basic clock frequency for our system.

In order to obtain pulses occuring at a 1-s rate (or alternatively a 1-Hz square wave) it is necessary to divide the 60-Hz power source by 60. If this 1-Hz square wave is again divided by 60, a 1 cycle per minute square wave is the result. Dividing this signal by 60 then provides a 1 cycle per hour square wave. This, then, is the basic idea to be used in forming a digital clock.

A block diagram showing the functions to be performed is given in Fig. 9-15. The first ÷60 counter simply divides the 60-Hz power signal down to a 1-Hz square wave. The second ÷60 counter changes state once each second and has 60 discrete states, and it can therefore be decoded to provide signals to display seconds. This counter is then referred to as the seconds counter. The third ÷60 counter changes state once each minute and has 60 discrete states. It can therefore be decoded to provide the necessary signals to display minutes. This counter is then the minutes counter. The last counter changes state once each 60 min (or once per hour). Thus, if it is a ÷12 counter it will have 12 states, and it can be decoded to provide signals to display the hour. It is therefore the hours counter.

As you know, there are a number of ways to implement a counter. What is desired here is to design the counters in such a way as to minimize the hardware required. The first counter must divide by 60, and it need not be decoded. Therefore, it should be constructed in the easiest manner with the minimum number of flip-flops.

To obtain 60 states requires at least six flip-flops, since $2^6 = 64$. The simplest configuration is a straight binary ripple counter, and this first ÷60 counter will therefore be a simple binary ripple counter using six flip-flops. Six flip-flops provide 64 states, and we only want a total of 60 states. It is therefore necessary to provide feedback to skip four states. The counter to provide this function is shown in Fig. 9-16.

Fig. 9-16. ÷60 ripple counter to provide a 1-Hz square wave.

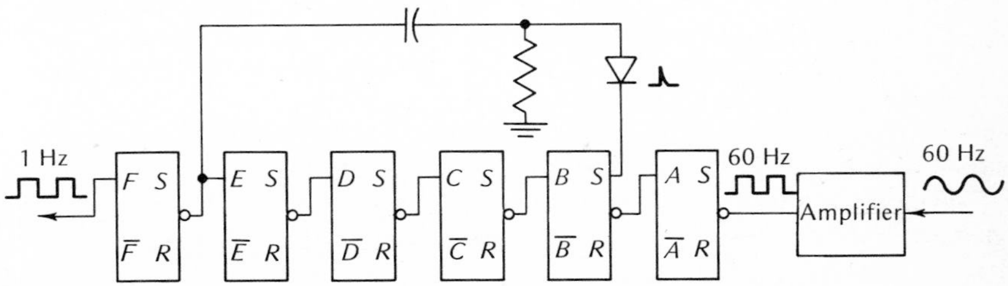

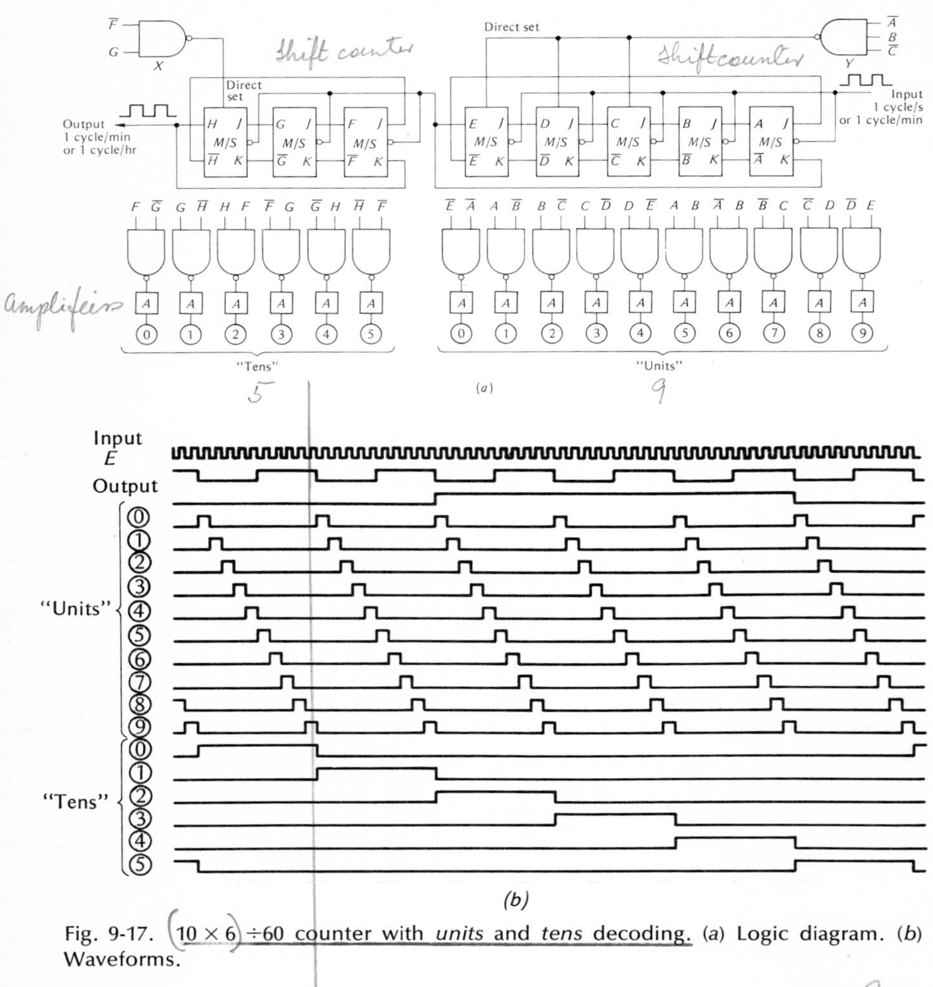

Fig. 9-17. 10 × 6 ÷60 counter with *units* and *tens* decoding. (*a*) Logic diagram. (*b*) Waveforms.

The amplifier at the input of the counter provides a 60-Hz square wave of the proper amplitude to drive the first flip-flop. The required feedback is from flip-flop *E* to flip-flop *B*. In the normal counting sequence without feedback, the counter progresses through 64 states, and *E* goes high twice during the sequence. Thus if *E* is fed back to the *S* input of flip-flop *B*, two counts are added to the count sequence each time *E* goes positive. Since this happens twice during a complete cycle, a total of four counts are added to the counter during one complete cycle. Thus the counter essentially skips four states and has a total of 60 states.

The second counter in the system must also divide by 60 and could be implemented in the same way. However, the seconds counter must be decoded, and to decode this binary ripple counter would require 60 six-input AND gates. This is a total of 360 diodes! There must be a better way.

Notice that $60 = 2 \times 2 \times 3 \times 5$. We can therefore use counters of these moduli, or products of these moduli, to construct the desired ÷60 counter. We are interested in decoding this counter to represent each of the 60 seconds in 1 minute. This can most easily be accomplished by constructing a mod-10 counter in series with a mod-6 counter to form the ÷60. The mod-10 counter can then be decoded to represent the *units* digit of seconds, and the mod-6 counter can be decoded to represent the *tens* digit of seconds. These two counters could be constructed by using feedback around ripple counters to obtain the correct counts. One such con-

Fig. 9-18. Logic diagram and waveforms for the ÷12 hours counter.

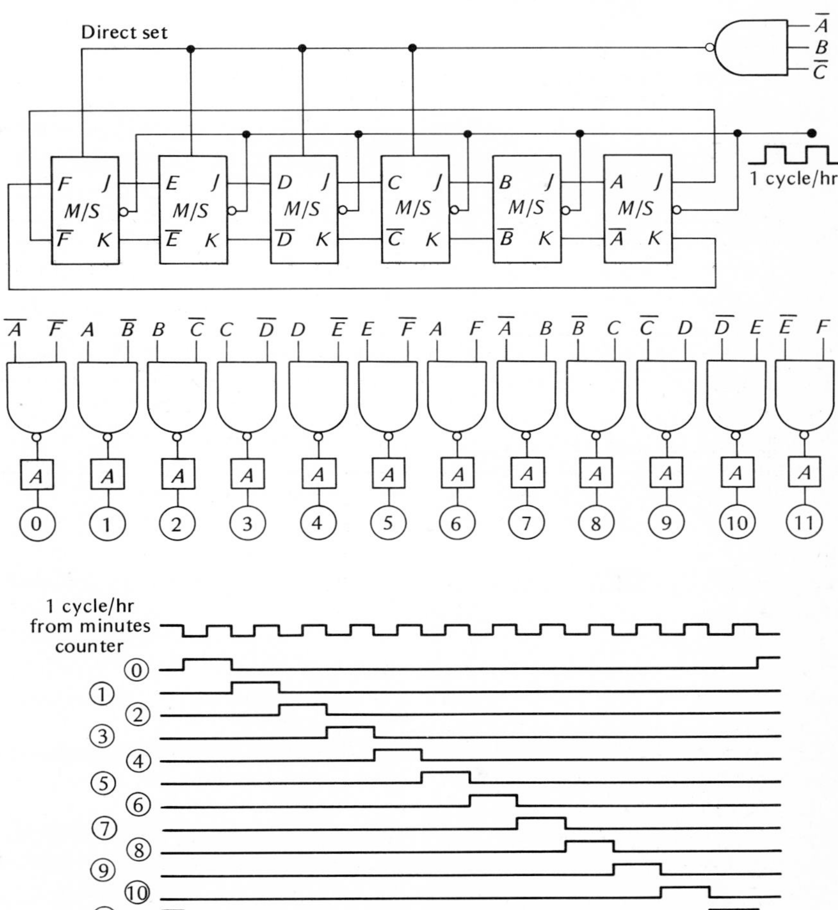

figuration requires 48 diodes for decoding. On the other hand, in the previous section it was shown that one of the more important advantages of a shift counter is the reduction in diodes necessary to decode the counter. Therefore, the seconds counter will be formed by connecting serially a mod-10 shift counter and a mod-6 shift counter. The entire counter is shown in Fig. 9-17.

Decoding of the counters is accomplished as discussed in the previous section, and the proper decoding gates are shown below the counters. A total of 32 diodes are required for decoding this counter. Thus, the shift-counter configuration requires one more flip-flop but provides a one-third reduction in decoding diodes. The amplifiers at the output of each of the gates provide the power necessary to drive the lamps, which are represented by the circles with numbers in them. The waveforms in the counter are shown in the figure, and the operation of the counter can be verified by examining these waveforms.

For example, if G and $\overline{H}$ are high and F is low, the lamp representing 1 in the *tens* column will be lit; if at the same time, B and $\overline{C}$ are high (all others are low) the lamp representing 2 in the *units* column will be lit. Thus, in this state the counter will display 12 s. NAND gates X and Y ensure that the counter always operates in the desired mode (see Prob. 9-16). The minutes counter is exactly the same as the seconds counter, except it is driven by the 1 cycle per minute square wave from the output of the seconds counter, and its output is a 1 cycle per hour square wave. The 16 output lamps now, of course, represent *units* and *tens* of minutes.

The ÷12 hours counter must be decoded into 12 states to display the hours. Again, for ease of decoding, a six-flip-flop shift counter will be used. The complete counter, along with the proper decoding gates and waveforms, is shown in Fig. 9-18.

The complete block diagram for the digital clock is shown in Fig. 9-19. Notice that the entire counter could have been constructed by using simple ripple counters throughout, and this would require only 22 flip-flops. This, however, would require over 700 diodes to decode the required states. Thus, the six extra flip-flops used to implement the shift counters seem well justified, since it now requires only 88 diodes for decoding. And the clock is much simpler to construct.

Finally, some means must be found to set the clock; for when the power is turned off and then turned back on again the flip-flops will turn on in random states. Setting the clock can be quite easily accomplished by means of the *set* push buttons shown in Fig. 9-19. Depressing the *set hours* button causes the hours counter to advance at a 1-s rate, and thus this counter can be set to the desired hour. The minutes counter can be similarly set by depressing the *set minutes* button. Depressing the *set seconds* button removes the signal from the seconds counter, and the clock can thus be brought into synchronization.

By means of large-scale integration (LSI), it is possible to construct a digital clock entirely on one semiconductor chip. Such units are commercially available, and they perform essentially the function shown in the logic diagram in Fig. 9-19 (the indicator lamps are, of course, separate). One such commercially available LSI digital clock comes in a 24-pin dual in-line package measuring 0.54 in by 1.25 in.

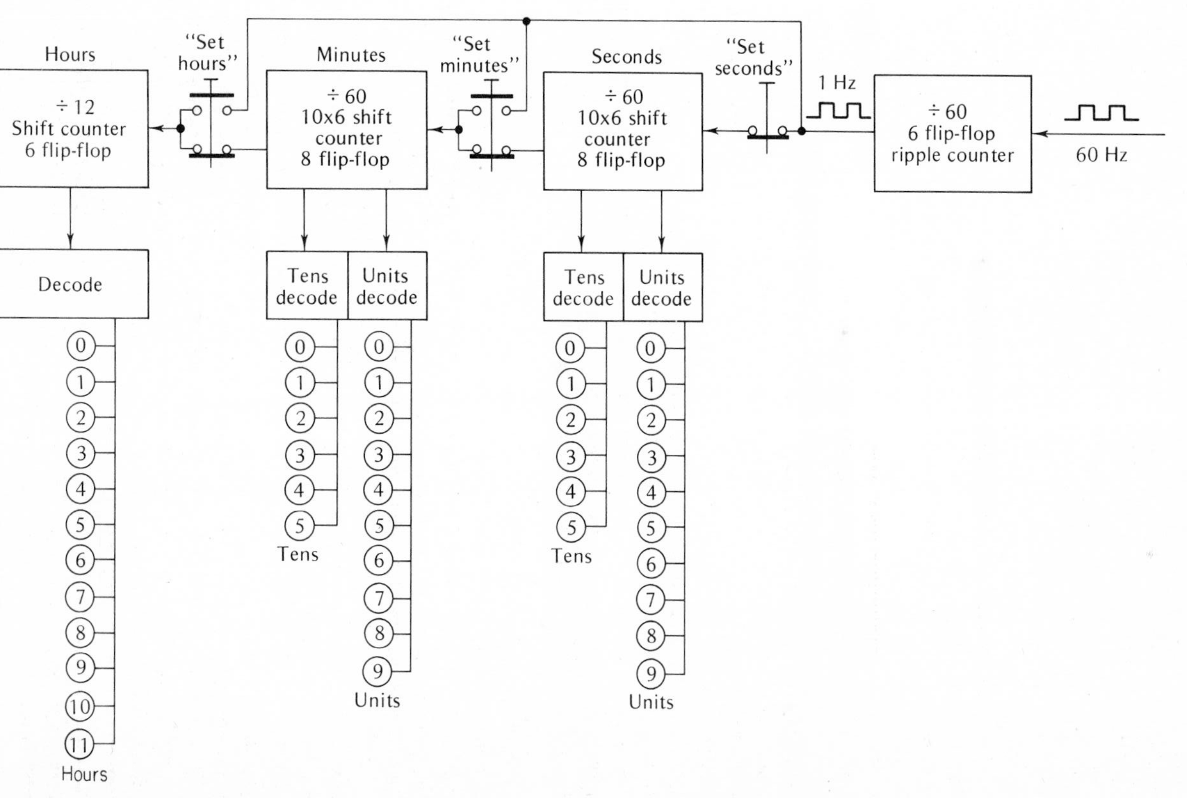

Fig. 9-19. Digital-clock logic diagram.

9-6 UP-DOWN COUNTER

Up to this point, we have considered a number of different counters, all of which count in an upward sequence. That is, the count sequences are 0-1-2-3-4 . . . *n*-0. It is sometimes useful to have a counter which can count in a downward sequence. Any of the binary counters previously discussed can be used to implement a down counter.

The three-flip-flop ripple counter discussed in the previous chapter is redrawn in Fig. 9-20. The waveforms for the false sides of the flip-flops have been added, along with a new truth table. The normal count sequence, 0-1-2-3-4-5-6-7, is shown under *count* in the table, and the corresponding state of the true side of each flip-flop is shown under *A*, *B*, and *C*. The false sides of the flip-flops are simply the negatives of the true sides, and these are shown under $\overline{A}$, $\overline{B}$, and $\overline{C}$. The binary states represented by $\overline{A}$, $\overline{B}$, and $\overline{C}$ are shown under $\overline{count}$.

Notice that the count sequence under $\overline{count}$ progresses in a downward fashion (7-6-5-4-3-2-1-0). Thus this counter could be decoded to provide a count-down sequence of waveforms. The proper decoding gates and the count-down waveforms are shown in Fig. 9-21.

Notice that these decoding gates are exactly the same set of gates that would be required to produce a set of count-up waveforms. It is only necessary to change the number labels on the output of each gate (i.e., 7 → 0, 6 → 1, 5 → 2, 4 → 3, 3 → 4, 2 → 5, 1 → 6, and 0 → 7). Thus we have not truly formed a down counter yet; we have simply produced a set of decoded waveforms which correspond to a down count by rearranging the decoding gates.

A true down counter can be formed by triggering the input of each flip-flop with

Fig. 9-20. Three-flip-flop ripple counter. (*a*) Logic diagram. (*b*) Waveforms. (*c*) Truth table.

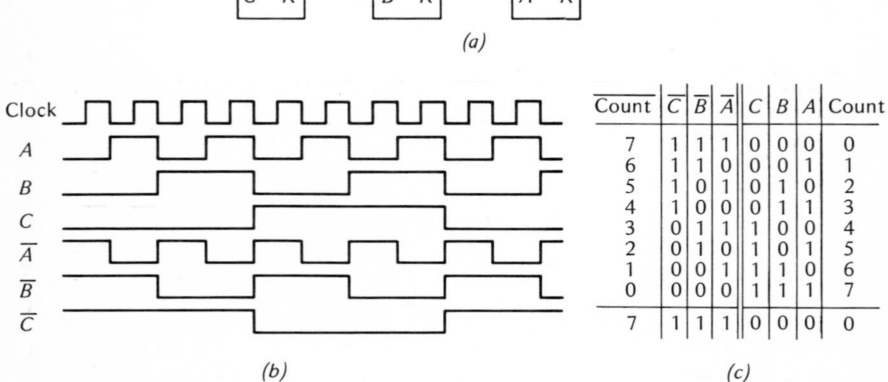

$\overline{Count}$	$\overline{C}$	$\overline{B}$	$\overline{A}$	C	B	A	Count
7	1	1	1	0	0	0	0
6	1	1	0	0	0	1	1
5	1	0	1	0	1	0	2
4	1	0	0	0	1	1	3
3	0	1	1	1	0	0	4
2	0	1	0	1	0	1	5
1	0	0	1	1	1	0	6
0	0	0	0	1	1	1	7
7	1	1	1	0	0	0	0

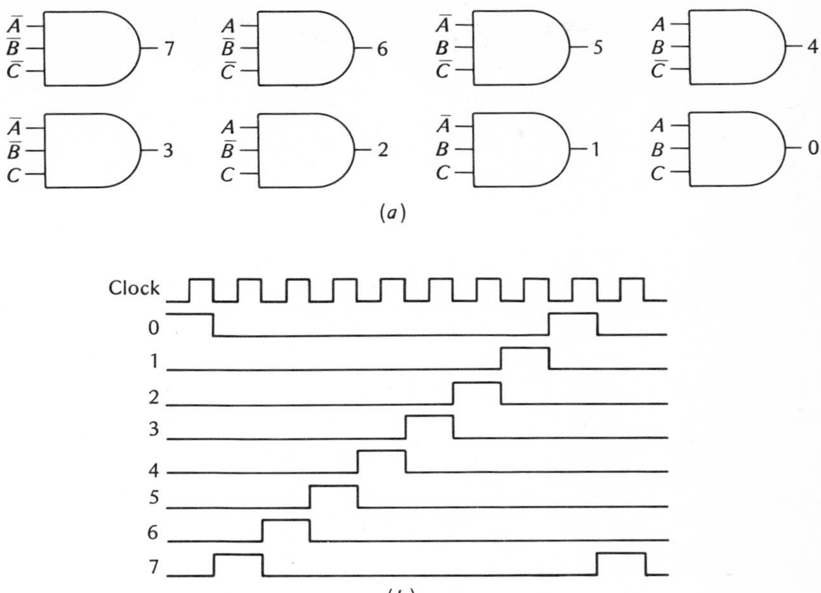

Fig. 9-21. (*a*) Decoding gates to provide count-down waveforms. (*b*) Count-down waveforms.

the false side of the previous flip-flop instead of the true side. Such a counter is shown in Fig. 9-22. Notice that A changes state each time the clock goes low, as before. However, B now changes state when A goes high since this is the time when $\bar{A}$ goes low. Similarly, C changes state when B goes high, since this is the time when B goes low.

Fig. 9-22. Down counter. (*a*) Logic diagram. (*b*) Waveforms.

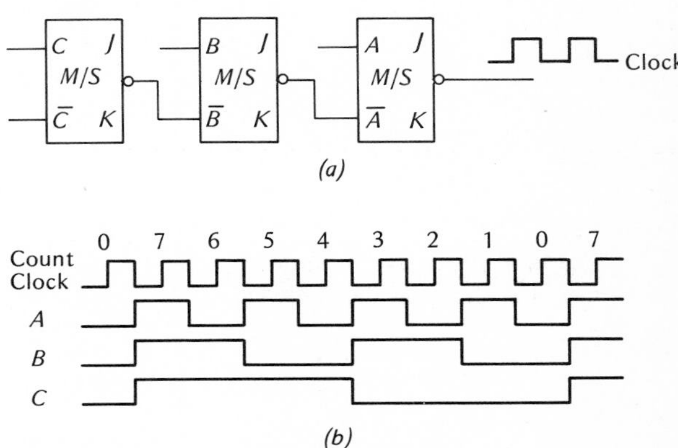

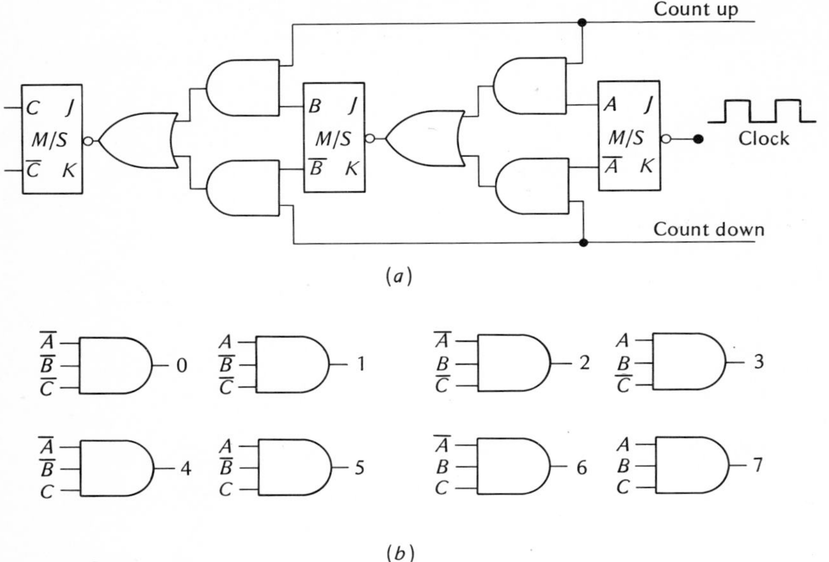

Fig. 9-23 Up-down counter. (a) Logic diagram. (b) Decoding gates.

In order to unify this idea, a three-flip-flop *up-down counter* is shown in Fig. 9-23. In order that the counter will progress through a count-up sequence, it is necessary to trigger each flip-flop with the true side of the previous flip-flop. If the *count-up* line is high and the *count-down* line is low, this will be the case, and the counter will progress through the waveforms shown in Fig. 9-23*c*.

In order to cause the counter to progress through a count-down sequence, it is necessary to hold the *count-up* line low and the *count-down* line high. This causes each flip-flop to be triggered from the false side of the previous flip-flop and causes the counter to progress through the count-down sequence in Fig. 9-23*d*.

This process can be continued to other flip-flops down the line to form an up-down counter of larger moduli. It should be noted, however, that the gates introduce additional delays that must be taken into account when determining the maximum rate at which the counter can operate.

Example 9-5

Show how to implement the up-down counter shown in Fig. 9-23 using NAND/NOR gates instead of AND gates and OR gates.

Solution

The necessary control gate can be implemented quite simply by using the configuration shown in Fig. 9-24. When the *count-down* line is low and the *count-up* line is high, gate *X* is disabled and gate *Y* is enabled. Therefore, *A* will not pass

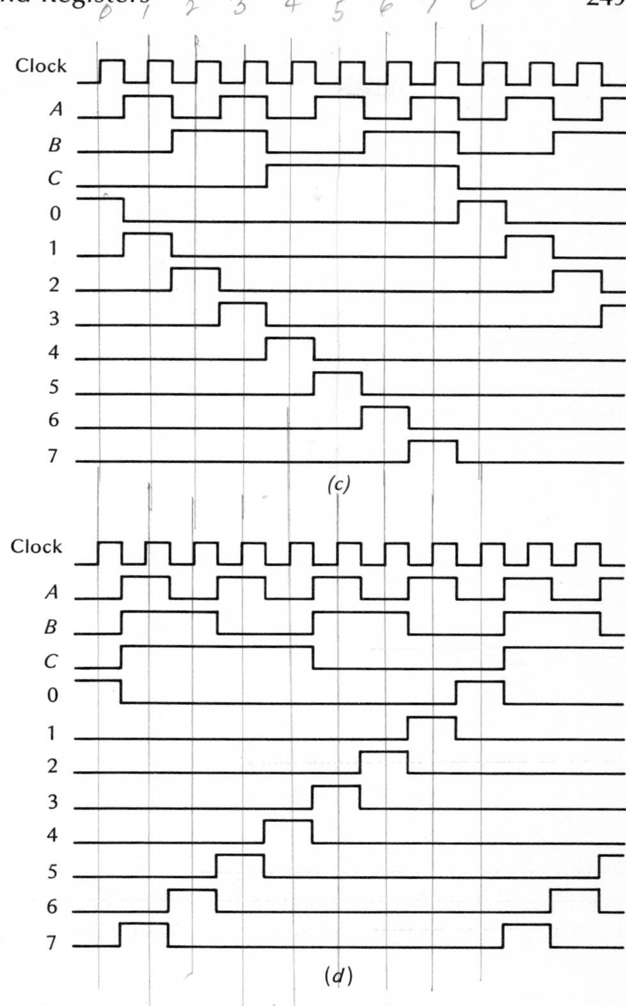

Fig. 9-23 (continued). Up-down counter. (c) Count-up waveforms. (d) Count-down waveforms.

through X, but $\overline{A}$ will pass through Y. In passing through gate Y, $\overline{A}$ inverted and thus appears at the input of the next flip-flop as A. With the *count-down* line high and the *count-up* line low, Y is disabled and X is enabled. Thus A is inverted as it passes through gate X and appears at the input of the next flip-flop as $\overline{A}$.

Care should be exercised in ORing the outputs of NAND/NOR gates (connecting the outputs together) in this fashion, since some NAND/NOR gates have active pull-up transistors. This means that if one of the gates is on while the other is off, there is nearly a direct short between the supply voltage and ground.

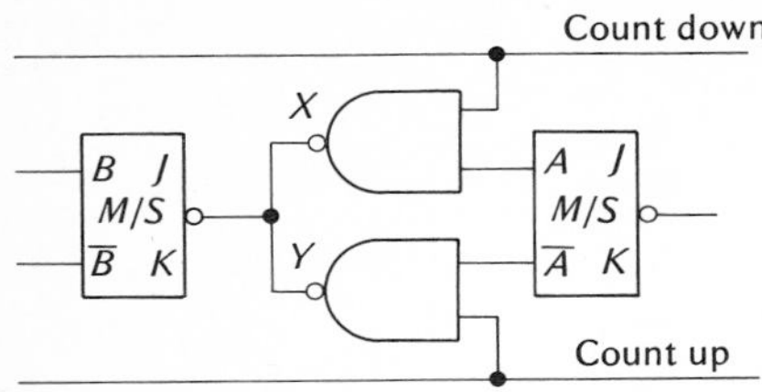

Fig. 9-24. Up-down counter control using NAND/NOR gates.

A *parallel up-down counter* can be formed in a similar manner, as shown in Fig. 9-25. This counter works in an inhibit mode, since the flip-flops change state when their *J* and *K* inputs are high, and do not change state when these inputs are low. Recall that in a parallel counter the time at which any flip-flop changes state is determined by the states of all previous flip-flops in the counter. To determine the logic necessary to implement this counter, it is convenient to refer to the truth table shown in Fig. 9-25*b*.

A is required to change state each time the clock goes low, and flip-flop *A* therefore has both its *J* and *K* inputs held in a high state. This is true in both the count-up and count-down modes, and therefore no other logic is necessary for this flip-flop.

In the count-up mode, *B* is required to change state each time *A* goes from a 1 to a 0. Any time the *count-up* line and *A* are both high, the output of gate X_1 is low. Any time either of the inputs to Z_1 is low, its output is high. Therefore, the *J* and *K* inputs to flip-flop *B* are high whenever both *count-up* and *A* are high; *B* then changes state each time *A* goes from a 1 to a 0.

In the count-down mode, *B* must change state each time *A* goes from a 0 to a 1. The output of gate Y_1 is low, and therefore the *J* and *K* inputs to flip-flop *B* are high, any time $\overline{A}$ and *count-down* are high. Thus, in the count-down mode, *B* changes state every time *A* goes from a 0 to a 1.

In the count-up mode, *C* is required to change state every time both *A* and *B* change from a 1 to a 0 (i.e., the transitions 3 to 4, 7 to 8, 11 to 12, and 15 to 0). The output of gate X_2 is low any time both *A* and *B* are high and the *count-up* line is high. Therefore, the *J* and *K* inputs to flip-flop *C* are high during these times and *C* changes state during the desired transitions.

In the count-down mode, *C* is required to change state whenever both *A* and *B* change from a 0 to a 1. The output of gate Y_2 is low any time both $\overline{A}$ and $\overline{B}$ are high and the *count-down* line is high. Thus, the *J* and *K* inputs to flip-flop *C* are high during these times and *C* then changes state during the required transitions.

In the count-up mode, *D* must change state every time *A* and *B* and *C* change from a 1 to a 0. The output of gate X_3 is low, and thus the *J* and *K* inputs to flip-flop *D* are high, any time *A* and *B* and *C* and *count-up* are all high. Thus *D* changes states during the desired transitions.

In the count-down mode, *D* must change state any time *A* and *B* and *C* go from a 0 to a 1. The output of gate Y_3 is low, and thus the *J* and *K* inputs to flip-flop *D* are high, whenever $\overline{A}$ and $\overline{B}$ and $\overline{C}$ and *count-down* are high. Thus *D* changes state during the desired transitions.

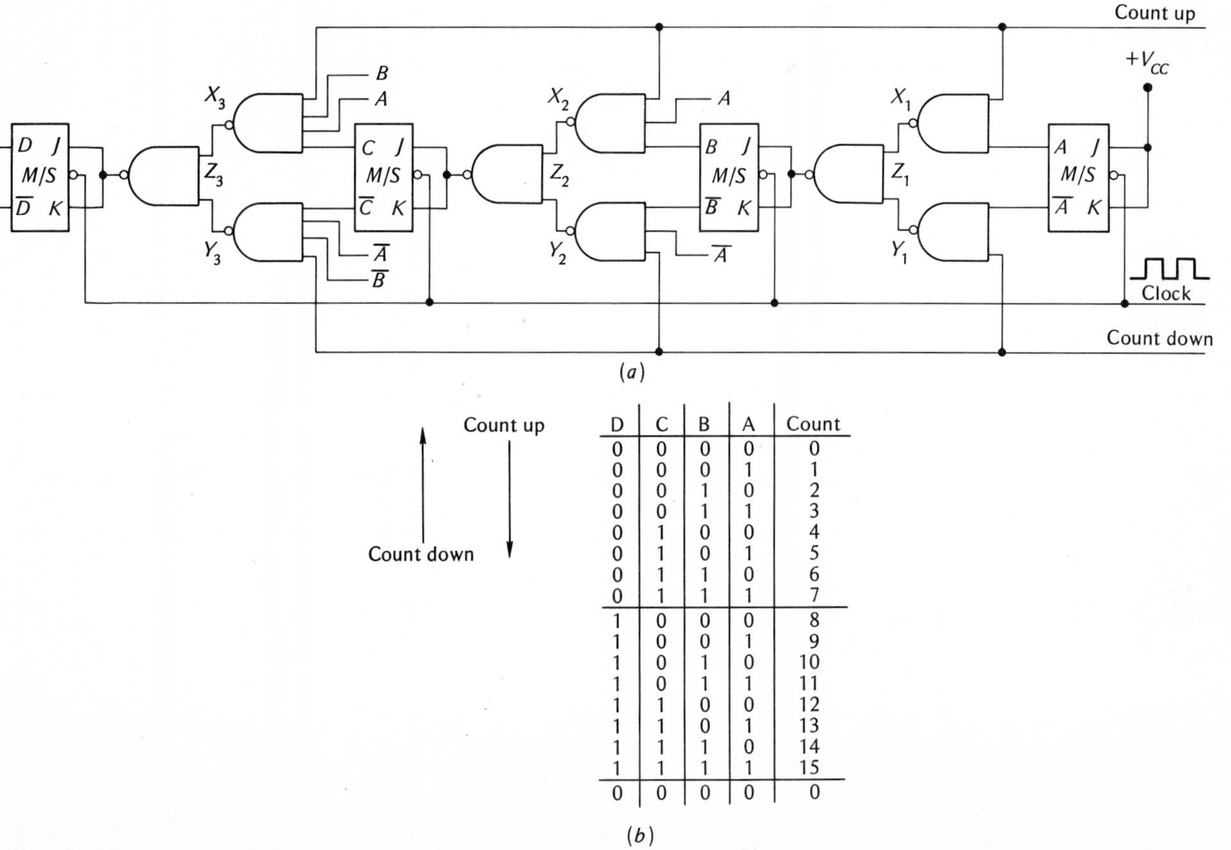

D	C	B	A	Count
0	0	0	0	0
0	0	0	1	1
0	0	1	0	2
0	0	1	1	3
0	1	0	0	4
0	1	0	1	5
0	1	1	0	6
0	1	1	1	7
1	0	0	0	8
1	0	0	1	9
1	0	1	0	10
1	0	1	1	11
1	1	0	0	12
1	1	0	1	13
1	1	1	0	14
1	1	1	1	15
0	0	0	0	0

(b)

Fig. 9-25. Parallel up-down counter. (a) Logic diagram. (b) Truth table.

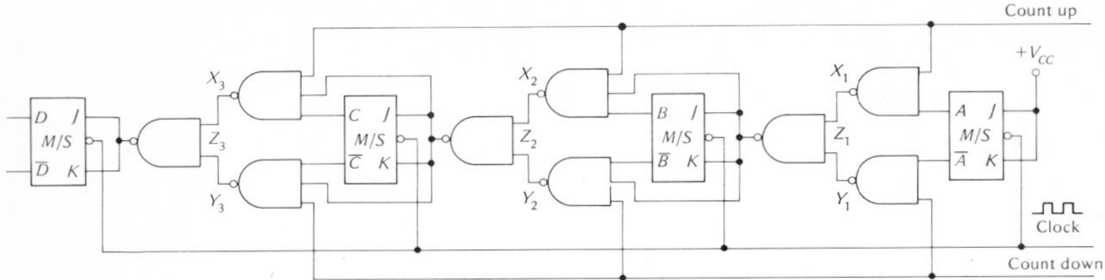

Fig. 9-26. Up-down counter with more efficient gating arrangement.

This process can be continued for *n* flip-flops to make larger-modulus counters. However, notice that the *X* and *Y* gates to each flip-flop require more and more inputs. In general, flip-flop *n* would be preceded by *X* and *Y* gates requiring *n* inputs.

This problem can be minimized by noting that a part of the information required in any one gate is present in the corresponding gate in the preceding flip-flop. For example, the required inputs to gate X_3 in Fig. 9-25*a* are *A, B, C,* and *count-up*. The information *A, B,* and *count-up* is present at the output of gate X_2, but it is of the wrong polarity. However, it appears with the proper polarity at the output of gate Z_2. Therefore, the output of Z_2 could be used at the input of X_3 in place of *A* and *B*. Similarly, Z_2 can be used at the input of Y_3 in place of $\overline{A}$ and $\overline{B}$. In this fashion, all the remaining gates down the line can be limited to three inputs.

The same counter, using this configuration, is shown in Fig. 9-26. Notice that the *count-up* line must still be used in every *X* gate since the *Z* outputs will be high at some time during both the up-count and down-count modes. For the same reason, the *count-down* line must be used in every *Y* gate.

9-7 SHIFT-REGISTER OPERATIONS

In the first section in this chapter, a basic shift register and the method for serially shifting a binary number into the register were discussed. This method requires one cycle of the clock for each bit in the word and is therefore relatively slow. The transfer of information into the register could be accomplished in a much shorter period of time if it were carried out in a parallel manner. This can be accomplished using the logic shown in Fig. 9-27. Each bit is shifted into the proper flip-flop by means of the AND gates shown. The 2^0 bit is shifted into *A*, the 2^1 bit is shifted into *B*, and so on, and the 2^n bit is shifted into flip-flop n - 1.

Entering data into the register requires two clock cycles, since the register must first be reset. Immediately after being reset, the shift pulse goes high, and if any of the data inputs to the AND gates (2^0, 2^1, 2^2, etc.) are high, the *set* input to the corresponding flip-flops will be high and a 1 will be set into these flip-flops. If any data input is low, that particular flip-flop will remain in the reset state. The proper control waveforms for this shifting operation are shown in Fig. 9-27*b*.

A similar method for entering information into a register is shown in Fig. 9-28. This method is twice as fast, since it requires no reset pulse. It does, however,

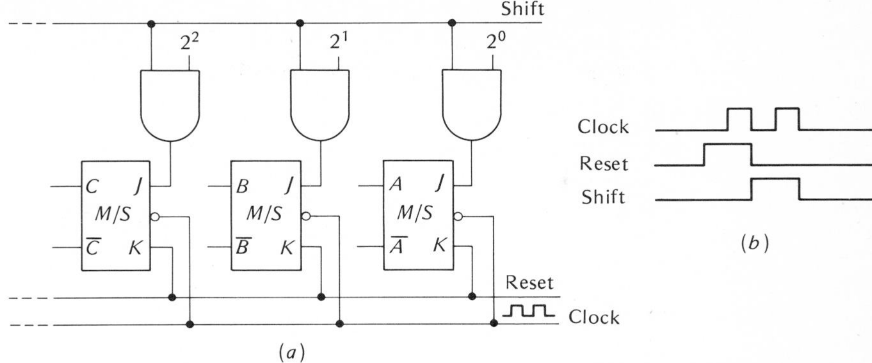

Fig. 9-27. (a) Shifting information into a shift register in parallel form. (b) Required control waveforms.

require twice the number of gates. The information is shifted into the flip-flops by means of the AND gates exactly as before. For example, if 2^0 is high, then $\overline{2^0}$ must be low. Therefore, the *J* input to flip-flop *A* must be high and the *K* input must be low, and *A* is then set high. Conversely, if 2^0 is low and $\overline{2^0}$ is high, *A* must be set low. The proper information will be shifted into each flip-flop regardless of the previous states of the flip-flops. Thus no reset pulse is required, and it requires only one clock period to shift the information into the register. Notice that the clock can continue to run since, when the *shift* and *reset* lines are both low, the flip-flops are prevented from changing state.

You will recall from Chap. 2 that binary arithmetic is sometimes carried out using the 1's complement of a number. It is quite easy to form the 1's complement in the register shown in Fig. 9-28. It is only necessary to hold both the *J* and *K* inputs to each flip-flop high for one cycle of the clock. Since the *master/slave* flip-flop will simply change state (complement) when both inputs are high, the contents of the register will be complemented. The register will then contain the 1's complement of the prior number.

Fig. 9-28. Method for shifting data into a register in parallel in one clock period.

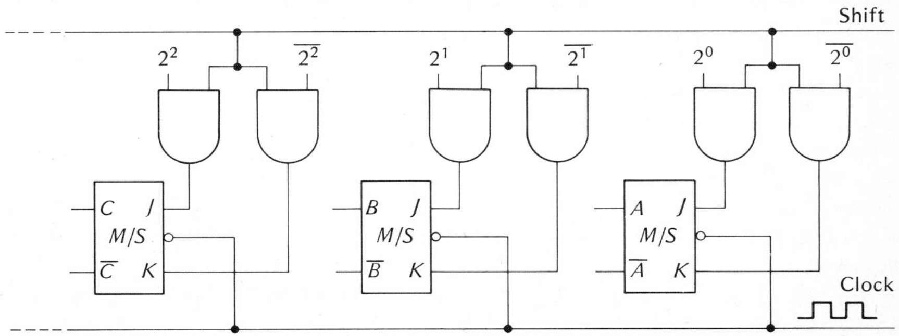

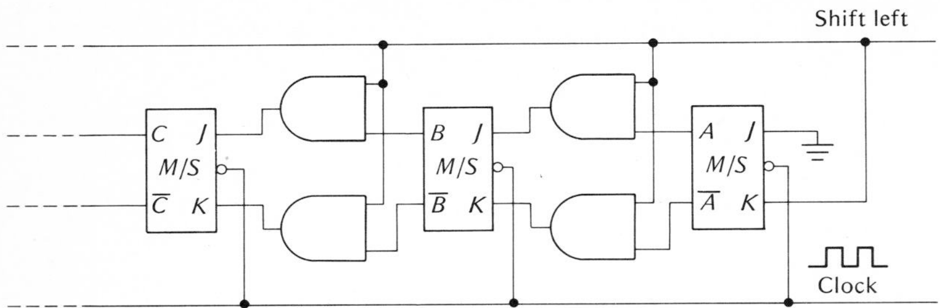

Fig. 9-29. Shift-left logic.

Two other very useful operations which can be performed with the basic shift register are *shift right* and *shift left*. The *shift left* operation can be performed by the register shown in Fig. 9-29 by simply holding the *shift left* line high for one cycle of the clock.

In this register, shifting the data left one place is equivalent to multiplying by 2. For example, suppose the register contained the number $011_2 = 3_{10}$. If the contents of the register are shifted left one place, the register now contains $110_2 = 6_{10}$. Thus the original number stored in the register has been effectively multiplied by 2.

As another example, suppose the number $101_2 = 5_{10}$ is stored in the register and the *shift left* operation is performed. After the shift left, the number $010_2 = 2_{10}$ is in the register. Something has gone wrong, since 2 is certainly not 2×5. Recall that a 3-bit register is capable of storing numbers from 0_{10} to 7_{10}. This defines the full capacity of the register. In the previous example, we were trying to double the number 5_{10}, which would have given 10_{10}. But this exceeds the capacity of the register, and we therefore obtained an erroneous result. What actually happened was that the MSB was shifted out of the left end of the register and lost. If there were one more flip-flop to accept this bit, which represents 8_{10}, the register would then contain the number $1010_2 = 10_{10}$. Thus, in the shifting operation it is always necessary to be aware of the capacity of the register.

In order to eliminate the necessity for stopping the clock, and to provide for a proper *shift left* control, the logic configuration shown in Fig. 9-29 can be used to shift left. The contents of the register will be shifted left one place each time the clock completes one cycle and the *shift left* line is high. Thus, shifting left two places can be accomplished by holding the *shift left* line high for two clock periods, and so on.

Notice that shifting left two places is equivalent to multiplying by 4, three places by 8, etc. Notice also that *A* must be set low for any *shift left* operation. This is true since only 0s can be shifted into the lower significant-bit positions, for if 1s were allowed the resulting number in the register would be more than twice the previous number. The *J* input to flip-flop *A* is shown grounded, but it is actually held low by other logic gates in the register (e.g., the gate used to enter information into the register serially).

The *shift right* operation is just the opposite of shifting left and results in a division by 2 of the number stored in the register. Shifting right two places results in division

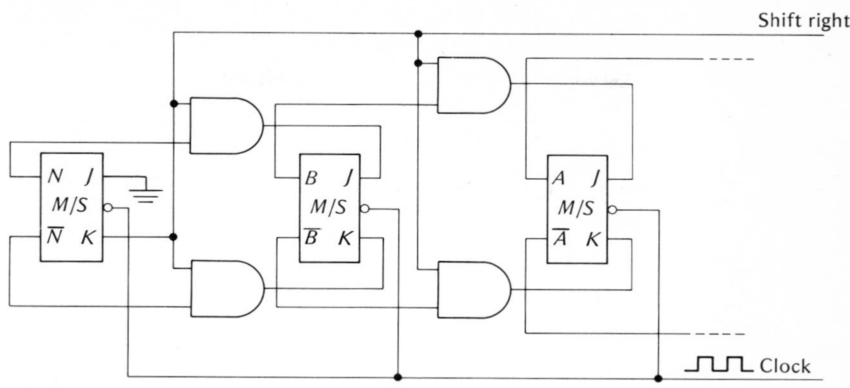

Fig. 9-30. Shift-right logic.

by 4, and so on. The logic necessary to effect a shift right is shown in Fig. 9-30.

Notice in this case that flip-flop N, which represents the MSB, must be reset and have only 0s shifted into it in order to effect a proper divide by 2.

9-8 SSI AND MSI INTEGRATED CIRCUITS

The various counters and shift registers we have discussed can be constructed using discrete ICs—that is, flip-flops, AND gates, OR gates, NAND gates, etc. However, with the advent of small-scale integration (SSI) and medium-scale integration (MSI), many of these counters are commercially available in a single package.

The 54/7400 series of TTL ICs is widely used in industry, and these units are available from a number of semiconductor manufacturers. Included in Table 9-1 is

Table 9-1

TTL ICs

Number	Title
7490	Decade counter
7491	Eight-bit shift register
7492	Divide-by-12 counter
7493	Binary counter
7494	Four-bit shift register
7495	Four-bit right/left shift register
7496	Five-bit shift register
74160	Decade counter
74161	Binary counter (presetable)
74190	Up/down decade counter
74191	Up/down binary counter
74195	Four-bit universal shift register

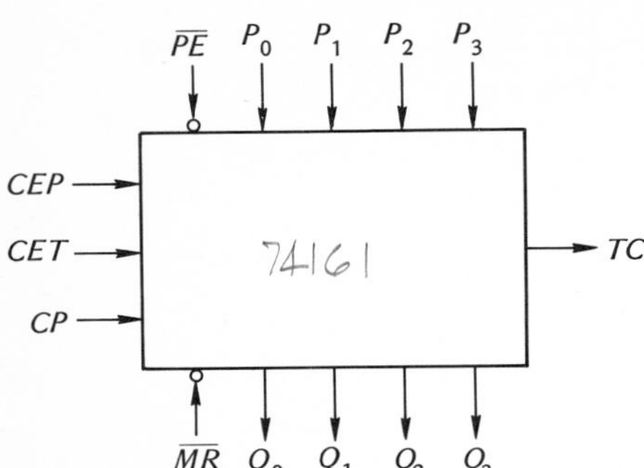

Fig. 9-31. Four-bit binary counter (presetable).

a partial listing of available SSI and MSI counters and registers. In most cases, the title of each unit is self-explanatory, and further information can be found by consulting specific manufacturers' data sheets.

The presetable counter (74161) is a most versatile unit, and we shall examine its characteristics in more detail. The 74161 is a four-flip-flop binary counter, and therefore has 16 possible states-0, 1, 2, 3, . . . , 14, 15. The logic diagram is shown in Fig. 9-31. The outputs are the four *binary digits* (Q_0, Q_1, Q_2, Q_3) and the *terminal-count* output (TC) which is high only when the counter is in state 15. The inputs are the *clock* (CP), the *parallel inputs* (P_0, P_1, P_2, P_3) which are used to preset the counter to any desired state, the *parallel enable* ($\overline{PE}$) which activates the parallel inputs, and three other special control inputs (CEP, CET, $\overline{MR}$) which we shall not discuss here. For our purposes, CEP, CET, and $\overline{MR}$ will be held at a high logic level($+V_{cc}$) to perform the preset counter operation.

Here is how the counter works. With $\overline{PE}$ held high ($+V_{cc}$), the counter progresses

Fig. 9-32.

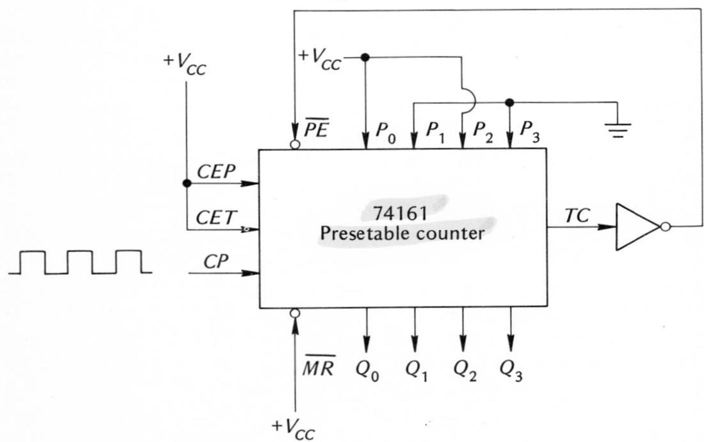

through a normal binary count sequence, advancing one count at the outputs (Q_0, Q_1, Q_2, Q_3) each time the clock goes positive. In this mode, it simply behaves as an ordinary 8421 binary counter.

Recall that the TC output is high each time the counter is in state 15 ($Q_0 = 1$, $Q_1 = 1$, $Q_2 = 1$, $Q_3 = 1$). This output can then be used to activate the parallel enable input ($\overline{PE}$) in order to preset the counter to the binary number present at the parallel inputs (P_0, P_1, P_2, P_3). For example, the counter can be preset to state 5 ($P_0 = 1$, $P_1 = 0$, $P_2 = 1$, $P_3 = 0$) by applying $1 = +V_{cc}$ and $0 = 0$ Vdc as shown in Fig. 9-32. The TC output is connected to the parallel enable input ($\overline{PE}$) through an inverter in order to obtain the correct logic level (recall that $\overline{PE}$ is active when *low,* and TC is *high* during count 15). Connected in this fashion, $\overline{PE}$ is activated in state 15, and when the clock goes high the number 0101 is shifted (preset) into the flip-flops. The counter then progresses through states 5, 6, 7, . . . , 14, 15 and then presets to 5 again. Notice the states 0, 1, 2, 3, and 4 are omitted and we have obtained a mod-11 counter. In this fashion, a counter of any modulus lower than 16 can be constructed.

Example 9-6

Show how to implement a mod-7 counter using a 74161. List the states omitted.

Solution

Preset the counter to state 9 ($P_0 = 1$, $P_1 = 0$, $P_2 = 0$, $P_3 = 1$). The omitted states are 0, 1, 2, 3, 4, 5, 6, 7, 8, and the counter will progress through states 9, 10, 11, 12, 13, 14, and 15.

STUDY AIDS

Summary

A basic binary register is a group of flip-flops (or some other binary elements) which is used to store binary information. Information can be shifted into the register in either serial or parallel fashion. The parallel method is much faster but requires considerably more hardware.

Information stored in a register can be shifted left or right, and this corresponds to binary multiplication or division by 2. The register capacity must be taken into account during the *shift right* and *shift left* operations. It is also quite easy to find the complement of the information stored in the register.

Direct feedback around the basic shift register leads to the formation of a ring counter. This counter will be quite useful later on in devising control waveforms (Chap. 14). Cross-coupled feedback around the basic register leads to the formation of the shift counter. Shift counters of any modulus can be formed by taking the feedback from the proper flip-flops. The shift counter has the great advantage of simplified decoding. This type of counter does, however, have undesired states, and these must be provided for in the counter design.

Two methods for implementing up-down counters have been discussed. This type of counter will be found quite useful in such applications as digital voltmeters and analog converters (Chap. 11).

The digital clock is an interesting application that illustrates some of the methods employing counters and decoders.

Glossary

register capacity Determined by the number of flip-flops in the register. There must be one flip-flop for each binary bit; the register capacity is 2^n, where n is the number of flip-flops.

ring counter A basic shift register with direct feedback such that the contents of the register simply circulate around the register when the clock is running.

shift counter A basic shift register with inverse feedback such that a cyclic counter is formed.

shift register A group of flip-flops connected in such a way that a binary number can be shifted into or out of the flip-flops.

up-down counter A basic counter, either serial or parallel, which is capable of counting in either an upward or a downward direction.

Review Questions

1. How many flip-flops are required to store a 10-bit binary number in a register?

2. How many clock pulses are required to shift a 16-bit binary number into a 16-flip-flop serial shift register?

3. How many flip-flops would be required to construct a register capable of storing the number 2561_{10} in binary form?

4. If it is desired to shift information into the register shown in Fig. 9-2, does the register have to be cleared (reset) before the shifting operation?

5. Describe the difference between the serial and parallel methods of entering data into a register.

6. How is the complement operation performed on a register, and what is the result?

7. How many flip-flops are required to construct shift counters of the following moduli?
(*a*) 5.
(*b*) 8.
(*c*) 9.
(*d*) 10.
(*e*) 21.

8. How many diodes would be required to decode a seven-flip-flop shift counter?

Problems

√ **9-1.** Demonstrate that it is possible to construct the six-bit shift register in Fig. 9-2 using *master/slave RS* flip-flops by drawing the logic diagram and the resulting waveforms.

√ **9-2.** How could you implement the controls for the *J* and *K* inputs to the first flip-flop in Fig. 9-2 using a single flip-flop?

9-3. Show the method for wiring the counter in Fig. 9-3 in order to preset it with a 1 in flip-flop F and 0s in all the other flip-flops.

9-4. What will happen if the counter in Fig. 9-3 has a 1 in every flip-flop just after power is applied?

9-5. How would you preset the counter in Fig. 9-3 to develop the waveform shown in Fig. 9-33?

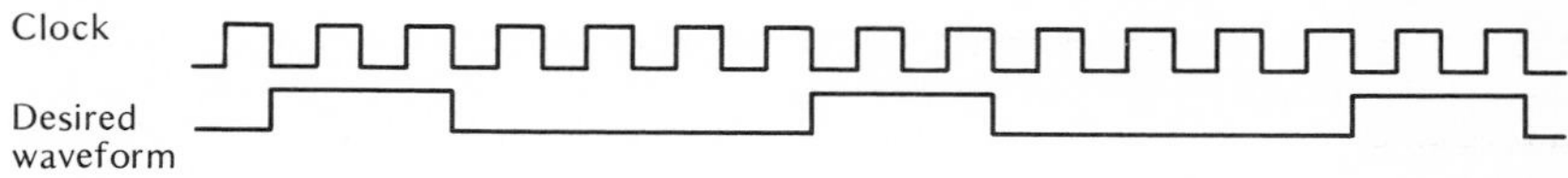

Fig. 9-33. Waveforms for Prob. 9-5.

9-6. Draw a two-flip-flop shift counter and its waveforms. Make a truth table and check for any illegal states.

9-7. Will the counter in Fig. 9-6 be forced into the proper mode if $\overline{A}$ is omitted from the NAND gate of Fig. 9-8?

9-8. In the worst case, how many clock periods will be required for the counter in Fig. 9-9 to get back into the proper count sequence?

9-9. Will the counter in Fig. 9-9 be forced into the proper mode if $\overline{C}$ is omitted from the NAND gate of Fig. 9-8? In the worst case, how many counts will be required to get the counter into the proper sequence?

9-10. List the 22 illegal states for the decade counter in Fig. 9-11, and verify the two illegal count sequences. Also verify the results of the counter existing in either count 10 or count 21.

9-11. Verify that the NAND gate in Fig. 9-8 will force the decade counter in Fig. 9-11 into the desired mode of operation. On what illegal counts will the counter be corrected? How many clock pulses are required in the worst case to get the counter back into the proper sequence? Is the $\overline{C}$ input to the NAND gate necessary?

9-12. Draw the logic diagram, waveforms, and truth table to form a mod-9 counter out of the decade counter shown in Fig. 9-11.

9-13. Do the complete design for a mod-7 shift counter. Draw the waveforms and truth table, and list the illegal states. Check the operation of the counter if it were to appear in any of the illegal states. Design a cure to place the counter back into the proper count sequence if one is needed.

9-14. Demonstrate that the mod-5 counter in Fig. 9-12 will always operate in the desired mode. Do the same for a mod-3 counter.

9-15. What states are skipped in the ÷60 counter in Fig. 9-16? Will the waveform at E be symmetrical? Would it affect the operation of the digital clock if it were not symmetrical?

9-16. Determine the method whereby NAND gates X and Y ensure proper operation of the counters in Fig. 9-17.

9-17. Draw the complete waveforms showing the results of allowing the up-down counter in Fig. 9-23 to progress through the following sequence: Starting at 0, count up five states, then down three states, then up four states, and then down six states.

9-18. Draw a three-flip-flop shift register capable of the following operations:
(*a*) Serial entry of data.
(*b*) Complement operation.
(*c*) Count up.
(*d*) Count down.

9-19. Use a 74161 presetable counter to implement a mod-8 counter. List the omitted states and the normal count sequence, and draw a complete logic diagram. Draw the set of waveforms you would expect, showing the clock (CP), and the four binary outputs (Q_0, Q_1, Q_2, Q_3). Remember that output transitions occur when the clock rises.

Input-Output Devices

In any digital system it is necessary to have a link of communication between man and machine. This communication link is often called the "man-machine interface" and it presents a number of problems. Digital systems are capable of operating on information at speeds much greater than man's, and this is one of their most important attributes. For example, a large-scale digital computer is capable of performing more than 500,000 additions per second.

The problem here is to provide data input to the system at the highest possible rate. At the same time, there is the problem of accepting data output from the system at the highest possible rate. The problem is further magnified since most digital systems do not speak English, or any other language for that matter, and some system of symbols must therefore be used for communication (there is at present a considerable amount of research in this area, and some systems have been developed which will accept spoken commands and give oral responses on a limited basis).

Since digital systems operate in a binary mode, a number of code systems which are binary representations have been developed and are being used as the language of communication between man and machine. In this chapter we discuss a number of these codes and, at the same time, consider the necessary input-output equipment.

The primary objective of this chapter is to acquire the ability to

1. Explain how Hollerith code and ASCII code are used in input/output media.
2. Discuss techniques for magnetic recording of digital information, including RZ, RZI, and NRZI.
3. Describe the limitations of a number of different digital input/output units.
4. Draw the logic diagrams for a simple tree decoder and a balanced multiplicative decoder.

10-1 PUNCHED CARDS

One of the most widely used media for entering data into a machine, or for obtaining output data from a machine, is the punched card. Some common examples of these cards are college registration cards, government checks, monthly oil company statements, and bank statements. It is quite simple to use this medium to represent binary information, since only two conditions are required. Typically, a hole in the card represents a 1 and the absence of a hole represents a 0. Thus, the card provides the means of presenting information in binary form, and it is only necessary to develop the code.

The typical punched card used in large-scale data-processing systems is 7⅜ in long, 3¼ in wide, and 0.007 in thick. Each card has 80 vertical columns, and there are 12 horizontal rows, as shown in Fig. 10-1. The columns are numbered 1 through 80 along the bottom edge of the card. Beginning at the top of the card, the rows are designated 12, 11, 0, 1, 2, 3, 4, 5, 6, 7, 8, and 9. The bottom edge of the card is the *9 edge,* and the top edge is the *12 edge*. Holes in the 12, 11, and 0 rows are called *zone punches,* and holes in the 0 through 9 rows are called *digit punches.* Notice that row 0 is both a zone-and a digit-punch row. Any number, any letter in the alphabet, or any of several special characters can be represented on the card by punching one or more holes in any one column. Thus, the card has the capacity of 80 numbers, letters, or combinations.

Probably the most widely used system for recording information on a punched card is the *Hollerith code.* In this code the numbers 0 through 9 are represented by a single punch in a vertical column. For example, a hole punched in the fifth row of column 12 represents a 5 in that column. The letters of the alphabet are represented by two punches in any one column. The letters A through I are represented by a zone punch in row 12 and a punch in rows 1 through 9. The letters J through R are represented by a zone punch in row 11 and a punch in rows 1 through 9. The letters S through Z are represented by a zone punch in row 0 and a punch in rows 2 through 9. Thus, any of the 10 decimal digits and any of the 26 letters of the alphabet can be represented in a binary fashion by punching the proper holes in the card. In addition, a number of special characters can be represented by punching combinations of holes in a column which are not used for the numbers or letters of the alphabet. These characters are shown with the proper punches in Fig. 10-1.

An easy device for remembering the alphabetic characters is the phrase "JR. is 11." Notice that the letters J through R have an 11 punch, those before have a 12 punch, and those after have a 0 punch. It is also necessary to remember that S begins on a 2 and not a 1.

Example 10-1

Decode the information punched in the card in Fig. 10-2.

Solution

Column 1 has a zone punch in row 0 and a punch in row 3. It is therefore the letter T. Column 2 has a zone punch in row 12 and another punch in row 8. It is

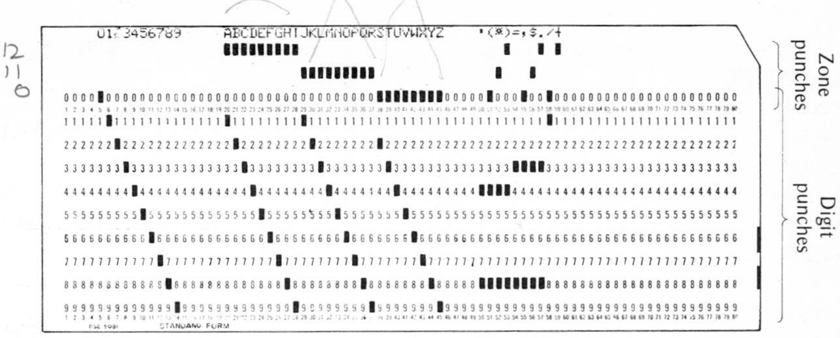

Fig. 10-1. Standard punched card using Hollerith code.

therefore the letter H. Continuing in this fashion, you should see that the complete message reads, "THE QUICK BROWN FOX JUMPED OVER THE LAZY DOGS BACK."

With this card code, any *alphanumeric* (alphabetic and numeric) information can be used as input to a digital system. On the other hand, the system is capable of delivering alphanumeric output information to the user. In scientific disciplines, the information might be missile flight number, location, or guidance information such as pitch rate, roll rate, and yaw rate. In business disciplines, the information could be account numbers, names, addresses, monthly statements, etc. In any case, the information is punched on the card with one character per column, and the card is then capable of containing a maximum of 80 characters.

Each card is considered as one block or unit of information. Since the machine operates on one card at a time, the punched card is often referred to as a "unit record." Moreover, the digital equipment used to punch cards, read cards into a system, sort cards, etc., is referred to as "unit-record equipment."

Occasionally, the information used with a digital system is entirely numeric; that is, no alphabetic or special characters are required. In this case, it is possible to input the information to the system by punching the cards in a straight binary fashion. In this system, the absence of a punch is a binary 0, and a punch is a

Fig. 10-2. Example 10-1.

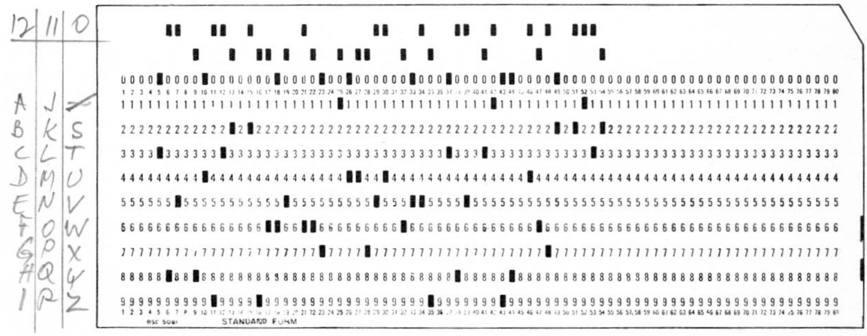

binary 1. It is then possible to punch 80 × 12 = 960 bits of binary information on one card.

Many large-scale data-processing systems use binary information in blocks of 36 bits. Each block of 36 bits is called a "word." You will recall from the previous chapter that a register capable of storing a 36-bit word must contain 36 flip-flops. There is nothing magical about the 36-bit word, and there are in fact other systems which operate with other word lengths. Even so, let's see how binary information arranged in words of 36 bits might be punched on cards.

There are two methods. The first method stores the information on the card horizontally by punching across the card from left to right. The first 36-bit word is punched in row 9 in columns 1 through 36. The second word is also in row 9, in columns 37 through 72. The third word is in row 8, columns 1 through 36, and so on. Thus a total of twenty-four 36-bit words can be punched in the card in straight binary form. It is then possible to store 864 bits of information on the card.

The second method involves punching the information vertically in columns rather than rows. Beginning in row 12 of column 1, the first 12 bits of the word are punched in rows 12, 11, 0, . . . , 9. The next 12 bits are punched in column 2, and the remaining 12 bits are punched in column 3. Thus, a 36-bit word can be punched in every three columns. The card is then capable of containing twenty-six 36-bit words.

The most common method of entering information into punched cards initially is by means of the key-punch machine. This machine operates very much the same as a typewriter, and the speed and accuracy of the operation depend entirely on the operator. The information on the punched cards can then be read into the digital system by means of a card reader. The information can be entered into the system at the rate of 100 to 1,000 cards per minute, depending on the type of card reader used.

The basic method for changing the punched information into the necessary electrical signals is shown in Fig. 10-3. The cards are stacked in the *read* hopper and are drawn from it one at a time. Each card passes under the *read* heads, which are either brushes or photocells. There is one *read* head for each column on the card, and when a hole appears under the *read* head an electrical signal is generated.

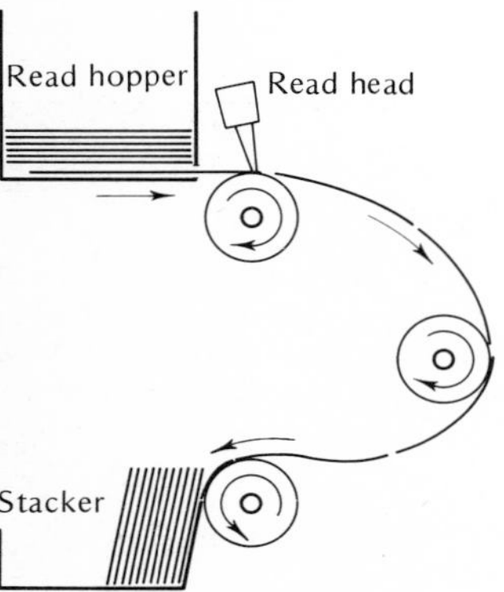

Fig. 10-3. Card-reading operation.

Thus, each signal from the *read* heads represents a binary 1, and this information can be used to set flip-flops which form the input storage register. The cards then pass over other rollers and are placed in the stacker. There is quite often a second *read* head which reads the data a second time to provide a validity check on the reading process.

Example 10-2

Suppose a deck of cards has binary data punched in them. Each card has twenty-four 36-bit words. If the cards are read at a rate of 600 cards per minute, what is the rate at which data are entering the system?

Solution

Since each card contains 24 words, the data rate is $24 \times 600 = 14{,}400$ words per minute. This is equivalent to $36 \times 14{,}400 = 518{,}400$ bits per minute, or $518{,}400/60 = 8{,}640$ bits per second.

Punched cards can also be used as a medium for accepting data output from a digital system. In this case, a stack of blank cards (having no holes punched in them) are held in a hopper in a card punch which is controlled by the digital system. The blank cards are drawn from the hopper one at a time and punched with the proper information. They are then passed under *read* heads, which check the validity of the punching operation, and stacked in an output hopper. Card punches are capable of operating at 100 to 250 cards per minute, depending on the system used.

Punched cards present a number of important advantages, the first of which is the fact that the cards represent a means of storing information permanently. Since the information is in machine code, and since this information can be printed on the top edge of the card, this is a very convenient means of communication between man and machine, and between machine and machine. There is also a wide variety of peripheral equipment which can be used to process information stored on cards. The most common are sorters, collators, calculating punches, reproducing punches, and accounting machines. Moreover, it is very easy to correct or change the information stored, since it is only necessary to remove the desired card(s) and replace it (them) with the corrected one(s). Finally, these cards are quite inexpensive.

10-2 PAPER TAPE

Another widely used input-output medium is punched paper tape. It is used in much the same way as punched cards. Paper tape was developed initially for the purpose of transmitting telegraph messages over wires. It is now used extensively for storing information and for transmitting information from machine to machine. Paper tape differs from cards in that it is a continuous roll of paper; thus, any amount of information can be punched into a roll. It is possible to record any alphabetic or numeric character, as well as a number of special characters, on paper tape by punching holes in the tape in the proper places.

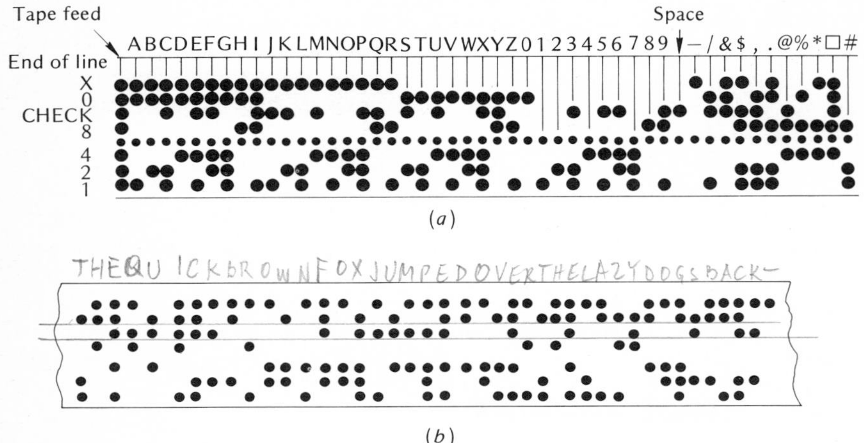

Fig. 10-4. Punched paper tape. (*a*) Eight-hole code. (*b*) Example 10-3.

There are a number of codes for punching data in paper tape, but one of the most widely used is the *eight-hole code* in Fig. 10-4*a*. Holes, representing data, are punched in eight parallel channels which run the length of the tape. (The channels are labeled 1, 2, 4, 8, parity, 0, *X*, and end of line.) Each character,—numeric, alphabetic, or special,—occupies one column of eight positions across the width of the tape.

Numbers are represented by punches in one or more channels labeled 0, 1, 2, 4, and 8, and each number is the sum of the punch positions. For example, 0 is represented by a single punch in the 0 channel; 1 is represented by a single punch in the 1 channel; 2 is a single punch in channel 2; 3 is a punch in channel 1 and a punch in channel 2, etc. Alphabetic characters are represented by a combination of punches in channels *X*, 0, 1, 2, 4, and 8. Channels *X* and 0 are used much as the zone punches in punched cards. For example, the letter *A* is designated by punches in channels *X*, 0, and 1. The special characters are represented by combinations of punches in all channels which are not used to designate either numbers or letters. A punch in the end-of-line channel signifies the end of a block of information, or the end of record. This is the only time a punch appears in this channel.

As a means of checking the validity of the information punched on the tape, the parity channel is used to ensure that each character is represented by an *odd* number of holes. For example, the letter C is represented by punches in channels *X*, 0, 1, and 2. Since an odd number of holes is required for each character, the code for the letter C also has a punch in the parity channel, and thus a total of five punches is used for this letter.

Example 10-3

What information is held in the perforated tape in Fig. 10-4*b*?

Solution

The first character has punches in channels 0, 1, and 2, and this is the letter T. The second character is the letter H, since there are punches in channels X, 0, and 8. Continuing, you should see that the message is the same as that punched on the card in Example 10-1.

The row of smaller holes between channels 4 and 8 are guide holes, used to guide and drive the tape under the *read* positions. The information on the tape can be sensed by brushes or photocells as shown in Fig. 10-5. The method for reading information from the paper tape and inputting it into the digital system is very similar to that used for reading punched cards. Depending on the type of reader used, information can be read into the system at a rate of 150 to 1,000 characters per second. You will notice that this is only slightly faster than reading information from punched cards.

Paper tape can be used as a means of accepting information output from a digital system. In this case the system drives a tape punch which enters the data on the tape by punching the proper holes. Typical tape punches are capable of operating at rates of 15 characters per second, and the data are punched with 10 characters to the inch. The number of characters per inch is referred to as the "data density," and in this case the density is 10 characters per inch. Recording density is one of the important features of magnetic-tape recording which will be discussed in the next session.

Paper tape can also be perforated by a manual tape punch. This unit is very similar to an electric typewriter, and indeed in some cases electric typewriters with special punching units attached are used. The accuracy and speed of this method are again a function of the machine operator. One advantage of this method is that

Fig. 10-5. Paper-tape drive and reading mechanism.

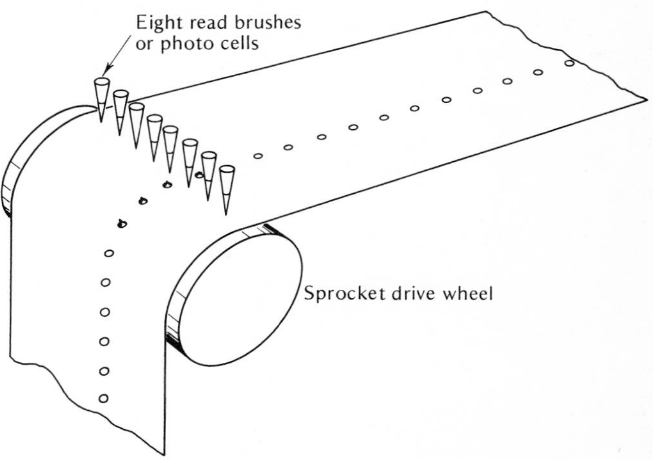

the typewriter provides a written copy of what is punched into the tape. This copy can be used for verification of the punched information.

10-3 MAGNETIC TAPE

Magnetic tape has become one of the most important methods for storing large quantities of information. Magnetic tape offers a number of advantages over punched cards and punched paper tape. One of the most important is the fact that magnetic tape can be erased and used over and over. Reading and recording are much faster than with either cards or paper tape. However, they require the use of a tape-drive unit which is much more expensive than the equipment used with cards and paper tape. On the other hand, it is possible to store up to 20 million characters on one 2,400-ft reel of magnetic tape, and if a high volume of data is one of the system requirements, the use of magnetic tape is well justified. Most commonly, magnetic tape is supplied on 2,400-ft reels. The tape itself is a ½-in-wide strip of plastic with a magnetic oxide coating on one side.

Data are recorded on the tape in seven parallel channels along the length of the tape. The channels are labeled 1, 2, 4, 8, *A*, *B*, and *C* as shown in Fig. 10-6. Since the information recorded on the tape must be digital in form, that is, there must be two states, it is recorded by magnetizing spots on the tape in one of two directions.

A simplified presentation of the *write* and *read* operations is shown in Fig. 10-7. The magnetic spots are recorded on the tape as it passes over the *write* head as shown in Fig. 10-7*a*. If a positive pulse of current is applied to the *write*-head coil, as shown in the figure, a magnetic flux is set up in a clockwise direction around the *write* head. As this flux passes through the record gap, it spreads slightly and passes through the oxide coating on the magnetic tape. This causes a small area on the tape to be magnetized with the polarity shown in the figure. If a current pulse of the opposite polarity is applied, the flux is set up in the opposite direction, and a spot magnetized in the opposite direction is recorded on the tape. Thus, it is possible to record data on the tape in a digital fashion. The spots shown in the figure are greatly exaggerated in size to show the direction of magnetization clearly.

In the *read* operation shown in Fig. 10-7*b*, a magnetized spot on the tape sets up a flux in the *read* head as the tape passes over the *read* gap. This flux induces a small voltage in the *read*-head coil which can be amplified and used to set or reset a flip-flop. Spots of opposite polarities on the tape induce voltages of opposite

Fig. 10-6. Magnetic-tape code.

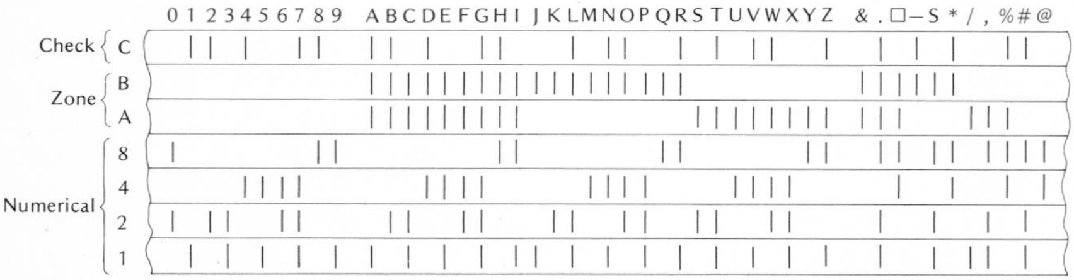

0 1 2 3 4 5 6 7 8 9 A B C D E F G H I J K L M N O P Q R S T U V W X Y Z & . □ – S * / , % # @

Check: C
Zone: B, A
Numerical: 8, 4, 2, 1

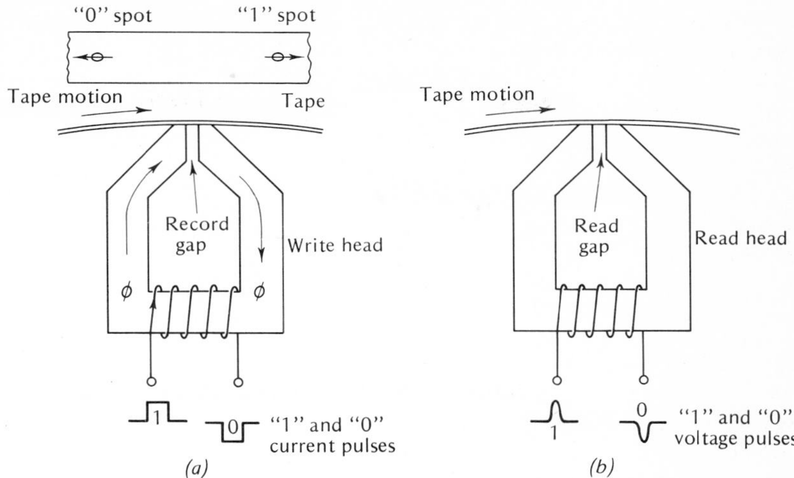

Fig. 10-7. Magnetic-tape recording and reading. (*a*) *Write* operation. (*b*) *Read* operation.

polarities in the *read* coil, and thus both 1s and 0s can be sensed. There is one *read/write* head for each of the seven channels on the tape. Typically, *read/write* heads are constructed in pairs as shown in Fig. 10-8. Thus, the *write* operation can be set up as a self-checking operation. That is, data recorded on tape are immediately read as they pass over the *read* gap and can be checked for validity.

A coding system similar to that used to punch data on cards is used to record alphanumeric information on tape. Each character occupies one column of seven bits across the width of the tape. The code is shown in Fig. 10-6. There are two independent systems for checking the validity of the information stored on the tape.

The first system is a vertical parity bit which is written in channel *C* of the tape. This is called a "character-check bit" and is written in channel *C* to ensure that all characters are represented by an *even* number of bits. For example, the letter A is

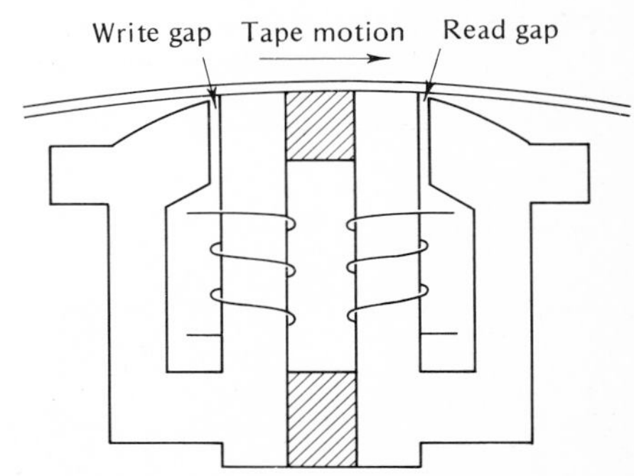

Fig. 10-8. Magnetic-tape *read/write* heads.

represented by spots in channels 1, *A*, and *B*. Since this is only three spots, an additional spot is recorded in channel *C* to maintain even parity for this character.

The second system is the *horizontal parity-check bit*. This is sometimes referred to as the longitudinal parity bit, and it is written, when needed, at the end of a block of information or record. The total number of bits recorded in each channel is monitored, and at the end of a record, a parity bit is written if necessary to keep the total number of bits an *even* number. These two systems form an even-parity system. They could, of course, just as easily be implemented to form an odd-parity system. Information can also be recorded on the tape in straight binary form. In this case, a 36-bit word is written across the width of the tape in groups of six bits. Thus it requires six columns to record one 36-bit word.

The vertical spacing between the recorded spots on the tape is fixed by the positions of the *read/write* heads. The horizontal spacing is a function of the tape speed and the recording speed. Tape speeds vary from 50 to 200 in/s, but 75 and 112.5 in/s are quite common.

The maximum number of characters recorded in 1 in of tape is called the "recording density," and it is a function of the tape speed and the rate at which data are supplied to the *write* head. Typical recording densities are 200, 556, and 800 bits per inch. Thus it can be seen that a total of $800 \times 2{,}400 \times 12 = 23.02 \times 10^6$ characters can be stored on one 2,400-ft reel of tape. This would mean that the data would have to be stored with no gaps between characters or groups of characters.

For purposes of locating information on tape, it is most common to record information in groups or blocks called "records." In between records there is a blank space of tape called the "interrecord gap." This gap is typically a 0.75-in space of blank tape, and it is positioned over the *read/write* heads when the tape stops. The interrecord gap provides the space necessary for the tape to come up to the proper speed before recording or reading of information can take place. The total number of characters recorded on a tape is then also a function of the record length (or the total number of interrecord gaps, since they represent blank space on the tape).

The data as recorded on the tape, including records (actual data) and interrecord gaps, can be represented as shown in Fig. 10-9. If there were no interrecord gaps, the total number of characters recorded could be found by multiplying the length of the tape in inches by the recording density in characters per inch. If the record were exactly the same length as the interrecord gap, the total storage would be cut in half. Thus, it is desirable to keep the records as long as possible in order to use the tape most efficiently.

Fig. 10-9. Recording data on magnetic tape.

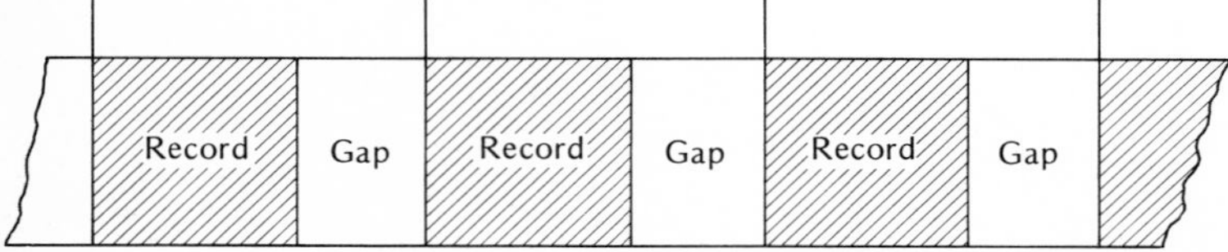

Given any one tape system and the recording density, it is a simple matter to determine the actual storage capacity of the tape. Consider the length of tape composed of one record and one record gap as shown in Fig. 10-9. This length of tape is repeated over and over down the length of the tape. The total number of characters that could be stored in this length of tape is the sum of the characters in the record R and the characters which could be stored in the record gap. The number of characters which could be stored in the gap is equal to the recording density D multiplied by the gap length G. Thus the total number of characters which could be stored in this length of tape is given by $R + GD$. The ratio of the characters actually recorded R to the total possible could be called a tape-utilization factor F and is given by

$$F = \frac{R}{R + GD} \tag{10-1}$$

Examination of the tape-utilization factor shows that if the total number of characters in the record is equal to the number of characters which could be stored in the gap, the utilization factor reduces to 0.5. This utilization factor can be used to determine the total storage capacity of a magnetic tape if the recording density and the record length are known. Thus the total number of characters stored on a tape $CHAR$ is given by

$$CHAR = LDF \tag{10-2}$$

where L = length of tape, in
D = recording density, characters per inch.

For a standard 2,400-ft reel of tape having a 0.75-in record gap, the formula in Eq. (10-2) reduces to

$$\boxed{CHAR = \frac{2{,}400 \times 12 \times DR}{R + 0.75D}} \tag{10-3}$$

Example 10-4

What is the total storage capacity of a 2,400-ft reel of magnetic tape if data are recorded at a density of 556 characters per inch and the record length is 100 characters?

Solution

The total number of characters can be found using Eq. (10-3).

$$CHAR = \frac{2{,}400 \times 12 \times 556 \times 100}{100 + (0.75 \times 556)} = 3.10 \times 10^6$$

This result can be checked by calculating the tape-utilization factor.

$$F = \frac{100}{100 + (0.75 \times 556)} = \frac{1}{5.17} \cong 0.19$$

The maximum number of characters that can be stored on the tape is $2{,}400 \times 12 \times 556 = 16.0128 \times 10^6$. Multiplying this by the utilization factor gives

$$CHAR = 16.0128 \times 10^6 \times \frac{1}{5.17} = 3.10 \times 10^6$$

10-4 DIGITAL RECORDING METHODS

There are a number of methods for recording data on a magnetic surface. The methods fall into two general categories, called "return-to-zero" and "non-return-to-zero," and they apply to magnetic-tape recording as well as recording on magnetic disk and drum surfaces (magnetic-disk and magnetic-drum storage will be discussed in a later chapter).

In the previous section, it was stated that digital information could be recorded on magnetic tape by magnetizing spots on the tape with opposite polarities. This type of recording is known as return-to-zero, or *RZ* for short, recording. The technique for recording data on tape using this method is to apply a series of current pulses to the *write*-head winding as shown in Fig. 10-10. The current pulses set up corresponding fluxes in the *write* head, as shown in the figure. The spots magnetized on the tape have polarities corresponding to the direction of the flux waveform, and it is only necessary to change the direction of the input current to write 1s or 0s. Notice that the input current and the flux waveform return to a zero reference level between individual bits. Thus the term "return to zero."

When it is desired to read the recorded information from the tape, the tape is passed over the *read* heads and the magnetized spots induce voltages in the *read*-coil winding as shown in the figure. Notice that there is somewhat of a problem here, since all the pulses have both positive and negative portions. One method of detecting these levels properly is to strobe the output waveform. That is, the output-

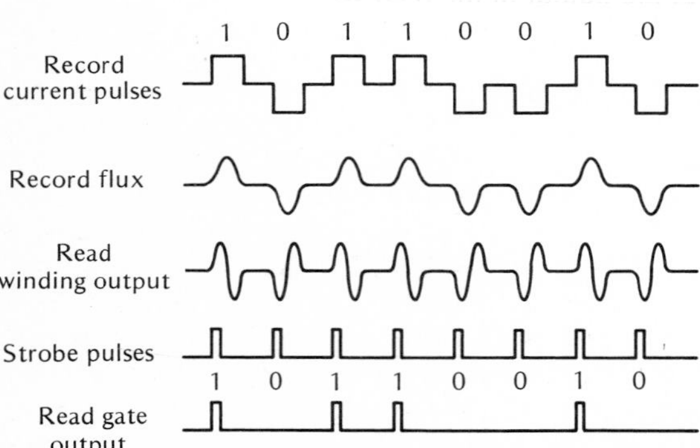

Fig. 10-10. Return-to-zero recording and reading.

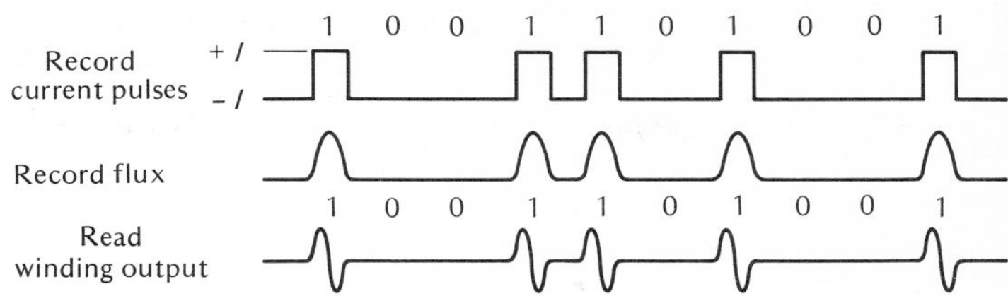

Fig. 10-11. Biased return-to-zero recording and reading.

voltage waveform is applied to one input of an AND gate (after being amplified), and a clock or strobe pulse is applied to the other input to the gate. The strobe pulse must be very carefully timed to ensure that it samples the output waveform at the proper time. This is one of the major difficulties of this type of recording, and it is therefore seldom used except on magnetic drums. On a magnetic drum, the strobe waveform can be recorded on one track of the drum, and thus the proper timing is achieved.

A second difficulty with this type of recording is the fact that between bits there is no record current, and thus between the spots on the tape the magnetic surface is randomly oriented. This means that if a new recording is to be made over old data, the new data have to be recorded precisely on top of the old data. If they are not, the old data will not be erased, and the tape will contain a conglomeration of information. The tape could be erased by installing another set of *erase* heads, but this is costly and unnecessary.

A method for curing these problems is to bias the record head with a current which will saturate the tape in either one direction or the other. In this system, a current pulse of positive polarity is applied only when it is desired to write a 1 on the tape as shown in Fig. 10-11. At all other times the flux in the *write* heads is sufficient to magnetize the entire track in the 0 direction. Now, recording data over old data is not a problem since the tape is effectively erased as it passes over the record heads. Moreover, the timing is not so critical since it is not necessary to record exactly over the previous data. When data are recorded in this fashion and then played back, a pulse appears at the output of the *read* winding only when a 1 has been recorded on the tape. This makes reading the information from the tape much simpler.

The *non-return-to-zero*, or NRZ, recording technique is a variation of the RZ technique where the *write* current pulses do not return to some reference level between bits. The NRZ recording technique can be best explained by examining the record-current waveform shown in Fig. 10-12. Notice that the current is at $+I$ while recording 1s and at $-I$ while recording 0s. Since the current levels are always at either $+I$ or $-I$, the recording problems of the first RZ system do not exist here.

Notice that the voltage at the *read*-winding output has a pulse only when the recorded data change from a 1 to a 0 or vice versa. Therefore, some means of sensing the recorded data is necessary for the *read* operation. If the *read*-winding voltage is amplified and used to set or reset a flip-flop as shown in the figure, the *A* side of the flip-flop is high during each time that a 1 is being read. It is low during

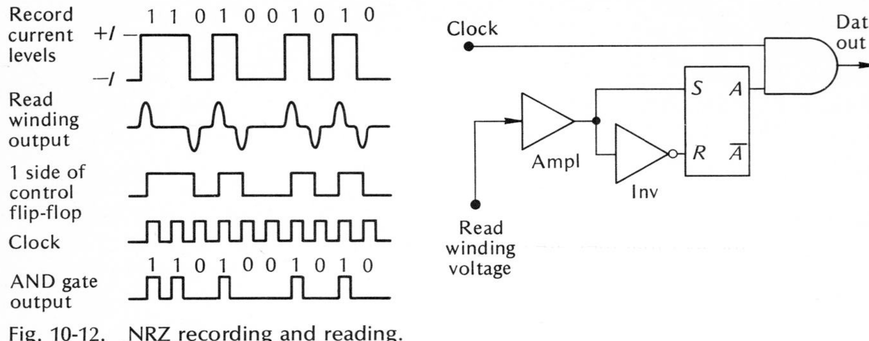

Fig. 10-12. NRZ recording and reading.

any time when the data being read is a 0. Thus if the *A* output of the flip-flop is used as a control signal at one input of an AND gate, while the other input is a clock, the output of the AND gate is an exact replica of the digital data being read. Notice that the clock must be carefully synchronized with the data train from the *read*-head winding. Notice also that the maximum rate of flux changes occurs when recording (or reading) alternate 1s and 0s.

In comparing this with the RZ recording methods, you can see that the NRZ method offers the distinct advantage that the maximum rate of flux changes is only one-half that for RZ recording. Thus the *read/write* heads and associated electronics can have reduced requirements for operation at the same rates, or they are capable of operating at twice the rate for the same specifications.

A variation on this basic form of NRZ recording is shown in Fig. 10-13. This technique is quite often called "non-return-to-zero-inverted," NRZI, since both 1s and 0s are recorded at both the high and low saturation-current levels. The key to this method of recording is that a 1 is sensed whenever there is a flux change, whether it be positive or negative. If the *read*-winding output voltage is amplified and presented to the OR gate as shown in the figure, the output of the gate will be the desired data train. The upper Schmitt trigger is sensitive only to positive pulses, while the lower one is sensitive only to negative pulses. Both outputs of the Schmitt triggers are low until a pulse arrives. At this time the output goes positive for a fixed duration and generates the desired output pulse.

Fig. 10-13. NRZI recording and reading.

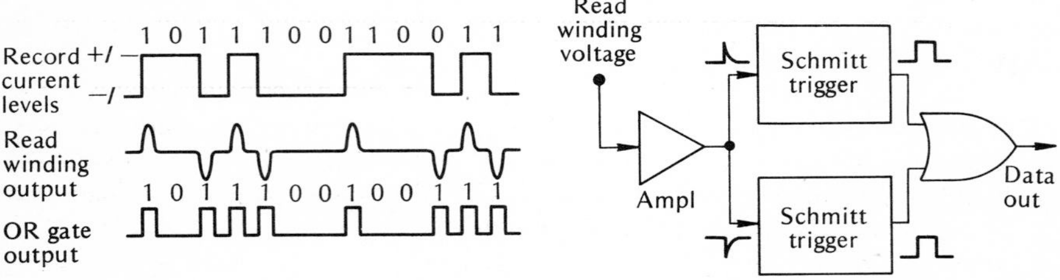

10-5 OTHER PERIPHERAL EQUIPMENT

A wide variety of peripheral equipment has been developed for use with digital systems. Only a cursory description of some of the various equipment will be given here, and the reader is encouraged to study equipment of particular interest by consulting the data manuals of the various manufacturers.

One of the simplest means of inputting information into a digital system is by the use of switches. These switches could be push-button, toggle, etc., but the important thing is the fact that they are capable of representing binary information. A row of 10 switches could, for example, be switched to represent the 10 binary bits in a 10-bit word.

Similarly, one of the simplest means of reading data out of a digital system is to put lights on the outputs of the flip-flops in a storage register. Admittedly, this is a rather slow means of communication, since the operator must convert the displayed binary data into something more meaningful. Nevertheless, this represents an inexpensive and practical means of communication between man and machine.

A much more sophisticated method for reading data out of a digital system is by means of a cathode-ray tube. One type of cathode-ray tube used is very similar to the tube used in oscilloscopes, and the operation of the tube is nearly the same. The unit is generally used to display curves representing information which has been processed by the system, and a camera can be attached to some units to photograph the display for a permanent record. The information displayed might be the transient response of an electrical network or a guided-missile trajectory.

A second type of cathode-ray tube for display is called a "charactron." It has the ability to display alphanumeric characters on the face of the screen. This tube operates by shooting an electron beam through a matrix (mask) which has each of the characters cut in it. As the beam passes through the matrix it is shaped in the form of the character through which it passes, and this shaped beam is then focused on the face of the screen. Since the operation of the electron beam is very fast, it is possible to write information on the face of the tube, and the operator can then read the display.

Some tubes of this type which are used in large radar systems have matrices with the proper characters to display map coordinates, friendly aircraft, unfriendly aircraft, etc. The operator thus sees a display of the surrounding area complete with all aircraft, properly designated, in the vicinity. These systems usually have an additional accessory called a "light pen" which enables the operator to input information into the digital system by placing the light pen on the surface of the tube and activating it. The operator can do such things as expand an area of interest, request information on an unidentified flying object, and designate certain aircraft as targets.

A somewhat more common piece of equipment, but nevertheless useful when large quantities of data are being handled, is the printer. Printers are available which will print the output data in straight binary form, octal form, or all the alphanumeric characters. The typical printer has the ability to print information on a 120-space line at rates from a few hundred lines up to over 1,200 lines per minute. The simplest printers are converted, or specially made, electric typewriters known as "character-at-a-time printers." They are relatively slow and operate at speeds of 10 to 30 characters per second.

A more sophisticated printer is known as the "line-at-a-time printer" since an entire line of 120 characters is printed in one operation. This type of printer is capable of operation at rates of around 250 lines per minute.

Somewhat faster operation is possible with machines which use a print wheel. The print-wheel printer is composed of 120 wheels, one for each position on the line to be printed. These wheels rotate continuously, and when the proper character is under the print position a hammer strikes an inked ribbon against the paper, which contacts the raised character on the print wheel. Wheel printers are capable of operation at the rate of 1,250 lines per minute and have a maximum capacity of 160 characters per line.

One other very important piece of peripheral equipment is the digital plotter. These units are being used more and more in a wide variety of tasks, including automatic drafting, numerical control, production artwork masters (used to manufacture integrated circuits), charts and graphs for management information, maps and contours, biomedical information, and traffic analysis, as well as a host of other applications. A somewhat hybrid form of digital plotting is used when the digital output of a system is converted to analog form (digital-to-analog conversion is the subject of the next chapter) to drive servomotors which position a cursor or pen. A piece of graph paper is positioned on a flat plotting surface, and the pen is caused to move across the paper in response to numbers received from the digital system.

Another digital plotting system, which is more truly a digital plotter, makes use of bidirectional stepping motors to position the pen and thus plot the information on graph paper. In this system, which is known as a "digital incremental plotter," the necessity for digital-to-analog conversion is eliminated, and these systems are usually less expensive and smaller in size. Digital incremental plotters are capable of plotting increments as small as 0.0025 in and offer much greater accuracies than the hybrid model. Furthermore, these plotters are capable of plotting at the rate of $4\frac{1}{2}$ in/s and providing a complete system of annotation and labeling.

10-6 TELETYPEWRITER TERMINALS

The teletypwriter (TTY) is presently one of the most popular *input/output* units. A TTY is an important and versatile link between man and computer, whether the computer is of the small-scale general-purpose type, or a large-scale model used on a time-share basis. It is common practice to use a TTY as a *remote* terminal connected to a large-scale general-purpose computer via telephone lines. The two binary logic levels (1 and 0) used in the TTY and the computer can be represented as two distinct audio frequencies which are then transmitted over telephone lines. An *acoustic tone coupler* is used in conjunction with the TTY to translate data from audio frequencies to logic levels, and vice versa. The central computer can be placed in a convenient site, and access to the computer via a TTY terminal is limited only by the requirement for a telephone line.

A TTY console consists of a basic keyboard for typing in information, and a printing mechanism for printing information output from the computer. Many TTYs are also equipped with a paper-tape punch, and thus either input data or output data can be recorded on punched paper tape.

Most modern TTYs use an eight-hole punched paper tape. There has been an attempt to standardize on an alphanumeric code, and the American Standard Code for Information Interchange (ASCII) is widely used. An eight-hole code has $2^8 = 256$ combinations, sufficient to provide for both uppercase and lowercase alphabets, the 10 numerals, and a number of special characters and control signals. The ASCII code is shown in Table 10-1.

10-7 ENCODING AND DECODING MATRICES

Encoding and decoding matrices are often used to alter the form of the data being entered into or taken out of a system. A decoding matrix is used to decode the binary information in a digital system by changing it into some other number system. For example, in a previous chapter the binary output of a register was decoded into decimal form by means of AND gates, and the decoded output was used to drive nixie tubes. Encoding information is just the reverse process and could, for example, involve changing decimal signals into equivalent binary signals for entry into a digital system.

The most straightforward way of decoding information is simply to construct the necessary AND gates, as was done for the nixie tubes. Decoding in this fashion is quite simple and is most easily accomplished by using the truth table or waveforms for the signals involved. The decoding of a four-flip-flop counter would, for example, require 16 four-input AND gates, since there are 16 possible states determined by the four flip-flops. This type of decoding then requires $n \times 2^n$ diodes, where n is the number of flip-flops, for the complete decoding network.

Example 10-5

Draw the 16 gates necessary to decode a four-flip-flop counter.

Solution

The necessary gates can best be implemented by using a truth table to determine the necessary gate connections. The gates are shown in Fig. 10-14.

There is a second method of decoding which can be used to realize a savings in diodes. This method is referred to as "tree decoding," and it results in a reduction of the number of required diodes by grouping the states to be decoded. Decoding of the four-flip-flop counter discussed in the previous example can be accomplished by separating the counts into four groups. These groups are 0,1,2,3; 4,5,6,7; 8,9,10,11; and 12,13,14,15. Notice that the first group can be distinguished by an AND gate whose output is $\overline{D}\overline{C}$, the second group by $\overline{D}C$, the third group by $D\overline{C}$, and the last group by DC. Each of these four groups can then be divided in half by using B or $\overline{B}$. These eight subgroups can then be further divided into the 16 counts by using A and $\overline{A}$. The complete decoding network is shown in Fig. 10-15.

Table 10-1
THE AMERICAN STANDARD CODE FOR INFORMATION EXCHANGE*

	000	001	010	011	100	101	110	111
0000	*NULL*	① DC_0	ƀ	0	@	*P*		
0001	*SOM*	DC_1	!	1	*A*	*Q*		
0010	*EOA*	DC_2	"	2	*B*	*R*		
0011	*EOM*	DC_3	#	3	*C*	*S*		
0100	*EOT*	DC_4 (Stop)	$	4	*D*	*T*		
0101	*WRU*	ERR	%	5	*E*	*U*		
0110	*RU*	*SYNC*	&	6	*F*	*V*		
0111	*BELL*	*LEM*	'	7	*G*	*W*		
1000	FE_0	S_0	(	8	*H*	*X*	Unassigned	
1001	*HT* / *SK*	S_1	)	9	*I*	*Y*		
1010	*LF*	S_2	*	:	*J*	*Z*		
1011	V_{TAB}	S_3	+	;	*K*	[		
1100	*FF*	S_4	,	<	*L*	\		ACK
1101	*CR*	S_5	—	=	*M*	]		②
1110	*SO*	S_6	*	>	*N*	↑		*ESC*
1111	*SI*	S_7	/	?	*O*	←		*DEL*

Example | 100 | 0001 | = *A*

b_1 - - - - - - - - - b_1

The abbreviations used in the figure mean:

NULL	Null idle	*CR*	Carriage return
SOM	Start of message	*SO*	Shift out
EOA	End of address	*SI*	Shift in
EOM	End of message	DC_0	Device control ① Reserved for data Link escape
EOT	End of transmission	$DC_1 - DC_2$	Device control
WRU	"Who are you?"	*ERR*	Error
RU	"Are you . . .?"	*SYNC*	Synchronous idle
BELL	Audible signal	*LEM*	Logical end of media
FE	Format effector	SO_0 - SO_7	Separator (information)
HT	Horizontal tabulation		Word separator (blank, normally non-printing)
SK	Skip (punched card)	*ACK*	Acknowledge
LF	Line feed	②	Unassigned control
V/TAB	Vertical tabulation	*ESC*	Escape
FF	Form feed	*DEL*	Delete idle

* Reprinted from *Digital Computer Fundamentals* by Thomas C. Bartee. Copyright 1960, 1966 by McGraw-Hill, Inc. Used with permission of McGraw-Hill Book Company.

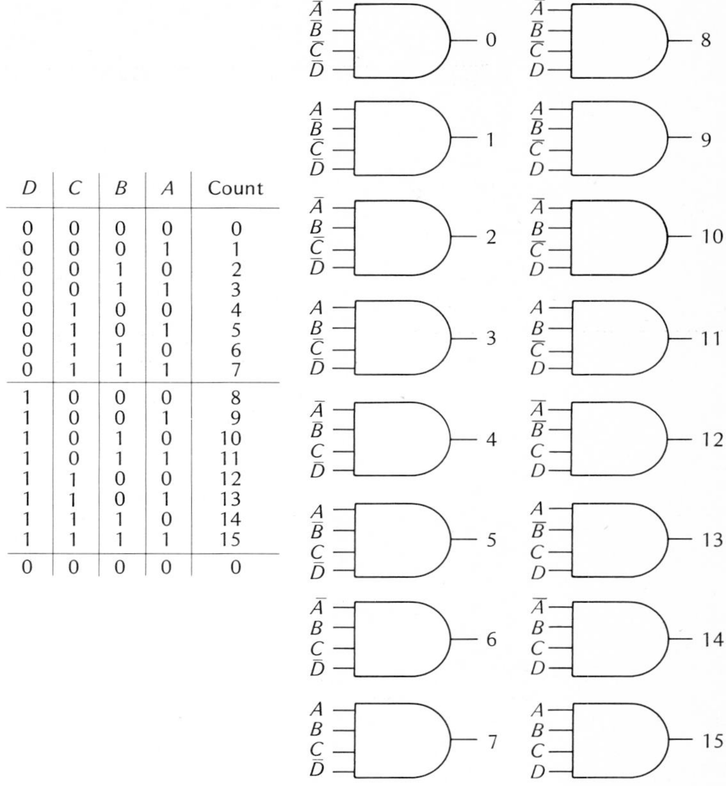

D	*C*	*B*	*A*	Count
0	0	0	0	0
0	0	0	1	1
0	0	1	0	2
0	0	1	1	3
0	1	0	0	4
0	1	0	1	5
0	1	1	0	6
0	1	1	1	7
1	0	0	0	8
1	0	0	1	9
1	0	1	0	10
1	0	1	1	11
1	1	0	0	12
1	1	0	1	13
1	1	1	0	14
1	1	1	1	15
0	0	0	0	0

Fig. 10-14. Four-flip-flop counter decoding.

A saving of 8 diodes has been achieved, since the previous decoding scheme required 64 diodes and this method only requires 56. The saving in diodes here is not very spectacular, but the construction of a matrix in this manner to decode five flip-flops would result in a saving of 40 diodes. As the number of flip-flops to be decoded increases, the saving in diodes increases very rapidly.

This type of decoding matrix does have the disadvantage that the decoded signals must pass through more than one level of gates (in the previous method the signal passes through only one gate). The output signal level may therefore suffer considerable reduction in amplitude. Furthermore, there may be a speed limitation due to the number of gates through which the decoded signals must pass.

A third type of decoding network is known as a "balanced multiplicative decoder." This always results in the minimum number of diodes required for the decoding process. The idea is much the same as a tree decoder, since the counts to be decoded are divided into groups. However, in this system the flip-flops to be decoded are divided into groups of two, and the results are then combined to give

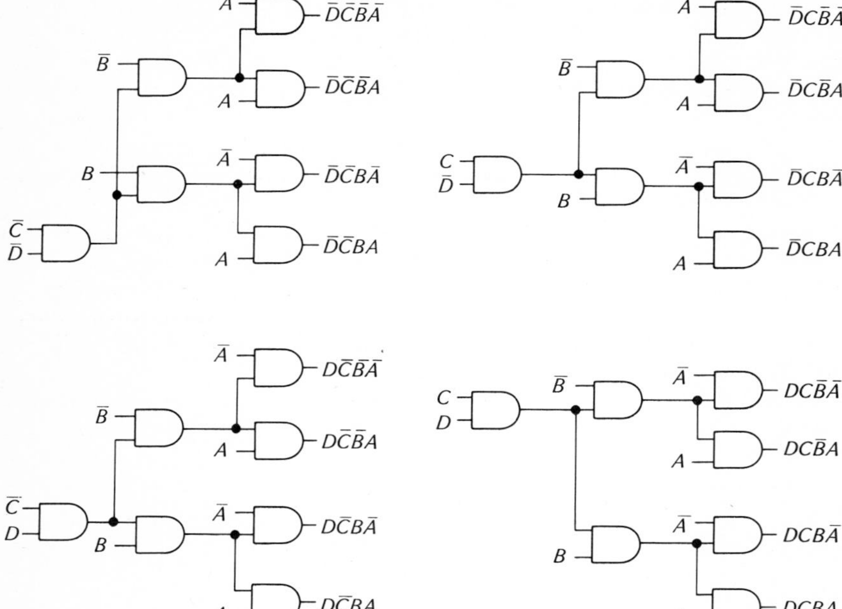

Fig. 10-15. Tree decoding matrix.

Fig. 10-16. Balanced multiplicative decoder.

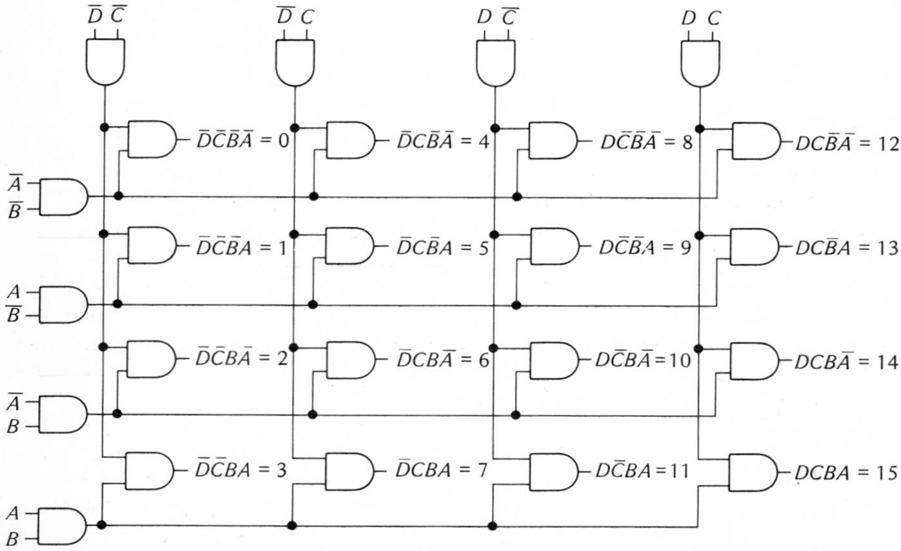

the desired output signals. To decode the four flip-flops discussed previously, four groups are formed by combining flip-flops *C* and *D* just as before. In addition, flip-flops *B* and *A* are combined in a similar arrangement. The outputs of these eight gates are then combined in 16 AND gates to form the 16 output signals. The results are shown in Fig. 10-16. It can be seen that a total of 48 diodes are required; a saving of 16 diodes is then realized over the first method, while a saving of 8 diodes is realized over the tree method. This scheme again has the same disadvantages of signal-level degradation and speed limitation as the tree decoder.

Encoding a number is just the reverse of decoding. One of the simplest examples of encoding would be the use of a thumb-wheel switch (a 10-position switch) which is used to enter data into a digital system. The operator can set the switch to any one of 10 positions which represent decimal numbers. The output of the switch is then transformed by a proper encoding matrix which changes the decimal number to an equivalent binary number.

An encoding matrix which changes a decimal number to an equivalent binary number and stores it in a register is shown in Fig. 10-17. Setting the switch to a

Fig. 10-17. Decimal encoding matrix.

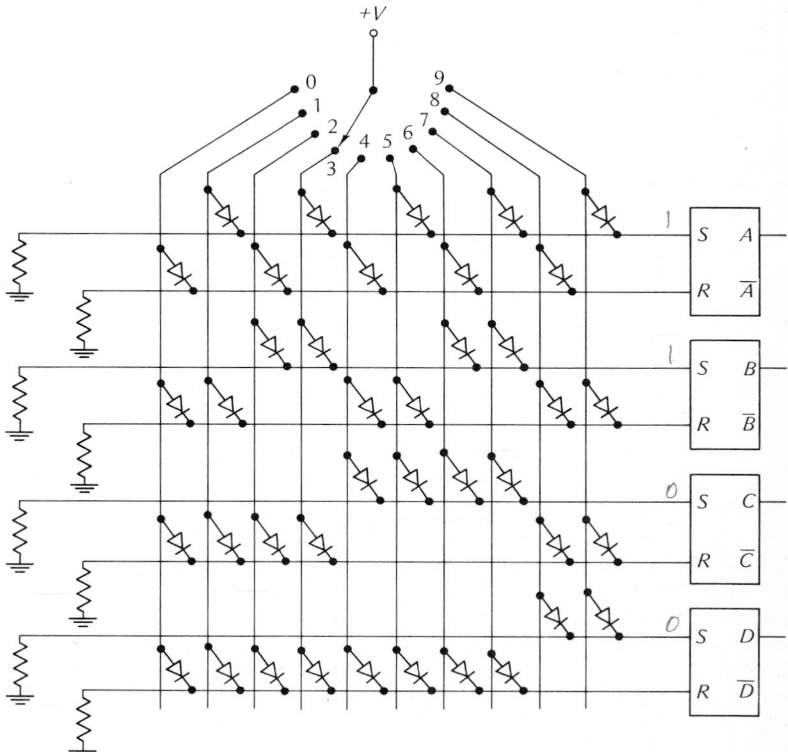

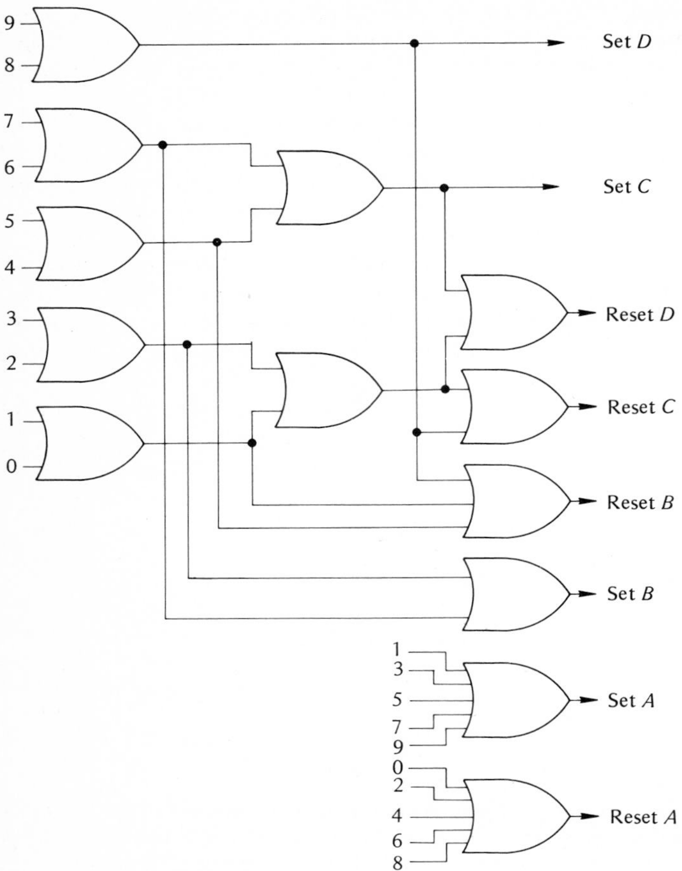

Fig. 10-18. Another decimal encoding matrix.

position places a positive voltage on the line connected to that position. Notice that the *R* and *S* input to each flip-flop is essentially the output of an OR gate.

For example, if the switch is set to position 1, the diodes connected to that line have a positive voltage on their plates (they are therefore forward-biased). Thus the *set* input to flip-flop *A* goes high while the *reset* inputs to flip-flops *B, C,* and *D* go high. This sets the binary number 0001 in the flip-flops, where *A* is the least significant bit. Notice that this encoding matrix requires 40 diodes. As might be expected, it is possible to reduce the number of diodes required by combining the input functions as was done with decoding matrices. One method of doing this is shown in Fig. 10-18; it represents a saving of 7 diodes, since this scheme requires only 33 diodes.

Any encoder or decoder can be constructed from basic gates as shown in this section, and when only one or two functions are needed this may provide the best technique. However, as shown in Chap. 3, many of the more common decoding functions are available as MSI ICs. Examples are the 7441 (or 74141) BCD-to-decimal decoder driver, the 7443 excess-3-to-decimal decoder, the 7446 BCD-to-seven-segment decoder driver, and the 74145 1-of-10 decoder driver. There are numerous others, and you are urged to consult manufacturers' data sheets for specific information.

There are also a few encoders available as MSI ICs—for example, the Fairchild 9318 eight-input priority encoder. This unit accepts eight inputs and produces a binary weighted code of the highest-order output. Again, you should consult specific manufacturers' data sheets for detailed information on encoders.

STUDY AIDS

Summary

Punched cards provide one of the most useful and widely used media for storing binary information. Each card is considered as a block or unit of information and is therefore referred to as a "unit record." Furthermore, punched-card equipment (punches, sorters, readers, etc.) is commonly called "unit-record equipment."

Alphanumeric information, as well as special characters, can be punched into cards by means of a code. The most common code in use is the Hollerith code.

A similar medium for information storage is punched paper tape. Alphanumeric and special characters are recorded by perforating the tape according to a code. There are a number of codes, but the one most commonly used is the eight-hole code. A perforated role of paper tape is a continuous record and is thus distinct from the unit record (punched card).

For handling large quantities of information, magnetic tape is a most convenient recording medium. Magnetic tape offers the advantages of much higher processing rate and much greater recording densities. Moreover, magnetic tape can be erased and used over and over.

The three most common methods for recording on magnetic tape are the return-to-zero (RZ), the non-return-to-zero (NRZ), and the non-return-to-zero-inverted (NRZI). The NRZ and NRZI methods effectively erase or clean the tape automatically during the record operation and thus eliminate one of the problems of RZ recording. These two methods also lend themselves to higher recording rates.

Encoding and decoding matrices form an important part of input-output equipment. These matrices are generally used to change information from one form to another, for example, binary to octal, or binary to decimal, or decimal to binary.

There is a wide variety of digital peripheral equipment including unit-record equipment, printers, cathode-ray-tube displays, and plotters. The choice of peripheral equipment to be used with any system is a major engineering decision. The decision involves establishing the system requirements, studying the available equipment, meeting with the equipment manufacturers, and then making the decision based on operational characteristics, delivery time, and cost.

Glossary

alphanumeric information Information composed of the letters of the alphabet, the numbers, and special characters.
bit One binary digit.
character A number, letter, or symbol represented by a combination of bits.
decoding matrix A matrix used to alter the format of information taken from the output of a system.
encoding matrix A matrix used to alter the format of information being entered into a system.
Hollerith code The system for representing information by punching holes in a prescribed manner in a punched card.
interrecord gap A blank piece of tape between recorded information.
NRZ Non-return-to-zero recording.
NRZI Non-return-to-zero inverted recording.
parity The method of using an additional punched hole (or magnetic spot for magnetic recording) to ensure that the total number of holes (or spots) for each character is even or odd.
recording density The number of characters recorded per inch of tape.
tape-utilization factor The ratio of the number of characters actually recorded to the maximum number of characters that could be recorded.
unit record A punched card represents a unit record since each card contains a unit or block of information.

Review Questions

1. Describe some of the problems of the man-machine interface.

2. Describe a typical punched card (size, number of columns, number of rows).

3. Which rows are the zone punches on a punched card?

4. Which rows are the digit punches on a punched card?

5. What is the Hollerith code? What does "JR. is 11" signify?

6. How is binary information represented on a card; i.e., what does a hole represent, and what does the absence of a hole represent?

7. What is the meaning of unit record?

8. Name three pieces of unit-record peripheral equipment, and give a brief description of how they are used.

9. Describe the eight-hole code used to punch information into paper tape.

10. Describe how 1s and 0s are recorded on magnetic tape by means of a magnetic record head.

11. How is alphanumeric information recorded on magnetic tape?

12. How is binary information recorded on magnetic tape?

13. Explain the dual-parity system used in magnetic-tape recording.

14. What is the purpose of an interrecord gap on magnetic tape?

15. How can the tape-utilization factor be used to determine the total number of characters stored on a magnetic tape?

16. Describe the operation of the RZ recording method. What are some of the difficulties with this system?

17. Describe the operation of the NRZ recording method. What advantages does this method offer over RZ recording?

18. Describe the NRZI recording technique.

19. Why is a digital incremental plotter a true digital plotting system?

20. What is the difference between an encoding and a decoding matrix?

Problems

10-1. Make a sketch of a punched card and code your name, address, and social security number using the Hollerith code. Use a dark spot to represent a hole.

10-2. Change your social security number to the equivalent binary number. Make a sketch of a punched card, and record this number on the card in the horizontal binary fashion.

10-3. Repeat Prob. 10-2, but record the number on the card in the vertical fashion.

10-4. Assume that alphanumeric information is being punched into cards at the rate of 250 cards per minute. If the cards have an average of 65 characters each, at what rate in characters per second is the information being processed?

10-5. Make a sketch of a length of paper tape. Using the eight-hole code, record your name, address, and social security number on the tape. Use a dark spot to represent a hole.

10-6. What length of paper tape is required for the storage of 60,000 characters of alphanumeric information using the eight-hole code? Assume no record gaps.

10-7. What length of magnetic tape would be required to store the information in Prob. 10-6 if the recording density is 500 bits per inch? Assume no record gaps.

10-8. Assume that data are recorded on magnetic tape at a density of 200 bits per inch. If the record length is 200 characters, and the interrecord gap is 0.75 in, what is the tape-utilization factor? Using this scheme, how many characters can be stored in 1,000 ft of tape?

10-9. Verify the solution to Prob. 10-8 above by using Eq. (10-3). Notice that the 2,400 in the equation must be replaced by 1,000, since this is the tape length.

10-10. Repeat Probs. 10-8 and 10-9 for a density of 800 bits per inch.

10-11. What length of magnetic tape is required to store 10^5 characters recorded at a density of 800 bits per inch with a record length of 500 characters?

10-12. Can you explain why it is best to keep the record length as long as possible in magnetic-tape recording?

10-13. Draw the eight three-input AND gates necessary to decode a three-flip-flop counter. Now draw the gates necessary to decode this counter using the tree decoding scheme. How many diodes are saved?

10-14. Draw the gates necessary to change a three-bit binary number to an octal number.

10-15. How would you change a four-bit binary number to an octal number? Draw the necessary gates.

10-16. Draw the gates necessary to form an octal-to-binary encoder.

10-17. See if you can simplify the gates in Prob. 10-16 by grouping the eight inputs.

10-18. What are the advantages and disadvantages of a tree and a balanced multiplicative decoder?

D/A and A/D Conversion

11

Digital-to-analog (D/A) and analog-to-digital (A/D) conversion form two very important aspects of digital data processing. D/A conversion involves translating digital information into equivalent analog information. As an example, the output of a digital system might be changed to analog form for the purpose of driving a pen recorder. Similarly, an analog signal might be required for the servomotors which drive the cursor arms of a plotter, as discussed in the preceding chapter. In this respect, a D/A converter is sometimes considered a decoding device since it operates on the output of a digital system.

Quite often the opposite conversion is needed. The process of changing an analog signal to an equivalent digital signal is accomplished by the use of an A/D converter. For example, an A/D converter might be used to change the analog output signals from transducers (measuring temperature, pressure, vibration, etc.) into equivalent digital signals. These signals would then be in a form suitable for entry into a digital system for processing. An A/D converter is often referred to as an encoding device since it is commonly used to encode signals for entry into a digital system.

D/A conversion is a straightforward process and is considerably easier than A/D conversion. In fact, a D/A converter is usually an integral part of any A/D converter. For this reason, we shall consider the digital-to-analog conversion process first.

After completing this chapter, you should be able to

1. Show how to use a binary ladder to accomplish D/A conversion.
2. Explain the operation of a counter type A/D converter.
3. Determine the accuracy and the resolution of an A/D or a D/A converter.

11-1 VARIABLE-RESISTOR NETWORK

The basic problem in converting a digital signal into an equivalent analog signal is to change the *n* digital voltage levels into one equivalent analog voltage. This can be most easily accomplished by designing a resistive network which will change

2^2	2^1	2^0
0	0	0
0	0	1
0	1	0
0	1	1
1	0	0
1	0	1
1	1	0
1	1	1

Fig. 11-1.

each of the digital levels into an *equivalent binary weighted voltage* (or current).

As an example of what is meant by equivalent binary weight, consider the truth table for the three-bit binary signal shown in Fig. 11-1. Suppose we want to change the eight possible digital signals in this figure into equivalent analog voltages. The smallest number represented is 000; let us make this equal to 0 V. The largest number is 111; let us make this equal to +7 V. This then establishes the range of the analog signal which will be developed. (There is nothing special about the voltage levels chosen; they were simply selected for convenience.)

Now, notice that between 000 and 111 there are seven discrete levels to be defined. Therefore, it will be convenient to divide the analog signal into seven levels. The smallest incremental change in the digital signal is represented by the least significant bit (LSB) 2^0. Thus we would like to have this bit cause a change in the analog output which is equal to $1/7$ of the full-scale analog output voltage. The resistive divider will then be designed such that a 1 in the 2^0 position will cause $+7 \times 1/7 = +1$ V at the output.

Since $2^1 = 2$ and $2^0 = 1$, it can be clearly seen that the 2^1 bit represents a number which is twice the size of the 2^0 bit. Therefore, a 1 in the 2^1 bit position must cause a change in the analog output voltage which is twice the size of the LSB. The resistive divider must then be constructed such that a 1 in the 2^1 bit position will cause a change of $+7 \times 2/7 = +2$ V in the analog output voltage.

Similarly, $2^2 = 4 = 2 \times 2^1 = 4 \times 2^0$, and thus the 2^2 bit must cause a change in the output voltage which is four times that of the LSB. The 2^2 bit must then cause an output-voltage change of $+7 \times 4/7 = +4$ V.

The process can be continued, and it will be seen that each successive bit must have a value which is twice that of the preceding bit. Thus the LSB is given a binary equivalent weight of $1/7$ or 1 part in 7. The next least significant bit is given a weight of $2/7$, which is twice the LSB, or 2 parts in 7. The MSB (in the case of this three-bit system) is given a weight of $4/7$, which is 4 times the LSB or 4 parts in 7. Notice that the sum of the weights must equal 1. Thus $1/7 + 2/7 + 4/7 = 7/7 = 1$. In general, the binary equivalent weight assigned to the LSB is $1/(2^n - 1)$, where n is the number of bits. The remaining weights are found by multiplying 2, 4, 8, ...

Example 11-1

Find the binary equivalent weight of each bit in a four-bit system.

Bit	Weight
2^0	1/7
2^1	2/7
2^2	4/7
Sum	7/7

(a)

Bit	Weight
2^0	1/15
2^1	2/15
2^2	4/15
2^3	8/15
Sum	15/15

(b)

Fig. 11-2. Binary equivalent weights.

Solution

The LSB has a weight of $1/(2^4 - 1) = 1/(16 - 1) = 1/15$, or 1 part in 15. The second LSB has a weight of $2 \times 1/15 = 2/15$. The third LSB has a weight of $4 \times 1/15 = 4/15$ and the MSB has a weight of $8 \times 1/15 = 8/15$. As a check, the sum of the weights must equal 1. Thus, $1/15 + 2/15 + 4/15 + 8/15 = 15/15 = 1$. The binary equivalent weights for a three-bit and a four-bit system are summarized in Fig. 11-2.

What is now desired is a resistive divider having three digital inputs and one analog output as shown in Fig. 11-3*a*. Assume that the digital input levels are 0 = 0 V and 1 = +7 V. Now for an input of 001, the output will be +1 V. Similarly, an input of 010 will provide an output of +2 V and an input of 100 will provide an output of +4 V. The digital input 011 is seen to be a combination of the signals 001 and 010. If the +1 V from the 2^0 bit is added to the +2 V from the 2^1 bit, the desired +3 V output for the 011 input is achieved. The other desired voltage levels are shown in Fig. 11-3*b*; they too are additive combinations of voltages.

Thus the resistive divider must do two things in order to change the digital input into an equivalent analog output voltage:

1. The 2^0 bit must be changed to +1 V; the 2^1 bit must be changed to +2 V; the 2^2 bit must be changed to +4 V.
2. These three voltages representing the digital bits must be summed together to form the analog output voltage.

Fig. 11-3.

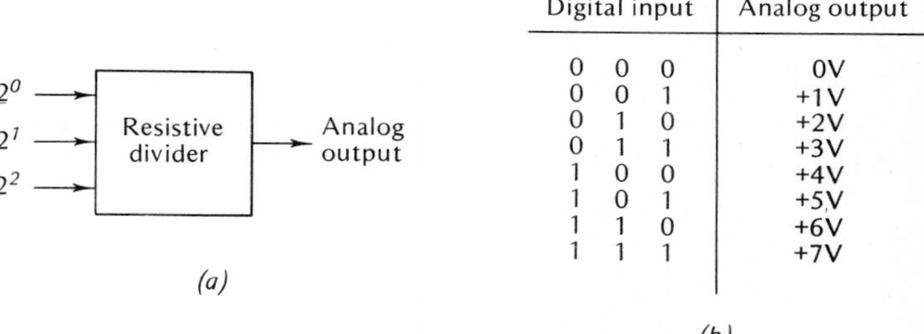

Digital input			Analog output
0	0	0	0V
0	0	1	+1V
0	1	0	+2V
0	1	1	+3V
1	0	0	+4V
1	0	1	+5V
1	1	0	+6V
1	1	1	+7V

(b)

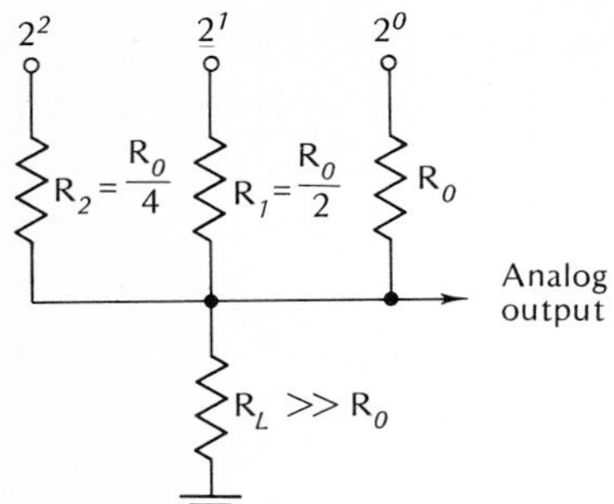

Fig. 11-4. Resistive ladder.

A resistive divider which performs the above functions is shown in Fig. 11-4. The resistors R_0 R_1, and R_2 form the divider network. R_L represents the load to which the divider is connected and is considered to be large enough so that it does not load the divider network.

Assume that the digital input signal 001 is applied to this network. Recalling that $0 = 0$ V and $1 = +7$ V, the equivalent circuit shown in Fig. 11-5 can be drawn. R_L is considered large and is neglected. The analog output voltage V_A can be most easily found by use of Millman's theorem. Millman's theorem states that the voltage appearing at any node in a resistive network is equal to the summation of the currents entering the node (found by assuming the node voltage is zero) divided by the summation of the conductances connected to the node. In equation form, Millman's theorem is

$$V = \frac{E_1/R_1 + E_2/R_2 + E_3/R_3 + \cdots}{1/R_1 + 1/R_2 + 1/R_3 + \cdots}$$

Applying Millman's theorem to Fig. 11-5 yields

$$V_A = \frac{V_0/R_0 + V_1/(R_0/2) + V_2/(R_0/4)}{1/R_0 + 1/(R_0/2) + 1/(R_0/4)}$$

$$= \frac{7/R_0}{1/R_0 + 2/R_0 + 4/R_0} = 7/7 = +1 \text{ V}$$

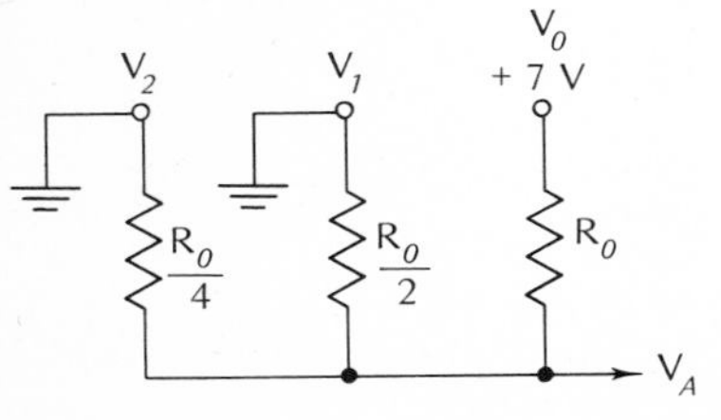

Fig. 11-5.

Drawing the equivalent circuits for the other seven input combinations and applying Millman's theorem will lead to the table of voltages shown in Fig. 11-3 (see Prob. 11-3).

Example 11-2

For a four-input resistive divider ($0 = 0$ V, $1 = +10$ V) find the following:

(a) The full-scale output voltage
(b) The output-voltage change due to the least significant bit
(c) The analog output voltage for a digital input of 1011

Solution

(a) The maximum output voltage occurs when all the inputs are at +10 V. If all four inputs are at +10 V, the output must also be at +10 V (ignoring the effects of R_L).

(b) For a four-bit digital number, there are 16 possible states. There are 15 steps between these 16 states, and the LSB must be equal to one-fifteenth of the full-scale output voltage. Therefore, the change in output voltage due to the LSB is $+10 \times 1/15 = +2/3$ V.

(c) Using Millman's theorem, the output voltage for a digital input of 1011 is

$$V_A = \frac{10/R_0 + 10/(R_0/2) + 0/(R_0/4) + 10/(R_0/8)}{1/R_0 + 1/(R_0/2) + 1/(R_0/4) + 1/(R_0/8)}$$

$$= \frac{110}{15} = 22/3 = +7\tfrac{1}{3}\text{ V}$$

To summarize, a resistive divider can be built to change a digital voltage into an equivalent analog voltage. The following criteria can be applied to this divider:

1. There must be one input resistor for each digital bit.
2. Beginning with the LSB, each following resistor value is one-half the size of the previous resistor.
3. The full-scale output voltage is equal to the + voltage of the digital input signal (the divider would work equally well with input voltages of 0 and $-V$ V).
4. The LSB has a weight of $1/(2^n - 1)$, where n is the number of input bits.
5. The change in output voltage due to a change in the LSB is equal to $V/(2^n - 1)$, where V is the digital input-voltage level.
6. The output voltage V_A can be found for any digital input signal by using the following modified form of Millman's theorem:

$$V_A = \frac{V_0 2^0 + V_1 2^1 + V_2 2^2 + V_3 2^3 + \cdots + V_{n-1} 2^{n-1}}{2^n - 1} \tag{11-1}$$

where $V_0, V_1, V_2, V_3, \ldots, V_n$ are the digital input voltage levels (0 or V V) and n is the number of input bits.

Example 11-3

For a five-bit resistive divider, determine the following:

(*a*) The weight assigned to the LSB.

(*b*) The weight assigned to the second and third LSB.

(*c*) The change in output voltage due to a change in the LSB, the second LSB, and the third LSB.

(*d*) The output voltage for a digital input of 10101. Assume $0 = 0$ V and $1 = +10$ V.

Solution

(*a*) The LSB weight is $1/(2^5 - 1) = 1/31$.

(*b*) The second LSB weight is $2/31$, and the third LSB weight is $4/31$.

(*c*) The LSB causes a change in the output voltage of $10/31$ V. The second LSB causes an output-voltage change of $20/31$ V, and the third LSB causes an output-voltage change of $40/31$ V.

(*d*) The output voltage for a digital input of 10101 is

$$V_A = \frac{10 \times 2^0 + 0 \times 2^1 + 10 \times 2^2 + 0 \times 2^3 + 10 \times 2^4}{2^5 - 1}$$

$$= \frac{10(1 + 4 + 16)}{32 - 1} = \frac{210}{31} = +6.77 \text{ V}$$

This resistive divider has two serious drawbacks. The first is the fact that each resistor in the network is a different value. Since these dividers are usually constructed using precision resistors, the added expense becomes unattractive. Moreover, the resistor used for the MSB is required to handle a much greater current than the LSB resistor. For example, in a 10-bit system, the current through the MSB resistor is approximately 500 times as large as the current through the LSB resistor (see Prob. 11-5). For these reasons, a second type of resistive network, called a "ladder," has been developed.

11-2 BINARY LADDER

The binary ladder is a resistive network whose output voltage is a properly weighted sum of the digital inputs. Such a ladder, designed for four bits, is shown in Fig. 11-6. It is constructed of resistors having only two values and thus overcomes one of the objections to the resistive divider previously discussed. The left end of the ladder is terminated in a resistance of $2R$, and we shall assume for the moment that the right end of the ladder (the output) is open-circuited.

Let us now examine the resistive properties of the network, assuming that all the digital inputs are at ground. Beginning at node A, the total resistance looking into the terminating resistor is $2R$. The total resistance looking out toward the 2^0 input is

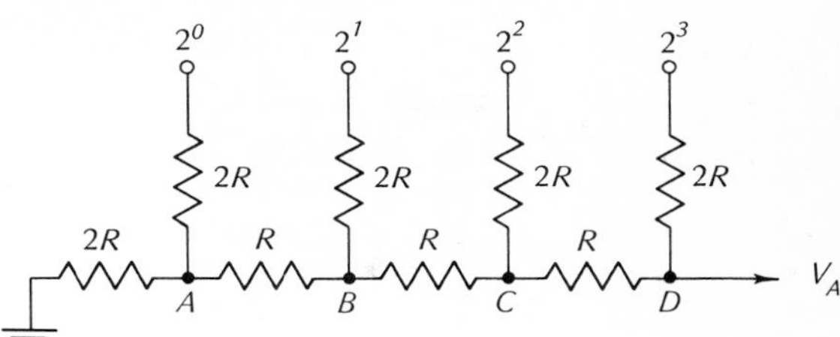

Fig. 11-6. Binary ladder.

also $2R$. These two resistors can be combined to form an equivalent resistor of value R as shown in Fig. 11-7*a*.

Now, moving to node *B*, we see the total resistance looking into the branch toward node *A* is $2R$, as is the total resistance looking out toward the 2^1 input. These resistors can be combined to simplify the network as shown in Fig. 11-7*b*.

From this figure, it can be seen that the total resistance looking from node *C* down the branch toward node *B* or out the branch toward the 2^2 input is still $2R$. The circuit in Fig. 11-7*b* can then be reduced to the equivalent shown in Fig. 11-7*c*.

From this equivalent circuit, it is clear that the resistance looking back toward node *C* is $2R$, as is the resistance looking out toward the 2^3 input.

From the above, we can conclude that the total resistance looking from any node back toward the terminating resistor or out toward the digital input is $2R$. Notice that this is true regardless of whether the digital inputs are at ground or $+V$ V. The justification for this statement is the fact that the internal impedance of an ideal voltage source is 0Ω, and we are assuming that the digital inputs are ideal voltage sources.

Fig. 11-7.

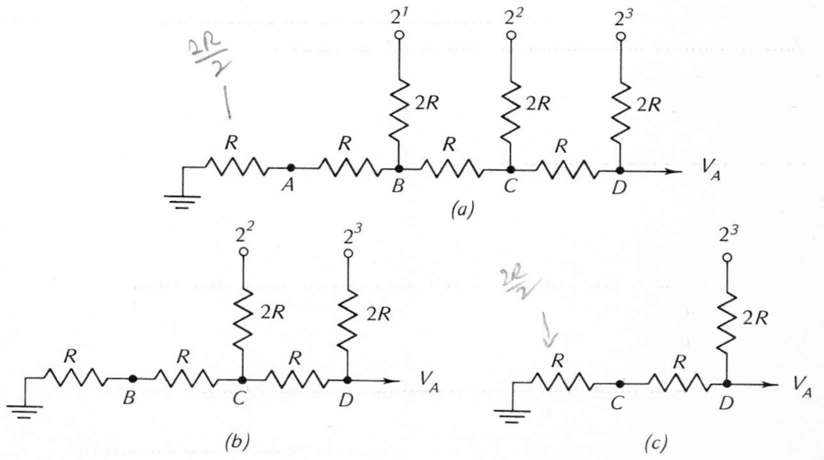

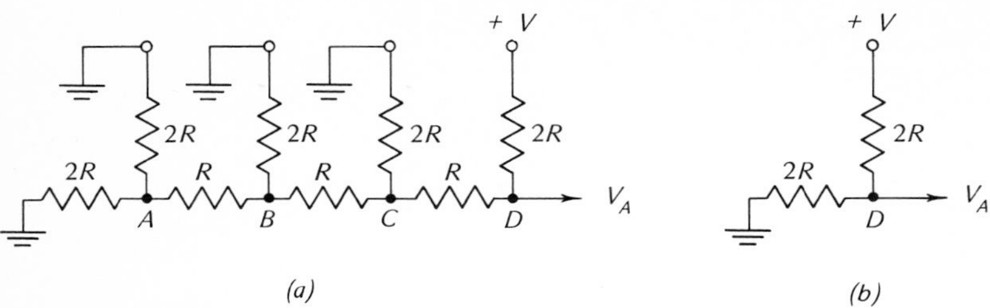

Fig. 11-8. (a) Binary ladder for a digital input of 1000. (b) Equivalent circuit for digital input of 1000.

We can use the resistance characteristics of the ladder to determine the output voltages for the various digital inputs. First, assume that the digital input signal is 1000. With this input signal, the binary ladder can be drawn as shown in Fig. 11-8a. Since there are no voltage sources to the left of node D, the entire network to the left of this node can be replaced by a resistance of $2R$ to form the equivalent circuit shown in Fig. 11-8b. From this equivalent circuit, it can be easily seen that the output voltage is

$$V_A = +V \frac{2R}{2R + 2R} = \frac{+V}{2}$$

Thus a 1 in the MSB position will provide an output voltage of $+V/2$.

To determine the output voltage due to the second most significant bit, assume a digital input signal of 0100. This can be represented by the circuit shown in Fig. 11-9a. Since there are no voltage sources to the left of node C, the entire network to the left of this node can be replaced by a resistance of $2R$, as shown in Fig. 11-9b. Let us now replace the network to the left of node C with its Thévenin equivalent by cutting the circuit on the jagged line shown in Fig. 11-9b. The Thévenin equivalent is clearly a resistance R in series with a voltage source $+V/2$. The final equivalent circuit with the Thévenin equivalent included is shown in Fig. 11-9c. From this circuit, the output voltage is clearly

$$V_A = \frac{+V}{2} \frac{2R}{R + R + 2R} = \frac{+V}{4}$$

Thus the second MSB provides an output voltage of $+V/4$.

This process can be continued, and it can be shown that the third MSB provides an output voltage of $+V/8$, the fourth MSB provides an output voltage of $+V/16$, and so on. The output voltages for the binary ladder are summarized in Fig. 11-10; note that each digital input is transformed into a properly weighted binary output voltage.

Example 11-4

What are the output voltages caused by each bit in a five-bit ladder if the input levels are $0 = 0$ V and $1 = +10$ V?

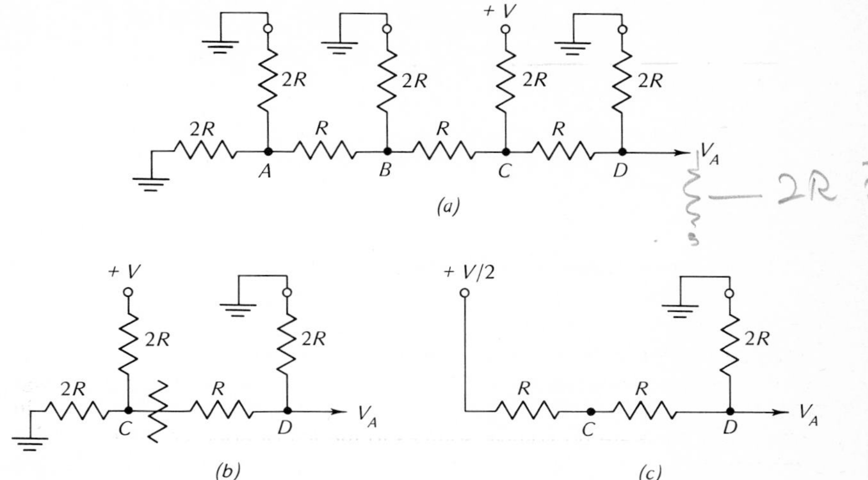

Fig. 11-9. (*a*) Binary ladder for digital input of 0100. (*b*) Partially reduced equivalent circuit. (*c*) Final equivalent circuit using Thévenin's theorem.

Solution

The output voltages can be easily calculated using Fig. 11-10. They are

$$\text{MSB} \qquad V_A = \frac{V}{2} = \frac{+10}{2} = +5 \text{ V}$$

$$\text{Second MSB} \qquad V_A = \frac{V}{4} = \frac{+10}{4} = +2.5 \text{ V}$$

$$\text{Third MSB} \qquad V_A = \frac{V}{8} = \frac{+10}{8} = +1.25 \text{ V}$$

$$\text{Fourth MSB} \qquad V_A = \frac{V}{16} = \frac{+10}{16} = +0.625 \text{ V}$$

$$\text{LSB} = \text{Fifth MSB} \qquad V_A = \frac{V}{32} = \frac{+10}{32} = +0.3125 \text{ V}$$

Bit position	Binary weight	Output voltage
MSB	1/2	V/2
2nd MSB	1/4	V/4
3rd MSB	1/8	V/8
4th MSB	1/16	V/16
5th MSB	1/32	V/32
6th MSB	1/64	V/64
7th MSB	1/128	V/128
.	.	.
.	.	.
.	.	.
*N*th MSB	$1/2^N$	$V/2^N$

Fig. 11-10. Binary-ladder output voltages.

Since this ladder is composed of linear resistors, it is a linear network and the principle of superposition can be used. This means that the total output voltage due to a combination of input digital levels can be found by simply taking the sum of the output levels caused by each digital input individually.

In equation form, the output voltage is given by

$$V_A = \frac{V}{2} + \frac{V}{4} + \frac{V}{8} + \frac{V}{16} + \cdots + \frac{V}{2^n} \qquad (11\text{-}2)$$

where n is the total number of bits at the input.

This equation can be simplified somewhat by factoring and collecting terms. The output voltage can then be given in the form

$$V_A = \frac{V_0 2^0 + V_1 2^1 + V_2 2^2 + V_3 2^3 + \cdots + V_{n-1} 2^{n-1}}{2^n} \qquad (11\text{-}3)$$

where $V_0, V_1, V_2, V_3, \ldots, V_{n-1}$ are the digital input-voltage levels. Equation (11-3) can be used to find the output voltage from the ladder for any digital input signal.

Example 11-5

Find the output voltage from a five-bit ladder which has a digital input of 11010. Assume $0 = 0$ V and $1 = +10$ V.

Solution

By Eq. (11-3),

$$V_A = \frac{0 \times 2^0 + 10 \times 2^1 + 0 \times 2^2 + 10 \times 2^3 + 10 \times 2^4}{2^5}$$

$$= \frac{10(2 + 8 + 16)}{32} = \frac{10 \times 26}{32} = +8.125 \text{ V}$$

This solution can be checked by adding the individual bit contributions calculated in Example 11-4 above.

Notice that Eq. (11-3) is very similar to Eq. (11-1), which was developed for the resistive divider. They are in fact identical with the exception of the denominators. This is a subtle but very important difference. Recall that the full-scale voltage for the resistive divider is equal to the voltage level of the digital input 1. On the other hand, examination of Eq. (11-2) shows that the full-scale voltage for the ladder is given by

$$V_A = V(1/2 + 1/4 + 1/8 + 1/16 + \cdots + 1/2^n)$$

The terms inside the brackets form a geometric series whose sum approaches 1, given enough terms. However, it never quite reaches 1. Therefore, the full-scale output voltage of the ladder approaches V in the limit, but never quite reaches it.

Example 11-6

What is the full-scale output voltage of the five-bit ladder in Example 11-4?

Solution

The full-scale output voltage is simply the sum of the individual bit voltages. Thus

$$V_A = 5 + 2.5 + 1.25 + 0.625 + 0.3125 = +9.6875 \text{ V}$$

To keep the ladder in perfect balance and to maintain symmetry, the output of the ladder should be terminated in a resistance of $2R$. This will result in a lowering of the output voltage, but if the $2R$ load is maintained constant, the output voltages will still be a properly weighted sum of the binary input bits. If the load is varied, the output voltage will not be a properly weighted sum and care must be exercised to ensure that the load resistance is constant.

Terminating the output of the ladder with a load of $2R$ also ensures that the input resistance to the ladder seen by each of the digital voltage sources is constant. With the ladder balanced in this manner, the resistance looking into any branch from any node has a value of $2R$. Thus, the input resistance seen by any input digital source is $3R$. This is a definite advantage over the resistive divider, since the digital voltage sources can now all be designed for the same load.

Example 11-7

Suppose the value of R for the five-bit divider described in Example 11-4 is 1,000 Ω. Determine the current that each input digital voltage source must be capable of supplying. Also determine the full-scale output voltage if the ladder is terminated with a load resistance of 2,000 Ω.

Solution

The input resistance into the ladder seen by each of the digital sources is $3R = 3{,}000\ \Omega$. Thus, for a voltage level of +10 V, each source must be capable of supplying $I = 10/(3 \times 10^3) = 3\frac{1}{3}$ mA (without the $2R$ load resistor, the resistance looking into the MSB terminal is actually $4R$). The no-load output voltage of the ladder has already been determined in Example 11-6. This open-circuit output voltage along with the open-circuit output resistance can be used to form a Thévenin equivalent circuit for the output of the ladder. The resistance looking back into the ladder is clearly $R = 1{,}000\ \Omega$. Thus the Thévenin equivalent is as shown in Fig. 11-11. From this figure, the output voltage is

$$V_A = +9.6875 \frac{2R}{2R + R} = +6.4583 \text{ V}$$

1-30-77

11-3 D/A CONVERTER

Either the resistive divider or the ladder can be used as the basis for a digital-to-analog converter. It is in the resistive network that the actual translation from a

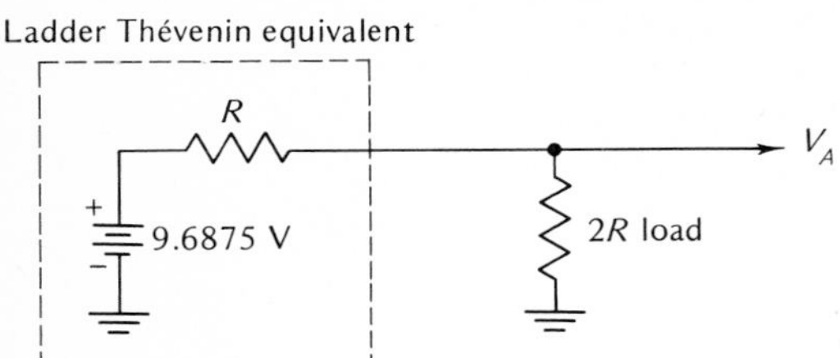

Fig. 11-11. Example 11-7.

digital signal to an analog voltage takes place. There is, however, the need for additional circuitry to complete the design of the D/A converter.

As an integral part of the D/A converter there must be a register which can be used to store the digital information. This register could be any one of the many types discussed in the previous chapter. The simplest register is formed using *RS* flip-flops, with one flip-flop per bit. There must also be level amplifiers between the register and the resistive network to ensure that the digital signals presented to the network are all of the same level and are constant. Finally, there must be some form of gating on the input of the register such that the flip-flops can be set with the proper information from the digital system. A complete D/A converter in block-diagram form is shown in Fig. 11-12.

Let us expand on the block diagram shown in this figure by drawing the complete schematic for a four-bit D/A converter as shown in Fig. 11-13. You will recognize that the resistor network used is of the ladder type.

The level amplifiers each have two inputs: one input is the +10 V from the precision voltage source, and the other is from a flip-flop. The amplifiers work in such a way that when the input from a flip-flop is high, the output of the amplifier is at +10 V. When the input from the flip-flop is low, the output is 0 V.

The four flip-flops form the register necessary for storing the digital information.

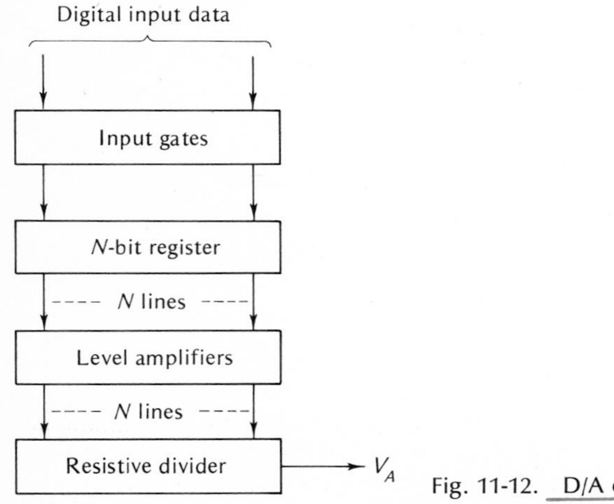

Fig. 11-12. D/A converter.

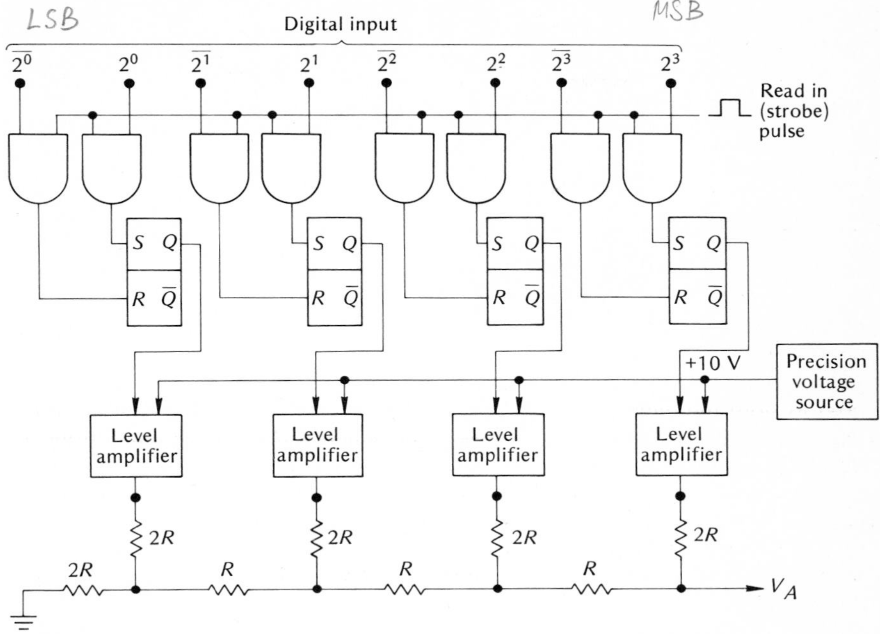

Fig. 11-13. Four-bit D/A converter.

The flip-flop on the right represents the MSB, and the flip-flop on the left represents the LSB. Each flip-flop is of the simple *RS* type and requires a positive level at the *R* or *S* inputs to reset or set it. The gating scheme for entering information into the register was discussed in the preceding chapter and should be familiar to you. Recall that with this gating scheme the flip-flops need not be reset (or set) each time new information is entered. When the *read in* line goes high, one of the two gates connected to each flip-flop is true, and the flip-flop is set or reset accordingly. Thus, data are entered into the register each time the *read in* (strobe) pulse occurs.

Quite often there is the necessity for decoding more than one signal, for example, the *X* and *Y* coordinates for a plotting board. In this event, there are two ways in which to decode the signals.

The first and most obvious method is simply to use one D/A converter for each signal. This method, shown in Fig. 11-14*a*, has the advantage that each signal to be decoded is held in its register and the analog output voltage is then held fixed. The digital input lines are connected in parallel to each converter. The proper converter is then selected for decoding by the select lines.

The second method involves using only one D/A converter and switching its output. This is called "multiplexing," and such a system is shown in Fig. 11-14*b*. The disadvantage here is that the analog output signal must be held between sampling periods, and the outputs must therefore be equipped with sample-and-hold amplifiers.

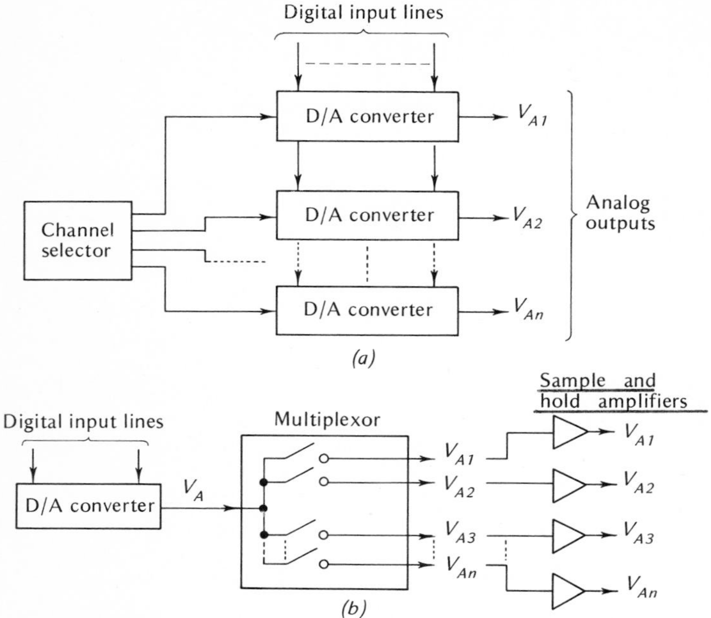

Fig. 11-14. Decoding a number of signals. (*a*) Channel-selection method. (*b*) Multiplex method.

A sample-and-hold amplifier can be approximated by a capacitor and high-gain amplifier (operational amplifier) as shown in Fig. 11-15. When the switch is closed, the capacitor charges to the D/A converter output voltage. When the switch is opened, the capacitor holds the voltage level until the next sampling time. The operational amplifier provides a large input impedance so as not to discharge the capacitor appreciably and at the same time offers gain to drive external circuits.

When the D/A converter is used in conjunction with a multiplexor, the maximum rate at which the converter can operate must be considered. Each time data are shifted into the register, transients appear at the output of the converter. This is due mainly to the fact that each flip-flop has different rise and fall times. Thus, a settling time must be allowed between the time data are shifted into the register and the time the analog voltage is read out. This settling time is the main factor in determining the maximum rate of multiplexing the output. The worst case is when all bits change (e.g., from 1000 to 0111).

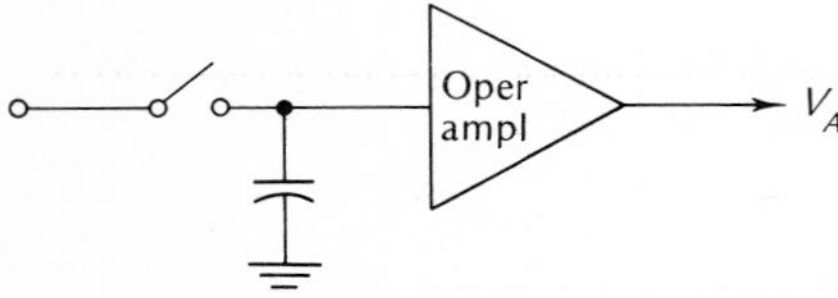

Fig. 11-15. Sample-and-hold amplifier.

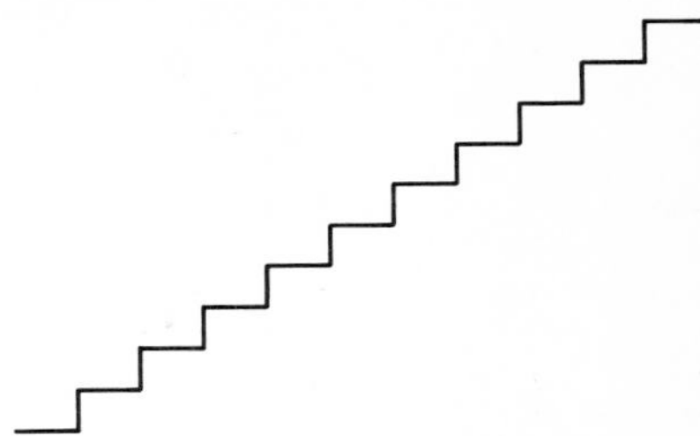

Fig. 11-16. Correct output-voltage waveform for monotonicity test.

Naturally, the capacitors on the sample-and-hold amplifiers are not capable of holding a voltage indefinitely, and the sampling rate must therefore be great enough to ensure that these voltages do not decay appreciably between samples. The sampling rate is a function of the capacitors as well as the frequency of the analog signal which is expected at the output of the converter.

Two simple but important tests which can be performed to check for proper operation of the D/A converter are the steady-state accuracy test and the monotonicity test.

The *steady-state accuracy* test involves setting a known digital number in the input register, measuring the analog output with an accurate meter, and comparing with the theoretical value.

Checking for *monotonicity* means checking that the output voltage increases regularly as the input digital signal increases. This can be accomplished by using a counter as the digital input signal and observing the analog output on an oscilloscope. For proper monotonicity, the output waveform should be a perfect staircase waveform as shown in Fig. 11-16. The steps on the staircase waveform must be equally spaced and of the exact same amplitude. Missing steps, steps of different amplitude, or steps in a downward fashion indicate malfunctions.

The monotonicity test does not check the system for accuracy, but if the system passes the test, it is relatively certain that the converter error is less than ± 1 LSB. Converter accuracy and resolution are the subjects of the next section.

Example 11-8

Suppose that in the course of a monotonicity check on the four-bit converter in Fig. 11-13, the waveform shown in Fig. 11-17 is observed. What is the probable malfunction in the converter?

Solution

There is obviously some malfunction since the actual output waveform is not continuously increasing as it should be. Directly below the waveform the actual digital inputs are shown. Notice that the converter functions correctly up to count 3. At count 4, however, the output should be 4 units in amplitude. Instead it drops to 0. It remains 4 units below the correct level until it reaches count 8. Then, from count 8 to count 11, the output level is correct. But again at count 12 the output falls 4 units below the correct level, and remains there for the next four levels. If you examine the waveform carefully, you will note that the output is 4 units below normal

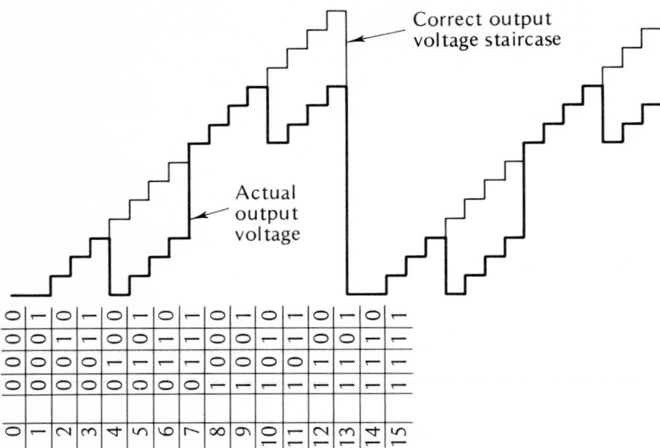

Fig. 11-17. Irregular output voltage for Example 11-8.

during the time when the 2^2 bit is supposed to be high. This then suggests that the 2^2 bit is being dropped (i.e., the 2^2 input to the ladder is not being held high). This means that the 2^2 level amplifier is malfunctioning, or the 2^2 flip-flop is not being set, or the 2^2 AND gate is not operating properly. In any case, the monotonicity check has clearly shown that the second MSB is not being used and that the converter is not operating properly.

11-4 D/A ACCURACY AND RESOLUTION

Two very important aspects of the D/A converter are the resolution and the accuracy of the conversion. There is a definite distinction between the two, and you should clearly understand the differences.

The accuracy of the D/A converter is primarily a function of the accuracy of the precision resistors used in the ladder and the precision of the reference voltage supply used. Accuracy is a measure of how close the actual output voltage is to the theoretical output voltage.

For example, suppose that the theoretical output voltage for a particular input should be +10 V. An accuracy of ± 10 percent means that the actual output voltage must be somewhere between +9 and +11 V. Similarly, if the actual output voltage were somewhere between + 9.9 and +10.1 V, this would imply an accuracy of ± 1 percent.

Resolution, on the other hand, defines the smallest increments in voltage that can be discerned. Resolution is primarily a function of the number of bits in the digital input signal. That is to say, the smallest increment in output voltage is determined by the LSB.

In a four-bit system using a ladder, for example, the LSB has a weight of $1/16$. This means that the smallest increment in output voltage is $1/16$ of the input voltage. To make the arithmetic easy, let us assume that this four-bit system has input voltage levels of +16 V. Since the LSB has a weight of $1/16$, a change in the LSB results in a change of 1 V in the output. Thus the output voltage changes in steps (or increments) of 1 V. The output voltage of this converter is then the staircase shown in Fig. 11-16 and ranges from 0 to +15 V in 1 V steps. This converter can be used to represent analog voltages from 0 to +15 V but it cannot resolve voltages into increments smaller than 1 V. Thus, if we desired to produce +4.2 V using this converter, the actual output voltage would be +4.0 V. Similarly, if we desired a voltage of +7.8 V the actual output voltage would be +8.0 V. It is clear that this converter is not capable of distinguishing voltages finer than 1 V . . . which is the resolution of the converter.

If we wanted to represent voltages to a finer resolution, we would have to use a converter with more input bits. As an example, the LSB of a 10-bit converter has a weight of $1/1024$. Thus the smallest incremental change in the output of this converter is approximately $1/1000$ of the full-scale voltage. If this converter has a +10-V full-scale output, the resolution is approximately $+10 \times 1/1000 = 10$ mV. This converter is then capable of representing voltages to within ± 10 mV.

Example 11-9

What is the resolution of a nine-bit D/A converter which uses a ladder network? What is this resolution expressed as a percent? If the full-scale output voltage of this converter is +5 V, what is the resolution in volts?

Solution

The LSB in a nine-bit system has a weight of $1/512$. Thus this converter has a resolution of 1 part in 512. The resolution expressed as a percentage is $1/512 \times 100$ percent $\cong 0.2$ percent. The voltage resolution is obtained by multiplying the weight of the LSB by the full-scale output voltage. Thus the resolution in volts is $1/512 \times 5 \cong 10$mV.

Example 11-10

How many bits are required at the input of a converter if it is required to resolve voltages to 5 mV and the ladder has +10 V full scale?

Solution

The LSB of an 11-bit system has a resolution of $1/2048$. This would provide a resolution at the output of $1/2048 \times (+10) \cong +5$ mV.

It is important to realize that resolution and accuracy in a system should be compatible. For example, in the four-bit system previously discussed, the resolution was found to be 1 V. Clearly it would be unjustified to construct such a system to an accuracy of 0.1 percent. This would mean that the system would be accurate to ± 16 mV but would be capable of distinguishing only to the nearest volt.

Similarly, it would be wasteful to construct the 11-bit system described in Example 11-10 to an accuracy of only ± 1 percent. This would mean that the output voltage would be accurate only to ± 100 mV whereas it is capable of distinguishing to the nearest 5 mV.

11-5 A/D CONVERTER— SIMULTANEOUS CONVERSION

The process of converting an analog voltage into an equivalent digital signal is known as analog-to-digital conversion. This operation is somewhat more complicated than the converse operation of D/A conversion. A number of different methods have been developed, the simplest of which is probably the simultaneous method.

The simultaneous method of A/D conversion is based on the use of a number of comparator circuits. One such system using three comparator circuits is shown in Fig. 11-18. The analog signal to be digitized serves as one of the inputs to each comparator. The second input is a standard reference voltage. The reference voltages used are $+V/4$, $+V/2$, and $+3V/4$. The system is then capable of accepting an analog input voltage between 0 and $+V$ V.

If the analog input signal exceeds the reference voltage to any comparator, that comparator turns on (let's assume this means that the output of the comparator goes high). Now, if all the comparators are off, the analog input signal must be between 0 and $+V/4$ V. If C_1 is high (comparator C_1 is on), and C_2 and C_3 are low, then the input must be between $+V/4$ and $+V/2$ V. If C_1 and C_2 are high while C_3 is low, the input must be between $+V/2$ and $+3V/4$. Finally, if all comparator outputs are high, the input signal must be between $+3V/4$ and $+V$ V. The comparator output levels for the various ranges of input voltages are summarized in Fig. 11-18*b*.

Examination of Fig. 11-18 shows that there are four voltage ranges that can be detected by this converter. Four ranges can be effectively discerned by two digital bits. The three comparator outputs can then be fed into a coding network to provide two bits which are equivalent to the input analog voltage. The bits of the coding network can then be entered into a flip-flop register for storage. The complete block diagram for such an A/D converter is shown in Fig. 11-19.

In order to gain a clear understanding of the operation of the simultaneous A/D

Fig. 11-18. Simultaneous A/D conversion. (*a*) Logic circuit. (*b*) Comparator outputs for input voltage ranges.

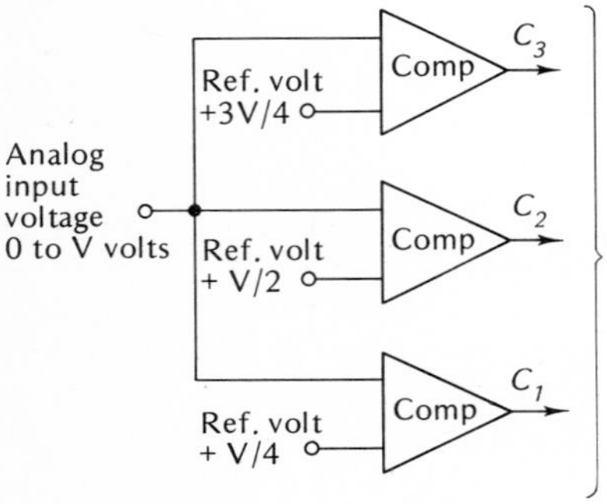

(*a*)

Input voltage	Comparator output C_1	C_2	C_3
0 to + V/4	Low	Low	Low
+V/4 to + V/2	High	Low	Low
+V/2 to + 3V/4	High	High	Low
+3V/4 to + V	High	High	High

(*b*)

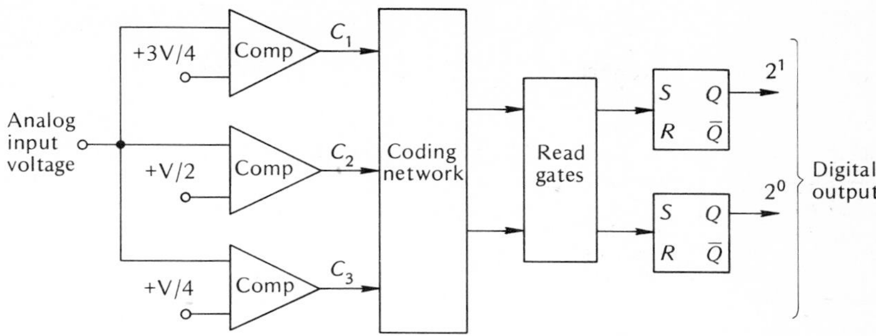

Fig. 11-19. 2-bit simultaneous A/D converter.

converter, let us investigate the three-bit converter shown in Fig. 11-20. Notice that to convert the input signal to a digital signal having three bits, it is necessary to have seven comparators (this allows a division of the input into eight ranges). Recall that, for the two-bit converter, it required three comparators to define four ranges. In general it can be said that $2^n - 1$ comparators are required to convert to a digital signal having n bits. Some of the comparators have inverters at their outputs since both C and $\overline{C}$ are needed for the encoding matrix.

Fig. 11-20. Three-bit simultaneous A/D converter.

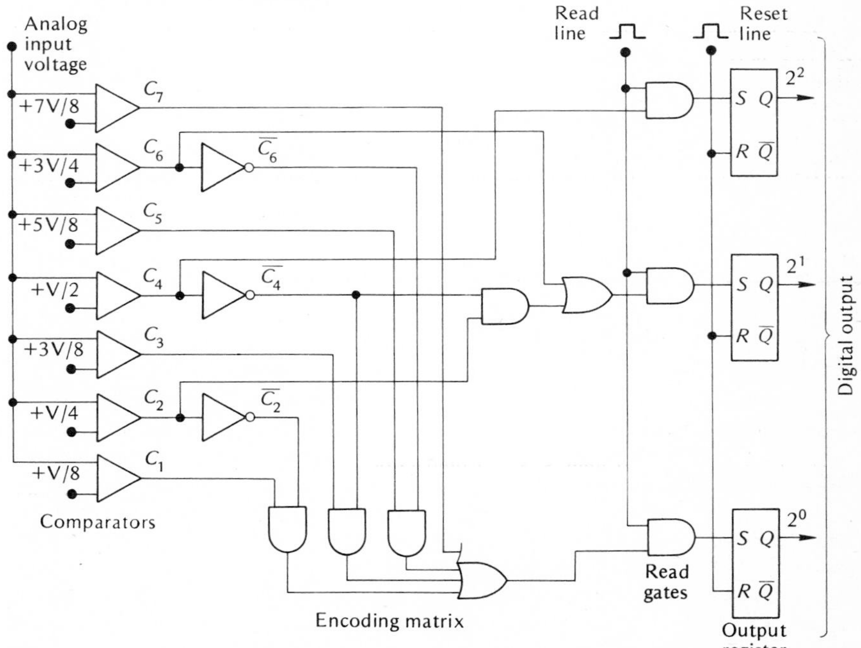

Input voltage	Comparator for level							Binary output		
	C_1	C_2	C_3	C_4	C_5	C_6	C_7	2^2	2^1	2^0
0 to V/8	Low	Low	Low	Low	Low	Low	Low	0	0	0
V/8 to V/4	High	Low	Low	Low	Low	Low	Low	0	0	1
V/4 to 3V/8	High	High	Low	Low	Low	Low	Low	0	1	0
3V/8 to V/2	High	High	High	Low	Low	Low	Low	0	1	1
V/2 to 5V/8	High	High	High	High	Low	Low	Low	1	0	0
5V/8 to 3V/4	High	High	High	High	High	Low	Low	1	0	1
3V/4 to 7V/8	High	High	High	High	High	High	Low	1	1	0
7V/8 to V	High	High	High	High	High	High	High	1	1	1

Fig. 11-21. Logic table for the simultaneous converter in Fig. 11-20.

The encoding matrix must accept seven input levels and encode them into a three-bit binary number (having eight possible states). Operation of the encoding matrix can be most easily understood by examination of the table of outputs in Fig. 11-21.

The 2^2 bit is easiest to determine since it must be high (the 2^2 flip-flop must be set) whenever C_4 is high.

The 2^1 line must be high whenever C_2 is high and $\overline{C}_4$ is high, or whenever C_6 is high. In equation form, we can write $2^1 = C_2\overline{C}_4 + C_6$.

The logic equation for the 2^0 bit can be found in a similar manner; it is $2^0 = C_1\overline{C}_2 + C_3\overline{C}_4 + C_5\overline{C}_6 + C_7$.

The transfer of data from the encoding matrix into the register must be carried out in two steps. First, a positive reset pulse must appear on the *reset* line to reset all the flip-flops low. Then, a positive *read* pulse allows the proper *read* gates to go high and thus transfer the digital information into the flip-flops.

The construction of a simultaneous A/D converter is quite straightforward and relatively easy to understand. However, as the number of bits in the desired digital number increases, the number of comparators increases very rapidly ($2^n - 1$), and the problem soon gets out of hand. Even though this method is simple and is indeed capable of extremely fast conversion rates, there are preferable methods for digitizing numbers having more than three or four bits.

11-6 A/D CONVERTER—COUNTER METHOD

A higher-resolution A/D converter using only one comparator could be constructed if a variable reference voltage were available. This reference voltage could then be applied to the comparator, and when it became equal to the input analog voltage the conversion would be complete.

To construct such a converter, suppose we begin with a simple binary counter. The digital output signals will be taken from this counter, and therefore we want it to be an *n*-bit counter, where *n* is the desired number of bits. Now let us connect the output of this counter to a standard binary ladder to form a simple D/A converter. If a clock is now applied to the input of the counter, the output of the binary ladder is the familiar staircase waveform shown in Fig. 11-16. This waveform is exactly the reference voltage signal we would like to have for the comparator! With a minimum of gating and control circuitry, this simple D/A converter can be changed into the desired A/D converter.

Figure 11-22 shows the block diagram for a counter-type A/D converter. The operation of the counter is as follows. First, the counter is reset to all 0s. Then, when a convert signal appears on the *start* line, the gate opens and clock pulses are allowed to pass through to the input of the counter. The counter advances through its normal binary count sequence, and the staircase waveform is generated at the output of the ladder. This waveform is applied to one side of the comparator, and the analog input voltage is applied to the other side. When the reference voltage equals (or exceeds) the input analog voltage, the gate is closed, the counter stops, and the conversion is complete. The number stored in the counter is now the digital equivalent of the analog input voltage.

Notice that this converter is composed of a D/A converter (the counter, level amplifiers, and binary ladder), one comparator, a clock, and the gate and control circuitry. This can really be considered as a closed-loop control system. An error signal is generated at the output of the comparator by taking the difference between the analog input signal and the feedback signal (staircase reference voltage). The error is detected by the control circuit, and the clock is allowed to advance the counter. The counter advances in such a way as to reduce the error signal by increasing the feedback voltage. When the error is reduced to zero, the feedback voltage is equal to the analog input signal, the control circuitry stops the clock from advancing the counter, and the system comes to rest.

The counter-type A/D converter provides a very good method for digitizing to a high resolution. This method is much simpler than the simultaneous method for high resolution, but the conversion time required is longer. Since the counter always begins at zero and counts through its normal binary sequence, it may require as many as 2^n counts before conversion is complete. The average conversion time is, of course, $2^n/2$ or 2^{n-1} counts.

Fig. 11-22. Counter-type A/D converter.

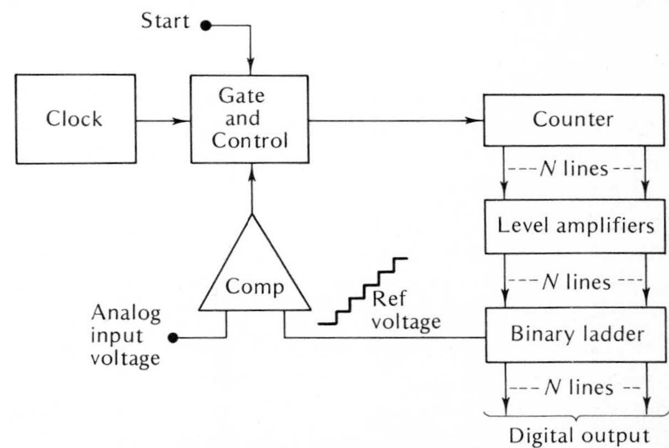

The counter advances one count for each cycle of the clock, and the clock therefore determines the conversion rate. Suppose, for example, we have a 10-bit converter. It requires 1,024 clock cycles for a full-scale count. If we are using a 1-MHz clock, the counter advances 1 count every microsecond. Thus, to count full scale requires $1024 \times 10^{-6} \cong 1.024$ ms. The converter reaches one-half full scale in half this time, or in 0.512 ms. The time required to reach one-half full scale can be considered the average conversion time for a large number of conversions.

Example 11-11

Suppose the converter shown in Fig. 11-22 is an eight-bit converter driven by a 500-kHz clock. Find:

(*a*) The maximum conversion time
(*b*) The average conversion time
(*c*) The maximum conversion rate

Solution

(*a*) An eight-bit converter has a maximum of $2^8 = 256$ counts. With a 500-kHz clock, the counter advances at the rate of 1 count each 2 μs. To advance 256 counts requires $256 \times 2 \times 10^{-6} = 512 \times 10^{-6} = 0.512$ ms.

(*b*) The average conversion time is one-half the maximum conversion time. Thus it is $\frac{1}{2} \times 0.512 \times 10^{-3} = 0.256$ ms.

(*c*) The maximum conversion rate is determined by the longest conversion time. Since the converter has a maximum conversion time of 0.512 ms, it is capable of making at least $1/(0.512 \times 10^{-3}) \cong 1{,}953$ conversions per second.

Figure 11-23 shows one method of implementing the control circuitry for the converter shown in Fig. 11-22. The waveforms for one conversion are also shown. A conversion is initiated by the receipt of a *start* signal. The positive edge of the *start* pulse is used to reset all the flip-flops in the counter and to trigger the one-shot. The output of the one-shot sets the *control* flip-flop, which makes the AND gate true and allows clock pulses to advance the counter.

The delay between the *reset* pulse to the flip-flops and the beginning of the clock pulses (ensured by the one-shot) is to ensure that all flip-flops are indeed reset before counting begins. This is a definite attempt to avoid any racing problems.

With the *control* flip-flop set, the counter advances through its normal count sequence until the staircase voltage from the ladder is equal to the analog input voltage. At this time, the comparator output changes state, generating a positive pulse which resets the *control* flip-flop. Thus the AND gate is closed and counting ceases. The counter now holds a digital number which is equivalent to the analog input voltage. The converter remains in this state until another conversion signal is received.

If a new start signal is generated immediately after each conversion is completed, the converter will operate at its maximum rate. The converter could then be used to digitize a signal as shown in Fig. 11-24*a*. Notice that the conversion times in digitizing this signal are not constant but depend on the amplitude of the input signal.

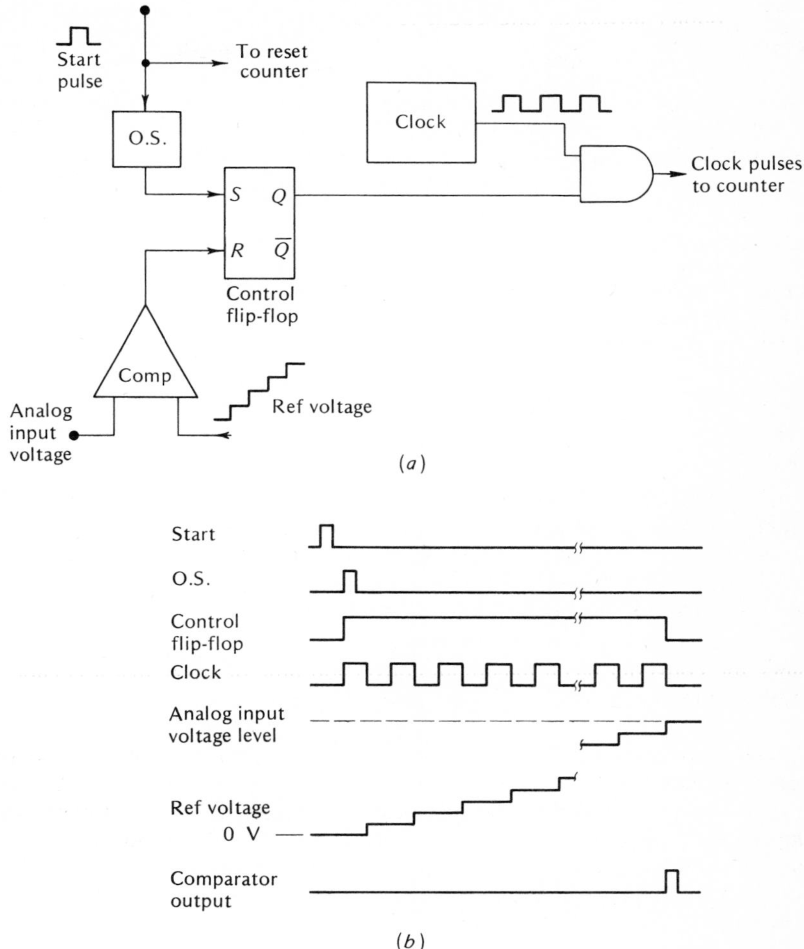

Fig. 11-23. Operation of the A/D converter in Fig. 11-22. (*a*) Control logic. (*b*) Waveforms.

The analog input signal can be reconstructed from the digital information by drawing straight lines from each digitized point to the next. Such a reconstruction is shown in Fig. 11-24*b*; it is indeed a reasonable representation of the original input signal. Take careful note of the fact that in this case the conversion times are small compared with the transient time of the input waveform.

On the other hand, if the transient time of the input waveform approaches the conversion time, the reconstructed output signal is not quite so accurate. Such a situation is shown in Fig. 11-25*a* and *b*. In this case, the input waveform changes at a

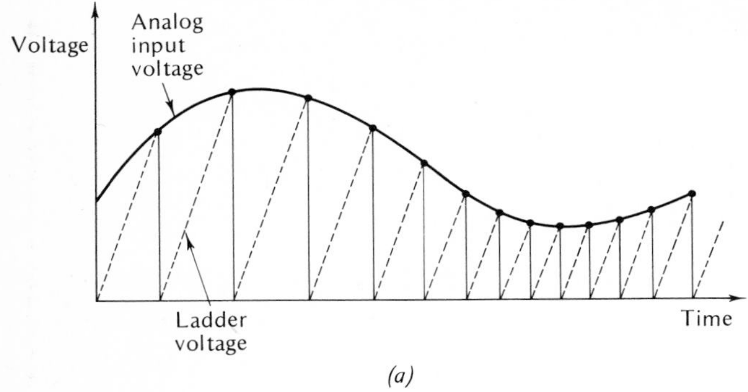

(a)

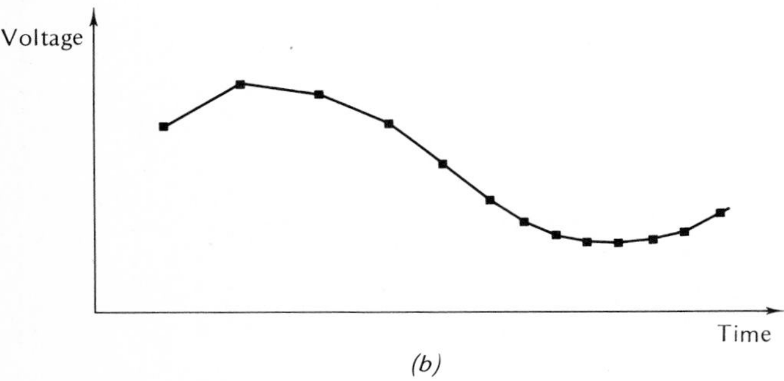

(b)

Fig. 11-24. (a) Digitizing an output voltage. (b) Reconstructed signal from digital data.

rate faster than the converter is capable of recognizing. Thus the need for reducing conversion time is apparent.

11-7 ADVANCED A/D TECHNIQUES

An obvious method for speeding up the conversion of the signal as shown in Fig. 11-25 is to eliminate the need for resetting the counter each time a conversion is made. If this were done, the counter would not begin at zero each time, but instead would begin at the value of the last converted point. This means that the counter would have to be capable of counting either up or down. This is no problem; we are already familiar with the operation of up-down counters (Chap. 9).

There is, however, the need for additional logic circuitry, since we must decide whether to count up or down by examining the output of the comparator. An A/D converter which uses an up-down counter is shown in Fig. 11-26. This method is known as continuous conversion, and thus is called a "continuous-type A/D converter."

The D/A portion of this converter is the same as those previously discussed, with

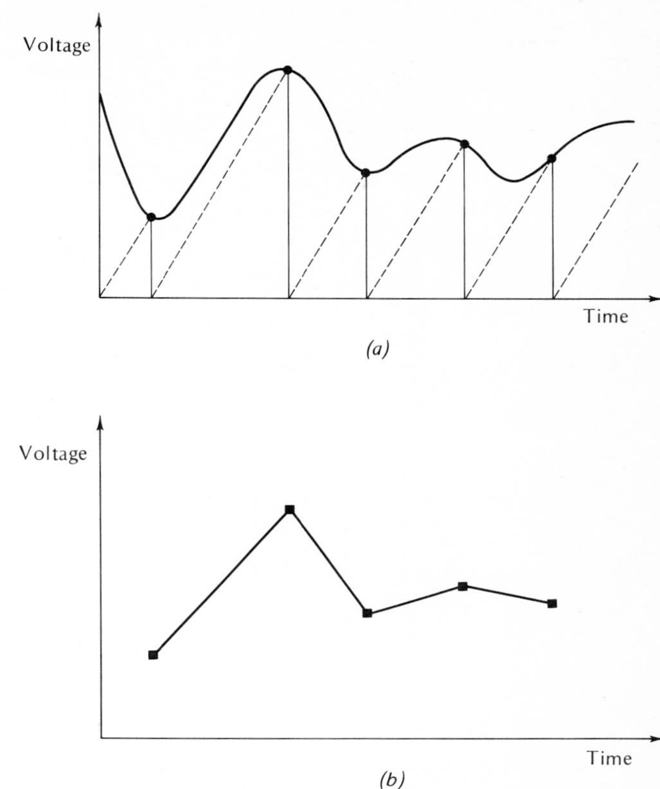

Fig. 11-25. (*a*) Digitizing an output voltage. (*b*) Reconstructed signal from digital data.

the exception of the counter. It is an up-down counter and has the *up* and *down count* control lines in addition to the *advance* line at its input.

The output of the ladder is fed into a comparator which has two outputs instead of one as before. When the analog voltage is more positive than the ladder output, the *up* output of the comparator is high. When the analog voltage is more negative than the ladder output, the *down* output is high.

If the *up* output of the comparator is high, the AND gate at the input of the *up* flip-flop is open, and the first time the clock goes positive, the *up* flip-flop is set. If we assume for the moment that the *down* flip-flop is reset, the AND gate which controls the *count up* line of the counter will be true and the counter will advance one count. The counter can advance only one count since the output of the one-shot resets both the *up* and *down* flip-flops just after the clock goes low. This can then be considered as one *count up* conversion cycle.

Notice that the AND gate which controls the *count up* line has inputs of *up* and $\overline{down}$. Similarly, the *count down* line AND gate has inputs of *down* and $\overline{up}$. This could be considered an exclusive-OR arrangement, and ensures that the *count down* and *count up* lines cannot both be high at the same time.

As long as the *up* line out of the comparator is high, the converter continues to operate one conversion cycle at a time. At the point where the ladder voltage becomes more positive than the analog input voltage, the *up* line of the comparator goes low and the *down* line goes high. The converter then goes through a *count down* conversion cycle. At this point, the ladder voltage is within one LSB of the analog voltage, and the converter oscillates about this point. This is not desirable since we want the converter to cease operation and not jump around the final value. The trick here is to adjust the comparator such that its outputs do not change at the same time.

Fig. 11-26. Continuous-type A/D converter. (*a*) Logic diagram. (*b*) Typical waveforms.

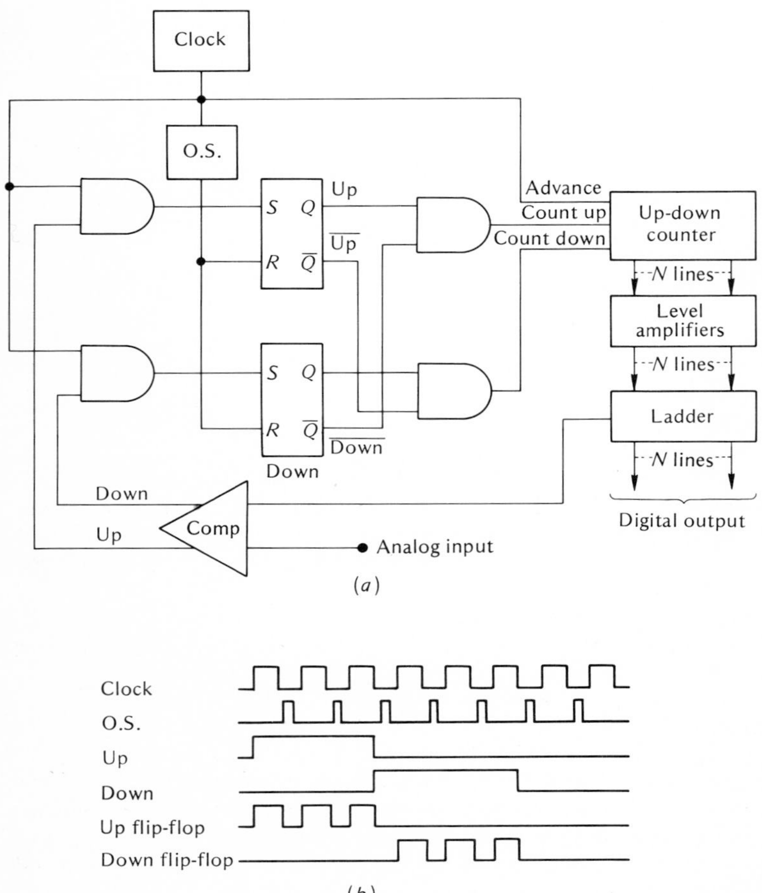

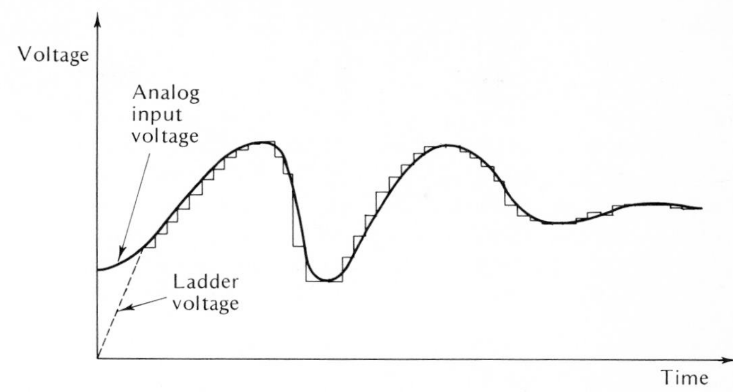

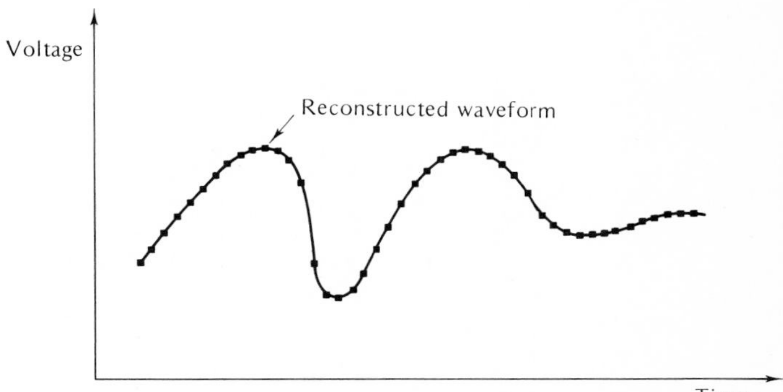

Fig. 11-27. Continuous A/D conversion.

We can accomplish this by adjusting the comparator such that the *up* output will not go high unless the ladder voltage is more than ½ LSB below the analog voltage. Similarly the *down* output will not go high unless the ladder voltage is more than ½ LSB above the analog voltage. This is called "centering on the LSB" and provides a digital output which is within ±½ LSB.

A waveform typical of this type of converter is shown in Fig. 11-27. You can see that this converter is capable of following input voltages which change at a much faster rate.

Example 11-12

Quite often, additional circuitry is added to a continuous converter to ensure that it cannot count off scale in either direction. For example, if the counter contained all 1s, it would be undesirable to allow it to progress through a *count up* cycle since the next count would advance it to all 0s. We would like to design the logic necessary to prevent this.

Solution

The two limit points which must be detected are all 1s and all 0s in the counter. Suppose we construct an AND gate having the 1 sides of all the counter flip-flops as its inputs. The output of this gate will be true whenever the counter contains all 1s. If the gate is then connected to the *reset* side of the *up* flip-flop, the counter will be unable to count beyond all 1s.

Similarly we can construct an AND gate having the 0 sides of all the counter flip-flops as its inputs. The output of this gate can be connected to the *reset* side of the *down* flip-flop, and the counter will then be unable to count beyond all 0s. The gates are shown in Fig. 11-28.

There are a variety of other methods for digitizing analog signals—too many to discuss in detail. Nevertheless, we shall take the time to examine some of the methods and the reasons for their importance.

Probably the most important single reason for investigating other methods of conversion is to determine ways to reduce the conversion time. Recall that the simultaneous converter has a very fast conversion time but becomes unwieldy for more than a few bits of information. The counter converter is simple logically but has a relatively long conversion time. The continuous converter has a very fast conversion time once it is locked on the signal but loses this advantage when multiplexing inputs.

If multiplexing is required, the successive-approximation converter is most useful. The block diagram for this type of converter is shown in Fig. 11-29*a*. The converter operates by successively dividing the voltage ranges in half. The counter is first reset to all 0s, and the MSB is then set. The MSB is then left in or taken out (by resetting the MSB flip-flop) depending on the output of the comparator. Then the second MSB is set in, and a comparison is made to determine whether or not to reset the second MSB flip-flop. The process is repeated down to the LSB, and at this time the desired number is in the counter. Since the conversion involves operating on one flip-flop at a time, beginning with the MSB, a ring counter may be used for flip-flop selection.

The successive-approximation method then is the process of approximating the analog voltage by trying one bit at a time beginning with the MSB. The operation is shown in diagram form Fig. 11-29*b*. It can be seen from this diagram that each conversion takes the same time and requires one conversion cycle for each bit. Thus the total conversion time is equal to the number of bits *n* times the time required for one conversion cycle. One conversion cycle normally requires one cycle of the

Fig. 11-28. Count-limiting gates for the converter in Fig. 11-26.

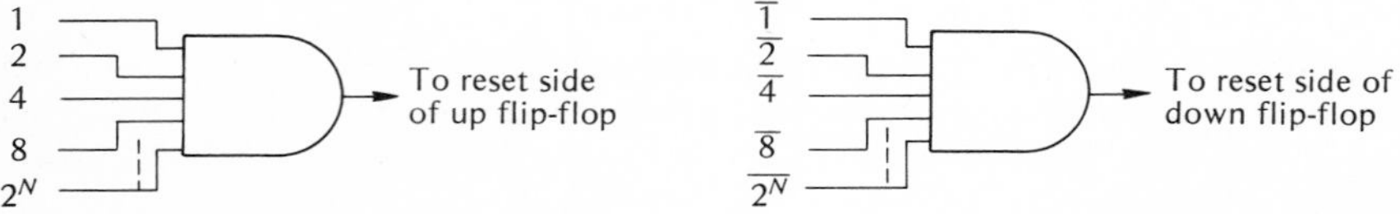

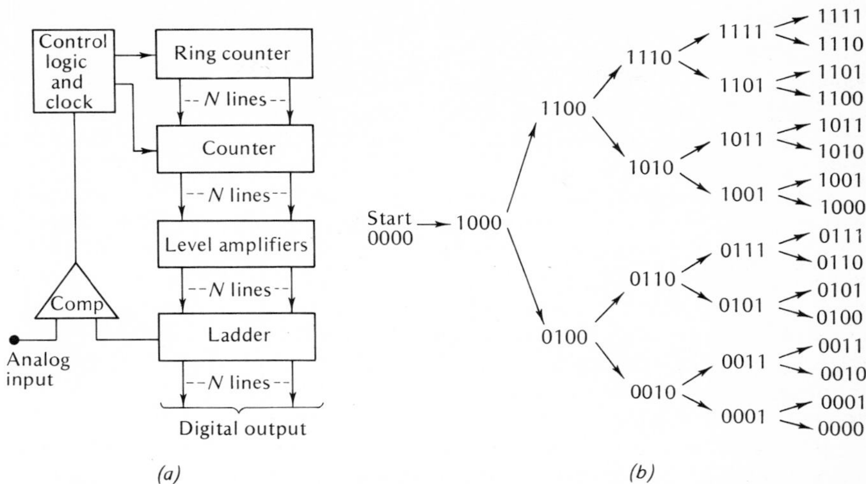

Fig. 11-29. Successive-approximation converter. (*a*) Logic diagram. (*b*) Operation diagram.

clock. As an example, a 10-bit converter operating with a 1-MHz clock has a conversion time of $10 \times 10^{-6} = 10^{-5} = 10\ \mu s$.

When dealing with conversion times this short, it is usually necessary to take into account the other delays in the system (e.g., switching time of the multiplexor, settling time of the ladder network, comparator delay, and settling time).

Another method for reducing the total conversion time of a simple counter converter is to divide the counter into sections. Such a configuration is called a "section counter." To determine how the total conversion time might be reduced by this method, assume we have a standard eight-bit counter. If this counter is divided into two equal counters of four bits each, we have a section converter. The converter operates by setting the section containing the four LSBs to all 1s and then advancing the other section until the ladder voltage exceeds the input voltage. At this point the four LSBs are all reset, and this section of the counter is then advanced until the ladder voltage equals the input voltage.

Notice that it requires a maximum of $2^4 = 16$ counts for each section to count full scale. Thus this method requires only $2 \times 2^4 = 2^5 = 32$ counts to reach full scale. This is a considerable reduction over the $2^8 = 256$ counts required for the straight eight-bit counter. There is, of course, some extra time required to set the counters initially and to switch from counter to counter during the conversion. This logical operation time is very small, however, compared with the total time saved by this method.

This type of converter is quite often used for digital voltmeters since it is very convenient to divide the counters by counts of 10. Each counter is then used to represent one of the digits of the decimal number appearing at the output of the voltmeter.

11-8 A/D ACCURACY AND RESOLUTION

Since the A/D converter is a closed-loop system involving both analog and digital systems, the overall accuracy must include errors from both the analog and digital positions. In determining the overall accuracy it is easiest to separate the two sources of error.

If we assume that all components are operating properly, the source of the digital error is simply determined by the resolution of the system. In digitizing an analog voltage, we are trying to represent a continuous analog voltage by an equivalent set of digital numbers. When the digital levels are converted back into analog form by the ladder, the output is the familiar staircase waveform. This waveform is a representation of the input voltage but is certainly not a continuous signal. It is, in fact, a discontinuous signal composed of a number of discrete steps. In trying to reproduce the analog input signal, the best we can do is to get on the step which most nearly equals the input voltage in amplitude.

The simple fact that the ladder voltage has steps in it leads to the digital error in the system. The smallest digital step, or *quantum,* is due to the LSB and can be made smaller only by increasing the number of bits in the counter. This inherent error is often called the "quantization error" and is commonly ± 1 bit. If the comparator is centered, as with the continuous converter, the quantization error can be made $\pm 1/2$ LSB.

The main source of analog error in the A/D converter is probably the comparator. Other sources of error are the resistors in the ladder, the reference-voltage supply ripple, and noise. These can, however, usually be made secondary to the sources of error in the comparator.

The sources of error in the comparator are centered around variations in the dc switching point. The dc switching point is the difference between the input-voltage levels which causes the output to change state. Variations in switching are primarily due to offset, gain, and linearity of the amplifier used in the comparator. These parameters usually vary slightly with input-voltage levels and quite often with temperature. It is these changes which give rise to the analog error in the system.

An important measure of converter performance is given by the differential linearity. *Differential linearity* is a measure of the variation in voltage-step size which causes the converter to change from one state to the next. It is usually expressed as a percent of the average step size. This performance characteristic is also a function of the conversion method and is best for the converters having counters which count continuously. The counter-type and continuous-type converters usually have better differential linearity than the successive-approximation-type converters. This is true since, in the one case, the ladder voltage is always approaching the analog voltage from the same direction. In the other case, the ladder voltage is first on one side of the analog voltage and then on the other. The comparator is then being used in both directions, and the net analog error from the comparator is thus greater.

The next logical question which might be asked is, what should be the relative order of magnitudes of the analog and digital errors? As mentioned previously, it would be hard to justify constructing a 15-bit converter which has an overall error of ± 1 percent. On the other hand, it would be equally difficult to justify building a six-bit converter to an accuracy of 0.1 percent. In general, it is considered good practice to construct converters having analog and digital errors of approximately the same magnitudes. There are many arguments pro and con for this, and any final argument would have to depend on the situation. As as example, an eight-bit converter would have a quantization error of $1/256 \cong 0.4$ percent. It would then seem reasonable to construct this converter to an accuracy of 0.5 percent in an effort to achieve an overall accuracy of 1.0 percent. This might mean constructing the ladder to an accuracy of 0.1 percent, the comparator to an accuracy of 0.2 percent, etc., since these errors are all accumulative.

Example 11-13

What overall accuracy could one reasonably expect from the construction of a 10-bit A/D converter?

Solution

A 10-bit converter has a quantization error of $1/1024 \cong 0.1$ percent. If the analog portion can be constructed to an accuracy of 0.1 percent, it would seem reasonable to strive for an overall accuracy of 0.2 percent.

11-9 ELECTROMECHANICAL A/D CONVERSION

There is another area of application in which A/D conversion is very important. This involves the translation of the angular position of a shaft into digital information. A very common application of this type of conversion is found in large radar installations where the azimuth and elevation information are determined directly from shaft position. There are many other examples in aircraft and aerospace applications. The method is not necessarily limited to rotational information since rectilinear information can be translated into rotational information by means of a gearing arrangement.

In any case, the task involves changing position information (which can usually be considered analog information since it is continuous) into equivalent digital information. This is most generally accomplished by the use of a code wheel as shown in Fig. 11-30. This particular wheel is coded in straight binary fashion and represents three bits. The wheel is divided into three concentric bands, each band representing one bit. The innermost band is divided into two equal segments and is the MSB. The middle band is divided into four equal segments and represents the second MSB. The outer band has eight equal segments and is the LSB.

If the light areas on the wheel are transparent and the dark areas are opaque, the digital information can be obtained by placing light sources and photosensors on opposite sides of the disk as shown in the figure. The output of a photosensor is high if light is sensed and low if no light is sensed. These outputs then represent 1s and 0s and can be amplified, passed through logic circuits, and used to set flip-flops. Such a system is called an "optical encoder." The sensing could also be ac-

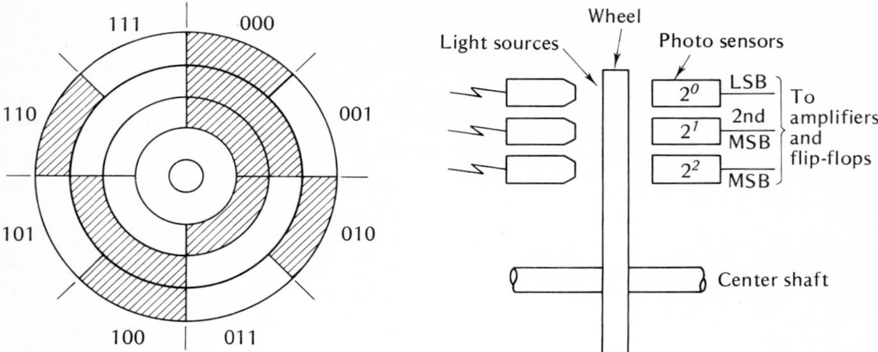

Fig. 11-30. Three-bit binary-code wheel.

complished by placing a brush on each band and making the light areas a conducting material and the dark areas an insulating material.

Now let us suppose that the wheel is positioned under the sensors such that the output reads 011. Further, let us assume that the wheel is very near the dividing line between 100 and 011. If there is an ambiguity in reading the LSB (i.e., the sensor cannot decide whether to read a 1 or a 0 since it is right on the dividing line), the output may vary between 011 and 010. This is not too bad since the output is jumping only from one adjacent position to the next. However, if the ambiguity is in the MSB, the results could be disatrous since the output is jumping from 011 to 111, which is 180° of error. Consider the results of this error if this wheel were the source of azimuth information which directs the firing of a large missile installation! Obviously something must be done to overcome this ambiguity.

Since it might be quite difficult to resolve the problem of sensing when the wheel

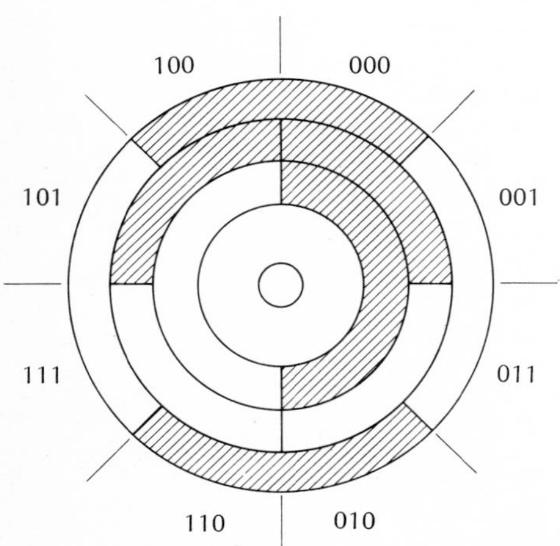

Fig. 11-31. Three-bit Gray-code wheel.

stops on the dividing line, it might be more fruitful to investigate a different code. What we would like is a code which changes only one bit at a time when going from any position to the next. This would then ensure an ambiguity of only one position, and would eliminate the 180° ambiguity previously demonstrated. Such a code, as you might recall from Chap. 3, is the Gray code. A code wheel constructed using Gray code is shown in Fig. 11-31. By examination of the wheel you can see that the greatest error caused by a reading ambiguity is one segment of rotation.

The three-bit wheel discussed has the ability to digitize shaft position into an equivalent three-bit binary number. This implies eight positions around the wheel, and thus each position represents a 45° segment. To determine shaft position to the nearest 45° is of course not very accurate, and we would like a better measurement.

To obtain a closer reading, it is only necessary to add extra bands to the code wheel and thus extra bits to the digital number. In general, the degree of resolution obtained is given by $360°/2^n$, where n is the number of bits in the binary number (or bands on the wheel).

Example 11-14

What degree of resolution can be obtained using an eight-bit optical encoder?

Solution

The resolution is $360° / 2^8 = 360° / 256 \cong 1.4°$.

The optical encoder described converts an analog position signal into an equivalent digital signal in Gray code. If this digital information is going to be entered into a digital system for processing, it might be more convenient to have it in straight binary form. One method of performing this conversion is to use the Gray code as the input to a series of gates wired to perform a Gray-to-binary conversion. The outputs of these gates will then be the equivalent binary number.

The logic gates necessary to perform a Gray-to-binary conversion for the three-bit encoder are shown in Fig. 11-32. The 2^2 bit (binary) is the same as the A bit (Gray). Thus the 2^2 bit is simply wired directly to the A bit.

The 2^1 bit must be high whenever $\overline{A}$ and B are high and when A and $\overline{B}$ are high. Thus the 2^1 bit is given by $2^1 = \overline{A}B + A\overline{B}$.

The 2^0 bit can be encoded by making use of the 2^1 bit. Notice that 2^0 must be high whenever 2^1 and $\overline{C}$ are high and when $\overline{2}^1$ and C are high. Thus the 2^0 bit can be formed by $2^0 = 2^1\overline{C} + \overline{2}^1C$. Recall that exclusive-OR gates can be used to perform Gray-to-binary conversion.

11-10 D/A CONVERTER CONTROL

The D/A converter shown in Fig. 11-13 can be redrawn as shown in Fig. 11-33a. In this figure we have expanded the basic four-bit converter to a 10-bit converter. This simply means that we need 10 flip-flops in the converter, and the ladder is ex-

Gray code			Binary code		
A	B	C	2^2	2^1	2^0
0	0	0	0	0	0
0	0	1	0	0	1
0	1	1	0	1	0
0	1	0	0	1	1
1	1	0	1	0	0
1	1	1	1	0	1
1	0	1	1	1	0
1	0	0	1	1	1

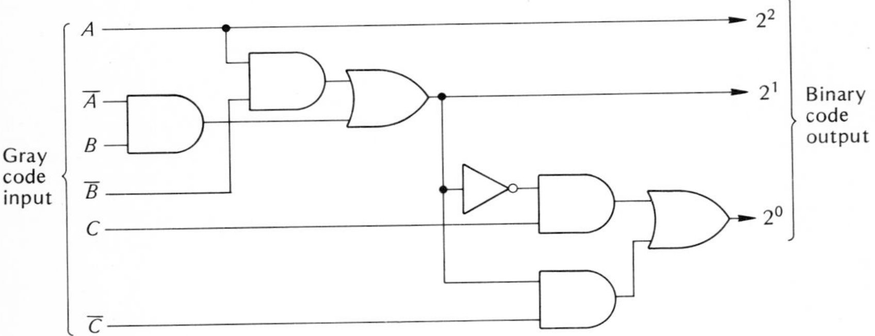

Fig. 11-32. Gray-to-binary encoder.

panded to accommodate 10 inputs. The storage register represents the flip-flops holding the data to be converted to analog form. Let's suppose that data to be converted appear in the storage register once every 10 μs and we wish to shift these data from the storage register into the D/A converter every 10 μs. Thus we shall be performing a D/A conversion every 10 μs. We need to develop the proper *read in* or strobe pulse for the D/A converter. Assume that the storage register is composed of *master/slave* flip-flops which are driven by a clock of 1 MHz. The desired strobe pulse can be developed using the logic shown in Fig. 11-33*b*. The basic 1-MHz clock is divided by 10 using a five-flip-flop shift counter. We can then form the control waveform *A* shown in Fig. 11-33*c* by decoding one of the 10 states of the shift counter (decoding of the shift counter is discussed in Chap. 9). This might at first glance seem to be all we need, since this waveform *A* is a pulse which occurs once every 10 μs as we desire. We must, however, consider the race problem, which might occur if we use this waveform. It is clear that the control waveform *A* changes state during the time the clock goes low. However, the storage register also changes state at this time; we do not therefore want to initiate a shift at this time, since we would certainly run the risk of racing at the shift-gate inputs. We need then to develop a strobe pulse to initiate the shift of data from the storage register into the converter. A convenient time for this to occur would be exactly in the middle of the positive portion of the control waveform *A*, since the clock is going high at this time and therefore the storage-register flip-flop outputs are static. The desired strobe pulse can be easily formed by ANDing the differentiated clock with the control waveform *A* as shown in Fig. 11-33*b*. It is clear that the D/A strobe pulse occurs at 10-μs intervals and at a time when the storage-register flip-flop outputs are static.

11-11 MULTIPLE D/A CONVERSION

We have discussed two methods for obtaining the analog equivalent of a number of digital signals. The two methods involve channel selection using a separate D/A converter for each signal to be decoded. The second method involves the use of only one D/A converter and multiplexing to switch the output of the converter to a number of sample-and-hold amplifiers.

In the channel-selection method all the D/A converter inputs are connected in parallel with the input storage register. Let us assume that we have 10 such converters, each of which is similar to the converter discussed in the previous section and shown in Fig. 11-33. The 10 signals to be converted appear in the storage register in a time sequence and remain in the register for 1 μs. In order to decode

Fig. 11-33. Use of a strobe pulse with a D/A converter. (*a*) 10-bit D/A converter. (*b*) Logic to develop a strobe pulse. (*c*) Control waveforms.

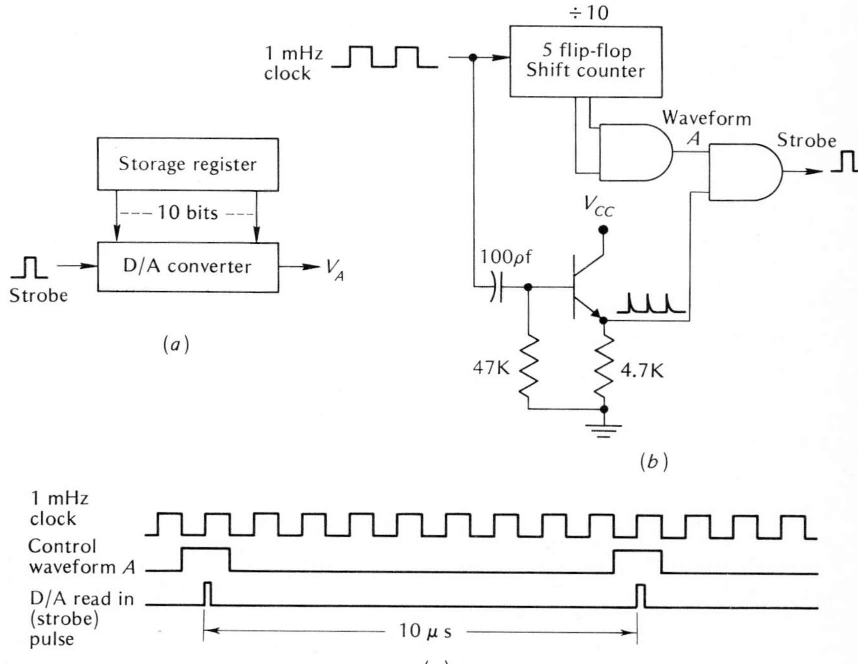

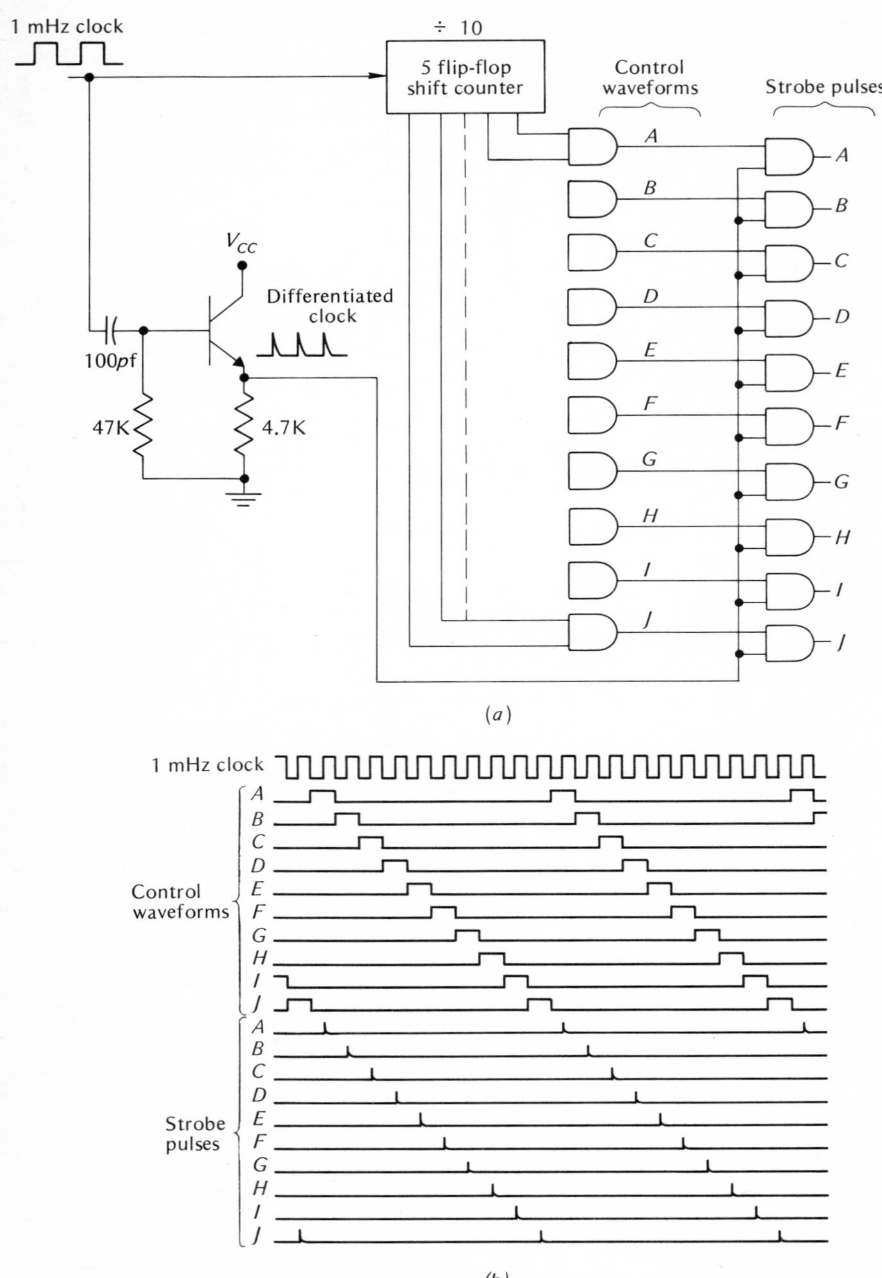

Fig. 11-34. Multiple D/A conversion by channel selection. (*a*) Logic diagram. (*b*) Waveforms.

these signals we must develop a series of 10 strobe pulses similar to the one developed in Fig. 11-33. The proper strobe signals can be quite easily developed by decoding all 10 states of the five-flip-flop shift counter in that figure. This then provides 10 control waveforms (*A, B, C, D, E, F, G, H, I,* and *J*) as shown in Fig. 11-34*b*. The correct strobe pulses can then be formed by differentiating the clock and ANDing the signals as before. The complete logic diagram and the resulting waveforms are shown in Fig. 11-34. It is clear from the strobe waveforms developed that the data are shifted into the converters at 1-μs intervals and that each converter is selected every 10 μs in a time sequence.

Example 11-15

What would be one method for altering the system shown in Fig. 11-34 to accommodate five D/A converters?

Solution

If symmetry is desired, it is quite easy to form the control signals by retaining only the gates required to form the strobe pulses *A, C, E, G,* and *I*.

11-12 A/D CONVERTER CONTROL

The operation of the A/D converter discussed previously is summarized by the waveforms for one conversion cycle shown in Fig. 11-23. These waveforms are self-contained in the converter, with the exception of the clock and the *start* pulse. The *start* pulse could be obtained via a push-button switch; the converter would make one conversion each time the push button is depressed. More often, however, it is desirable to have the converter perform conversions at a predetermined rate. The *start* pulse could then be derived from the output of a free-running multivibrator, and a conversion would begin each time the multivibrator output went high. It is only necessary to ensure that the period of the multivibrator waveform is greater than maximum conversion time of the converter. If the frequency of the multivibrator is variable, the number of conversions per second can be varied. This principle is quite often used in commercial counters and digital voltmeters.

Example 11-16

The D/A converter previously discussed operates with a basic system clock of 1 MHz. What is the maximum frequency at which the *start* pulse multivibrator can operate if the full-scale count of the converter is 500 counts?

Solution

With a 1-MHz clock, the basic clock cycle time is 1 μs. Therefore, the full-scale count requires 500 μs, which corresponds to the maximum conversion time. The *start* pulse multivibrator must then have a period greater than 500 μs. Therefore, the multivibrator must operate at a frequency less than $1/(500 \times 10^{-6}) = 2$ kHz.

Example 11-17

The output of the D/A converter in the previous example is used to drive the grids of nixie tubes. If the *start* pulse multivibrator runs at 10 Hz, for what percentage of the time will the nixie tubes be illuminated?

Solution

The converter conversion time is 500 μs, or 0.5 ms. During the conversion time, the converter output is changing, and the nixie tubes do not respond properly. However, a conversion is initiated only once every tenth of a second, or once every 100 ms. Thus, the output of the converter is static with the desired output signal for $100 - 0.5 = 99.5$ ms out of every 100 ms. Therefore, the nixie tubes are illuminated with the proper output signal for $(99.5/100) \times 100 = 99.5$ percent of the total operating time.

STUDY AIDS

Summary

D/A conversion, the process of converting digital input levels into an equivalent analog output voltage, is most easily accomplished by the use of resistance networks. The binary ladder was found to have definite advantages over the resistance divider. The complete D/A converter consists of a binary ladder (usually) and a flip-flop register to hold the digital input information.

The simultaneous method for A/D conversion is very fast but becomes cumbersome for more than a few bits of resolution. The counter-type A/D converter is somewhat slower but represents a much more reasonable solution for digitizing high-resolution signals. The continuous-converter method, the successive-approximation method, and the section-counter method are all variations of the basic counter-type A/D converter which lead to a much faster conversion time.

Optical encoders provide a convenient means for digitizing shaft-position information (and possibly rectilinear-position information). Gray code is used to code the optical encoder wheel to eliminate the large ambiguity error found in binary-coded wheels.

Glossary

A/D conversion The process of converting an analog input voltage to a number of equivalent digital output levels.

D/A conversion The process of converting a number of digital input signals to one equivalent analog output voltage.

differential linearity A measure of the variation in size of the input voltage to an A/D converter which causes the converter to change from one state to the next.

equivalent binary weight The value assigned to each bit in a digital number, expressed as a fraction of the total. The values are assigned in binary fashion according to the sequence 1,2,4,8, . . . , 2^n, where n is the total number of bits.

Millman's theorem A theorem from network analysis which states that the voltage at any node in a resistive network is equal to the sum of the currents entering the node divided by the sum of the conductances connected to the node, all determined by assuming the voltage at the node is zero.

quantization error The error inherent in any digital system due to the size of the LSB.

Review Questions

1. Why is a D/A converter usually considered a decoder?
2. Why is an A/D converter usually considered an encoder?
3. How is a binary equivalent weight determined?
4. Describe how Millman's theorem is used to find the output voltage of a resistive divider.
5. What is the principle of superposition? How can it be used to find the output voltage of a binary ladder?
6. What are some of the advantages of the binary ladder over the resistive divider?
7. What is monotonicity?
8. What is the difference between accuracy and resolution?
9. What is meant by the statement "resolution and accuracy should be compatible"?
10. How does a simultaneous A/D converter operate?
11. Describe the operation of a counter-type A/D converter.
12. Why is the counter-type A/D converter better than the simultaneous type for high-resolution conversions?
13. Explain why the counter-type A/D converter is slower than the simultaneous type.
14. Where does the staircase waveform appear in a counter-type A/D converter?
15. What does the staircase waveform have to do with monotonicity?
16. How does a continuous-type A/D converter aid conversion time?
17. Explain how it is possible to perform A/D conversion to $\pm\frac{1}{2}$ LSB.
18. Why is a section-counter-type A/D converter useful for digital voltmeters?
19. Describe how Gray code eliminates large ambiguity errors in reading optical encoders.
20. Explain why you would or would not make the innermost ring of the code wheel for an optical encoder the LSB.

Problems

11-1. What is the binary equivalent weight of each bit in a six-bit resistive divider?

11-2. Draw the schematic for a six-bit resistive divider.

✓ **11-3.** Verify the voltage-output levels for the network of Fig. 11-5 using Millman's theorem. Draw the equivalent circuits. see p. 291

✓ **11-4.** Assume the divider in Prob. 11-2 has +10 V full-scale output, and find the following:

(*a*) The change in output voltage due to a change in the LSB.
(*b*) The output voltage for an input of 110110.

✓ **11-5.** A 10-bit resistive divider is constructed such that the current through the LSB resistor is 100 μA. Determine the maximum current that will flow through the MSB resistor.

✓ **11-6.** What is the full-scale output voltage of a six-bit binary ladder if 0 = 0 V and 1 = +10 V? What is it for an eight-bit ladder?

✓ **11-7.** Find the output voltage of a six-bit binary ladder with the following inputs:

(*a*) 101001.
(*b*) 111011.
(*c*) 110001.

✓ **11-8.** Check the results of Prob. 11-7 by adding the individual bit contributions.

✓ **11-9.** What is the resolution of a 12-bit D/A converter which uses a binary ladder? If the full-scale output is +10 V, what is the resolution in volts?

✓ **11-10.** How many bits are required in a binary ladder to achieve a resolution of 1 mV if full scale is +5 V?

✓ **11-11.** How many comparators are required to build a five-bit simultaneous A/D converter?

11-12. Redesign the encoding matrix and *read* gates of Fig. 11-20 using NAND gates.

11-13. Find the following for a 12-bit counter-type A/D converter using a 1-MHz clock:

(*a*) Maximum conversion time.
(*b*) Average conversion time.
(*c*) Maximum conversion rate.

11-14. What clock frequency must be used with a 10-bit counter-type A/D converter if it must be capable of making at least 7,000 conversions per second?

11-15. What is the conversion time of a 12-bit successive-approximation-type A/D converter using a 1-MHz clock?

11-16. What is the conversion time of a 12-bit section-counter-type A/D converter using a 1-MHz clock? The counter is divided into three equal sections.

11-17. What overall accuracy could you reasonably expect from a 12-bit A/D converter?

11-18. What degree of resolution can be obtained using a 12-bit optical encoder?

11-19. Redesign the Gray-to-binary encoder in Fig. 11-32 using NAND gates.

11-20. Redesign the Gray-to-binary encoder in Fig. 11-32 using exclusive-OR gates.

Magnetic Devices and Memories

There is a large class of devices and systems which are useful as digital elements because of their magnetic behavior. A ferromagnetic material can be magnetized in a particular direction by the application of a suitable magnetizing force (a magnetic flux resulting from a current flow). The material remains magnetized in that direction after the removal of the excitation. Application of a magnetizing force of the opposite polarity will switch the material, and it will remain magnetized in the opposite direction after removal of the excitation. Thus the ability to store information in two different states is available, and a large class of binary elements has been devised using these principles. In this chapter we investigate a number of these devices and systems that make use of them.

After studying this chapter you should be able to

1. Illustrate how magnetic cores are used to store binary information.
2. Explain the fundamental principles of a coincident-current memory.
3. Describe the operation of a semiconductor memory using either bipolar or MOS devices.

12-1 MAGNETIC CORES

One of the most widely used magnetic elements is the magnetic core. The typical core is toroidal (doughnut-shaped), as shown in Fig. 12-1, and is usually constructed in one of two ways. The metal-ribbon core is constructed by winding a very thin metallic ribbon on a ceramic-core form. A popular ribbon is ⅛-mil-thick 4-79 molybdenum-permaloy (known as ultrathin ribbon), and a typical core might consist of 20 turns of this ribbon wound on a 0.2-in-diameter ceramic form.

Ferrite cores are constructed from a finely powdered mixture of magnetite, various bivalent metals such as magnesium or maganese, and a binder material. The powder is pressed into the desired shape and fired. During firing, the powder is fused into a solid, homogeneous, polycrystalline form. Ferrite cores such as this are commonly constructed with 50 mil outside diameters and 30 mil inside diameters.

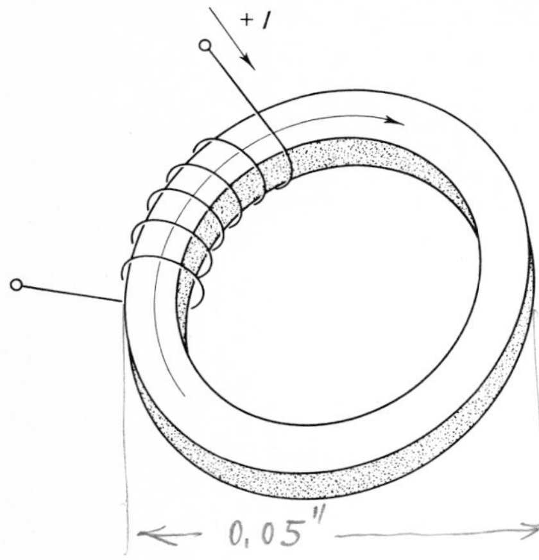

Fig. 12-1. Magnetic core.

Ferrite cores can be constructed in smaller dimensions than metal-ribbon cores and usually have better uniformity and lower cost. Furthermore, ferrite cores typically have resistivities greater than 10^5 Ω-cm, which means eddy-current losses are negligible and thus core heating is reduced. For these reasons, they are widely used as the principal memory or storage elements in large-scale digital computers.

Metal-ribbon cores, on the other hand, have very good magnetic characteristics and generally require a smaller driving current for switching. They are somewhat better for the construction of logic circuits and shift registers.

The binary characteristics of a core can be most easily seen by examining the *hysteresis curve* for a typical core. Hysteresis comes from the Greek word *hysterein,* which means to lag behind. A magnetic core exhibits a lag-behind characteristic in the hysteresis curve shown in Fig. 12-2*a*. In this figure, the *magnetic flux density* **B** is plotted as a function of the *magnetic force* **H**. However, since the flux density **B** is directly proportional to the flux ϕ, and since the magnetic field **H** is directly proportional to the current I producing it, a plot of ϕ versus I is a curve of the same

Fig. 12-2 Ferrite-core hysteresis curves. (*a*) Magnetic flux density **B** versus magnetic field **H.** (*b*) Magnetic flux ϕ versus current I.

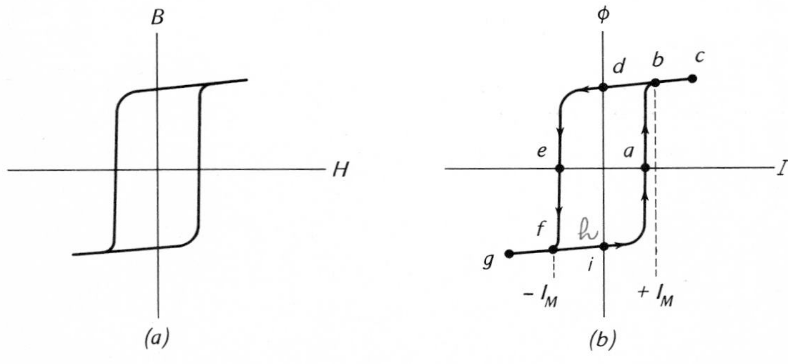

general shape. A plot of flux in the core ϕ versus driving current I is shown in Fig. 12-2*b*. We shall base our discussion on this curve since it is generally easier to talk in terms of these quantities.

Now, suppose that a current source is attached to the windings on the core shown in Fig. 12-1, and a positive current is applied (current flows into the upper terminal of the winding). This creates a flux in the core in the clockwise direction shown in the figure (remember the *right-hand rule*). If the drive current is just slightly greater than I_m shown in Fig. 12-2, the operating point of the core is somewhere between points *b* and *c* on the ϕI curve. The magnitude of the flux can then be read from the ϕ axis in this figure.

If the drive current is now removed, the operating point moves along the ϕI curve through point *b* to point *d*. The core is now storing energy with no input signal, since there is a remaining or *remanent* flux in the core at this point. This property is known as *remanence,* and this point is known as a *remanent point.*

The repeated application of positive current pulses simply causes the operating point to move between points *d* and *c* on the ϕI curve. Notice that the operating point always comes to rest at point *d* when all drive current is removed.

If a negative drive current somewhat greater than $-I_m$ is now applied to the winding (in a direction opposite to that shown in Fig. 12-1), the operating point moves from *d* down through e and stops at a point somewhere between *f* and g on the ϕI curve. At this point the flux has switched in the core and is now directed in a counterclockwise direction in Fig. 12-1. If the drive current is now removed, the operating point comes to rest at point *h* on the ϕI curve of Fig. 12-2. Notice that the flux has approximately the same magnitude but is the negative of what it was previously. This indicates that the core has been magnetized in the opposite direction.

Repeated application of negative drive currents will simply cause the operating point to move between points g and *h* on the ϕI curve, but the final resting place with no applied current will be point *h*. Point *h* then represents a second remanent point on the ϕI curve.

By way of summary, a core has two remanent states: point *d* after the application of one or more positive current pulses, point *h* after the application of one or more negative current pulses. For the core in Fig. 12-1, point *d* corresponds to the core magnetized with flux in a clockwise direction, and point *h* corresponds to magnetization with flux in the counterclockwise direction.

Example 12-1

Cores can be magnetized by utilizing the magnetic field surrounding a current-carrying wire by simply *threading* the cores on the wire. For the two possible current directions in the wire shown in Fig. 12-3, what are the corresponding directions of magnetization for the core?

Solution

According to the right-hand rule, a current of $+I$ magnetizes the core with the flux in a clockwise direction around the core. A current of $-I$ magnetizes the core with flux in a counterclockwise direction around the core.

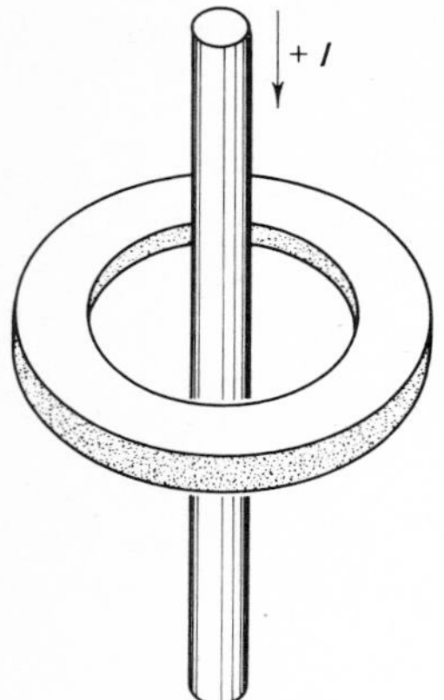

Fig. 12-3.

It is now quite easy to see how a magnetic core is used as a binary storage device in a digital system. The core has two states, and we can simply define one of the states as a 1 and the other state as a 0. It is perfectly arbitrary which is which, but for discussion purposes let us define point *d* as a 1 and point *h* as a 0. This means that a positive current will record a 1 and result in clockwise flux in the core in Fig. 12-1. A negative current will record a 0 and result in a counterclockwise flux in the core.

We now have the means for recording or writing a 1 or a 0 in the core but we do not as yet have any means of detecting the information stored in the core. A very simple technique for accomplishing this is to apply a current to the core which will switch it to a known state and detect whether or not a large flux change occurs. Consider the core shown in Fig. 12-4. Application of a drive current of $-I$ will switch the core to the 0 state. If the core has a 0 stored in it, the operating point will move between points *g* and *h* on the ϕI curve (Fig. 12-2), and a very small flux change will occur. This small change in flux will induce a very small voltage across the sense-winding terminals. On the other hand, if the core has a 1 stored in it, the operating point will move from point *d* to point *h* on the ϕI curve, resulting in a much larger flux change in the core. This change in flux will induce a much larger voltage in the sense winding, and we can thus detect the presence of a 1.

To summarize, we can detect the contents of a core by applying a *read* pulse which resets the core to the 0 state. The output voltage at the sense winding is

Fig. 12-4. Sensing the contents of a core.

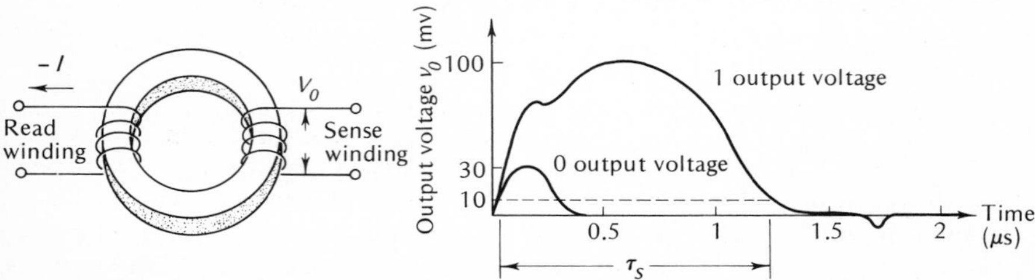

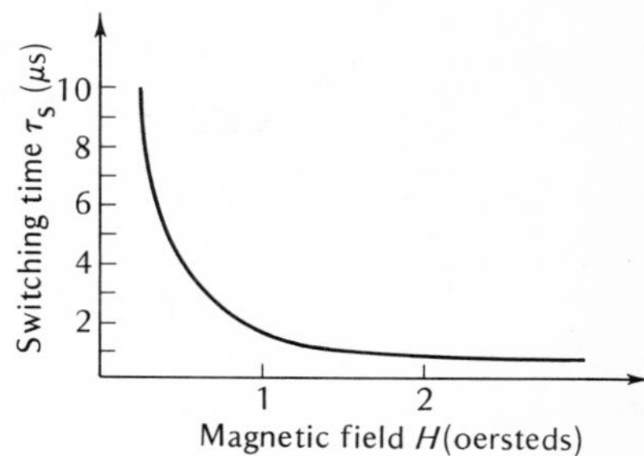

Fig. 12-5. Magnetic-core switching time characteristics.

much greater when the core contains a 1 than when it contains a 0. We can therefore detect a 1 by distinguishing between the two output-voltage signals. Notice that we could set the core by applying a *read* current of $+I$ and detect the larger output voltage at the sense winding as a 0.

The output voltage appearing at the sense winding for a typical core is also shown in Fig. 12-4. Notice that there is a difference of about 3 to 1 in output-voltage amplitude between a 1 and a 0 output. Thus a 1 can be detected by using simple amplitude discrimination in an amplifier. In large systems where many cores are used on common windings (such as the large memory systems in digital computers) the 0 output voltage may become considerably larger because of additive effects. In this case, amplitude discrimination is quite often used in combination with a strobing technique. Even though the amplitude of the 0 output voltage may increase because of additive effects, the width of the output will not increase appreciably. This means that the 0 output-voltage signal will have decayed and will be very small before the 1 output voltage has decayed. Thus if we strobe the *read* amplifiers some time after the application of the *read* pulse (for example, between 0.5 and 1.0 μs in Fig. 12-4), this should improve our detection ability.

The *switching time* of the core is commonly defined as the time required for the output voltage to go from 10 percent up through its maximum value and back down to 10 percent again (see Fig. 12-4). The switching time for any one core is a function of the drive current as shown in Fig. 12-5. It is evident from this curve that an increased drive current results in a decreased switching time. In general, the switching time for a core depends on the physical size of the core, the type of core, and the materials used in its construction, as well as the manner in which it is used. It will be sufficient for our purposes to know that cores are available with switching times from around 0.1μs up to milliseconds, with drive currents of 100 mA to 1 A.

12-2 MAGNETIC-CORE LOGIC

Since a magnetic core is a basic binary element, it can be used in a number of ways to implement logical functions. Because of its inherent ruggedness, the core is a particularly useful logical element in applications where environmental extremes are experienced, for example, the temperature extremes and radiation exposure experienced by space vehicles.

Since the core is essentially a storage device and its content is detected by resetting the core to the 0 state, any logic system using cores must necessarily be a

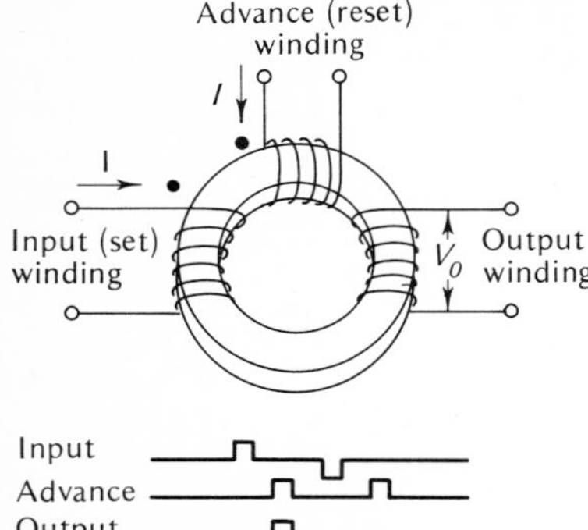

Fig. 12-6. Basic magnetic-core logic element.

dynamic system. The basis for using the core as a logical element is shown in Fig. 12-6. A 1 input to the core is represented by a current of $+I$ at the *input* winding; this sets a 1 in the core (magnetizes it in a clockwise direction). An *advance* pulse occurs sometime after the *input* pulse has disappeared. Logical operations are carried out during the time the *advance* pulse appears at the *advance* (*reset*) winding. At this time the core is forced into the 0 state and a pulse appears at the output winding only if the core previously stored a 1. The current in the output winding can then be used as the input for other cores or other logical elements.

There is some energy loss in the core during switching. For this reason, the output winding normally has more turns than either the *input* or *advance* windings, so that the output will be capable of driving one or more cores.

Notice that a 0 can be set in the core by application of a current of $-I$ at the *input* winding. Alternatively, a 0 could be stored by a current of $+I$ into the undotted side of the *input* winding. The important thing to notice is that either a 1 or a 0 can be stored in the core by application of a current to the proper terminal of the *input* winding.

To simplify our discussion and the logic diagrams, we shall adopt the symbols for the core and its windings shown in Fig. 12-7. A pulse at the 1 input sets a 1 in the core; a pulse at the 0 input sets a 0 in the core; during the *advance* pulse, a pulse appears at the output only if the core previously held a 1. Let us now consider some of the basic logic functions using the symbol shown in Fig. 12-7*b*.

A method for implementing the OR function is shown in Fig. 12-8*a*. A current pulse at either the X or Y inputs sets a 1 in the core. Sometime after the input pulse(s) have been terminated, an *advance* pulse occurs. If the core has been set to the 1 state, a pulse appears at the output winding. Notice that this is truly an OR function since a pulse at either the X or Y input or *both* sets a 1 in the core.

The method shown in Fig. 12-8*b* provides the means for obtaining the complement of a variable. The *set input* winding to the core has a 1 input. This means that during the *input* pulse time this winding always has a *set input* current. If there is no current at the X input (signifying $X = 0$), the core is set. Then, when the *advance* pulse occurs, a 1 appears at the output, signifying that $\overline{X} = 1$. On the other hand, if

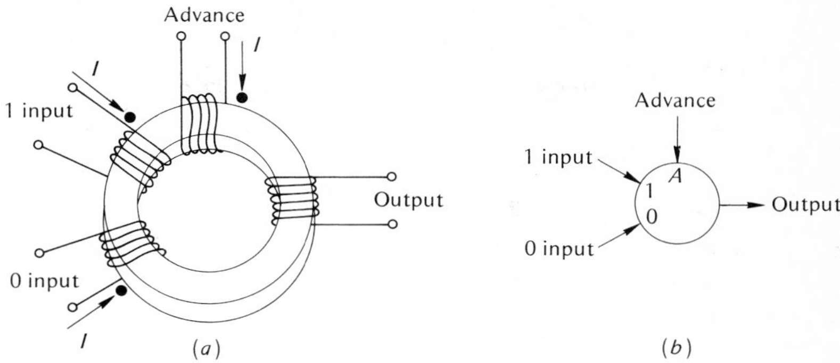

Fig. 12-7. Magnetic-core logic element. (*a*) Core windings. (*b*) Logic symbol.

$X = 1$, a current appears at the X input during the *set* time, and the effects of the X input current and the 1 input current cancel one another. The core then remains in the reset state (recall that the core is reset during the *advance* pulse). In this case no pulse appears at the output during the *advance* pulse since the core previously contained a 0. Thus the output represents $\overline{X} = 0$.

The AND function can be implemented using a core as shown in Fig. 12-8*c*. The two inputs to the core are X and $\overline{Y}$, and there are four possible combinations of these two inputs. Let's examine these input combinations in detail.

Fig. 12-8. Basic core logic functions. (*a*) OR. (*b*) Complement. (*c*) AND. (*d*) Exclusive-OR.

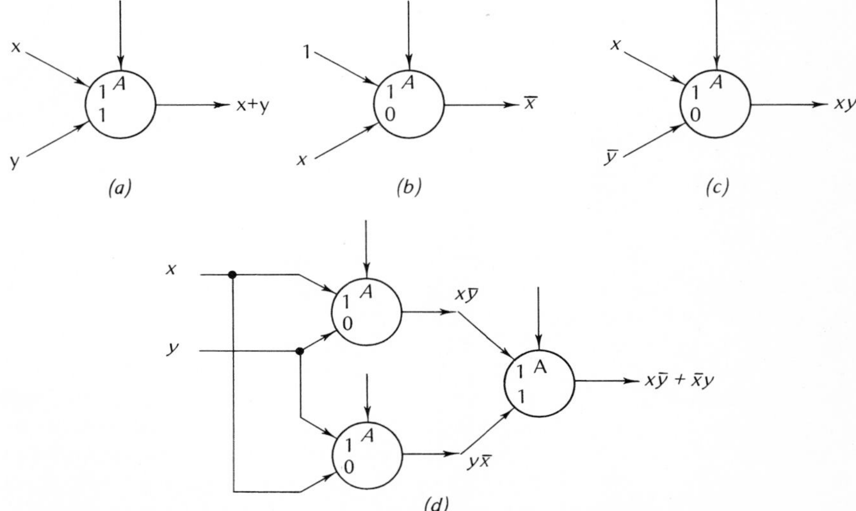

1. $X=0$, $Y=0$. Since $X=0$, the core cannot be set. Since $Y=0$, $\overline{Y}=1$ and the core will then be reset. Thus this input combination resets the core and it stores a 0.
2. $X=0$, $Y=1$. Since $X=0$, the core still cannot be set. $Y=1$ and therefore $\overline{Y}=0$. In this input combination, there is no input current in either winding and the core cannot change state. Thus the core remains in the 0 state because of the previous *advance* pulse.
3. $X=1$, $Y=0$. The current in the *X winding* will attempt to set a 1 in the core. However, $\overline{Y}=1$ and this current will attempt to reset the core. These two currents offset one another, and the core does not change states. It remains in the 0 state because of the previous advance pulse.
4. $X=1$, $Y=1$. The current in the *X winding* will set a 1 in the core since $\overline{Y}=0$ and there is no current in the *Y winding*. Thus this combination stores a 1 in the core.

In summary, the input X AND $\overline{Y}$ is the only combination which results in a 1 being stored in the core. Thus this is truly an AND function.

An exclusive-OR function can be implemented as shown in Fig. 12-8*d* by ORing the outputs of two AND-function cores.

Example 12-2

Make a truth table for the exclusive-OR function shown in Fig. 12-8*d*.

Solution

X	Y	$X\overline{Y}$	$\overline{X}Y$	$X\overline{Y}+\overline{X}Y$
0	0	0	0	0
0	1	0	1	1
1	0	1	0	1
1	1	0	0	0

One of the major problems of core logic becomes apparent in the operation of the exclusive-OR shown in Fig. 12-8*d*. This is the problem of the time required for the information to shift down the line from one core to the next. For the exclusive-OR, the inputs X and Y appear at time t_1, and the AND cores are *set* or *reset* at this time. At time t_2 an *advance* pulse is applied to the AND cores and their outputs are used to *set* the OR core. Then at time t_3 an *advance* pulse is applied to the OR core and the final output appears. It should be obvious from this discussion that the operation time for more complicated logic functions may become excessively long.

A second difficulty with this type of logic is the fact that the *input* pulses must be of exactly the same width. This is particularly true for functions such as the COMPLEMENT and the AND, since the input signals are at times required to cancel one another. It is apparent that if one of the input signals is wider than the other, the core may contain erroneous data after the *input* pulses have disappeared.

You will recall that in order to switch a core from one state to another a certain minimum current I_m is required. This is sometimes referred to as the *select current.* The core arrangement shown in Fig. 12-8*a* can be used to implement an AND function if the X and Y inputs are each limited to one-half the select current $\frac{1}{2}I_m$. In this way, the only time the core can be set is when both X and Y are present, since this is the only time the core receives a full select current I_m. Core logic functions can be constructed using the half-select current idea. This idea is quite important; it forms the basis of one type of large-scale memory system which we discuss later in this chapter.

12-3 MAGNETIC-CORE SHIFT REGISTER

A review of the previous section will reveal that a magnetic core exhibits at least two of the major characteristics of a flip-flop: first, it is a binary device capable of storing binary information; second, it is capable of being set or reset. Thus it would seem reasonable to expect that the core could be used to construct a shift register or a ring counter. Cores are indeed frequently used for these purposes, and in this section we consider some of the necessary precautions and techniques.

The main idea involves connecting the output of each core to the input of the next core. When a core is reset (or set), the signal appearing at the output of that core is used to set (or reset) the next core. Such a connection between two cores, called a "single-diode transfer loop," is shown in Fig. 12-9.

There are three major problems to overcome when using the single-diode transfer loop. The first problem is the gain through the core. This is similar to the problem discussed previously, and the solution is the same. That is, the losses in signal through the core can be overcome by constructing the output winding with more turns than the input winding. This ensures that the output signal will have sufficient amplitude to switch the next core.

The second problem concerns the polarity of the output signal. A signal appears at the output when the core is set or when the core is reset. These two signals have opposite polarities, and either is capable of switching the next core. In general, it is desirable that only one of the two output signals be effective, and this can be achieved by the use of the diode shown in Fig. 12-9. In this figure, the current produced in the output winding will go through the diode in the forward direction (and thus set the next core) when the core is reset from the 1 state to the 0 state. On

Fig. 12-9. Single-diode transfer loop. (*a*) Circuit. (*b*) Symbolic representation.

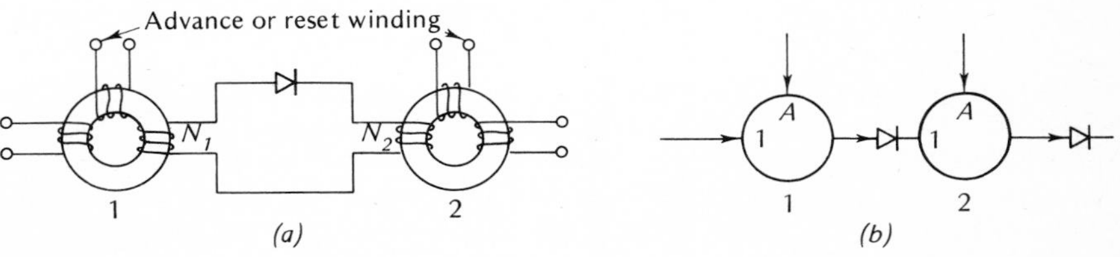

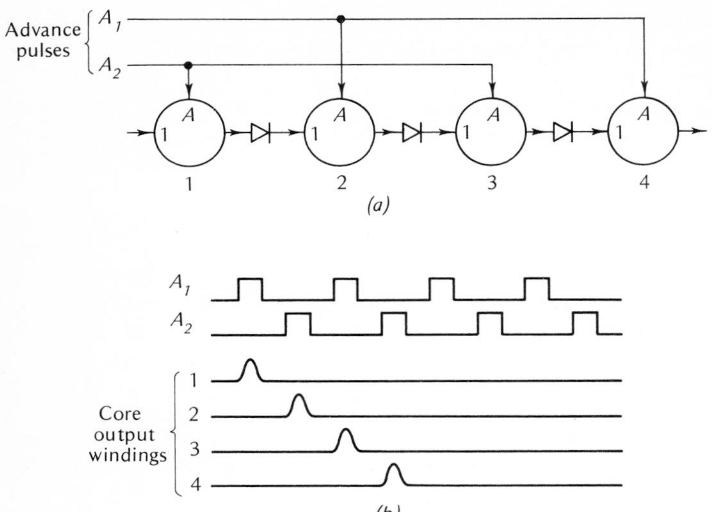

Fig. 12-10. Four-core shift register. (*a*) Symbolic circuit. (*b*) Waveforms.

the other hand, when the core is being set to the 1 state, the diode will prevent current flow in the output and thus the next core cannot be switched. Notice that the opposite situation could be realized by simply reversing the diode.

The third problem arises from the fact that resetting core 2 induces a current in winding N_2 which will pass through the diode in the forward direction and thus tend to set a 1 in core 1. This constitutes the transfer of information in the reverse direction and is highly undesirable. Fortunately, the solution to the first problem (that of gain) results in a solution for this problem as well. That is, since N_2 has fewer windings than N_1, this reverse signal will not have suffcient amplitude to switch core 1. With this understanding of the basic single-diode transfer loop, let us investigate the operation of a simple core shift register.

A basic magnetic-core shift register in symbolic form is shown in Fig. 12-10. Two sets of advance windings are necessary for shifting information down the line. The advance pulses occur alternately as shown in the figure. A_1 is connected to cores 1 and 3 and would be connected to all *odd*-numbered cores for a larger register. A_2 is connected to cores 2 and 4 and would be connected to all *even*-numbered cores. If we assume that all cores are reset with the exception of core 1, it is clear that the advance pulses will shift this 1 down the register from core to core until it is shifted "out the end" when core 4 is reset. The operation is as follows: the first A_1 pulse resets core 1 and thus sets core 2. This is followed by an A_2 pulse which resets core 2 and thus sets core 3. The next A_1 pulse resets core 3 and sets core 4, and the following A_2 pulse shifts the 1 "out the end" by resetting core 4. Notice that the two phases of advance pulses are required, since it is not possible to set a core while an advance (or reset) pulse is present.

The output of each core winding can be used as an input to an amplifier to produce the waveforms shown in Fig. 12-10*b*. Notice that after four advance pulses the 1 has been shifted completely through the register, and the output lines all remain low after this time.

The need for a two-phase clock or advance pulse system could be eliminated if some delay were introduced between the output of each core and the input of the next core. Suppose that a delay greater than the width of the advance pulses were introduced between each pair of cores. In this case, it would be possible to drive every core with the same advance pulse since the output of any core could not arrive at the input to the next core until after the advance pulse had disappeared.

One method for introducing a delay between cores is shown in Fig. 12-11. The advance-pulse amplitude is several times the minimum required to switch the cores and will reset all cores to the 0 state. If a core previously contained a 0, no switching occurs and thus no signal appears at the output winding. On the other hand, if a core previously contained a 1, current flows in the output winding and charges the capacitor. Some current flows through the set winding of the next core, but it is small because of the presence of the resistor; furthermore, it is overridden by the magnitude of the advance pulse. However, at the cessation of the advance pulse, *C* remains charged. Thus *C* discharges through the input winding and *R*, and sets core 2 to the 1 state.

In this system, the amplitude of the advance pulses is not too critical, but the width must be matched to the *RC* time constant of the loop. If the advance pulses are too long, or alternatively if the *RC* time constant is too short, the capacitor will discharge too much during the advance pulse time and will be incapable of setting the core at the cessation of the advance pulse. The *RC* time constant may limit the upper frequency of operation; it should be noted, however, that resetting a core induces a current in its input winding in a direction which tends to discharge the capacitor.

The arrangements we have discussed here are called *one-core-per-bit* registers. There are numerous other methods (too many to discuss here) for implementing

Fig. 12-11. Core shift register using a capacitor for delay between cores.

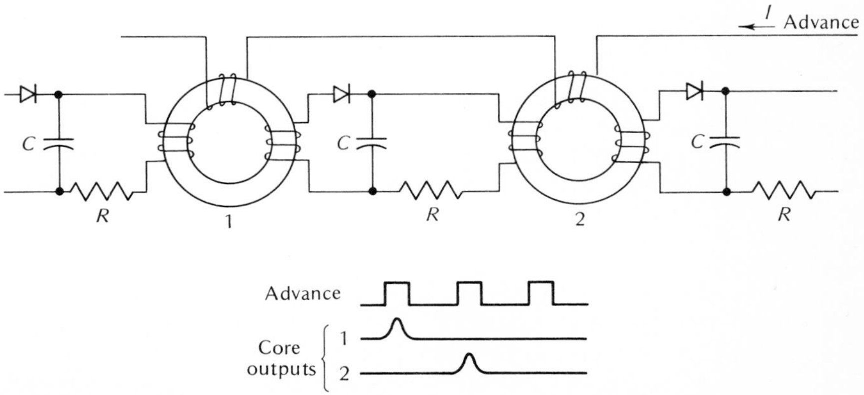

registers and counters, and the reader is referred to the references for more advanced techniques. Some of the other methods include *two-core-per-bit* systems, *modified-advance-pulse* systems, *modified-winding-core* systems, *split-winding-core* systems, and *current-routing-transfer* systems.

Example 12-3

Using core symbols and the capacitor-delay technique, draw the diagram for a four-stage ring counter. Show the expected waveforms.

Solution

A ring counter can be formed from a simple shift register by using the output of the last core as the input for the first core. Such a system, along with the expected waveforms, is shown in Fig. 12-12.

12-4 COINCIDENT-CURRENT MEMORY

The core shift register discussed in the previous section suggests the possibility of using an array of magnetic cores for storing words of binary information. For example, a 10-bit core shift register could be used to store a 10-bit word. The operation would be serial in form, much like the 10-bit flip-flop shift register discussed earlier. It would, however, be subject to the same speed limitations observed in the serial flip-flop register. That is, since each bit must travel down the register from core to core, it requires n clock periods to shift an n-bit word into or out of the register. This shift time may become excessively long in some cases, and a faster method must then be developed. Much faster operation can be achieved if the information is written into and read out of the cores in a parallel manner. Since all the bits are processed simultaneously an entire word can be transferred in only one

Fig. 12-12. Four-stage ring counter for Example 12-3.

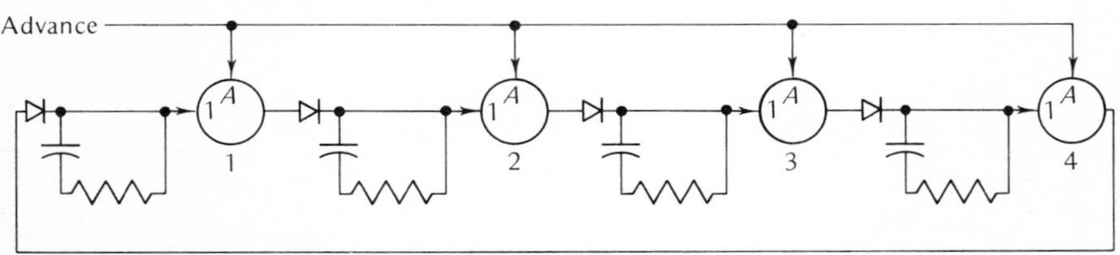

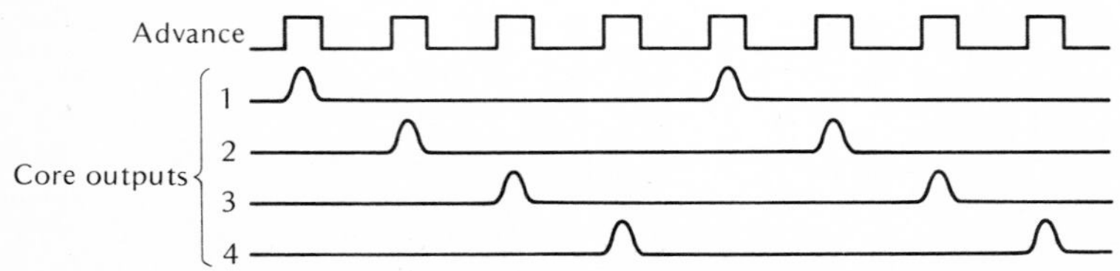

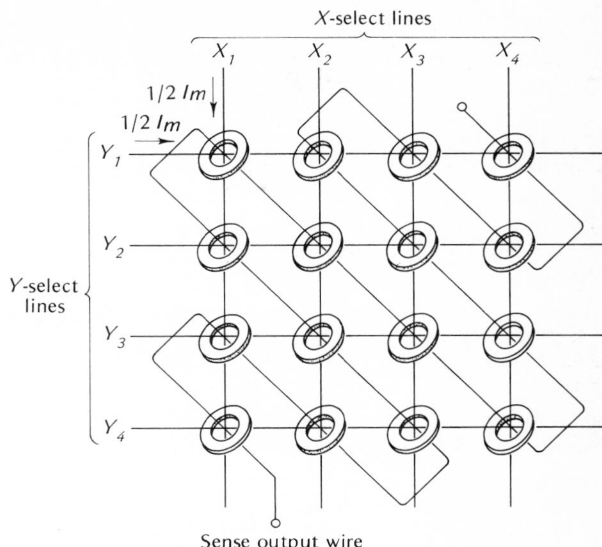

Fig. 12-13. Magnetic-core coincident-current memory.

clock period. A straight parallel system would, however, require one input wire and one output wire for each core. For a large number of cores the total number of wires makes this arrangement impractical, and some other form of core selection must be developed.

The most popular method for storing binary information in parallel form using magnetic cores is the *coincident-current drive* system. Such memory systems are widely used in all types of digital systems from small-scale special-purpose machines up to large-scale digital computers. The basic idea involves arranging cores in a matrix and using two *half-select currents;* the method is shown in Fig. 12-13.

The matrix consists of two sets of drive wires: the X drive wires (vertical) and the Y drive wires (horizontal). Notice that each core in the matrix is threaded by one X wire and one Y wire. Suppose one half-select current $\frac{1}{2}I_m$ is applied to line X_1 and one half-select current $\frac{1}{2}I_m$ is applied to line Y_1. Then the core which is threaded by both lines X_1 and Y_1 will have a total of $\frac{1}{2}I_m + \frac{1}{2}I_m = I_m$ passing through it, and it will switch states. The remaining cores which are threaded by X_1 or Y_1 will each receive only $\frac{1}{2}I_m$, and they will therefore not switch states. Thus we have succeeded in switching one of the 16 cores by selecting two of the input lines (one of the X lines and one of the Y lines). We designate the core that switched in this case as core X_1Y_1, since it was switched by selecting lines X_1 and Y_1. The designation X_1Y_1 is called the *address* of the core since it specifies its location. We can then switch any core X_aY_b located at address X_aY_b by applying $\frac{1}{2}I_m$ to lines X_a and Y_b. For example, the core located in the lower right-hand corner of the matrix is at the address X_4Y_4 and can be switched by applying $\frac{1}{2}I_m$ to lines X_4 and Y_4.

In order that the selected core will switch, the directions of the half-select currents through the X line and the Y line must be additive in the core. In Fig. 12-13, the X select currents must flow through the X lines from the top toward the bottom, while the Y select currents flow through the Y lines from left to right. Application of the *right-hand rule* will demonstrate that currents in this direction switch the core such that the core flux is in a clockwise direction (looking from the top). We define this as switching the core to the 1 state. It is obvious, then, that reversing the directions of both the X and Y line currents will switch the core to the 0 state. Notice that if the X and Y line currents are in a subtractive direction the selected core receives $\frac{1}{2}I_m - \frac{1}{2}I_m = 0$ and the core does not change state.

With this system we now have the ability to switch any one of 16 cores by selecting any two of eight wires. This is a saving of 50 percent over a direct parallel selection system. This saving in input wires becomes even more impressive if we enlarge the existing matrix to 100 cores (a square matrix with 10 cores on each side). In this case, we are able to switch any one of 100 cores by selecting any two of only 20 wires. This represents a reduction of 5 to 1 over a straight parallel selection system.

At this point we need to develop a method of sensing the contents of a core. This can be very easily accomplished by threading one *sense wire* through every core in the matrix. Since only one core is selected (switched) at a time, any output on the sense wire will be due to the changing of state of the selected core, and we will know which core it is since the core address is prerequisite to selection. Notice that the sense wire passes through half the cores in one direction and through the other half in the opposite direction. Thus the output signal may be either a positive or a negative pulse. For this reason, the output from the sense wire is usually amplified and rectified to produce an output pulse which always appears with the same polarity.

Example 12-4

From the standpoint of construction, the core matrix in Fig. 12-14 is more convenient. Explain the necessary directions of half-select currents in the X and Y lines for proper operation of the matrix.

Solution

Core X_1Y_1 is exactly similar to the previously discussed matrix in Fig. 12-13. Thus a current passing down through X_1 and to the right through Y_1 will set core X_1Y_1 to the 1 state. To set core X_1Y_2 to the 1 state, current must pass down through line X_1, but current must pass from the right to the left through line Y_2 (check with the *right-hand rule*). Proceeding in this fashion, we see that core X_1Y_3 is similar to X_1Y_1. Therefore, current must pass through line Y_3 from left to right. Similarly, core X_1Y_4 is similar to core X_1Y_2 and current must therefore pass through line Y_4 from right to left. In general, current must pass from *left to right* through the *odd*-numbered Y lines, and from *right to left* through *even*-numbered Y lines.

Now, since current must pass from left to right through line Y_1, it is easily seen that current must pass upward through line X_2 in order to set core X_2Y_1. By an argument similar to that given for the Y lines, current must pass *downward* through the *odd*-numbered X lines and *upward* through the *even*-numbered X lines.

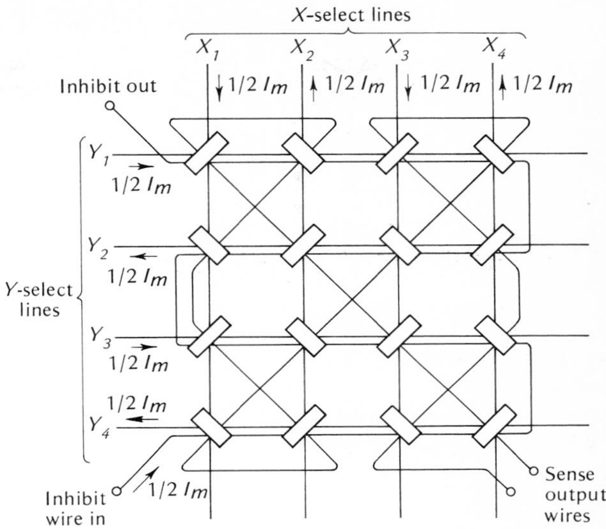

Fig. 12-14. Coincident-current memory matrix (one plane).

The matrix shown in Fig. 12-14 has one extra winding which we have not yet discussed. This is the *inhibit wire*. In order to understand its operation and function, let us examine the methods for writing information into the matrix and reading information from the matrix.

To write a 1 in any core (that is, to set the core to the 1 state), it is only necessary to apply $\frac{1}{2}I_m$ to the X and Y lines selecting that core address. If we desired to write a 0 in any core (that is, set the core to the 0 state), we could simply apply a current of $-\frac{1}{2}I_m$ to the X and Y lines selecting that core address. We can also write a 0 in any core by making use of the *inhibit* wire shown in Fig. 12-14. (We assume that all cores are initially in the 0 state.) Notice that the application of $\frac{1}{2}I_m$ to this wire in the direction shown on the figure results in a complete cancellation of the Y line select current (it also tends to cancel an X line current). Thus, to write a 0 in any core, it is only necessary to select the core in the same manner as if writing a 1, and at the same time apply an *inhibit* current to the *inhibit* wire. The major reason for writing a 0 in this fashion will become clear when we use these matrix planes to form a complete memory.

To summarize, we write a 1 in any core X_aY_b by applying $\frac{1}{2}I_m$ to the select lines X_a and Y_b. A 0 can be written in the same fashion by simply applying $\frac{1}{2}I_m$ to the *inhibit* line at the same time (if all cores are initially reset).

To read the information stored in any core, we simply apply $-\frac{1}{2}I_m$ to the proper X and Y lines and detect the output on the sense wire. The select currents of $-\frac{1}{2}I_m$ reset the core, and if the core previously held a 1, an output pulse occurs. If the core previously held a 0, it does not switch, and no output pulse appears.

This, then, is the complete coincident-current selection system for one plane. Notice that reading the information out of the memory results in a complete loss of

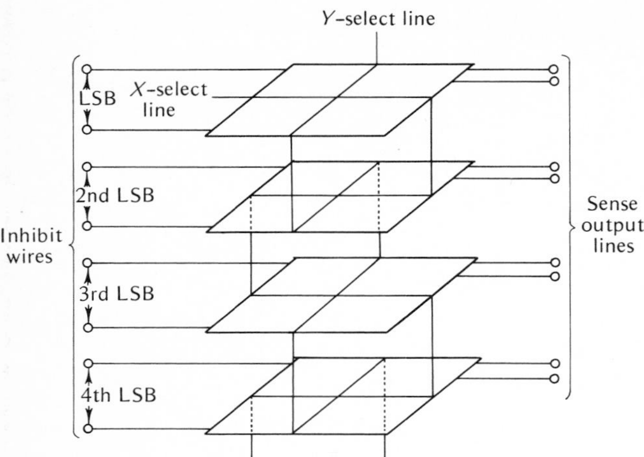

Fig. 12-15. Complete coincident-current memory system.

information from the memory, since all cores are reset during the *read* operation. This is referred to as a *destructive readout* or DRO system. This matrix plane is used to store one bit in a word, and it is necessary to use n of these planes to store an n-bit word.

A complete parallel coincident-current memory system can be constructed by stacking the basic memory planes in the manner shown in Fig. 12-15. All the X drive lines are connected in series from plane to plane as are all the Y drive lines. Thus the application of $\frac{1}{2}I_m$ to lines X_a and Y_b results in a selection of core X_aY_b in every plane. In this fashion we can simultaneously switch n cores, where n is the number of planes. These n cores represent one word of n bits. For example, the top plane might be the LSB, the next to the top plane would then be the second LSB, and so on; the bottom plane would then hold the MSB.

To read information from the memory, we simply apply $-\frac{1}{2}I_m$ to the proper address and sense the outputs on the n sense lines. Remember that readout results in resetting all cores to the 0 state, and thus that word position in the memory is cleared to all 0s.

To write information into the memory, we simply apply $\frac{1}{2}I_m$ to the proper X and Y select lines. This will, however, write a 1 in every core. So for the cores in which we desire a 0, we simultaneously apply $\frac{1}{2}I_m$ to the *inhibit* line. For example, to write 1001 in the upper four planes in Fig. 12-15, we apply $\frac{1}{2}I_m$ to the proper X and Y lines and at the same time apply $\frac{1}{2}I_m$ to the *inhibit* lines of the second and third planes.

This method of writing assumes that all cores were previously in the 0 state. For this reason it is common to define a *memory cycle*. One memory cycle is defined as a *read* operation followed by a *write* operation. This serves two purposes: first it ensures that all the cores are in the 0 state during the *write* operation; second, it provides the basis for designing a *nondestructive readout* (NDRO) system.

It is quite inconvenient to lose the data stored in the memory every time they are read out. For this reason, the NDRO has been developed. One method for accomplishing this function is to read the information out of the memory into a temporary storage register (flip-flops perhaps). The outputs of the flip-flops are then used to drive the *inhibit* lines during the *write* operation which follows (inhibit to write a 0 and do not inhibit to write a 1). Thus the basic memory cycle allows us to form an NDRO memory from a DRO memory.

Example 12-5

Describe how a coincident-current memory might be constructed if it must be capable of storing 1,024 twenty-bit words.

Solution

Since there are 20 bits in each word, there must be 20 planes in the memory (there is one plane for each bit). In order to store 1,024 words, we could make the planes square. In this case, each plane would contain 1,024 cores; it would be constructed with 32 rows and 32 columns since $(1024)^{1/2} = (2^{10})^{1/2} = 2^5 = 32$. This memory is then capable of storing $1,024 \times 20 = 20,480$ bits of information. Typically, a memory of this size might be constructed in a 3-in cube. Notice that in this memory we have the ability to switch any one of 20,480 cores by controlling the current levels on only 84 wires (32 *X* lines, 32 *Y* lines, and 20 *inhibit* lines). This is indeed a modest number of control lines.

Example 12-6

Devise a means for making the memory system in the previous example a NDRO system.

Solution

One method for accomplishing this is shown in Fig. 12-16. The basic core array consists of twenty 32-by-32 core planes. For convenience, only the three LSB planes and the MSB core plane are shown in the diagram. The wiring and operation for the other planes are the same. For clarity, the *X* and *Y select* lines have also been omitted. The output sense line of each plane is fed into a bipolar amplifier which rectifies and amplifies the output so that a positive pulse appears any time a set core is reset to the 0 state. A complete memory cycle consists of a *clear* pulse followed by a *read* pulse followed by a *write* pulse. The proper waveforms are shown in Fig. 12-17. The *clear* pulse first sets all flip-flops to the 0 state (this *clear* pulse can be generated from the trailing edge of the *write* pulse). When the *read* line goes high, all the AND gates driven by the bipolar amplifiers are enabled. Shortly after the rise of the *read* pulse, $-\frac{1}{2}I_m$ is applied to the *X* and *Y* lines designating the address of the word to be read out. This resets all cores in the selected word to the 0 state, and any core which contained a 1 will switch. Any core which switches generates a pulse on the *sense* line which is amplified and appears as a positive pulse at the output of one of the bipolar amplifiers. Since the *read* AND gates are enabled, a positive pulse at the output of any amplifier passes through the AND gate and sets the flip-flop. Shortly thereafter the half-select currents disappear,

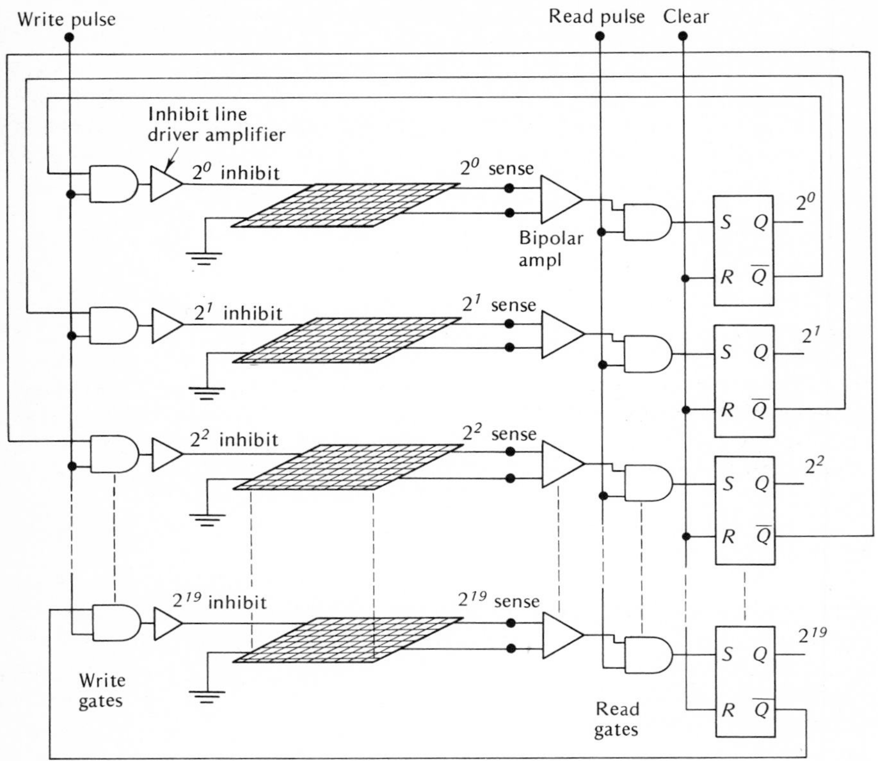

Fig. 12-16. NDRO system for Example 12-6.

the *read* line goes low, and the flip-flops now contain the data which were previously in the selected cores. Shortly after the *read* line goes low, the *write* line goes high, and this enables the *write* AND gates (connected to the *inhibit* line drivers). The 0 side of any flip-flop which has a 0 stored in it is high, and this enables the *write* AND gate to which it is connected. In this manner an *inhibit* current is applied to any core which previously held a 0. Shortly after the rise of the *write* pulse, positive half-select currents are applied to the same *X* and *Y* lines. These select currents set a 1 in any core which does not have an *inhibit* current. Thus the information stored in the flip-flops is written directly back into the cores from which it came. The half-select currents are then reduced to zero, and the *write* line goes low. The fall of the *write* line is used to reset the flip-flops, and the system is now ready for another *read/write* cycle.

The NDRO memory system discussed in the preceding example provides the means for reading information from the system without losing the individual bits stored in the cores. To have a complete memory system, we must have the

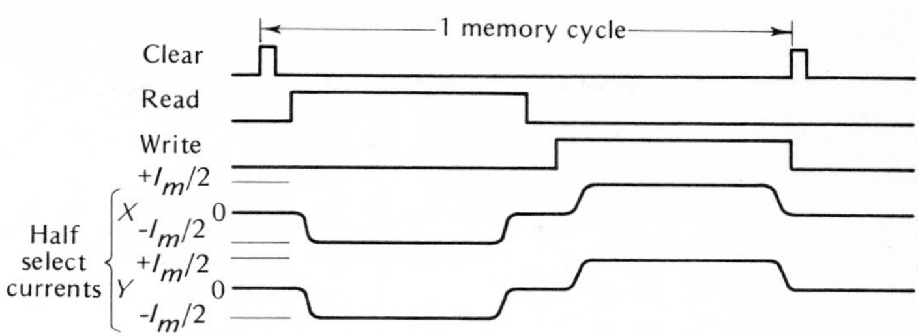

Fig. 12-17. NDRO waveforms for Fig. 12-16 (read from memory).

capability to write information into the cores from some external source (e.g., input data). The *write* operation can be realized by making use of the exact same NDRO waveforms shown in Fig. 12-17. We must, however, add some additional gates to the system such that during the *read* pulse the data set into the flip-flops will be the external data we wish stored in the cores. This could easily be accomplished by adding a second set of AND gates which can be used to set the flip-flops. The logic diagram for the complete memory system is shown in Fig. 12-18. For simplicity, only the LSB is shown since the logic for every bit is identical.

For the complete memory system we recognize that there are two distinct operations. They are *write into memory* (i.e., store external data in the cores) and *read from memory* (i.e., extract data from the cores to be used elsewhere). For these two operations we must necessarily generate two distinct sets of control waveforms. The waveforms for *read from memory* are exactly those shown in Fig. 12-17, and the events are summarized as follows:

1. The *clear* pulse resets all flip-flops.
2. During the *read* pulse, all cores at the selected address are reset to 0, and the data stored in them are transferred to the flip-flops by means of the read AND gates.
3. During the *write* pulse, the data held in the flip-flops are stored back in the cores by applying positive half-select currents (the *inhibit* currents are controlled by the 0 sides of the flip-flops and provide the means of storing 0s in the cores).

The *write into memory* waveforms are exactly the same as shown in Fig. 12-17 with one exception: that is, the *read* pulse is replaced with the *enter data* pulse. The events for *write into memory* are shown in Fig. 12-19, and are summarized as follows:

1. The *clear* pulse resets all flip-flops.
2. During the *enter data* pulse, the negative half-select currents reset all cores at the selected address. The core outputs are not used, however, since the *read* AND gates are not enabled. Instead, external data are set into the flip-flops through the enter AND gates.

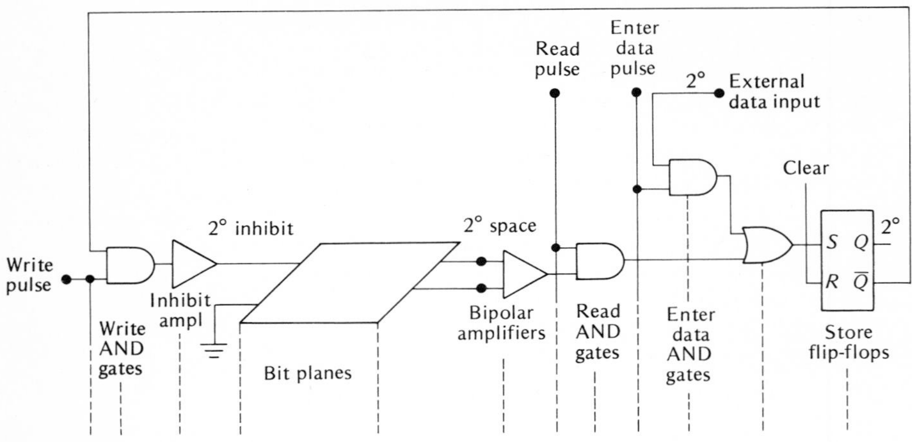

Fig. 12-18. Complete NDRO memory system (LSB plane only).

3. During the *write* pulse, data held in the flip-flops are stored in the cores exactly as before.

In conclusion, we see that *write into memory* and *read from memory* are exactly the same operations with the exception of the data stored in the flip-flops. The waveforms are exactly the same when the *read* and *enter data* pulses are used appropriately, and the same total cycle time is required for either operation.

It should be pointed out that a number of difficulties are encountered with this type of system. First of all, since the *sense* wire in each plane threads every core in that plane, a number of undesired signals will be on the *sense* wire. These undesired signals are a result of the fact that many of the cores in the plane receive a half-select current and thus exhibit a slight flux change.

The geometrical pattern of core arrangement and wiring shown in Fig. 12-13 represents an attempt to minimize the *sense*-line noise by cancellation. For example, the signals induced in the *sense* line by the *X* and *Y* drive currents would hopefully

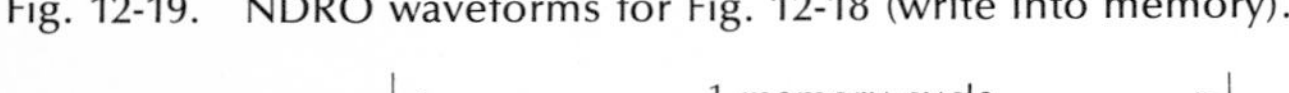
Fig. 12-19. NDRO waveforms for Fig. 12-18 (write into memory).

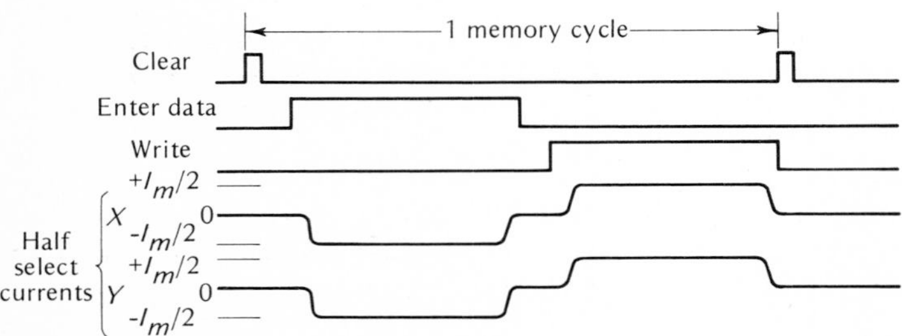

be canceled out since the *sense* line crosses these lines in the opposite direction the same number of times. Furthermore, the *sense* line is always at a 45° angle to the X and Y select lines. Similarly, the noise signals induced in the *sense* line by the partial switching of cores receiving half-select currents should cancel one another. This, however, assumes that all cores are identical, which is hardly ever true.

Another method for eliminating noise due to cores receiving half-select currents would be to have a core which exhibits an absolutely rectangular **BH** curve as shown in Fig. 12-20*a*. In this case, a half-select current would move the operating point of the core perhaps from point *a* to point *b* on the curve. However, since the top of the curve is horizontal, no flux change would occur, and therefore no undesired signal could be induced in the sense wire. This is an ideal curve, however, and cannot be realized in actual practice. A measure of core quality is given by the *squareness ratio,* which is defined as

$$\text{Squareness ratio} = \frac{\mathbf{B}_r}{\mathbf{B}_m}$$

This is the ratio of the flux density at the remanent point $\mathbf{B}_r$ to the flux density at the switching point $\mathbf{B}_m$ and is shown graphically in Fig. 12-20*b*. The ideal value is, of course, 1.0, but values between 0.9 and 1.0 are the best obtainable.

12-5 MEMORY ADDRESSING

In this section we investigate the means for activating the X and Y selection lines which supply the half-select currents for switching the cores in the memory. First of all, since it typically requires 100 to 500 mA in each select line (that is, I_m is typically between 100 and 500 mA), each select line must be driven by a current amplifier. A special class of transistors has been developed for this purpose; they are referred to as *core drivers* in data sheets. What is then needed is the means for activating the proper core-driver amplifier.

Up to this point, we have designated the X lines as X_1, X_2, X_3, . . . ,X_n, and the Y

Fig. 12-20. Hysteresis curves. (*a*) Ideal. (*b*) Practical (realizable).

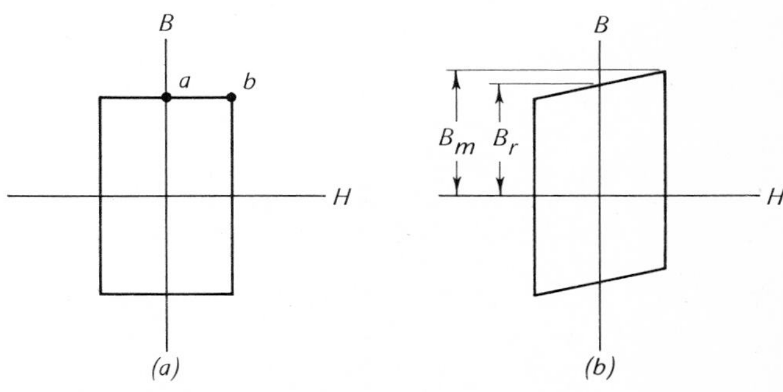

lines as $Y_1, Y_2, Y_3, \ldots, Y_n$. For a square matrix, n is the number of cores in each row or column, and there are then n^2 cores in a plane. When the planes are arranged in a stack of M planes, where M is the number of bits in a word, we have a memory capable of storing n^2, M-bit words. Any two select lines can then be used to read or write a word in memory, and the address of that word is X_aY_b, where a and b can be any number from 1 to n. For example, X_2Y_3 represents the column of cores at the intersection of the X_2 and Y_3 select lines, and we can then say that the address of this word is 23. Notice that the first digit in the address is the X line and the second digit is the Y line. This is arbitrary and could be reversed.

This method of address designation entails but one problem: in a digital system we can use only the numbers 1 and 0. The problem is easily resolved, however, since the address 23, for example, can be represented by 010 011 in binary form. If we use three bits for the X line position and three bits for the Y line position, we can then designate the address of any word in a memory having a capacity of 64 words or less. This is easy to see, since with three bits we can represent eight decimal numbers, which means we can define an $8 \times 8 = 64$ word memory. If we chose an eight-bit address, four bits for the X line and four bits for the Y line, we could define a memory having $2^4 \times 2^4 = 16 \times 16 = 256$ words. In general, an address of B bits can be used to define a square memory of 2^B words, where there are $B/2$ bits for the X lines and $B/2$ bits for the Y lines. From this discussion it is easy to see why large-scale coincident-current memory systems usually have a capacity which is an even power of 2.

Example 12-7

What would be the structure of the binary address for a memory system having a capacity of 1,024 words?

Solution

Since $2^{10} = 1{,}024$, there would have to be 10 bits in the address word. The first five bits could be used to designate one of the required 32 X lines, and the second five bits could be used to designate one of the 32 Y lines.

Example 12-8

For the memory system described in the previous example, what is the decimal address for the following binary addresses?

(*a*) 10110 00101
(*b*) 11001 01010
(*c*) 11110 00001

Solution

(*a*) The first five bits are the X line and correspond to the decimal number 22. The second five bits represent the Y line and correspond to the decimal number 5. Thus the address is $X_{22}Y_5$.

(*b*) $11001_2 = 25_{10}$ and $01010_2 = 10_{10}$. Therefore, the address is $X_{25}Y_{10}$.

(*c*) The address is $X_{30}Y_1$.

The B bits of the address in a typical digital system are stored in a series of flip-

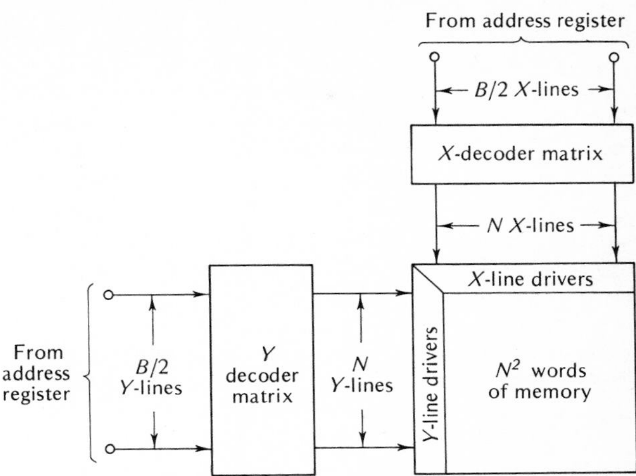

Fig. 12-21. Coincident-current memory addressing.

flops called the "address register." The address in binary form must then be decoded into decimal form in order to drive one of the X line drivers and one of the Y line driver amplifiers as shown in Fig. 12-21. The X and Y decoding matrices shown in the figure can be identical and are essentially binary-to-decimal decoders. Binary-to-decimal decoding and appropriate matrices were discussed in Chap. 10.

12-6 SEMICONDUCTOR MEMORIES—BIPOLAR

Reduced cost and size, improved reliability and speed of operation, and increased packing density are among the technological advances which have made semiconductor memories a reality in modern digital systems. A *bipolar* memory is constructed using the familiar bipolar transistor, while the MOS memory makes use of the MOSFET. In this section we consider the characteristics of bipolar semiconductor memories; MOS memories are considered in the next section.

A "memory cell" is a unit capable of storing binary information; the basic memory unit in a bipolar semiconductor memory is the flip-flop (latch) shown in Fig. 12-22. The cell is *selected* by raising the *X select* line and the *Y select* line; the *sense* lines are both returned through low-resistance *sense* amplifiers to ground. If the cell contains a 1, current is present in the 1 *sense* line. On the other hand, if the cell contains a 0, current is present in the 0 *sense* line.

To write information into the cell, the X and Y *select* lines are held high; holding the 0 *sense* line high ($+V_{cc}$) while the 1 *sense* line is grounded writes a 1 into the cell. Alternatively, holding the 1 *sense* line high ($+V_{cc}$) and the 0 *sense* line at ground during a select writes a 0 into the cell. The basic bipolar memory cell in Fig. 12-22 can be used to store one binary digit (bit), and thus many such cells are required to form a memory.

Sixteen of the *RS* flip-flop cells in Fig. 12-22 have been arranged in a 4-by-4 ma-

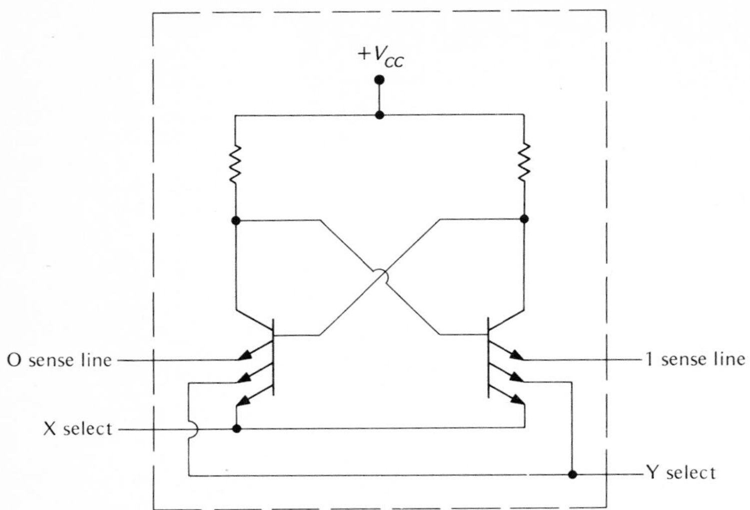

Fig. 12-22. Bipolar memory cell circuit.

trix to form a 16-word by one-bit memory in Fig. 12-23. It is referred to as a random access memory (RAM) since each bit is individually addressable by selecting one *X* line and one *Y* line. It is also a nondestructable readout since the *read* operation does not alter the state of the selected flip-flop. This memory comes on a single semiconductor chip (in a single package) as shown in Fig. 12-24*a*. To construct a 16-word memory with more than one bit per word requires stacking these basic units. For example, six of these chips can be used to construct a 16-word by six-bit memory as shown in Fig. 12-24*b*. The *X* and *Y address* lines are all connected in parallel. The units shown in Figs. 12-23 and 12-24 are essentially equivalent to the Texas Instruments 9033 and Fairchild 93407 (5033 or 9033).

Example 12-9

Using a 9033, explain how to construct a 16-word by 12-bit memory. What address would select the 12-bit word formed by the bits in column 1 and row 1 of each plane?

Solution

Connect twelve 16-word by one-bit memory planes in parallel. The address $X_0X_1X_2X_3Y_0Y_1Y_2Y_3 = 10001000$ selects the bit in the first column and the first row of each plane (a 12-bit word represented by the vertical column of 12 bits).

For larger memories, the appropriate address decoding, driver amplifiers, and *read/write* logic are all constructed in a single package. Such a unit, for example, is the Fairchild 93415—this is a 1,024-word by one-bit read/write RAM. The logic diagram is shown in Fig. 12-25. An address of 10 bits is required ($A_0A_1A_2A_3A_4A_5A_6A_7A_8A_9$) to obtain 1,024 words. That is, *x* bits provide 2^x word

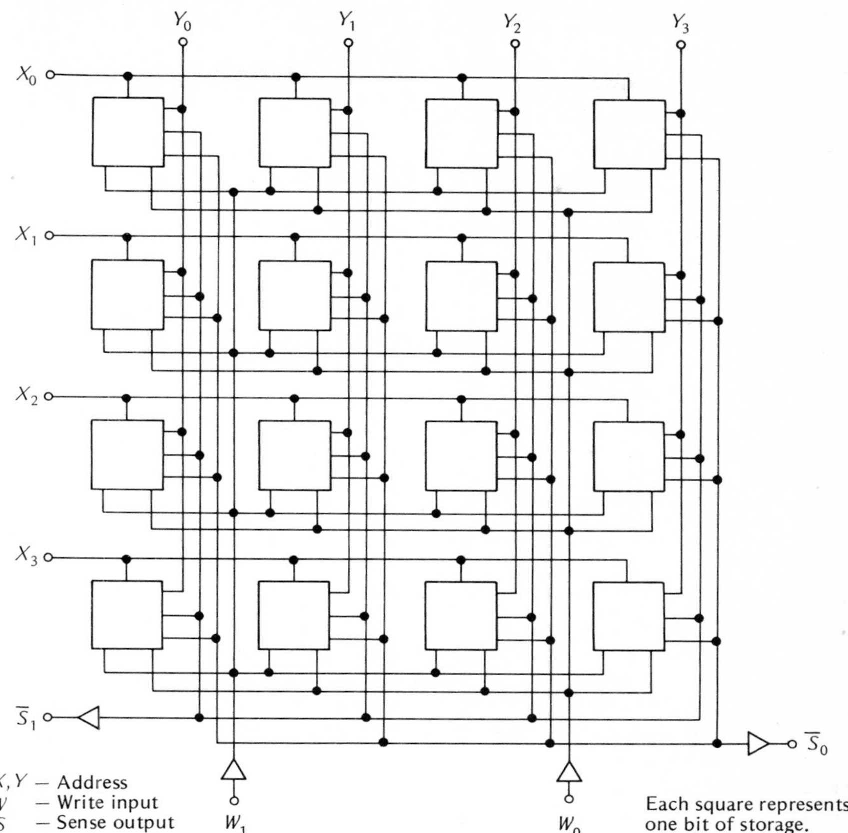

Fig. 12-23. 16-word 1-bit memory.

locations. In this case, the 10-bit address is divided into two groups of five bits each. The first five (A_0, A_1, A_2, A_3, A_4) select a unique group of 32 lines from the 32-by-32 array. The second five (A_5, A_6, A_7, A_8, A_9) select exactly one of the 32 preselected lines for reading or writing. These basic units are then stacked in parallel as shown previously; *n* units provide a memory having 1,024 words by *n* bits.

Another interesting and useful type of semiconductor memory is shown in Fig. 12-26. This is a bipolar TTL read-only memory (ROM). The information stored in a ROM can be read out, but new information cannot be written into it. Thus, the information stored is permanent in nature. ROMs can be used to store mathematical tables, code translations, and other fixed data. The logic required for a ROM is generally simpler than that required for a *read/write* memory, and the unit shown in Fig. 12-26 (equivalent to a TI 9034 or Fairchild 93434) provides an eight-bit output word for each five-bit input address. There are, of course, 32 words, since an address of five bits provides 32 words ($2^5 = 32$).

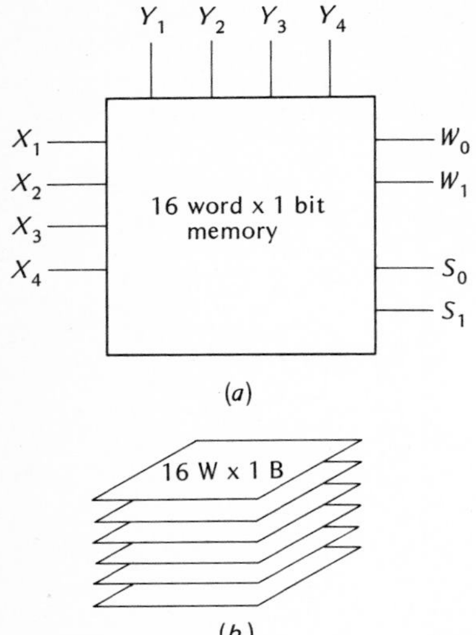

Fig. 12-24. (*a*) Logic diagram. (*b*) Six chips stacked to get a 16-word × 6-bit memory.

Fig. 12-25. 1024-word × 1-bit RAM.

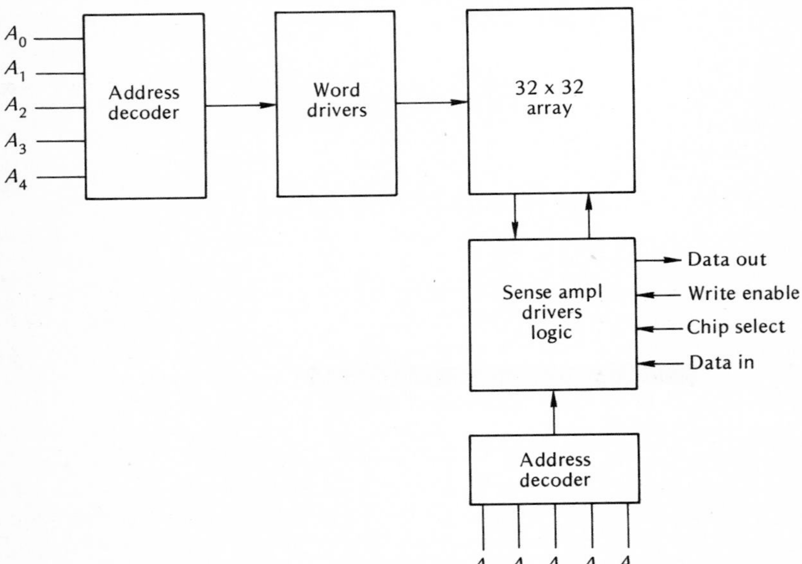

Example 12-10

How many address bits are required for a 123-word by four-bit ROM constructed similarly to the unit in Fig. 12-26? How many memory cells are there in such a unit?

Solution

It requires seven address bits, since $2^7 = 128$. There would be $128 \times 4 = 512$ memory cells.

12-7 SEMICONDUCTOR MEMORIES—MOS

The basic device used in the construction of an MOS semiconductor memory is the MOSFET. Both *p*-channel and *n*-channel devices are available. The *n*-channel memories have simpler power requirements, usually only $+V_{cc}$, and are quite compatible with TTL since they are usually referenced to ground and have positive signal levels up to $+V_{cc}$. The *p*-channel devices generally require two power-supply voltages and may require signal inversion in order to be compatible with TTL. MOS devices are somewhat simpler than bipolar devices; as a result, MOS memories can be constructed with more bits on a chip, and they are generally less expensive than bipolar memories. The intrinsic capacitance associated with an MOS device generally means that MOS memories are slower than bipolar units, but this capacitance can be used to good advantage, as we shall see.

Fig. 12-26. 256-bit (32-word × 8-bit) ROM.

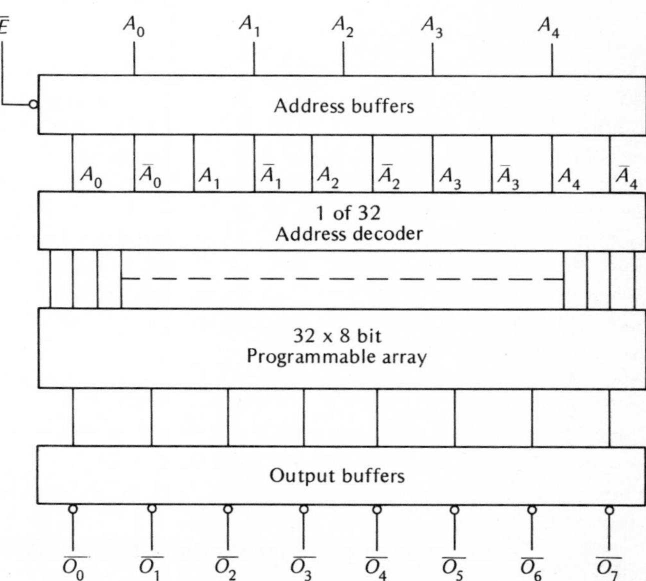

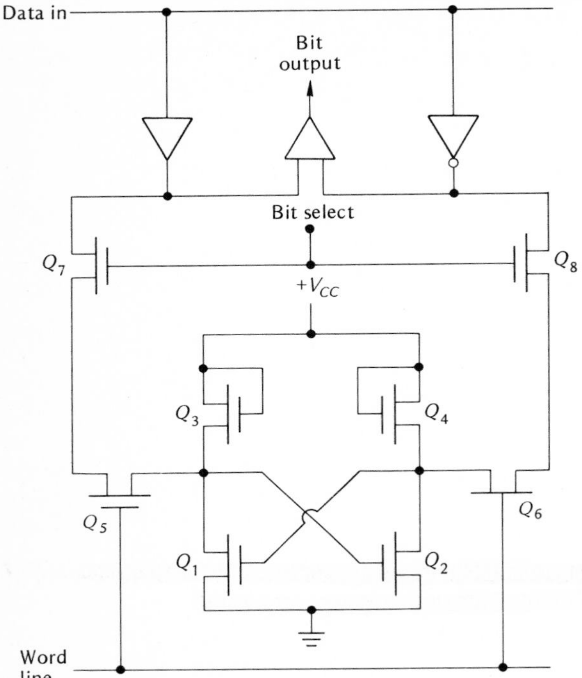

Fig. 12-27.

An *RS* flip-flop constructed using MOSFETs is shown in Fig. 12-27. It is a standard bistable circuit, with Q_1 and Q_2 as the two active devices, and Q_3 and Q_4 acting as active pull-ups (essentially resistances). Q_5 and Q_6 couple the flip-flop outputs to the two *bit lines*. This cell is constructed using *n*-channel devices, and selection is accomplished by holding both the *word* line and the *bit select* line high ($+V_{CC}$). The positive voltage on the *word* line turns on Q_5 and Q_6, and the positive voltage in the *bit select* line turns on Q_7 and Q_8. Under this condition, the flip-flop outputs are coupled directly to the *bit output* amplifier (one input side is high, and the other must be low). On the other hand, data can be stored in the cell when it is selected by applying 1 or 0 ($+V_{cc}$ or 0 V dc) at the *data input* terminal. The basic memory cell in Fig. 12-27 is used to construct a 1,024-bit RAM having a logic diagram similar to Fig. 12-25. This particular unit is a 2602 as manufactured by Signetics Corp.

A memory cell using *p*-channel MOSFETs is shown in Fig. 12-28. Q_1 and Q_2 are the two active devices forming the flip-flop, while Q_3 and Q_4 act as active load resistors. The cell is selected by a low logic level at the *bit select* input. This couples the contents of the flip-flop out to appropriate amplifiers (as in Fig. 12-27) through Q_5 and Q_6.

A *static memory* is composed of cells capable of storing binary information indefinitely. For example, the bipolar or MOSFET flip-flop remains set or reset as long

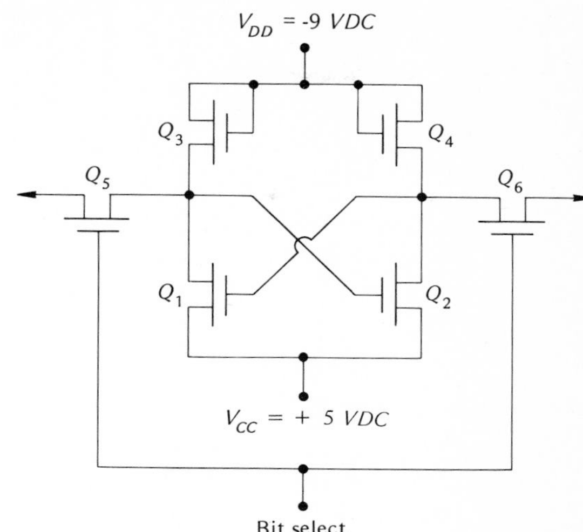

Fig. 12-28.

as power is applied to the circuit. Also, a magnetic core remains set or reset, even if power is removed. These basic memory cells are used to form a *static memory*. On the other hand, a *dynamic memory* is composed of memory cells whose contents tend to decay over a period of time (perhaps milliseconds or seconds); thus, their contents must be restored (refreshed) periodically. The leaky capacitance associated with a MOSFET can be used to store charge, and this is then the basic unit used to form a dynamic memory. (There are no dynamic bipolar memories because there is no suitable intrinsic capacitance for charge storage.) The need for extra

Fig. 12-29. Basic dynamic memory cell.

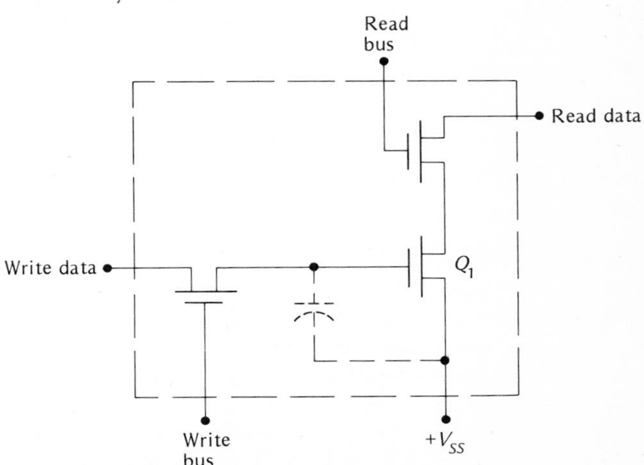

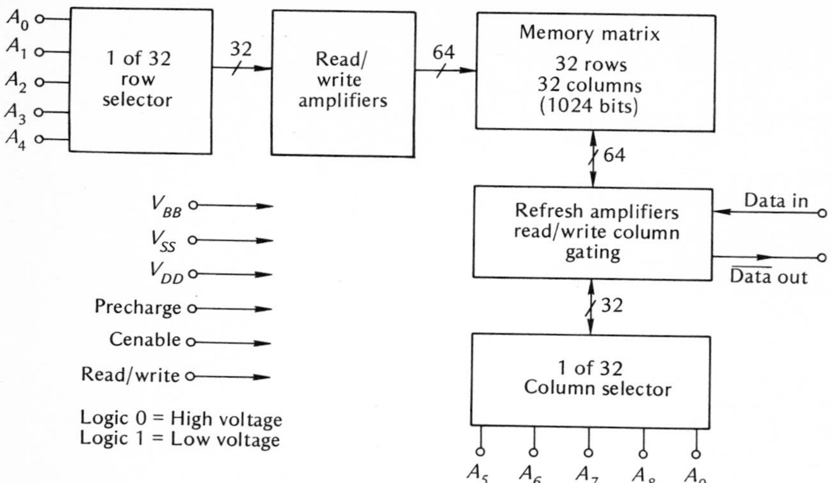

Fig. 12-30. 1103 Dynamic RAM logic diagram.

timing signals and logic to periodically refresh the dynamic memory is a disadvantage, but the higher speeds and lower power dissipation, and therefore the increased cell density, outweighs the disadvantages. Note that a dynamic memory dissipates energy only when reading, writing, or refreshing cells. A typical dynamic memory cell is shown in Fig. 12-29.

The dynamic memory cell in Fig. 12-29 is constructed from *p*-channel MOSFETs. The gate capacitance (shown as a dotted capacitor) is used as the basic storage element. To write into the cell requires holding the *write bus* at a low logic level; then a low level at the *write data* input charges the gate capacitance (stores a 1 in the cell). With the write bus held low, and a high logic level ($+V_{cc}$) at the *write data* input, the gate capacitance is discharged (a 0 is stored in the cell).

To read from the cell requires holding the *read bus* input at a low logic level. If the gate capacitance is charged (cell contains a 1), the *read data* line goes to $+V_{ss}$; if the cell contains a 0, the *read data* line remains low.

The memory cell in Fig. 12-29 is used by a number of manufacturers to construct the widely used 1103 1,024-bit dynamic RAM. The logic diagram is shown in Fig. 12-30. Refer to manufacturers' data sheets for more detailed operating information.

12-8 MAGNETIC-DRUM STORAGE

Magnetic cores and semiconductor devices arranged in three-dimensional form offer great advantages as memory systems. By far the most important advantage is the speed with which data can be written into or read from the memory system. This is called the *access time,* and for core memory systems it is simply the time of one *read/write* cycle. Thus the access time is directly related to the clock, and typical values are from less than 1 to a few microseconds. These types of memory systems are said to be *random-access* since any word in the memory can be selected at random. The primary disadvantage of this type of memory system is the cost of construction for the amount of storage available. As an example, recall that a magnetic tape is capable of storing large quantities of data at a relatively low cost per bit of storage. A typical tape might be capable of storing up to 20 million characters, which corresponds to 120 million bits (Chap. 10). To construct such a memory with magnetic cores requires about 3 million cores per plane, assuming we use a stack of 36 planes corresponding to a 36-bit word. It is quite easy to understand the impracticality of constructing such a system. What is needed, then, is a system capable of storing information with less cost per bit but having a greater capacity.

Such a system is the *magnetic-drum* storage system. The basis of a magnetic drum is a cylindrical-shaped drum, the surface of which has been coated with a magnetic material. The drum is rotated on its axis as shown in Fig. 12-31, and the *read/write* heads are used to record information on the drum or read information from the drum. Since the surface of the drum is magnetic, it exhibits a rectangular-hysteresis-loop property and can thus be magnetized. The process of recording on the drum is much the same as for recording on magnetic tape, as discussed in Chap. 10, and the same methods for recording are commonly used (i.e., RZ, NRZ, and NRZI). The data are recorded in tracks around the circumference of the drum, and there is one *read/write* head for each track. There are three major methods for storing information on the drum surface; they are *bit-serial, bit-parallel,* and *bit-serial-parallel*.

In bit-serial recording, all the bits in one word are stored sequentially, side by side, in one track of the drum. Bit-serial storage is shown in Fig. 12-32*a*. Storage densities of 200 to 1,000 bits per in are typical for magnetic drums. A typical drum might be 8 in in diameter and thus have the capacity to store $\pi \times 8$ in $\times$ 200 bits per in = 5,024 bits in each track. Drums have been constructed with anywhere from 15 to 400 tracks, and a spacing of 20 tracks to the inch is typical. If we assume this particular drum is 8 in wide and has a total of 100 tracks, we see immediately that it has a storage capacity of 5,024 bits per track $\times$ 100 tracks =

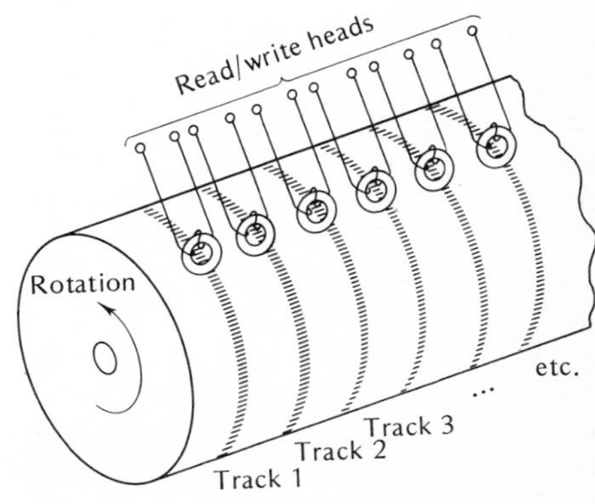

Fig. 12-31. Magnetic-drum storage.

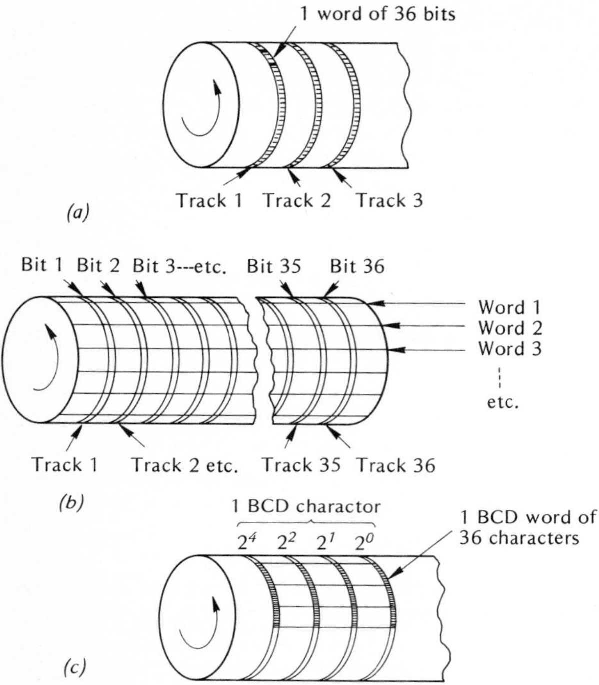

Fig. 12-32. Magnetic-drum organization. (*a*) Bit-serial storage. (*b*) Bit-parallel storage. (*c*) Bit-serial-parallel storage.

502,400 bits of information. Compare this capacity with that of a coincident core memory, which is 64 cores on a side (quite a large core system) with 64 core planes. This core memory has a capacity of $2^6 \times 2^6 \times 2^6 = 262{,}144$ bits. The drum described above is actually considered small, and much larger drums have been constructed and are now in use.

Example 12-11

A certain magnetic drum is 12 in in diameter and 12 in long. What is the storage capacity of the drum if there are 200 tracks and data are recorded at a density of 500 bits per in?

Solution

Each track has a capacity of $\pi \times 12$ in $\times$ 500 bits per in $\cong$ 18,840 bits. Since there are 200 tracks, the drum has a total capacity of $18{,}840 \times 200 = 3{,}768{,}000$ bits.

In the preceding example, each track has the ability to store about 18,840 bits. If we use a 36-bit word, we can store about 523 words in each track. Since the words are stored sequentially around the drum, and since there is only one *read/write* head for the track, it is easy to see that we may have to wait to read any one word. That is, the drum is rotating, and the word we want to read may not be under the *read* head at the time we choose to read it. It may in fact have just passed under the head, and we will have to wait until the drum completes nearly a full revolution before it is under the head again. This points out one of the major disadvantages of the drum compared with the core storage. That is the problem of access time. On the average, we can assume that we will have to wait the time required for the drum to complete one-half a revolution. A drum is thus said to have *restricted access*.

Example 12-12

If the drum in Example 12-11 rotates at a speed of 3,000 rpm, what is the average access time for the drum?

Solution

3,000 rpm = 50 rps. Thus the time for one revolution is 1/(50 rps) = 20 ms. Thus, the average access time is one-half the time of one revolution, which is 10 ms. Contrast this with a coincident-current core memory which has a direct access time of a few microseconds.

Notice in the previous example that it requires a short period of time to read the 36 bits of the word, since they appear under the *read* head one bit at a time in a serial fashion. The actual time required is small compared with the access time and is found to be (20 ms/r)/(523 words per track) $\cong$ 40 μs. This read time can be reduced by storing the data on the drum in a parallel manner, as shown in Fig. 12-32*b*.

The average access time for bit-parallel storage is the same as for bit-serial storage, but it is possible to read and record information at a much faster rate with the bit-parallel system. Let us use the drum in Example 12-11 once more. Since there are 523 words around each track, and since the drum rotates at 50 rps, we can read (or write) 523 words per revolution $\times$ 50 rps = 26,150 words per second. If the data were stored in parallel fashion, we could read (or write) at 36 times this rate, or at a rate of 18,840 words per revolution $\times$ 50 rps = 942,000 words per second. We would, of course, arrange to have the number of tracks on the drum an even multiple of the number of bits in a word. For example, with a 36-bit word we might use a drum having 36 or 72 or 108 tracks.

A third method for recording data on a drum is called "bit-serial-parallel." The method is shown in Fig. 12-32*c* and is commonly used for storing BCD information. The access and read (or write) times are a combination of the serial and parallel times. One BCD character occupies one bit in each of four adjacent tracks. Thus, every four tracks might be called a "band," and each BCD character occupies one space in the band. If there are 36 BCD characters in a word, we can store 523 words on the drum of Example 12-11.

Quite often the access time is speeded up by the addition of extra *read/write* heads around the drum. For example, we might use two sets of heads placed on opposite sides of the drum. This would obviously cut the access time in half. Alternatively, we might use three sets of heads arranged around the drum at 120° angles. This would reduce the access time by one-third.

Since writing on and reading from the drum must be very carefully timed, one track in the drum is usually reserved as a timing track. On this track, a series of timing pulses is permanently recorded and is used to synchronize the *write* and *read* operations. For the drum discussed in Example 12-11, there are 523 words in each track around the circumference of the drum. We might then record a series of 523 equally spaced timing marks around the circumference of the timing track. Each pulse would then designate the *read* or *write* position for a word on the drum.

STUDY AIDS

Summary

A wide variety of magnetic devices can be used as binary devices in digital systems. By far the most widely used is the magnetic core. Cores can be used to implement various logic functions such as AND, OR, and NOT, and more complicated functions can be formed from combinations of these basic circuits. Magnetic-core shift registers and ring counters can be constructed by using the single-diode transfer loop between cores. Magnetic-core logic is particularly useful in applications experiencing environmental extremes.

Direct-access memories with very fast access times can be conveniently constructed using either magnetic cores or transistors. The most popular method for constructing these memories is the coincident-current technique. Memories constructed using cores are inherently DRO-type memories but can be transformed into NDRO memories by the addition of external logic.

Semiconductor memories constructed from bipolar transistors or MOSFETs are available. Bipolar memories are static memories, but are available as random-access ROMs, or as complete *read/write* units. MOS memories can be either static or dynamic, and are available as RAMs.

Magnetic drums and disks provide larger storage capacities at a lower cost per bit than core-type memories. They do, however, offer the disadvantage of increased access time.

Glossary

access time For a coincident-current memory, it is the time required for one *read/write* cycle. In general, it is the time required to write one word into memory or to read one word from memory.

address A series of binary digits used to specify the location of a word stored in a memory.

coincident-current selection The technique of applying $\frac{1}{2}I_m$ on each of two lines passing through a magnetic device in such a way that the net current of I_m will switch the device.

DRO Destructive readout.

dynamic memory A memory whose contents must be restored periodically.

hysteresis Derived from the Greek word *hysterein,* which means to lag behind.

hysteresis curve Generally a plot of magnetic flux density **B** versus magnetic force

H. Can also refer to the plot of magnetic flux ϕ versus magnetizing current I.

memory cycle In a coincident-current memory system, a *read* operation followed by a *write* operation.

NDRO Nondestructive readout.

RAM Random-access memory.

ROM Read-only memory.

select current I_m The minimum current required to switch a magnetic device.

single-diode transfer loop A method of coupling the output of one magnetic core to the input of the next magnetic core.

squareness ratio A measure of core quality. From the hysteresis curve, it is the ratio $\mathbf{B}_r/\mathbf{B}_m$.

static memory A memory capable of storing binary information indefinitely.

Review Questions

1. Name one advantage of a ferrite core over a metal-ribbon core.
2. Name one advantage of a metal-ribbon core over a ferrite core.
3. Describe the method for detecting a stored 1 in a core.
4. Why is a strobing technique often used to detect the output of a switched core?
5. How is core switching time t_s affected by the switching current?
6. Explain why more complicated logic functions using cores can lead to excessive operating times.
7. What is the purpose of the diode in the single-diode transfer loop?
8. Why is a delay in signal transfer between cores desired?
9. Explain how the R and C in Fig. 12-11 introduce a delay in signal transfer between cores.
10. Explain the operation of the *sense* wire in a magnetic-core matrix plane. Why is it possible to thread every core in the plane with the same wire?
11. Explain how it is possible to store a 0 in a coincident-current memory core using the *inhibit* line.
12. Why is a basic coincident-current core memory inherently a DRO-type system?
13. In the basic memory cycle for a coincident-current core memory system, why must the *read* operation come before the *write* operation?
14. What is the difference between the *write into memory* and the *read from memory* cycles for a coincident-current core memory system?
15. Explain the meaning of the title "64-word by eight-bit static RAM."
16. Why are there no dynamic bipolar memories?
17. What does it mean to "refresh" a dynamic memory?

18. Describe the difference between random-access and restricted-access memories.

19. Describe the advantages of using a magnetic-drum storage system.

Problems

12-1. Draw a typical hysteresis curve for a core, and show the two remanent points.

12-2. Show graphically on a ϕI curve the path of the operating point as the core is switched from a 1 to a 0. Repeat for switching from a 0 to a 1.

12-3. Draw the symbol for a magnetic-core logic element, and explain the function of each winding.

12-4. Draw a set of waveforms showing how the exclusive-OR circuit of Fig. 12-8*d* must operate (notice it requires only two clocks which are spaced 180° out of phase).

12-5. Draw a single-diode transfer loop between two cores, and explain its operation (use waveforms if needed).

12-6. Draw a schematic and the waveforms for a core ring counter which provides seven output pulses.

12-7. Draw a sketch and explain how a core can be switched by the coincident-current method.

12-8. Make a sketch similar to Fig. 12-15 showing a three-dimensional core memory capable of storing 100 ten-bit words. Show all input and output lines clearly.

12-9. Describe the geometry of a coincident-current core memory capable of storing 4,096 thirty-six-bit words (i.e., how many planes, how many cores per plane, etc.).

12-10. How many bits can be stored in the memory in Prob. 12-9?

12-11. How many control lines are required for the memory in Prob. 12-9?

12-12. Show graphically the meaning of squareness ratio for a magnetic core, and explain its importance for magnetic-core memories.

12-13. Describe a structure for the address which could be used for the memory of Prob. 12-9.

12-14. If a certain core memory is composed of square matrices, what is the word capacity if the address is 12 binary digits?

12-15. How many bits are required in the address of a 256-word by one-bit *read/write* bipolar RAM?

12-16. Draw the polarity of the stored charge on the gate capacitance shown in the basic dynamic memory cell in Fig. 12-29.

12-17. What is the bit-storage capacity of a magnetic drum 10 in in diameter if data are stored with a density of 200 bits per in in 20 tracks?

12-18. What would be the diameter of a magnetic drum capable of storing 3,140 thirty-six-bit words if there are 10 tracks and data are stored bit-serial at 300 bits per in?

12-19. What is the average access time for the drum in Prob. 12-18 if it rotates at 36,000 rpm? What could be done to reduce this access time by a factor of 2?

12-20. For the drum in Prob. 12-18, at what bit rate must data be moved (i.e., read or write) if the drum rotates at 36,000 rpm?

Digital Arithmetic

13

In a previous chapter you learned how to perform various arithmetic operations using binary numbers. You learned that binary addition is carried out according to the rules of mod-2 arithmetic and that binary subtraction can be performed by using 1's and 2's complements. In later chapters various logic circuits were developed which were shown to be capable of carrying out binary arithmetic operations. These logic circuits—half- and full-adders, half- and full-subtractors, and shift registers—form the basis of the arithmetic unit. An arithmetic unit is usually included in a digital system whenever the capability to add, subtract, multiply, divide, etc., is desired. In a small special-purpose digital system the arithmetic requirements may be limited, and the arithmetic unit is then quite small. On the other end of the scale, in a large general-purpose digital computer with the capability of performing any number of arithmetic operations, the arithmetic unit is quite large and necessarily involved. In this chapter we investigate some of the methods and techniques for carrying out digital arithmetic using the basic arithmetic logic circuits previously discussed. It should be pointed out that there are almost as many ways of implementing arithmetic functions as there are designers, and this chapter is by no means a comprehensive coverage of the subject. The intent here is to give a solid foundation in the basic principles. The reader interested in more detail is urged to study advanced texts.

After studying this chapter you should be able to

1. Write positive or negative decimal numbers as either floating-point or fixed-point binary numbers.
2. Apply 1's and 2's complements to perform binary addition and subtraction.
3. Show how to use adders to accomplish both serial and parallel binary addition.
4. Discuss the characteristics of a BCD adder.

13-1 NUMBER REPRESENTATIONS

We saw earlier that we can store a binary number in a register in two ways, that is, with the MSB on the right or with the MSB on the left. For example, the number 0111_2 when stored in a register with the MSB on the left is equal to 7_{10}. In finding the decimal equivalent of this binary number we simply assumed that the binary

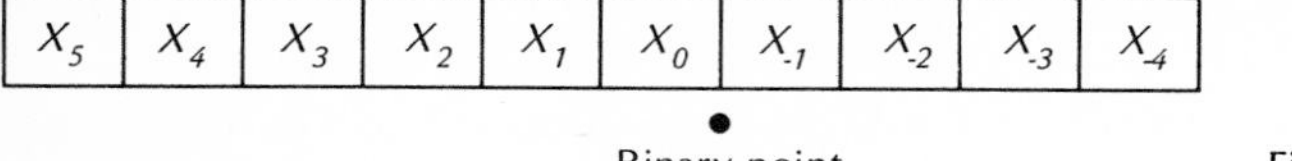

Fig. 13-1.

point was at the far right. Since no previous mention was made of the binary point, we could just as easily have assumed that it was on the far left. In this case the decimal equivalent would be 0.4375_{10}. For that matter we could have assumed the binary point to be at any number of other places. In any case, it must now be clear that the position of the binary point must be defined.

In trying to decide where to locate the binary point, if we intend to discuss logic capable of doing arithmetic in general, we must have the capability of handling fractions as well as integers. The most straightforward way of handling this is simply to define the position of the binary point in the word. For example, in Fig. 13-1, the binary-point location is defined. In this particular case, the digits to the left of the binary point represent the integers of the binary number, and the digits to the right represent the fractional part of the number. We can then represent numbers between 111111.1111 and 000000.0000, with the smallest fractional part being $0.0001 = 1/16_{10} = 0.0625_{10}$. We can then add or subtract binary numbers in this form and have no difficulties with the location of the binary point. For example, we might add 001110.0011 and 000100.1000 as follows:

$$\begin{array}{rr} & 001110.0011 \\ (+) & \underline{000100.1000} \\ & 010010.1011 \end{array} \qquad \begin{array}{rr} & 14.1875 \\ (+) & \underline{4.5000} \\ & 18.6875 \end{array}$$

Or we might add the following two numbers:

$$\begin{array}{rr} & 100100.0010 \\ (+) & \underline{011100.1000} \\ & 1000000.1010 \end{array} \qquad \begin{array}{rr} & 36.125 \\ (+) & \underline{28.500} \\ & 64.625 \end{array}$$

Notice that in this case we added two numbers whose sum exceeded the capacity of our register. That is to say, this register is capable of handling numbers only up to $111111.1111_2 = 63.9375_{10}$. The fact that the sum exceeded the capacity of the register can be easily detected by sensing the 1 in the leftmost position of the sum. This 1 is the result of a carry generated when the two MSBs were added, and we use this fact to detect this condition. Thus, when the contents of the register is exceeded, an *overflow* condition exists, and we sense this by detecting a 1 in the carry from the two MSBs. An overflow must be detected since the contents of the register after this last addition will be $000000.1010 = 0.625_{10}$. This obviously incorrect answer is a result of the fact that the register is capable of storing only 10 bits, and the eleventh bit (i.e., the carry from the two MSBs) is lost during the addition. Thus the overflow is an indication of incorrect operation.

Up to this point, we have shown that the selection of the binary point as shown in Fig. 13-1 is perfectly all right for addition so long as we detect any overflows. Let us see what happens when we multiply two numbers with this system. We might, for example, multiply two numbers as follows:

```
     100110.1001
    ×000001.0001
      1001101001
  1001101001
  10100011111001   = product
```

In the product we have no difficulty in finding the proper place for the binary point; we simply count eight places over from the right, and the answer is obviously 101000.11111001. This does not seem to offer any difficulties since there are still only six bits to the left of the binary point. We could simply omit the four extra digits to the right of the binary point. However, let us try one more example:

```
           100110.1001
          ×100001.0001
            1001101001
       1001101001
  1001101001
  100111111010111111001
```

In this case, we still have no trouble locating the binary point; again we count eight places from the right, giving 10011111010.11111001. We have now obviously greatly exceeded the capacity of the register on both ends. Furthermore, the register has no means for counting the number of places from the right to locate the binary point. In this case most machines would simply use the 10 MSBs to give the answer 100111.1101, which is obviously incorrect.

To overcome these difficulties, it is quite common practice to place the binary point to the left of the MSB. We would then represent all numbers as $.X_1X_2 \cdots X_n$, where n is the number of bits. For example, the 10-bit number previously discussed would be $.X_1X_2X_3 \cdots X_9X_{10}$. Let us see how this solves the problems previously described. First, it makes no difference where the binary point is located for addition (and therefore for subtraction), and this method is therefore acceptable. We may, however, still have an overflow problem and must still account for this possibility. Second, all numbers are represented as fractions, since the binary point is to the far left. This means that the product of any two numbers is always a number less than 1. Thus we can easily locate the binary point by simply using the 10 most significant digits. To clarify this point let us multiply two 10-bit numbers using this system:

```
        .1101100011
       ×.1000000010
         0000000000
        1101100011
 1101100011
.01101100110011000110
```

The arithmetic unit is designed such that it always generates 20 digits. The 10 most significant are stored in the register, and the other 10 are simply truncated (omitted). In this way the positioning of the binary point is solved. The fact that all numbers are represented as fractions is no problem since any number can be represented as a fraction multiplied by a power of 2. For example, $21.75_{10} = 10101.11_2 = .1010111 \times 2^5$. That is, the binary point is simply shifted to the left five places.

Any arithmetic unit which uses the system $.X_1X_2 \ldots X_n$ for representing numbers is called a *binary fractional machine* since all the numbers are fractions. Furthermore this is called a *fixed-point machine* since the location of the binary point is fixed. Using this system we have the capability of storing any number from $.000 \ldots 0$ to $(1 - 1/2^n)$.

Example 13-1

What range of numbers can be represented in a binary fractional machine having four bits?

Solution

The four bits are represented as $.X_1X_2X_3X_4$. The minimum number is, of course, .0000. The maximum number is $.1111_2 = 0.9375_{10}$. Using the above relationship, the maximum number is $1 - 1/2^n = 1 - 1/2^4 = 1 - 1/16 = 1 - 0.0625 = 0.9375$.

With the binary fractional system we have the capability to represent only positive numbers. In order to extend our capability to include negative numbers, it is common practice to include one more bit to the *left* of the binary point. This bit is used exclusively to designate the sign of the number (either positive or negative) and as such is called the *sign bit*. By convention, and for other reasons which will become clear later, a positive number is represented by a 0 in the sign bit, and a negative number is represented by a 1 in the sign bit. Thus we can represent a number by the following system:

$$\underbrace{X_0}_{\text{Sign bit}}.\underbrace{X_1X_2X_3 \cdots X_n}_{\text{Magnitude}}$$

With this system we can then define numbers between $-(1 - 1/2^n)$ and $+(1 - 1/2^n)$.

Example 13-2

What is the range of numbers defined by the *sign and magnitude system* if there are a maximum of five bits?

Solution

The system is $X_0.X_1X_2X_3X_4$. Notice that for a word of five bits, one bit is the sign and four bits are the magnitude. The most negative number is $1.1111 = -.1111 = -0.9375_{10}$. The most positive number is $0.1111 = +.1111 = +0.9375_{10}$.

Example 13-3

Represent the following decimal numbers in sign and magnitude form:

(*a*) +0.75
(*b*) −0.75
(*c*) +0.3125
(*d*) −0.4375

Solution

(*a*) +0.75 = 0.11
(*b*) −0.75 = 1.11
(*c*) +0.3125 = 0.0101
(*d*) −0.4375 = 1.0111

The representation of numbers with the system shown in Example 13-2 is more properly called the sign and *true-magnitude* form. You will recall from the chapter on binary arithmetic that the addition of a positive binary number and a negative binary number is carried out by adding the former to the complement of the latter (either the 1's or 2's complement). We shall find it convenient to represent all negative numbers in complement form. There are then three ways of representing a negative binary number:

1. Sign and true magnitude
2. Sign and 1's complement
3. Sign and 2's complement

Example 13-4

Represent the two negative numbers in Example 13-3 in the three different ways described.

Solution

Number	True magnitude	1's complement	2's complement
−.75	1.11	1.00	1.01
−.4375	1.0111	1.1000	1.1001

Notice that in each case the 1 in the sign bit shows that the number is negative. To find the 1's or 2's complement representation, only the magnitude portion of the number is complemented.

The next logical question which might arise is, how do we represent a number such as 831 in this sign and magnitude system? The answer is quite simple. In a fixed-point system such as this, the operator must ensure that all input information is in the proper form. Thus, in punching cards, paper tape, etc., the operator must change the number 831 to 0.831×10^3. The data which are entered into the machine are simply 0.831 (in binary form), and the operator must remember that the $\times 10^3$ is needed. This process is commonly called "scaling." As an example of this, suppose that one of the parameters to be entered into a system is temperature, and the range is from -100 to $+100$°C. The operator would simply enter this parameter as $X_0.X_1X_2X_3$. A value of -50°C would be entered as $-.050$. Output data from the machine representing temperature values would simply have to be multiplied by 1,000 to obtain the correct values.

Example 13-5

Describe the sign and magnitude system required to represent a parameter whose values vary between ± 200.

Solution

The input data will be scaled by a factor of 10^3. Thus all input data will be between ± 0.200, and all output data will then have to be multiplied by 10^3 for the correct value. We must now make a decision as to the resolution of the data. For example, if we choose to use six bits, the LSB is $1/2^6 = 1/64 = 0.0156$. For greater resolution we might choose a system having eight bits; the LSB is then $1/2^8 = 1/256 = 0.0039$. If we are then satisfied with an eight-bit system, the binary form will be $X_0.X_1X_2X_3X_4X_5X_6X_7X_8$. Notice that it actually requires a nine-bit word, one bit for the sign and eight bits for the magnitude.

Example 13-6

In the example above, what are the binary representation and resolution error for a value of $+125$? For a value of -25?

Solution

$+125$ will be represented as $+0.125$. The binary word for this value is 0.00100000. There is no resolution error in this case. -25 will be represented as -0.025. The binary word for this value is 1.00000110 (true magnitude) or 1.11111001 (in 1's complement) or 1.11111010 (in 2's complement). The decimal equivalent of the magnitude is 0.0234375. This represents a resolution error of about $(0.025 - 0.023)/0.025 = 8$ percent. It is quite obvious from this that the resolution error may be very large for small-magnitude numbers.

At this point some of the disadvantages of a fixed-point machine are apparent. To overcome the difficulties of fixed-point arithmetic, most large-scale digital computers use another method called "floating-point arithmetic." In this system numbers are represented with a sign bit, bits for the magnitude, and bits for the power of 2. One possible form of the word is shown as

$$\underbrace{X_0}_{\text{Sign bit}}.\underbrace{X_1X_2X_3}_{\text{Power of 2}}\underbrace{X_4X_5 \cdots X_n}_{\text{Magnitude}}$$

The sign and magnitude bits are used just as before. The exponent bits are used as follows: with three bits we can represent eight powers of 2. We would like to have negative powers of 2 as well as positive powers of 2, and we divide them evenly as shown in Table 13-1. For example, the number +7.5 would be represented as

$$7.5 = 111.1_2 = 0.1111_2 \times 2^3$$

The word for +7.5 is then 0.111 1111.

Table 13-1

EXPONENT BITS

	X_1	X_2	X_3	Power
Negative powers	0	0	0	2^{-4}
	0	0	1	2^{-3}
	0	1	0	2^{-2}
	0	1	1	2^{-1}
Positive powers	1	0	0	2^0
	1	0	1	2^1
	1	1	0	2^2
	1	1	1	2^3

As one more example, the number −6.125 is represented by the word 1.111 110001 (true magnitude).

13-2 FUNDAMENTAL PROCESSES

In the preceding section we developed two methods for representing numbers in binary form. The first is the binary fractional method which we choose to call sign and magnitude. This method is used in systems employing fixed-point arithmetic. The second method is sign and magnitude and exponent. This method is used in systems employing floating-point arithmetic. In the remainder of this chapter we shall discuss arithmetic units employing fixed-point arithmetic using sign and magnitude numbers. Machines using floating-point arithmetic are usually extensions of the basic operations with the added ability to handle exponents.

Using the sign and magnitude system, any number can be represented as $X_0.X_1X_2X_3 \cdots X_n$, where n is the number of bits in the magnitude. Since the sign bit

X_0 is not actually a part of the magnitude, we need to discover how to handle this bit during simple arithmetic operations. That is, must we treat the sign bit separately or can we simply operate on the entire number $X_0X_1X_2X_3 \cdots X_n$ as a whole? As it turns out, we can simply treat the entire number as a whole. This fortunate result is due to our choice of binary-point location and our choice to represent a + with a 0 in the sign bit and a − with a 1 in the sign bit.

Let us first investigate the addition of two numbers using sign and magnitude representation. We use a five-bit system for our investigation: $X_0.X_1X_2X_3X_4$. There are four cases to be considered:

1. The addition of two positive numbers. In this case, addition is straightforward; but we must check for overflow. For example:

Fractional notation	*Computer word*	*Fractional notation*	*Computer word*
3/16	0.0011	9/16	0.1001
(+) 7/16	0.0111	(+) 8/16	0.1000
10/16	0.1010	17/16	1.0001

   In adding 3/16 and 7/16, the answer 10/16 is correctly obtained. All three sign bits contain 0, and all is well. However, in adding 9/16 and 8/16, the answer 17/16 exceeds the capacity of our register and we obtain the incorrect answer of $1.0001_2 = -1/16$. This is an overflow error, detected by sensing the 1 in the sign bit.

2. Addition of a positive number and a negative number of smaller magnitude. You will notice first of all that it is impossible for an overflow to occur. We can then carry out the addition using either the 1's complement or the 2's complement for the negative number. For example,

Fractional notation	*True magnitude*	*1's complement*	*2's complement*
13/16	0.1101	0.1101	0.1101
(+) −5/16	1.0101	1.1010	1.1011
8/16		0.0111	0.1000
		→1	omit carry
		0.1000	

   You will notice that we cannot simply add the true-magnitude numbers. In the 1's complement, the end-around carry must be used. In the 2's complement, the end-around carry is disregarded as before. Notice in particular that the sign bit is simply used as a part of the number, and the proper carry as well as the proper sign in the sum is generated. This is always the case.

3. Addition of a positive number and a negative number of larger magnitude. As

in case 2 above, there cannot be any overflow condition. We therefore proceed with the addition, using complements.

Fractional notation	*True magnitude*	*1's complement*	*2's complement*
7/16	0.0111	0.0111	0.0111
(+) −12/16	1.1100	1.0011	1.0100
−5/16		1.1010	1.1011

Notice that there are no end-around carries generated; this is always the case. The 1's-complement answer is 1.1010, which is −5/16 in 1's-complement form. Similarly, the 2's-complement answer is −5/16 in 2's-complement form.

Notice the result of adding two numbers of the same magnitude but opposite sign.

Fractional notation	*True magnitude*	*1's complement*	*2's complement*
7/16	0.0111	0.0111	0.0111
(+) −7/16	1.0111	1.1000	1.1001
0		1.1111	0.0000 (omit carry)

From the 1's complement the sum is 1.1111, which is the 1's complement for −0.0. From the 2's complement the sum is 0.0000, which is + 0.0.

4. The addition of two negative numbers. We note first of all that the sum must necessarily be negative, and we must therefore end up with a 1 in the sign bit. Secondly, there is the definite possibility of an overflow. We will be able to detect this overflow by the presence of a 0 in the sign bit of the sum. A few examples best illustrate the methods.

Fractional notation	*True magnitude*	*1's complement*	*2's complement*
−7/16	1.0111	1.1000	1.1001
(+) −8/16	1.1000	1.0111	1.1000
−15/16		0.1111	1.0001 (omit carry)
		→1	
		1.0000	

In the 1's complement, the sum is 1.0000, which is the representation for −15/16. In the 2's complement, the sum is 1.0001, which is the representation for −15/16 in 2's-complement form.

Now let us examine an example where an overflow occurs:

Fractional notation	*True magnitude*	*1's complement*	*2's complement*
−9/16	1.1001	1.0110	1.0111
(+) −9/16	1.1001	1.0110	1.0111
−18/16		0.1100	0.1110
		→1	omit carry
		0.1101	

Both sums are indicated to be positive numbers, and thus we know an overflow has occurred giving incorrect results. We detect the overflow by detecting the change of sign in the sign bit.

At this point let us summarize in tabular form the basic addition operations for fixed-point arithmetic. We use numbers in binary fractional form, and we use sign and magnitude with either 1's or 2's complements for negative numbers. Table 13-2 summarizes and gives examples of binary-fractional-number representations using 1's and 2's complements.

Table 13-2

NUMBER REPRESENTATIONS, $X_0.X_1X_2X_3 \cdots X_n$

Fractional notation	True magnitude	1's complement	2's complement
3/16	0.0011		
−3/16	1.0011	1.1100	1.1101
35/64	0.100011		
−35/64	1.100011	1.011100	1.011101

Table 13-3 lists the methods for adding two numbers in the form of Table 13-2.

Table 13-3

ADDITION ($A > B$)

A	B	
+	+	Add numbers as they are. Overflow indicated by a 1 in the sign bit of the sum.
+	−	1. Use 1's complement for B. Add. No overflow possible. Must use carries to correct sum. 2. Use 2's complement for B. Add. No overflow possible. Disregard all carries.
−	+	1. Use 1's complement for A. Add. No overflow possible. 2. Use 2's complement for A. Add. No overflow possible. Disregard all carries.
−	−	1. Use 1's complement for A and B. 0 in sign bit of sum indicates overflow. Must use carries to correct sum. 2. Use 2's complement for A and B. 0 in sign bit of sum indicates overflow. Disregard all carries.

13-3 SERIAL BINARY ADDER

Using the sign and magnitude method developed in the previous sections, we now have the capability to represent any number. The fractional-binary-number representation which we are going to use is characteristic of fixed-point arithmetic units. What we are interested in then is to develop the logic necessary to perform fixed-point binary addition.

Let us first consider a binary addition as we would perform it mentally.

7/16	0.0111
(+) 4/16	0.0100
11/16	0.1011

We perform this simple addition in binary form in the following steps:

1. Add the two LSBs.
2. Add the next two bits.
3. Add the next two bits and place a carry over the MSB column.
4. Add the two MSBs and the carry generated from the previous step.
5. Add the two sign bits.

Notice that there are five distinct steps in the complete addition. If we performed the addition according to a time scale, we might do step 1 at time t_1, step 2 at time t_2, . . . , and step 5 at time t_5.

You will recall that a binary full-adder is capable of adding two bits and a carry present at its input terminals. If we could then present the two LSBs of the numbers to be added to the input terminals of a full-adder at time t_1, the LSB of the sum would appear at the output of the adder at time t_1. We would then present the next two bits to the adder at time t_2 and the second sum bit would appear at the output of the adder at time t_2. Continuing in this fashion, the third sum bit would appear at time t_3, the MSB at time t_4, and the sign bit at time t_5. A method for causing the bits of a number to appear at a pair of terminals in a time sequence is reminiscent of a shift register. We shall then use two shift registers and a full-adder to add two binary numbers in this fashion.

A *serial binary* adder is shown in Fig. 13-2. It will be used to find the sum of two

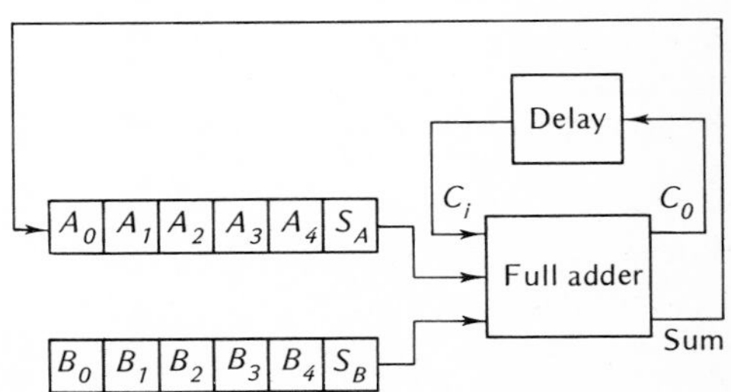

Fig. 13-2. Serial binary adder.

numbers: $A = A_0.A_1A_2A_3A_4$ and $B = B_0.B_1B_2B_3B_4$. It is called a serial adder since the sum is found by adding two bits at a time sequentially, beginning with the two LSBs. The two numbers are initially stored in the two registers as shown. Each register is shown simply as a series of five boxes, but it is actually five flip-flops forming a shift register. Each register contains one extra flip-flop at the right end. These two flip-flops are used as the inputs to the adder. The carry output of the adder is delayed one bit time. Thus, for example, a carry generated at time t_1 will not appear at the input to the adder until time t_2. The operation proceeds as follows:

1. At time t_o: all flip-flops have been previously reset to 0 and the numbers A and B are shifted into the registers from memory. Notice that the numbers can be entered into the registers in either serial or parallel form.
2. At time t_1: the contents of both registers are shifted right one place. Since the output of the adder was initially 0, 0 is shifted into A_0. The two LSBs are now in the sum flip-flops S_A and S_B (no carry appears) and the LSB of the sum appears at the output of the adder.
3. At time t_2: the contents of both registers are shifted right one place. The LSB of the sum is shifted into A_0. The two second LSBs are now in the sum flip-flops, any carry generated at time t_1 appears at C_i, and the second digit of the sum appears at the output of the adder.
4. This shifting process is continued up to and including time t_6. At the end of time t_6 the sum of the two numbers appears in the upper register. The number A initially stored in the upper register is lost. The number initially stored in the lower register is also lost since it has been shifted out the right end. We could have preserved this number by feeding the output of flip-flop S_B into the first flip-flop in the lower register. In this way, B remains in the lower register after the addition process is complete.

As an example of the functioning of the adder, the two numbers 0.0110 and 0.0010 are added. The waveforms for the system and the contents of the registers are shown in Fig. 13-3*a* and *b*, respectively. You should study these two figures until the operation is thoroughly understood. Notice that the 1 in the sum appears at the sum output of the adder at time t_4 (waveforms) and is shifted into bit position A_0 at time t_5 (registers).

We now have the basic requirements to form a serial binary adder. We must, however, consider a few details in order to complete the system. First of all let us consider the problem of overflows. You will recall that an overflow will occur only when adding two numbers of like sign. Furthermore, the overflow is detected when the sign of the sum is different from the sign of the original numbers. One method for detecting overflows is shown in Fig. 13-4. The operation of the circuit is as follows:

1. First of all, we examine for overflow only when the signs of A and B are the same. During time t_0 we compare the signs of the two numbers. If they are the same, the output of the exclusive-OR gate will be low and the check flip-flop CK will be set. If they are different, the check flip-flop will remain reset.

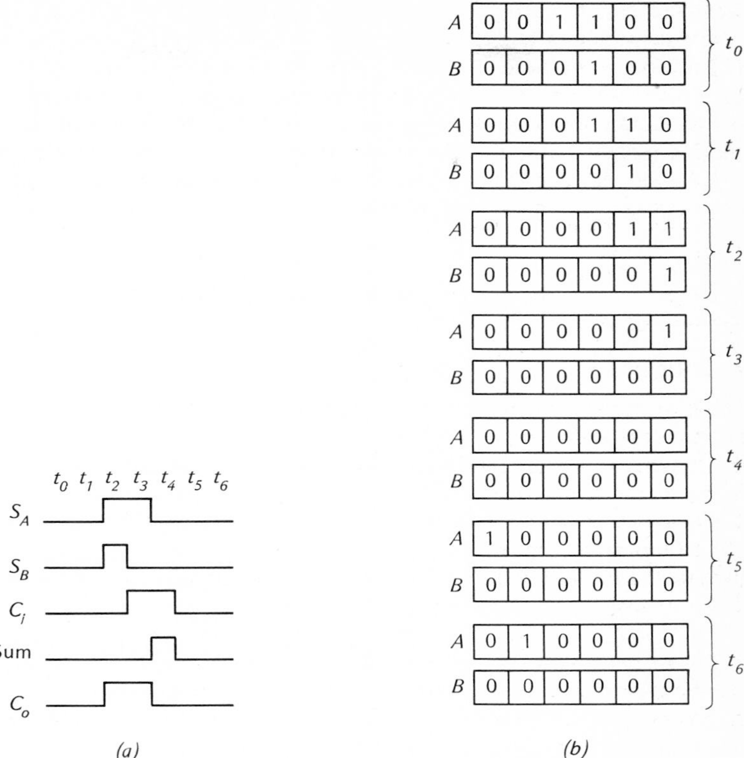

Fig. 13-3. Addition of 0.0110 and 0.0010 in the serial binary adder of Fig. 13-2.

Fig. 13-4. Overflow logic for serial adder.

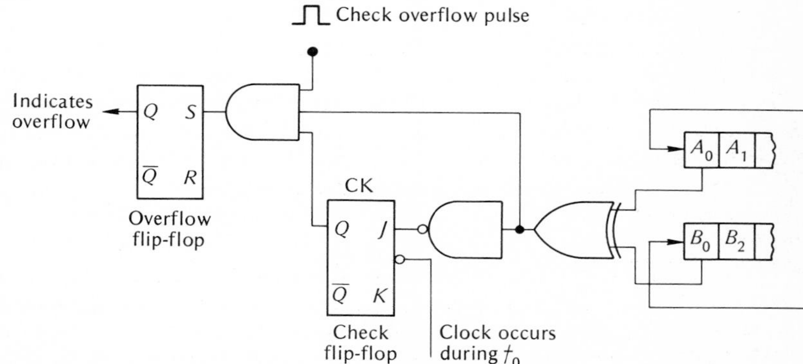

2. We circulate the contents of the lower register such that the number B appears in it at the end of the addition cycle (at time t_6). The sum will be in the upper register at this time.
3. Now during time t_6 a *check overflow pulse* occurs. If flip-flop CK has been set, and if the output of the exclusive-OR is high, the overflow flip-flop will be set. Notice that the output of the exclusive-OR will be high at this time only if the sign bit of the sum A_0 is different from the sign bit of one of the original numbers B_0.

Example 13-7

Draw the waveforms for the overflow logic shown in Fig. 13-4 assuming an overflow occurs.

Solution

The waveforms are shown in Fig. 13-5. Notice that at time t_0 the signs of A and B are both the same (indicating positive numbers). The CK flip-flop clock pulse sets this flip-flop during t_0 since the output of the NAND gate is high. At time t_6, the sign of the sum A_0 is now high (representing a negative number and thus an overflow). Thus the output of the exclusive-OR is high, as is CK, and the check overflow pulse will then pass through the AND gate and set the overflow flip-flop. In the usual case, the output of the overflow flip-flop would be used to stop machine operation, since this represents an error condition.

The second detail we must attend to is the handling of signed numbers. If the numbers are stored in memory in sign and true-magnitude form, the 1's or 2's complement of the negative numbers has to be formed before addition can take place. Forming the 1's complement of a number is quite simple. You will recall that the 1's complement of the number stored in a register can be formed by simply triggering each flip-flop in the register. Since the trigger pulse causes each flip-flop to change state, the register then contains the 1's complement of the previously

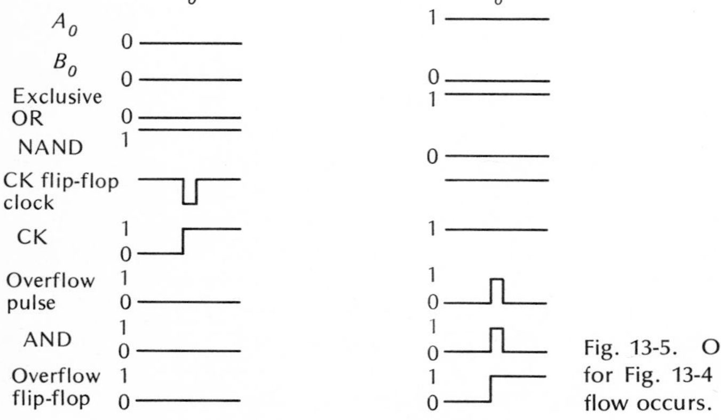

Fig. 13-5. Overflow waveforms for Fig. 13-4 assuming an overflow occurs.

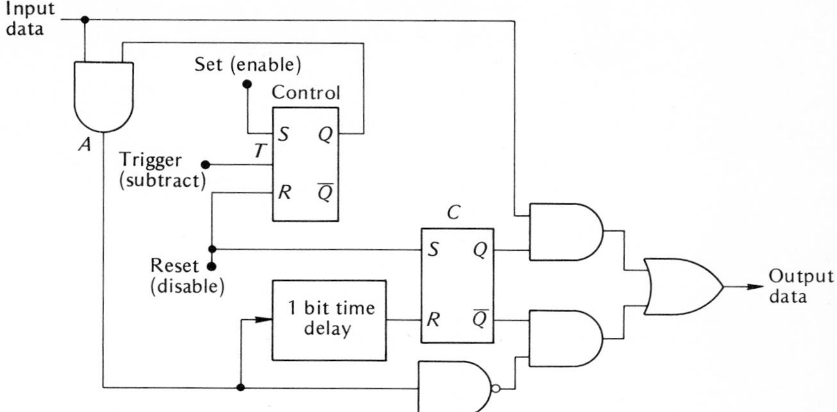

Fig. 13-6. Serial 2's-complement circuit.

stored number. In the serial adder, we can obtain the 1's complement more easily, however. Since the number appearing at the 0 side of the S_A or S_B flip-flops is the complement of the number appearing at the 1 side, we can simply use the 0 side when we are shifting out a negative number. However, if we choose to use 1's complements for negative numbers we have to include additional logic to handle the end-around carries. In the serial adder we are discussing, this involves setting a 1 in the lower register (in place of the number *B*) and going through the add cycle one more time. For this reason, the 1's-complement system is not widely used in serial adders.

We do better by representing negative numbers by their 2's complements. This eliminates the problem of end-around carries. On the other hand, it does not seem quite so easy to form the 2's complements. There is, however, a very simple circuit which can be used to form the 2's complement of a number in serial fashion. The circuit is shown in Fig. 13-6. To find the 2's complement of any number it is necessary to complement every bit beginning with the LSB which passes after, but not including, the first 1. The bits are shifted through the 2's complementer serially beginning with the LSB. Flip-flop *C* is initially set, and the first bit passes directly through the upper AND gate. The bits continue to pass through this AND gate until the first 1 appears. The first 1 passes through the AND gate unchanged, but it also resets flip-flop *C* after a delay of one bit time. Thereafter, the input bits pass through the lower AND gate after being inverted by the NAND gate and appear at the output in complement form. The control flip-flop is used to enable or disable the circuit. When it is set, AND gate *A* is true and the 2's complement can be formed. When it is reset, AND gate *A* is disabled and the circuit cannot function. In this case data simply pass through the upper AND gate and appear at the output unchanged.

Example 13-8

Show the waveforms for forming the 2's complement of the number 0110100.

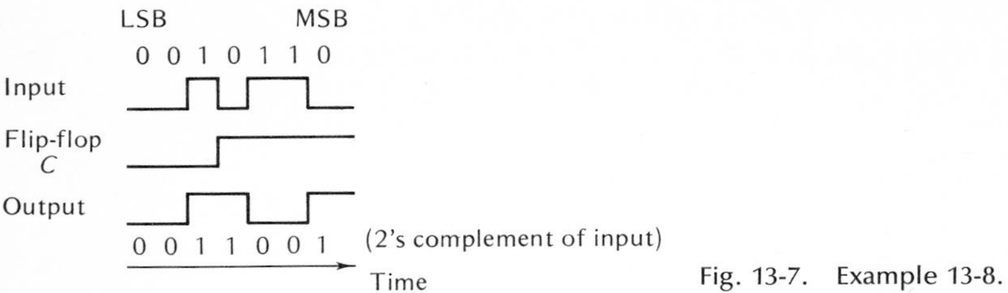

Fig. 13-7. Example 13-8.

Solution

The waveforms are shown in Fig. 13-7. Notice that the waveforms appear backward from the manner in which we would usually write the number. That is, the LSB is on the left and the MSB is on the right.

We can now use this 2's-complement circuit in our serial adder by placing one in series with each register directly in front of flip-flops S_A and S_B. The complete adder is shown in Fig. 13-8.

This adder is capable of adding positive or negative numbers which are stored in memory in sign and true-magnitude form. The overflow logic will handle any overflow problems, and the 2's-complement circuit will handle any negative numbers. We must take care to ensure that the 2's-complement circuit is allowed to operate for only n bit times (where n is the number of bits in the magnitude). In this fashion we use the 2's complement of the magnitude of any negative number, but we do not change the sign bit. Thus the control flip-flop in the 2's-complement circuit will be set at time t_0 if a negative sign is detected. It will then be reset at time t_n to ensure that the sign bit passes by unchanged.

A little thought reveals that the adder shown in Fig. 13-8 can also be used for

Fig. 13-8. Serial binary adder using numbers stored in sign and true-magnitude form.

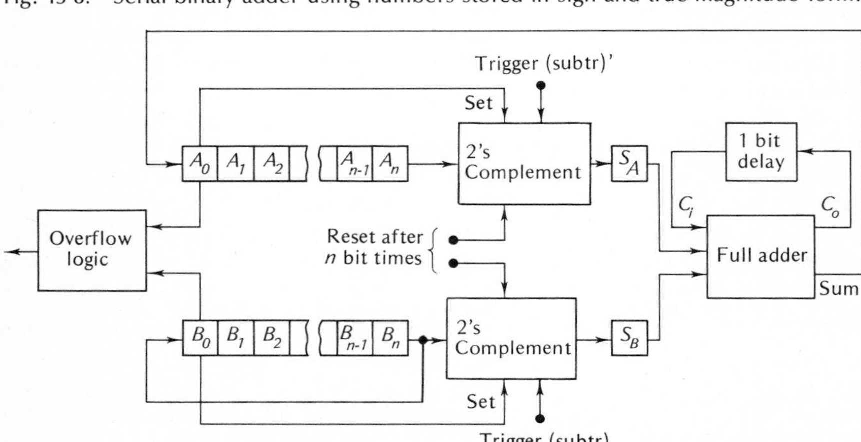

subtraction. We simply take the four cases ot two numbers, ++, +−, −+, and −−, and show the results of subtracting them instead of adding (as was done in a previous section). It can be shown (see Prob. 13-9) that these four cases reduce to the same problems of addition. To use the adder in Fig. 13-8 for subtraction then, it is only necessary to add a trigger input to the control flip-flop in the 2's-complement circuit of Fig. 13-6. Subtraction will then occur if we apply a pulse to this input just prior to time t_1. There are two cases to consider:

1. To subtract a positive number: the 2's-complement circuit will not have been enabled. The subtract control pulse appearing at the T input of the control flip-flop of the 2's-complement circuit will enable it, and the 2's complement will be used, resulting in a subtraction.
2. To subtract a negative number: the 2's-complement circuit will have been enabled and the subtract pulse will disable it. This will then result in a straight addition.

Example 13-9

Explain how the serial binary adder in Fig. 13-8 could be used to add or subtract binary numbers stored in the following form: sign and true magnitude for positive numbers, and sign and 2's complement for negative numbers.

Solution

(*a*) Since the negative numbers are stored in 2's-complement form, we need not use the 2's-complement circuits during addition. Therefore, the set inputs to the 2's-complement circuits are removed from the sign bits (A_0 and B_0). The 2's-complement circuits are held in the reset (disabled) state during addition. The sum will appear in register A.

(*b*) For subtraction we take the 2's complement of the number to be subtracted and add. Therefore, to subtract B from A we set the 2's-complement circuit in the B register at time t_0 and add (this 2's-complement circuit will be reset after n bit times). The 2's-complement circuit in the A register is held reset (disabled) during this cycle. The difference $(A - B)$ will appear in register A.

(*c*) To subtract A from B, we activate the 2's-complement circuit in register A for n bit times beginning at t_0 while holding the 2's-complement circuit in register B reset (disabled). The difference $(B - A)$ will appear in register A.

At this point let us examine the total time required to complete an addition or a subtraction using the serial adder of Fig. 13-8. Each word is composed of $n + 1$ bits, one sign bit and n bits of magnitude. A little thought will show that it requires $n + 2$ bit times to complete an addition (or subtraction) cycle. This time is defined at the *add-cycle* of the adder. Typically, one bit time is the time of one cycle of the clock since the adder is usually driven by the clock. Thus, one add-cycle time is equal to $n + 2$ clock periods.[1]

[1] If the add flip-flops (S_A and S_B) in the registers are omitted, it requires only $n + 1$ clock periods for the basic add-cycle time. These two flip-flops are not really necessary and were included to simplify the description of the operation of the adder.

Example 13-10

What is the add-cycle time of the serial adder in Fig. 13-8 if each word has 12 bits of magnitude and the clock is 100 kHz?

Solution

One clock period is $1/(100 \times 10^3) = 10\ \mu s$. The add-cycle time is then

$$10 \times 10^{-6}(12 + 2) = 140\ \mu s$$

The length of time required to perform an addition or subtraction using the serial binary adder is one of the major drawbacks to this method. Admittedly a serial adder offers the attraction of a minimum of hardware, but in the interest of accomplishing addition in a shorter period of time we shall investigate other methods. The parallel binary adder to be discussed in the next section offers a substantial reduction in add-cycle time, at the expense, however, of additional hardware.

13-4 PARALLEL BINARY ADDER

The primary advantage of a parallel binary adder over a serial adder is the reduction in add-cycle time (the main disadvantage is the increase in hardware required). You will recall that a serial adder requires $n + 1$ clock periods to add two binary numbers having one sign bit and n bits of magnitude. The parallel adder, on the other hand, provides the sum of two numbers in only one clock period. That is, the sum appears almost immediately after the two numbers are entered in the registers (allowing a short period of time for settling). Thus the actual addition time for a parallel adder could be considered one clock cycle.

A simple parallel adder is shown in Fig. 13-9 (at this point you might like to review the discussion of the parallel adder given in Chap. 5). This adder has the ability to add two numbers of the form $A = A_0.A_1A_2A_3A_4$ and $B = B_0.B_1B_2B_3B_4$. The number A is stored in register A, the number B is stored in register B, and the sum $S_0.S_1S_2S_3S_4$ appears at the adder outputs. To perform an addition using this adder requires two steps: (1) reset all flip-flops; (2) shift A into register A and B into register B. After a short settling time, the sum appears at the adder outputs. The add-cycle time can then be considered to be two clock periods. The reason for the short add-cycle time is now clear. The two LSBs are added in full-adder 4 and the carry C_o is connected to the C_i of full-adder 3. Simultaneously, the corresponding pairs of bits from each number are added and each carry is passed on to the next adder. Thus the two numbers are added all at once and the add time is actually the time required for the carries to propagate from one adder to the next down the line. This is the *settling time* required before the addition is complete.

In order to complete the discussion of this parallel adder, we must consider the four cases summarized in Table 13-3. First of all, however, we must decide how the numbers are to be stored in memory. If they are stored in sign and true-magnitude form, we have to complement the negative numbers in the registers (A or B) before addition can take place. This is easily accomplished for the 1's complement.

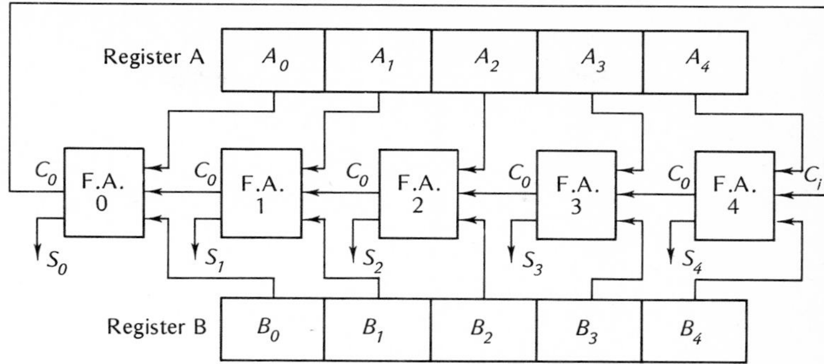

Fig. 13-9. Parallel binary adder.

It is only necessary to trigger each flip-flop in the register in order to form the 1's complement. This will, however, add one clock period to our add-cycle time, and we therefore do not choose this representation. The 2's complement is not easily formed, and we therefore do not choose this method. The only method remaining is the most convenient, and we shall use it. That is, we shall store positive numbers in sign and true magnitude and negative numbers in sign and 1's complement. Recall that the only disadvantage of this method in the serial adder is the end-around carries generated. This is no problem in the parallel adder, however, since the end-around carry (the C_o of adder 0 to the C_i of adder 4) is easily accommodated.

Now assuming we use sign and true magnitude for positive numbers, and sign and 1's complement for negative numbers, let us examine the four cases given in Table 13-3:

1. Sum of two positive numbers. The only problem here is the overflow, and we have to arrange a method for checking this.
2. A positive number added to a negative number of smaller magnitude. No overflow is possible in this case. The end-around carries are accommodated by connecting the C_o of adder 0 to the C_i of adder 4.
3. A positive number added to a negative number of larger magnitude. No overflow is possible, and there are no end-around carries.
4. The sum of two negative numbers. We must arrange for checking overflows. The end-around carries are taken care of as in case 2 above.

In summarizing the four cases, it is clear that the end-around carries will be taken care of and it is only necessary to add circuitry to check for overflows. A circuit which can be used to detect overflows for this parallel adder is shown in Fig. 13-10. There are two cases where overflows can occur: the addition of two positive numbers and the addition of two negative numbers. In Fig. 13-10, the output of exclusive-OR gate X will be low whenever A_0 and B_0 are the same (i.e., the numbers have the same sign). This will cause the top leg of the AND gate to be high. The output of exclusive-OR gate Y will be high whenever B_0 and S_0 are different (in-

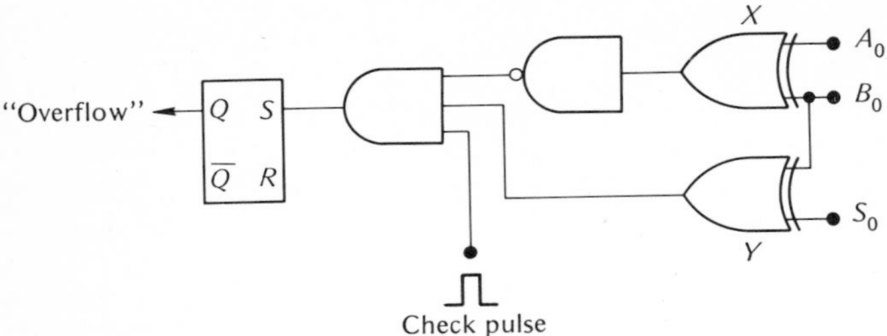

Fig. 13-10. Overflow circuit for parallel adder in Fig. 13-9.

dicating an overflow). This will cause the middle leg of the AND gate to be high. Thus when a *check* pulse occurs on the bottom leg of the AND gate, the *overflow* flip-flop will be set only when an overflow condition exists. The check pulse will occur some time after the addition is complete.

Example 13-11

Label the levels present on the gate legs of the circuit shown in Fig. 13-10 for an output of $A_0 = 1$, $B_0 = 1$, $S_0 = 0$.

Solution

The levels are shown on the circuit in Fig. 13-11. This case corresponds to the addition of two negative numbers with the sign of the sum indicating a positive sum — an overflow error.

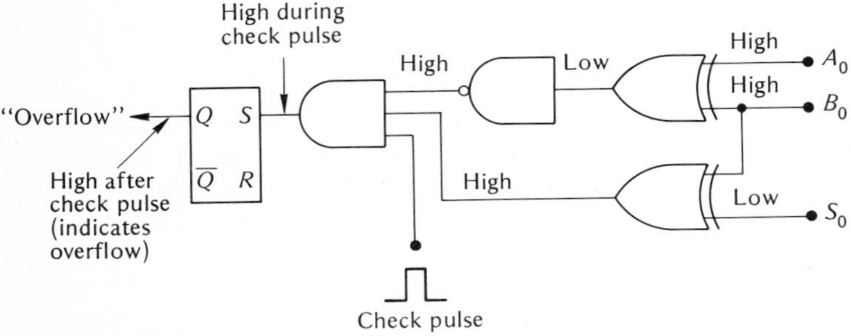

Fig. 13-11. Example 13-9.

The adder in Fig. 13-9 can also be used for subtraction. It can be shown, as in the previous section, that the four cases of subtraction reduce to the very same four cases of addition listed in Table 13-3. The subtraction process is carried out, however, by complementing the number to be subtracted and then adding. This simply requires one more clock period. There are two cases to consider:

1. $A - B$. During time t_1, the two numbers are shifted into registers A and B. During time t_2, the contents of register B are complemented by applying a pulse to the T input of every flip-flop in that register except the sign bit. The difference then appears at the adder outputs (after a small settling time).
2. $B - A$. Similar to the case for $A - B$ except that the contents of register A are complemented.

It should be noted that the parallel adder discussed could be used to add two numbers having n bits of magnitude by simply constructing the registers with the appropriate number of flip-flops and using $n + 1$ full-adders. The total add time is limited by the settling time due to the propagation of the carries from one adder to the next. There are a number of advanced methods for reducing this settling time, and they are discussed in more advanced texts.

13-5 BCD ADDITION

It is sometimes more desirable to perform digital arithmetic using numbers in BCD form. In Chap. 5, we discussed methods for adding numbers stored in 8421 BCD code and in excess-3 BCD code. The results of these studies are the 8421 adder and the excess-3 adder developed in that chapter. At this point we would like to form a general BCD adder using the basic adders developed in Chap. 5. We first of all need to consider the methods of representing numbers in BCD form and the four cases of addition investigated for adding numbers in straight binary form.

We shall soon discover that BCD arithmetic is very similar to arithmetic carried out in straight binary form, and for this reason we locate the binary point to the left of the MSB as in the previous section. Similarly, we use a 0 in the sign bit for a positive number and a 1 in the sign bit for a negative number. We therefore represent numbers in the following form:

$$\underbrace{X_0}_{\text{Sign bit}}.\underbrace{X_1X_2X_3X_4}_{\text{Tenths digit}}\ \underbrace{X_5X_6X_7X_8}_{\text{Hundredths digit}}$$

With this system we can represent numbers between $+.99_{10}$ and $-.99_{10}$. The system could be enlarged to include three-digit decimal numbers by adding four more bits to the magnitude, or four-digit decimal numbers by adding eight more bits, etc. Negative numbers can be represented in sign and true magnitude, sign and 9's complement, or sign and 10's complement.

Example 13-12

Show the binary representation for the following numbers in 8421 BCD code and in excess-3 BCD code:

(*a*) 34
(*b*) −34
(*c*) −81

Solution

	Number	Scaled	Sign and magnitude	True magnitude	9's complement	10's complement
8421	34	+.34	0.34	0.0011 0100		
BCD	−34	−.34	1.34	1.0011 0100	1.0110 0101	1.0110 0110
code	−81	−.81	1.81	1.1000 0001	1.0001 1000	1.0001 1001
Excess-3	34	+.34		0.0110 0111		
BCD	−34	−.34		1.0110 0111	1.1001 1000	0.1001 1001
code	−81	−.81		1.1011 0100	1.0100 1011	1.0100 1100

In this example, notice that the formation of the 9's and 10's complements of numbers in 8421 BCD code is not easily accomplished electronically. It can be done, but it requires additional logic (see Prob. 13-14), and we want to minimize the logic necessary to perform addition. On the other hand, the 9's complements of numbers in excess-3 code is easily formed. It is only necessary to complement the entire magnitude portion of the number. The 10's complement is also easily formed since it involves only the addition of a binary 1 to the rightmost position of the magnitude. With this in mind then, let us consider the four cases of adding two numbers in excess-3 BCD form.

1. Addition of two positive numbers. Overflows can obviously occur, and we have to check for them. Consider the addition of the following two numbers:

Numbers	*True magnitude*	*Excess-3*	
21	0.0010 0001	0.0101 0100	
+ 13	0.0001 0011	0.0100 0110	
34	0.0011 0100	0.1001 1010	occurs in BCD adder
		−11 −11	
		0.0110 0111	

The two numbers are first scaled and then changed into sign and true-magnitude form. The sum is then obtained by simply adding these two numbers. Under the excess-3 column, the excess-3 representations of the two numbers are added and the answer (34) appears in excess-3 code. The subtraction of 3 from each digit is accomplished automatically in the excess-3 adder (if you are not sure of the reasoning behind this, you should review the excess-3 BCD adder discussed in Chap. 5). Let us now consider the addition of two positive numbers when an overflow occurs.

Numbers	*True magnitude*	*Excess-3*	
		1	
34	0.0011 0100	0.0110 0111	
+ 86	0.1000 0110	0.1011 1001	
120	0.1011 1010		
	+110 +110	1.0001 0000	occurs in BCD adder
	1	+11 +11	
	1.0010 0000	1.0101 0011	

Under the true magnitude, the answer is −20, which is obviously incorrect. The negative sign (the 1 in the sign bit) is the indication that an overflow has occurred. The answer given under excess-3 is also −20 in excess-3 code. This obviously incorrect answer can also be detected by the 1 in the sign bit.

2. Addition of a positive number and a negative number of smaller magnitude.

		Excess-3		
Numbers	*True magnitude*	*9's complement*	*10's complement*	
21	0.0010 0001	0.0101 0100	0.0101 0100	
−13	1.0001 0011	1.1011 1001	1.1011 1010	
8	Cannot	0.0000 1101	0.0000 1110	occurs in
	add	+11 −11	+11 −11	adder
	directly	0.0011 1010	0.0011 1011	
		1		
		0.0011 1011	Discard carry	

In this case no overflow can occur. Notice that we use the end-around carry for the 9's complement and discard it for the 10's complement. The correct answer of +8 is given in excess-3 code in both cases.

3. Addition of a positive number and a negative number of greater magnitude.

		Excess-3	
Number	*True magnitude*	*9's complement*	*10's complement*
−21	1.0010 0011	1.1010 1011	1.1010 1100
13	0.0001 0011	0.0100 0110	0.0100 0110
−8	Cannot	1.1111 0001	1.1111 0010
	add	−11 +11	−11 +11
	directly	1.1100 0100	1.1100 0101

Again, no overflow can occur in this case. The answer under 9's complement is −8 in excess-3 code. The magnitude appears in complement form since the answer is negative. Similarly, the answer under 10's complement appears as the 10's complement of +8 since it is negative. The 10's-complement answer can be changed to true magnitude by subtracting 1 from the LSB of the magnitude and complementing it.

4. Addition of two negative numbers. We must of course examine for any overflow condition. Consider the following two numbers:

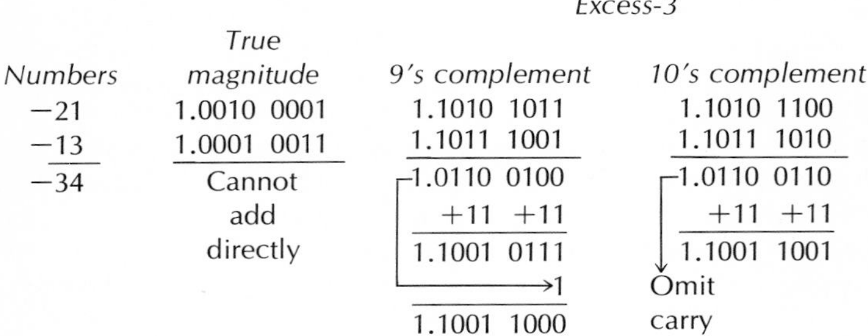

Numbers	True magnitude	Excess-3 9's complement	Excess-3 10's complement
−21	1.0010 0001	1.1010 1011	1.1010 1100
−13	1.0001 0011	1.1011 1001	1.1011 1010
−34	Cannot add directly	1.0110 0100	1.0110 0110
		+11 +11	+11 +11
		1.1001 0111	1.1001 1001
		→1	Omit carry
		1.1001 1000	

In this case, no overflow occurs and the correct sum appears in complement form. Now consider a case where an overflow occurs.

Numbers	True magnitude	Excess-3 9's complement	Excess-3 10's complement
−51	1.0101 0001	1.0111 1011	1.0111 1100
−51	1.0101 0001	1.0111 1011	1.0111 1100
−102	Cannot add directly	0.1111 0110	0.1111 1000
		−11 +11	−11 +11
		0.1100 1001	0.1100 1011
		→1	Omit carry
		0.1100 1010	

You will notice in these two cases that the sum appears as +2 in complement form. There is obviously an overflow, and the error can be detected by the 0 in the sign bit of the sum.

We can summarize these four cases of addition by noting the similarity between addition in excess-3 BCD using 9's and 10's complements for negative numbers and addition in straight binary form using 1's and 2's complements for negative numbers. That is to say, we must examine for overflow when adding two numbers of like sign, and we must provide for end-around carries when using 9's complements (similar to using the end-around carries when using the 1's complements). With this knowledge we can then proceed with the formulation of a BCD adder using excess-3 code and we can rest assured that the signs will take care of themselves.

13-6 A PARALLEL BCD ADDER

We will now discuss the operation of a parallel adder. This adder will operate in excess-3 BCD code, and numbers will be stored in sign and true magnitude for positive numbers and sign and 9's complement for negative numbers.

The basic adder is shown here in Fig. 13-12. It is capable of adding two two-digit decimal numbers, and it has the capacity of handling any numbers between $+99_{10}$ and -99_{10}. Stored in registers A and B, respectively, are the two numbers $A = A_0.A_1A_2A_3A_4A_5A_6A_7A_8$ and $B = B_0.B_1B_2B_3B_4B_5B_6B_7B_8$. The sum $S = S_0.S_1S_2S_3S_4S_5S_6S_7S_8$ appears at the outputs of the adders. The add-cycle time is quite similar to that for the parallel binary adder discussed previously. The numbers A and B are shifted into their respective registers at time t_1, and the sum S appears at the output after a small settling time. Since we are using 9's complements for negative numbers, we must account for the end-around carries. This is accomplished by connecting the carry out C_o of adder 3 to the carry in C_i of adder 1. Since overflows are detected in the same way as for parallel binary addition, we can use the overflow-detecting circuit of Fig. 13-10. Notice that we must use a BCD excess-3 adder for each four bits of the magnitude, but we use a straight binary full-adder for the sign bits. The adder could, of course, be extended to accommodate decimal numbers having n digits by simply using n BCD adders and $4n$ flip-flops for the magnitude portion of the registers.

Since the four cases of subtraction ultimately reduce to the four cases of addition

Fig. 13-12. Parallel BCD adder using excess-3 code.

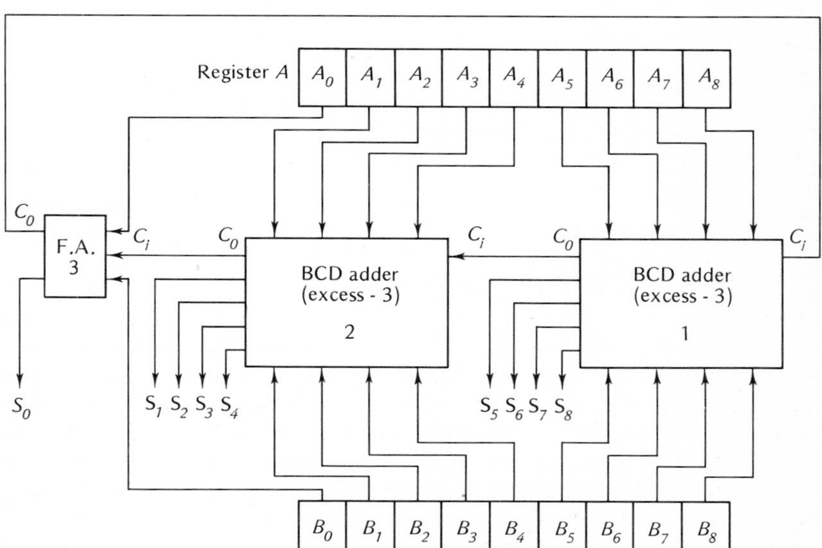

studied, the BCD adder in Fig. 13-12 can also be used to subtract BCD numbers in excess-3 code. It is only necessary to add one more clock period to the total during which the number to be subtracted is complemented. At this point the great value of using excess-3 code becomes apparent. Since subtraction is performed by finding the complement of a number and then adding, and since excess-3 code is *self-complementing,* we only need to complement the contents of the register containing the number to be subtracted and then execute an addition cycle. The complement operation is, of course, carried out in this case by applying a pulse to the T input of every flip-flop in the register (except the sign bit) containing the number to be complemented. The time required to perform a subtraction using this BCD adder is then two clock periods, and we could define the *subtract-cycle time* as being equal to two clock periods.

Because of the great similarity between this BCD adder and the parallel binary adder previously discussed, we might suppose that there would be a similarity between a serial BCD adder and a serial binary adder. This is indeed the case, and, as you might expect, it could be most easily accomplished using sign and true magnitude for positive numbers and sign and 10's complements for negative numbers (all in excess-3 code).

Example 13-13

Explain how the BCD adder in Fig. 13-12 could be used to add numbers stored in sign and true magnitude (excess-3 code).

Solution

There are three cases to consider:

(*a*) The addition of two positive numbers is carried out by simply shifting the numbers into the registers and reading the output.

(*b*) The addition of a positive number and a negative number is carried out by shifting the two numbers into the registers, complementing the negative number (magnitude portion only), and reading the output.

(*c*) The addition of two negative numbers is carried out by shifting the numbers into the registers, complementing both registers (except the signs), and reading the output.

13-7 BINARY MULTIPLICATION

Binary multiplication and division are both quite complex operations. There are many different methods for performing these operations, and a more comprehensive treatment can be found in more advanced textbooks. There are, however, one basic method for multiplication and one basic method for division, which we shall discuss at this time. The other methods are variations directed toward shortening the time required to perform a computation. We discuss multiplication in this section and division in the next.

We can easily discover a method for performing multiplication by carefully examining the multiplication process itself. First consider the multiplication of one digit by another. For example, the multiplication of 8 by 4 is really the same thing as adding the number 8 four times. That is, $8 \times 4 = 8 + 8 + 8 + 8 = 32$ (it is also the same thing as adding 4 eight times). Similarly the product of 5 and 3 could be written as $5 \times 3 = 5 + 5 + 5 = 3 + 3 + 3 + 3 + 3 = 15$. Thus it can be seen that the multiplication process is really the same thing as repeated addition. In the decimal system we may have to add any one number up to nine times. However, in the binary system, multiplication by repeated addition is much simpler since we have only two numbers, 1 and 0. Thus we have to add any one number only once. For example, consider the binary multiplication of the number 6. There are only two possibilities: $110 \times 0 = 0$, and $110 \times 1 = 110$.

We must now consider the multiplication of some number by a number having more than one digit. The binary multiplication of 6 and 5 would be carried out as follows:

```
  110 ← multiplicand
  101 ← multiplier
  ---
  110 ⎫
 000  ⎬ ← partial products
110   ⎭
-----
11110 ← product
```

The first partial product (the top one) is formed by multiplying the multiplicand by the LSB of the multiplier. The second partial product is formed by multiplying the multiplicand by the middle bit of the multiplier, and the third partial product is formed using the MSB of the multiplier. The product is then the sum of the partial products, properly shifted with respect to one another. You will notice that in longhand multiplication we add the partial products simultaneously; we could, however, add them individually and achieve the same result.

The formation of these partial products and their addition (and therefore multiplication) can be easily accomplished using two registers as shown in Fig. 13-13. The multiplicand is initially stored in the multiplicand register, and the multiplier is stored in the MQ register. The accumulator register is initially reset to all 0s. The multiplication proceeds as follows:

1. Examine the sign bits. If they are alike, the sign of the product will be positive; if unlike, the sign of the product will be negative. Store the sign of the product until multiplication is complete. Set sign bits to 0s (note that an exclusive-OR gate may be used as before).
2. Examine the 2^0 bit of the MQ register; if it is 1, add the multiplicand to the accumulator register; if it is 0, do not add.
3. Shift the accumulator and MQ register right one place.
4. The second LSB of the multiplier is now in the 2^0 position of the MQ register. Repeat steps 2 and 3 above.
5. The MSB of the multiplier is now in the 2^0 position of the MQ register. Repeat steps 2 and 3 above.

6. Multiplication is now complete, and the product is in the accumulator register. Place the sign determined in step 1 above in the sign bit of the accumulator.

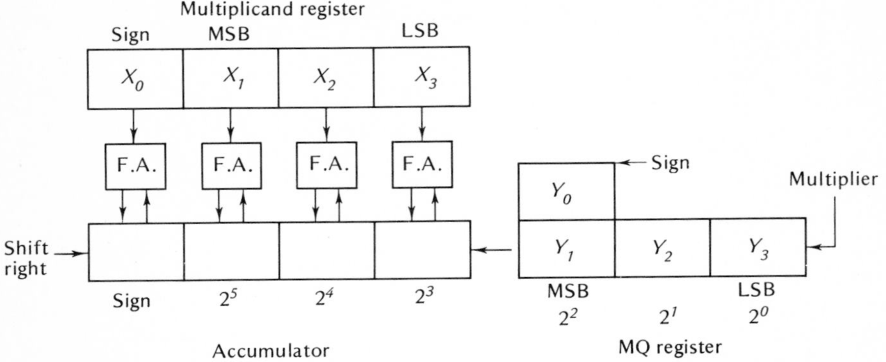

Fig. 13-13. Registers for multiplying two 3-bit numbers. Product appears in accumulator and MQ registers as sign, 2^5, 2^4, 2^3, 2^2, 2^1, and 2^0.

Example 13-14

Show the contents of the registers of Fig. 13-13 for the multiplication of 6 and 5.

Solution

0. Set numbers in registers:

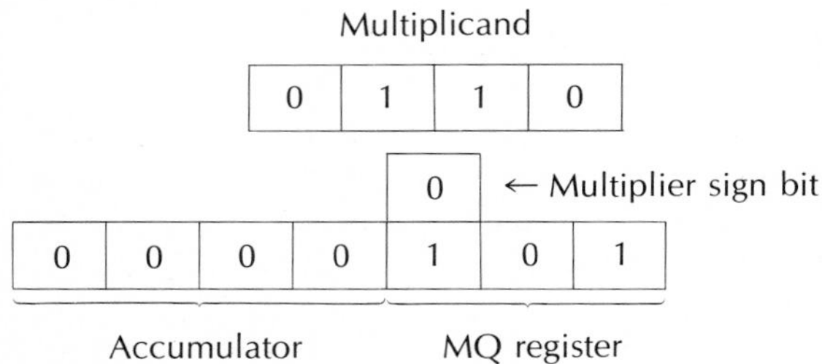

1. Determine sign:

0	0	0	0	1	0	1

2. Add multiplicand:

0	1	1	0	1	0	1

3. Shift right:

0	0	1	1	0	1	0

4. Do not add:

0	0	1	1	0	1	0

5. Shift right:

0	0	0	1	1	0	1

6. Add:

0	1	1	1	1	0	1

7. Shift right:

0	0	1	1	1	1	0

8. Set sign in accumulator:

0	0	1	1	1	1	0	← Product
↑ Sign	2^5	2^4	2^3	2^2	1^1	2^0	

Only the accumulator and MQ registers were shown, since the multiplicand register remains unchanged during the multiply operation. At the end of the eight steps, the product appears in the accumulator and MQ registers.

You will notice that in the preceding example the product of two three-bit numbers generated a six-bit product. This is necessary for the proper location of the binary point. We assume, of course, that the numbers are used in binary fractional notation with the binary point between the MSB and the sign bit. In multiplying these two three-bit numbers a total of eight steps were required after the numbers were shifted into the registers. Two of these steps deal with the sign bit and will always be needed. The remaining steps deal with the magnitudes only, and since there were three bits, we had to add and shift right three times for a total of six steps. This technique can, of course, be extended to handle n-bit numbers by simply increasing the size of the registers to $n + 1$ bits.

We can now make a general statement regarding the time required to multiply any two numbers using this technique: (1) It will always require two steps to handle the sign bits. (2) It will require $2n$ steps to handle the magnitude, when n is the number of bits in the magnitude. Thus the total time required to perform a multiplication in this way is $2n + 2 = 2(n + 1)$. Since each step is usually performed in one clock period, we can say that the time required to perform one multiplication is $2(n + 1)$ clock periods.

13-8 BINARY DIVISION

Like multiplication, a process for binary division can be best determined by examining longhand division. Let us consider the longhand method for carrying out the division $2_{10}/6_{10} = 010/110$. This problem is normally written as

```
                  0.0101 ← quotient
divisor → 110 | 010.0000 ← dividend
                  1 10
                  ------
                  0 1000 ← first remainder
                     110
                  ------
                     010 ← second remainder
```

The steps in this division are:

1. Place the divisor under the first three bits of the dividend. We notice that the divisor is too large to be divided into these three bits, and we then place a 0 to the left of the binary point. We determine that division is not possible by noting that the magnitude of the first three bits is smaller than the divisor.
2. We then move the divisor one place to the right and place it under the second, third, and fourth bits of the dividend. The magnitude of the dividend is still smaller than the divisor; so we place a 0 just after the binary point (in the first binary place).
3. We move the divisor one more place to the right (under the third, fourth, and fifth bits of the dividend). Now the magnitude of the dividend is greater than the divisor, and we can perform a subtraction. We note this by placing a 1 in the second binary place of the quotient. The divisor is then subtracted from the dividend to yield the first remainder.
4. We then bring down a 0 from the dividend, add it to the right side of the first remainder, and place the divisor under the three rightmost bits of the first remainder. We note that the magnitude of the first remainder is smaller than the divisor, and we therefore cannot make a subtraction. We note this by placing a 0 in the third binary place of the quotient.
5. We then bring down another 0 from the dividend, add it to the right end of the first remainder, and move the divisor to the right one place. Now the magnitude of the first remainder is larger than the divisor and we can make a subtraction; we note this by placing a 1 in the fourth binary place of the quotient. The subtraction is made, yielding the second remainder.

6. This process is repeated as many times as desired to generate the required number of bits in the quotient.

We notice from the above discussion that division is really a process of repeated subtractions. That is, we repeatedly subtract the divisor from the dividend or the remainders generated. We determine whether or not to subtract by comparing the magnitudes of the divisor and the dividend or remainders. For this reason, this method is sometimes called the "comparison method."

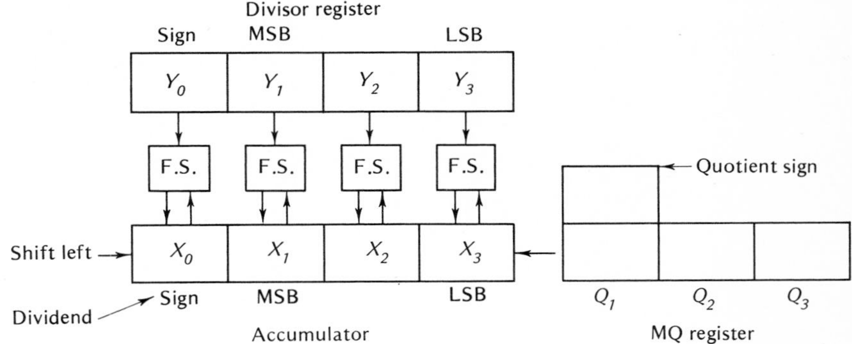

Fig. 13-14. Register for dividing two 3-bit numbers. Quotient appears in MQ register as Q_1, Q_2, and Q_3.

Division by the comparison method can be accomplished by means of the registers shown in Fig. 13-14. You will notice that the figure is the same as Fig. 13-13 with the exception of the full-subtractors. The divisor is held in the divisor register, the dividend is initially stored in the accumulator, and the quotient appears in the MQ register after the division is complete. The name MQ register is now obvious, since this register holds the *multiplier* at the beginning of a multiplication operation and contains the *quotient* at the end of a division operation. Division is carried out using the registers shown in Fig. 13-14 and the following steps:

1. Shift the divisor into the divisor register; shift the dividend into the accumulator, and set the MQ register to 0s.
2. Determine the sign of the quotient by comparing the signs of the divisor and dividend. It is positive if they have like signs and negative if they have unlike signs. Store the sign of the quotient, and set the sign bits in the registers to 0s.
3. Compare the magnitudes of the divisor and dividend; if the divisor is greater, proceed; if the dividend is greater, halt. Recall that we are using binary fractional notation and we therefore cannot generate a number larger than $0.111 \ldots 1_n$, where n is the number of bits in the magnitude. In other words, we cannot have a bit to the left of the binary point, other than the sign bit.
4. Shift the contents of the accumulator and MQ registers *left* one place.
5. Compare the magnitudes of the divisor and the accumulator register. If the divisor is smaller, subtract the divisor from the accumulator and place a 1 in

the rightmost bit of the MQ register. If the divisor is larger, do not subtract.

6. Shift the accumulator and MQ *left* one place.
7. Repeat steps 5 and 6 above until n bits have been generated in the MQ register (three bits in this case).
8. Place the sign in the sign bit of the MQ register. The quotient is now in the MQ register, and the division is complete.

Example 13-15

Show the contents of the accumulator and MQ registers while performing the division 010/110.

Solution

The divisor register is shown initially only since its contents remain unchanged during the divide operation.

0. Shift numbers into registers. Determine sign.

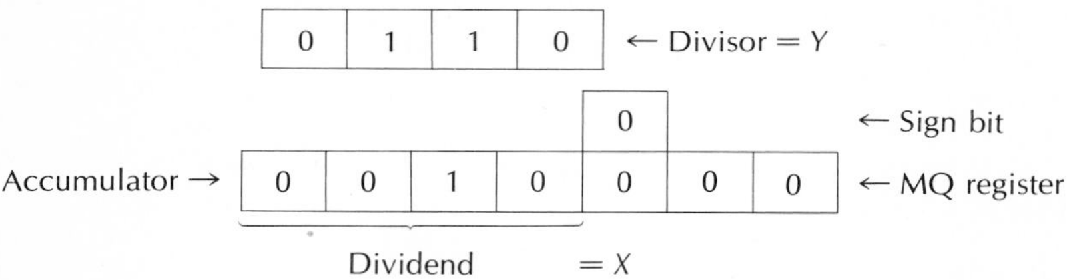

1. $X < Y$; proceed. Shift left.

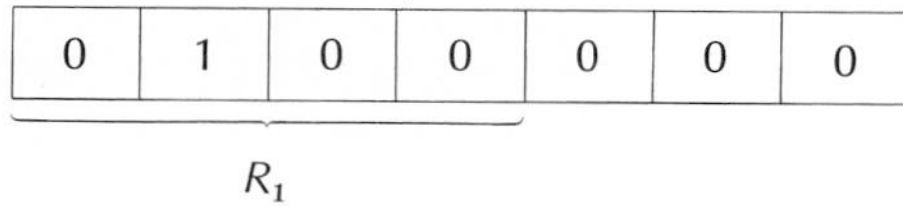

2. $R_1 < Y$, no subtraction. MSB of quotient = 0.

0	1	0	0	0	0	0
R_1						Q_1

3. Shift left.

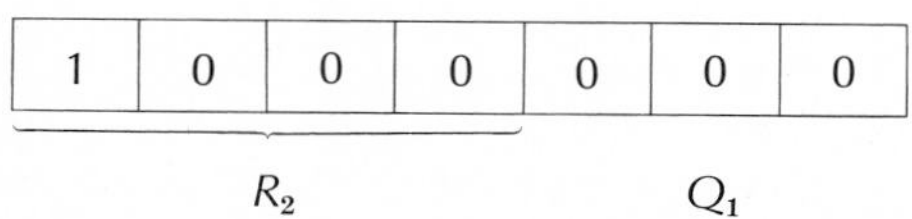

4. $R_2 > Y$; subtract divisor from accumulator. Place a 1 in LSB position of MQ register.

0	0	1	0	0	0	1
					Q_1	Q_2

5. Shift left.

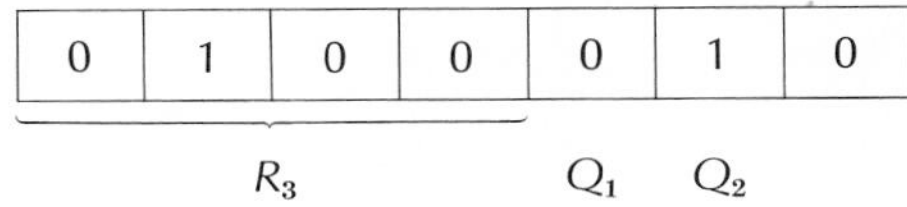

6. $R_3 < Y$; no subtraction. LSB of quotient = 0.

0	1	0	0	0	1	0
				Q_1	Q_2	Q_3

7. Place sign in sign bit of MQ register. Division complete; quotient in MQ register. Quotient = 0.010.

				0	← Sign of quotient	
0	1	0	0	0	1	0
				Q_1	Q_2	Q_3

Notice that we cease the division process when we have generated three bits of quotient since we have provision for only three-bit numbers in this system. Notice also that the answer to this problem is $(1/4)/(3/4) = 1/3 = 0.333 \ldots 3_{10}$. We have ended the division process at three places and thus obtain an approximate answer of $0.010 = 0.25_{10}$. We naturally do not expect great accuracy with a system using only three bits. Had we used a four-bit system we would have obtained an answer of $0.0101 = 0.31_{10}$, which is somewhat closer.

In Example 13-15 it required two clock periods to determine the sign of the quotient. We had to shift left once for each bit in the magnitude; this required three clock periods. We had to make a comparison once for each bit in the magnitude and subtract once; this required four clock periods. Thus the total division time required nine clock periods. It can be easily seen that the division time is variable since it depends on the number of times we have to subtract.

This division process can, of course, be extended to numbers having n bits of magnitude. The total division time is variable, but we can calculate the maximum division time as follows:

1. Two clock periods are required for the sign determination.
2. For a number having n bits of magnitude, we must shift left n times, which requires n clock periods.
3. We must make n comparisons, which may require n clock periods.
4. We may make a maximum of n subtractions, which requires n clock periods.

The total division time is then the sum of these individual times and is $2 + 3n$ clock periods.

Example 13-16

What is the maximum time required to perform a division using 10-bit numbers if the clock is 1 MHz?

Solution

It will require $2 + 3n = 2 + 30 = 32$ clock periods. Each clock period is 1 μs, and therefore the maximum division time is 32 μs.

STUDY AIDS

Summary

In this chapter we have discussed the basic methods of performing binary arithmetic. We began by establishing the methods of representing both positive and negative numbers in binary fractional notation. In this system we represent positive numbers in sign and true magnitude and negative numbers in sign and true magnitude, or sign and 1's complement, or sign and 2's complement. These numbers are then used to perform arithmetic in fixed-point machines. We then discussed one method for representing numbers in sign, magnitude, and exponents; this representation is used in machines designed for floating-point arithmetic.

The discussion of serial adders and parallel adders clearly shows that parallel addition is considerably faster. By examining the four cases of addition we showed that subtraction reduces to addition and therefore the adders discussed could be easily used for subtraction as well. During the discussion it became clear that the use of 2's complements is more desirable for serial adders while 1's complements are more desirable for parallel addition. The discussion of a parallel BCD adder showed clearly the convenience of using numbers in excess-3 BCD code.

The most basic method of multiplication by repeated addition was discussed as well as a simple method of division by repeated subtraction. Using these two methods for multiplication and division explains why these two operations are more complicated than addition and subtraction and therefore require longer execution times.

Glossary

add-cycle time The time required to complete an addition operation.
binary fractional machine A machine designed to perform arithmetic operations using numbers in binary fractional notation.
binary fractional notation The representation of binary numbers by placing the binary point to the left of the MSB.
divide-cycle time The time required to complete a division operation.
fixed-point machine A machine using binary numbers having a fixed position for the binary point and no provision for powers of 2.
floating-point machine A machine which uses binary numbers having a fixed position for the binary point and provision for powers of 2.
multiplication-cycle time The time required to complete a multiplication operation.
overflow The act of exceeding the capacity of a storage register.
parallel binary adder A system which adds all $n + 1$ bits of two binary numbers, having n bits of magnitude, simultaneously.
serial binary adder A system which adds two binary numbers by adding two bits at a time sequentially, beginning with the two LSBs.
settling time The time required for all logic elements in an arithmetic system to reach their final steady-state levels after a change in data.
subtract-cycle time The time required to complete a subtraction operation.

Review Questions

1. Why is it important to detect overflows? How are they detected?
2. Why is it common practice to locate the binary point to the left of the MSB?
3. Explain the use of the sign bit.
4. What is meant by scaling, and where is it used?
5. What are the three ways of representing a negative binary number using sign and magnitude?
6. What is meant by floating-point arithmetic?
7. What are the four cases of addition? When can overflows occur?
8. When adding two numbers, how are overflows detected?
9. Why is the 1's complement for negative numbers not used in a serial adder?
10. What is the major advantage of a parallel adder over a serial adder? What is the major disadvantage?
11. What is the primary cause of settling time in a parallel adder?
12. What are the similarities between binary addition and excess-3 BCD addition (consider the four cases)?
13. Why is excess-3 more useful in BCD addition than 8421 code?

14. Explain how multiplication is performed by repeated addition.

15. Explain how division is accomplished by repeated subtraction.

Problems

13-1. What range of numbers can be represented in a binary fractional machine having eight bits of magnitude?

13-2. Represent the following decimal numbers in sign and true-magnitude form:

(*a*) +0.65625.
(*b*) −0.65625.
(*c*) +0.53125.

13-3. What are the decimal equivalents of the following numbers given in sign and true-magnitude form:

(*a*) 1.00101.
(*b*) 0.00011.
(*c*) 1.01010.
(*d*) 0.101010.

13-4. Represent the number -0.625_{10} in sign and true magnitude, sign and 1's complement, and sign and 2's complement.

13-5. What would be the scale factor for the number 4,381 if it were used in a machine accepting binary fractional numbers only?

13-6. Make a sketch showing the format of the numbers in a binary fractional machine using six bits of magnitude and a sign bit. Using this format, what is the binary number equivalent to 4,381? What is the resolution error?

13-7. Make a sketch showing the format of a floating-point number using a sign bit, four bits of exponent, and eight bits of magnitude.

13-8. Using the system of Prob. 13-7, give the binary equivalent of the number 4381_{10}.

13-9. Demonstrate that the four cases of subtraction of two numbers reduce to the same four cases of addition of two numbers.

13-10. What is the add-cycle time of a serial binary adder having 12 bits of magnitude if the system uses a 1-MHz clock?

13-11. Estimate the settling time for a parallel binary adder having 12 bits of magnitude if the delay through each full-adder is 0.1 μs. Assume the flip-flop delay times are negligible.

13-12. Make a table showing the eight possible input conditions to the overflow detector in Fig. 13-10, and determine which cases give an overflow indication.

13-13. Show the binary representation of the following numbers in 8421 code and excess-3 code:

(*a*) 67_{10}.
(*b*) -34_{10}.
(*c*) -93_{10}.

13-14. Make a truth table and verify the operation of the 9's-complement circuit shown in Fig. 13-15.

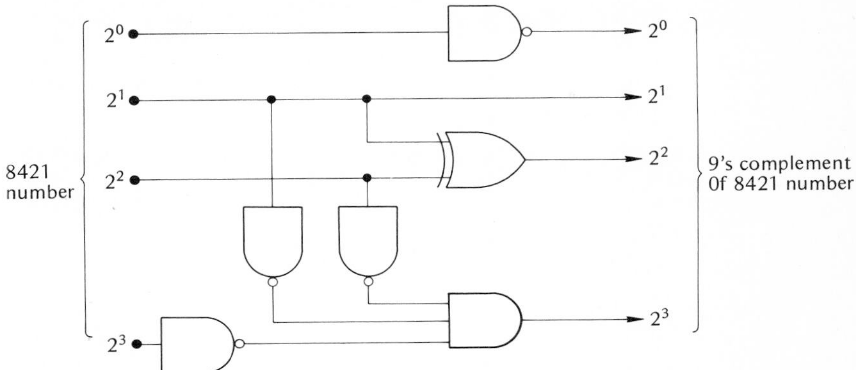

Fig. 13-15. Circuit to form the 9's complement of a 8421 BCD number (Prob. 13-14).

13-15. Draw a logic diagram for an excess-3 parallel BCD adder capable of handling numbers between ±0.999.

13-16. What multiply-cycle time is required for a system using numbers having 12 bits of magnitude and having a 2-MHz clock? What is the time required for numbers having 30 bits of magnitude?

13-17. What is the maximum division time required for two numbers having 12 bits of magnitude if the clock is 2 MHz? What is the maximum time for two numbers having 30 bits of magnitude?

13-18. Make a sketch similar to Example 13-14 showing the multiplication of +.111 and −.111 (these numbers are sign and true magnitude).

13-19. Make a sketch similar to Example 13-15 showing the division of +.101 by −.111 (these numbers are sign and true magnitude).

13-20. What happens if you attempt to divide +.111 by +.111?

Introduction to Digital Computers

14

The digital principles discussed in the previous chapters have been utilized to devise a great many different digital systems. The applications are many and varied. They include simple systems such as counters and digital clocks, and more complex applications such as digital voltmeters, A/D converters, frequency counters, and time-period measuring systems. Among the most sophisticated digital systems devised are digital computers, including special-purpose machines, small general-purpose computers (such as the Digital Equipment Corp. PDP-8/E), and large general-purpose computers (such as the IBM 360 and 370 systems). In this chapter we consider some of the basic principles common to digital computer systems.

After studying this chapter you should be able to

1. State the difference between a special purpose and a general purpose digital computer.
2. Discuss the 4 main blocks in a general purpose computer.
3. Write a simple computer program using mnemonic code.

14-1 BASIC CLOCKS

The operation or control of a digital system can be classified in two general categories—synchronous and asynchronous. In a *synchronous* system the flip-flops are controlled by the system clock and can therefore change states only when the clock changes state. Therefore, all the flip-flops and logic gates change levels in time (or in synchronism) with the clock. An example of such a synchronous system is the parallel counter constructed using the *master/slave* clocked flip-flops. In this counter, the flip-flops can change state only when the clock goes low and at no other time (notice that a system could be constructed such that the flip-flops would change state when the clock goes high). On the other hand, in an *asynchronous* system the flip-flops are controlled by events which occur at random times. Thus

Fig. 14-1. Basic system clock.

the flip-flops may change states at random and are not in synchronism with any timing signal such as a clock. An example of such a system might be the operation of a push button by a human operator. Depression of the push button would cause a flip-flop to change state. Since the operator can depress the button at any time he or she desires, the flip-flop would change states at some random time, and this is therefore an asynchronous operation. Most large-scale digital systems operate in the synchronous mode; if you give a little thought to the checkout and maintenance of such a system, it is easy to see why.

Since all logic operations in a synchronous machine occur in synchronism with a clock, the system clock becomes the basic timing unit. The system clock must provide a periodic waveform which can be used as a synchronizing signal. The square wave shown in Fig. 14-1*a* is a typical clock waveform used in a digital system. It should be noted that the clock need not be a perfectly symmetrical square wave as shown. It could simply be a series of positive pulses (or negative pulses) as shown in Fig. 14-1*b*. This waveform could, of course, be considered as an asymmetrical square wave. The main requirement is simply that the clock be perfectly periodic. Notice that the clock defines a basic timing interval during which logic operations must be performed. This basic timing interval is defined as a *clock cycle time* and is equal to one period of the clock waveform. Thus all logic elements, flip-flops, counters, gates, etc., must complete their transitions in less than one clock cycle time.

Example 14-1

What is the clock cycle time for a system which uses a 500-kHz clock? A 2-MHz clock?

Solution

A clock cycle time is equal to one period of the clock. Therefore, the clock cycle time for a 500-kHz clock is $1/(500 \times 10^3) = 2\ \mu s$. For a 2-MHz clock, the clock cycle time is $1/(2 \times 10^6) = 0.5\ \mu s$.

Example 14-2

The total propagation delay through a *master/slave* clocked flip-flop is given as 100 ns. What is the maximum clock frequency that can be used with this flip-flop?

Solution

An alternative way of expressing the question is, how fast can the flip-flop operate? The flip-flop must complete its transition in less than one clock cycle time. Therefore, the minimum clock cycle time must be 100 ns. So, the maximum clock frequency must be $1/(100 \times 10^{-9}) = 10$ MHz.

In many digital systems the clock is used as the basic standard for measurement. For example, the accuracy of the digital clock discussed in Chap. 9 is related directly to the frequency of the clock used to drive the counter. If the clock changes frequency, the accuracy is reduced. For this reason, it is necessary to ensure that the clock maintains a stable and predictable frequency. In many digital systems only short-term stability is required of the clock. This would be the case in a system where the clock could be monitored and adjusted periodically. For such a system, the basic clock might be derived from a free-running multivibrator or a simple sine-wave oscillator as shown in Fig. 14-2*a* and *b*. For the free-running multivibrator the clock frequency f is given by

$$f \cong \frac{1}{2RC \ln (1 + V_C/V_B)} \tag{14-1}$$

Fig. 14-2. Basic clock circuits. (*a*) Free-running multivibrator. (*b*) Wien-bridge oscillator.

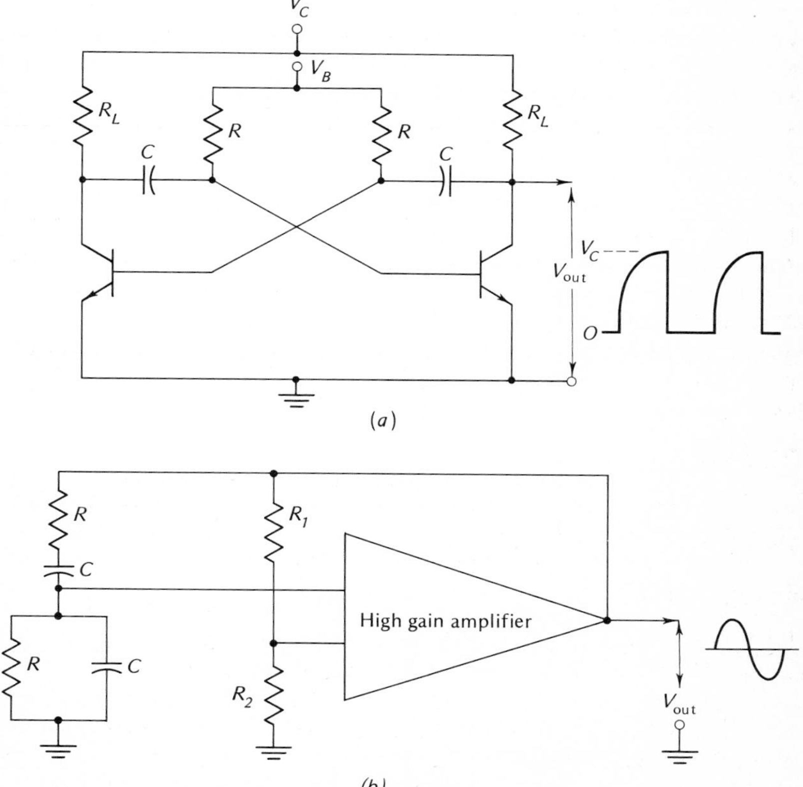

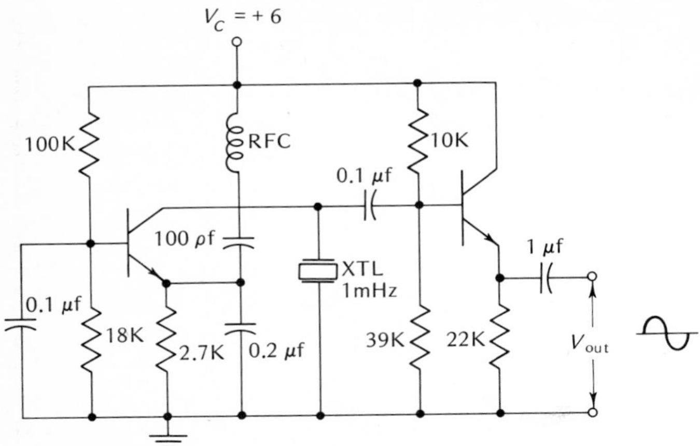

Fig. 14-3. Crystal oscillator.

From Eq. (14-1) it can be seen that the basic clock frequency is affected by the supply voltages as well as the values of the resistors R and capacitors C. Even so, it is possible to construct multivibrators such as this which have stabilities better than a few parts in 10^3 per day. The frequency of oscillation f for the Wien-bridge oscillator is given by

$$f \cong \frac{1}{2\pi RC} \tag{14-2}$$

Again it is not difficult to construct these oscillators with stabilities better than a few parts in 10^3 per day. If greater clock accuracy is desired, a crystal-controlled oscillator such as that shown in Fig. 14-3 might be used. This type of oscillator is quite often housed in an enclosure containing a heating element which maintains the crystal at a constant temperature. Such oscillators can have accuracies better than a few parts in 10^9 per day.

Example 14-3

The multivibrator in Fig. 14-2a is being used as a system clock and operated at a frequency of 100 kHz. If its accuracy is better than ± 2 parts in 10^3 per day, what are the maximum and minimum frequencies of the multivibrator?

Solution

One part in 10^3 can be thought of as 1 cycle in 1,000 cycles. Two parts in 10^3 can be thought of as 2 cycles in 1,000 cycles. Since the multivibrator runs at 100 kHz, two parts in 10^3 is equivalent to 200 cycles. Thus the maximum frequency would be 100 kHz + 200 cycles = 100.2 kHz, and the minimum frequency would be 100 kHz − 200 cycles = 99.8 kHz.

Fig. 14-4. Oscillator and output amplifier.

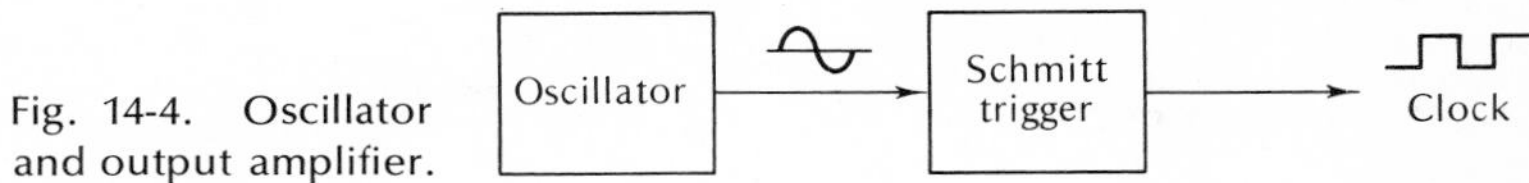

None of the oscillators shown in Figs. 14-2 and 14-3 has a square-wave output waveform, and it is therefore necessary to convert the basic frequency into a square wave before use in the system. The simplest way of accomplishing this is to use a Schmitt trigger on the output of the basic oscillator as shown in Fig. 14-4. This provides two advantages:

1. It provides a square wave of the basic clock frequency as desired.
2. It ensures that the clock-output amplifier (the Schmitt trigger in this case) has enough power to drive all the necessary circuits without loading the basic oscillator and thus changing the oscillating frequency.

14-2 CLOCK SYSTEMS

Quite often it is desirable to have clocks of more than one frequency in a system. Alternatively, it might be desirable to have the ability to operate a system at different clock frequencies. We might then begin with a basic clock which is the highest frequency desired and develop other basic clocks by simple frequency division using counters. As an example of this, suppose we desire a system which will provide basic clock frequencies of 3, 1.5, and 1 MHz. This could be accomplished by using the clock system shown in Fig. 14-5. We begin with a 3-MHz oscillator followed by a Schmitt trigger to provide the 3-MHz clock. The 3-MHz signal is then fed through one flip-flop which divides the signal by 2 to provide the 1.5-MHz clock. The 3 MHz signal is also fed through a divide-by-3 counter, which provides the 1-MHz clock. Systems having multiple clock frequencies can be provided by using this basic method.

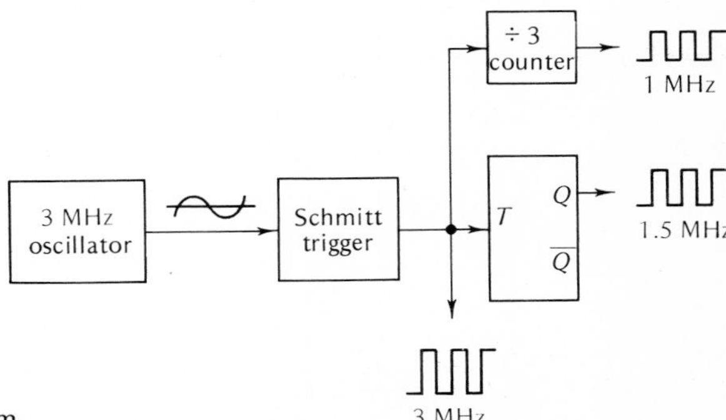

Fig. 14-5. Basic clock system.

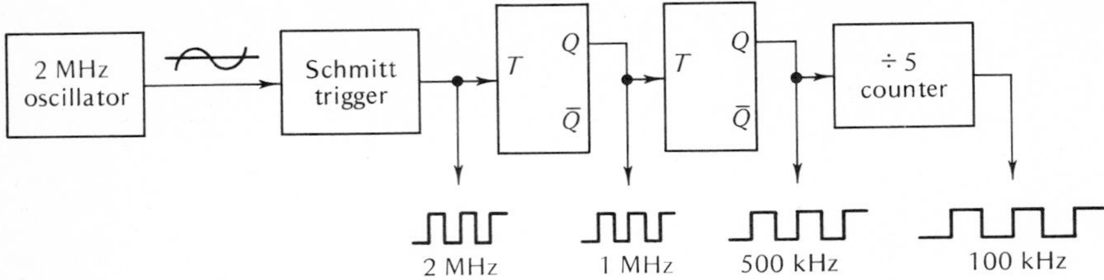

Fig. 14-6. Clock system.

Example 14-4

Show a clock system which will provide clock frequencies of 2 MHz, 1 MHz, 500 kHz, and 100 kHz.

Solution

The desired system is shown in Fig. 14-6. Beginning with a 2-MHz oscillator and a Schmitt trigger, the 2-MHz clock appears at the output of the Schmitt trigger. The first flip-flop divides the 2 MHz signal by 2 to provide the 1 MHz clock. The second flip-flop divides the 1-MHz clock by 2 to provide the 500-kHz clock. Dividing the 500-kHz clock by 5 provides the 100-kHz clock.

It is sometimes desirable to have a two-phase clock in a digital system. A *two-phase clock* simply means we have two clock signals of the same frequency which are 180° out of phase with one another. This can be accomplished with the outputs of a flip-flop. The Q output is one phase of the clock and the $\bar{Q}$ output is the other phase. These two signals are clearly 180° out of phase with one another, since one is the complement of the other. A system for developing a two-phase clock of 1 MHz is shown in Fig. 14-7. For distinction, the two clocks are sometimes referred to as phase A and phase B. You will recall that one use for a two-phase clock system is to drive the magnetic-core shift register discussed in Chap. 12 (Fig. 12-10). It is interesting to note that the two-phase clock system can be used to overcome the race problem encountered with the basic parallel counter discussed in Chap. 8 (Fig. 8-5). The race problem is solved by driving the *odd* flip-flops (i.e., flip-flops A, C, E, etc.) with phase A of the clock, and the *even* flip-flops (i.e., flip-flops B, D, F, etc.) with phase B of the clock (see Prob. 14-12).

The race problem as initially discussed in Chap. 8 can occur any time two or more signals at the inputs of a gate are undergoing changes at the same time. The

Fig. 14-7. 1-MHz two-phase clock.

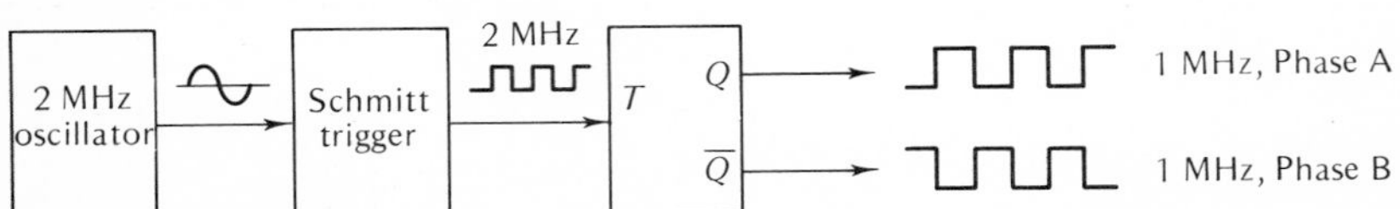

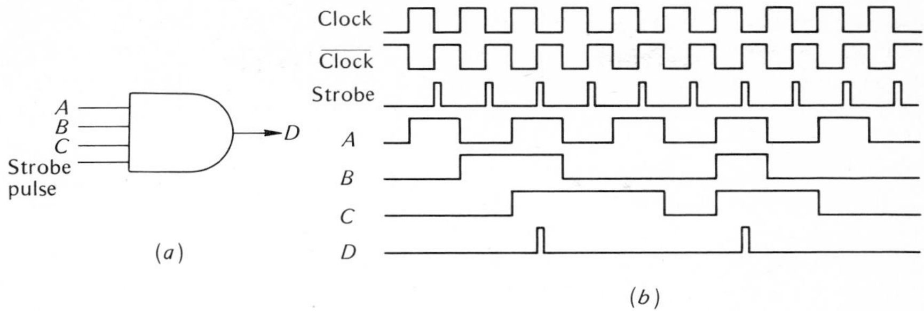

Fig. 14-8. The use of a strobe pulse. (*a*) Three-input AND interrogated by a strobe pulse. (*b*) Waveforms for the AND gate.

problem is therefore not unique in counters and can occur anywhere in a digital system. For this reason, a *strobe pulse* is quite often developed using the basic clock. This strobe pulse is used to interrogate the condition of a gate at a time when the input levels to the gate are not changing. If the gate levels render the gate in a true condition, a pulse appears at the output of the gate when the strobe pulse is applied. If the gate is false, no pulse appears. In Fig. 14-8, a strobe pulse is used to interrogate the simple three-input AND gate. The waveforms clearly show that outputs appear only when the three input levels to the gate are true. It is also quite clear that no racing can possibly occur since the strobe pulses are placed exactly midway between the input-level transitions. The strobe signal can be developed in a number of ways. One way is to differentiate the complement of the clock, $\overline{\text{clock}}$, and use only the positive pulses. A second method would be to differentiate the clock and feed it into an "off" transistor as shown in Fig. 14-9.

14-3 MPG COMPUTER

Up to this point we have covered quite a wide variety of the topics generally encountered in the study of digital systems. Some of the topics have been discussed in

Fig. 14-9. Developing a strobe pulse.

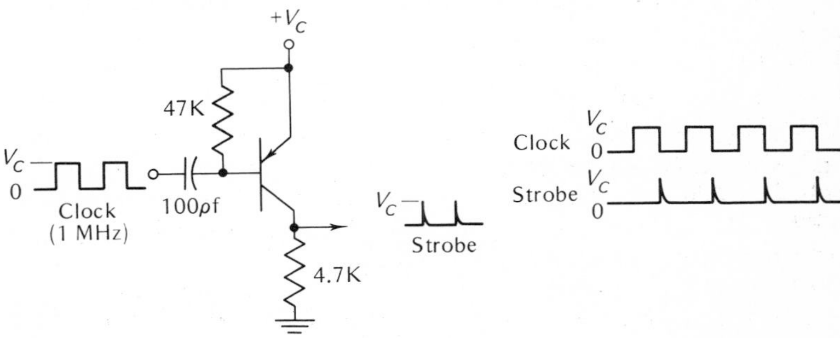

great detail, while others have been treated in a more general way. In any case you should now have the necessary background to study any digital system with good comprehension and a minimum of effort. Even so, you may be somewhat unsure about the overall organization of a digital system. In an effort to overcome this feeling and to attempt to tie together many of the topics discussed in the previous chapters, we shall at this time consider the implementation of a small special-purpose digital computer.

The special-purpose computer we shall consider will be used to calculate the *miles per gallon* of a motor vehicle, thus the name *MPG computer*. It is a *special-purpose* computer since this is the only use for which it is intended. A *general-purpose* computer would be a more complicated machine which might be used for a number of different applications.

The first step in the design of the MPG computer must necessarily be the determination of the system performance requirements. The first requirement might be that the system be capable of operating from a supply voltage of ±6 or ±12 V dc since the machine will be operated in a motor vehicle. The second requirement might be that the readout of the computer be in decimal form. Nixie tubes might be good for the readout, but they require an additional power supply of around +100 V to operate the tubes. Digital modules are commercially available which provide decimal readout, and they operate on +6 or +12 V dc. These modules do not require the +100 V, and might be a better choice in this case. The final decision will be one of economics. The third requirement is that the computer calculate the miles per gallon used by the vehicle to an accuracy of ±1 mile per gallon. The fourth requirement we shall impose is that the computer perform a calculation at least once every 15 s when the vehicle is traveling at a speed greater than 10 mph. In other words, we would like to sample the mileage performance of the vehicle at least once every 15 s (faster sampling rates are acceptable). The fifth requirement is that the computer be capable of operating in vehicles using fuel at rates between 10 and 40 miles per gallon. We can now summarize the five basic requirements of the MPG computer as follows:

1. Power-supply voltage is either ±6 or ±12 V dc.
2. The computer must provide a decimal readout in miles per gallon.
3. The computer must provide the readout to an accuracy of ±1 mile per gallon.
4. The computer must provide a readout of miles per gallon at least once every 15 s when the vehicle is traveling at a speed greater than 10 mph.
5. The computer must be capable of calculating miles per gallon between the limits of 10 and 40 miles per gallon.

It should be noted that the system requirements for the computer under study here are quite simple and somewhat less stringent than in the usual case. The requirements here are intentionally made simple in order to simplify the discussion. Nevertheless the principles are the same regardless of the severity of the system specifications, and the study is therefore instructive.

We assume that we have available two transducers which are to be used as an integral part of the MPG computer. The first transducer is used to measure the vol-

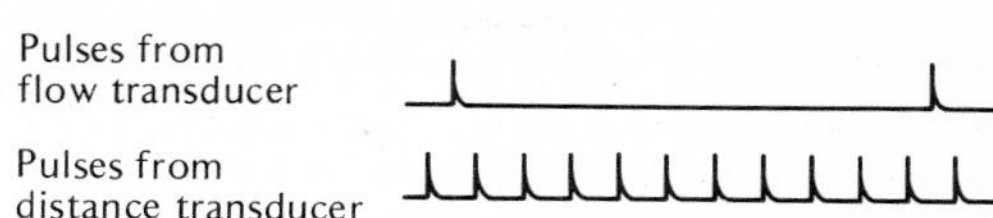

Fig. 14-10. Transducer pulses for the MPG computer when the rate is 10 miles per gallon.

ume of fuel flowing into the engine. This flow tranducer provides an electrical pulse each time 1/1000 of a gallon of fuel passes through it. The second transducer is used to measure the distance traveled and is driven by the speedometer cable. This distance transducer provides an electrical pulse each time the vehicle has traveled a distance of 1/1000 of a mile.

Now in order to implement the necessary logic for the computer, let us examine the outputs of the flow and distance transducers. Let us begin by assuming that we have a flow transducer which gives an output pulse each time 1 gallon is used, and we have a distance transducer which gives an output pulse each time the vehicle has traveled 1 mile. If our vehicle is obtaining a mileage slightly better than 10 miles per gallon, the transducer waveforms appear as shown in Fig. 14-10. Notice that the number of distance pulses appearing between two flow pulses is exactly equal to the miles per gallon we desire. Thus we can *calculate* the miles per gallon by simply counting the number of distance pulses occurring between two flow pulses. We can check this by noting that, if the vehicle were operating at 20 miles per gallon, there would be 20 distance pulses between two flow pulses. Notice that if the flow transducer supplied 10 pulses per gallon, and at the same time the distance transducer provided 10 pulses per mile, the basic waveform in Fig. 14-10 would remain unchanged. That is, the number of distance pulses appearing between two flow pulses would still be equal to the number of miles per gallon. From this it should be clear that we can choose any number of pulses per gallon from the flow transducer so long as we choose the same number of pulses per mile from the distance transducer. The transducers we are going to use in the MPG computer provide 1,000 pulses per gallon of flow and 1,000 pulses per mile of distance. Therefore, the number of miles per gallon can be obtained by simply counting the number of distance pulses between consecutive flow pulses.

The reason for using these transducers can be seen by examining the time between flow pulses. Let us first consider the flow transducer having one pulse per gallon and the distance transducer having one pulse per mile. If the vehicle were obtaining a rate of 10 miles per gallon, one flow pulse would occur every 10 miles. If the vehicle were traveling at a speed of 10 mph, the flow pulses would occur at a rate of one per hour. This is clearly not a fast enough sampling rate. On the other hand, with the specified transducers, the flow pulses occur at a rate of 1,000 pulses per gallon and at the rate of 1,000 pulses per hour under the same conditions. Thus the flow pulses occur every 1 hr/1000 = 3.6 s. This sampling time is clearly within the specified rate. The worst case occurs when the vehicle obtains the maximum miles per gallon. At 40 miles per gallon and 10 mph the flow pulses occur every $3.6 \times 4 = 14.4$ s. We have therefore met the minimum-sampling-time requirements.

The logic diagram for the MPG computer can now be drawn; it is shown in Fig. 14-11 along with the complete waveforms. The flow pulses are fed into a conditioning amplifier and then into a one-shot to develop the waveform OS_1 and $\overline{OS}_1$. The distance pulses are also fed into a conditioning amplifier. Since we desire to count the number of distance pulses occuring between two pulses, we use the distance pulses as one input to the *count* AND gate. If $\overline{OS}_1$ is used as the other input to this AND gate, it is enabled between flow pulses, and the distance pulses appear at its output. We use the pulses appearing at the output of the *count* AND gate to drive a counter. Since we desire to display the miles per gallon between the limits

Fig. 14-11. Complete MPG computer.

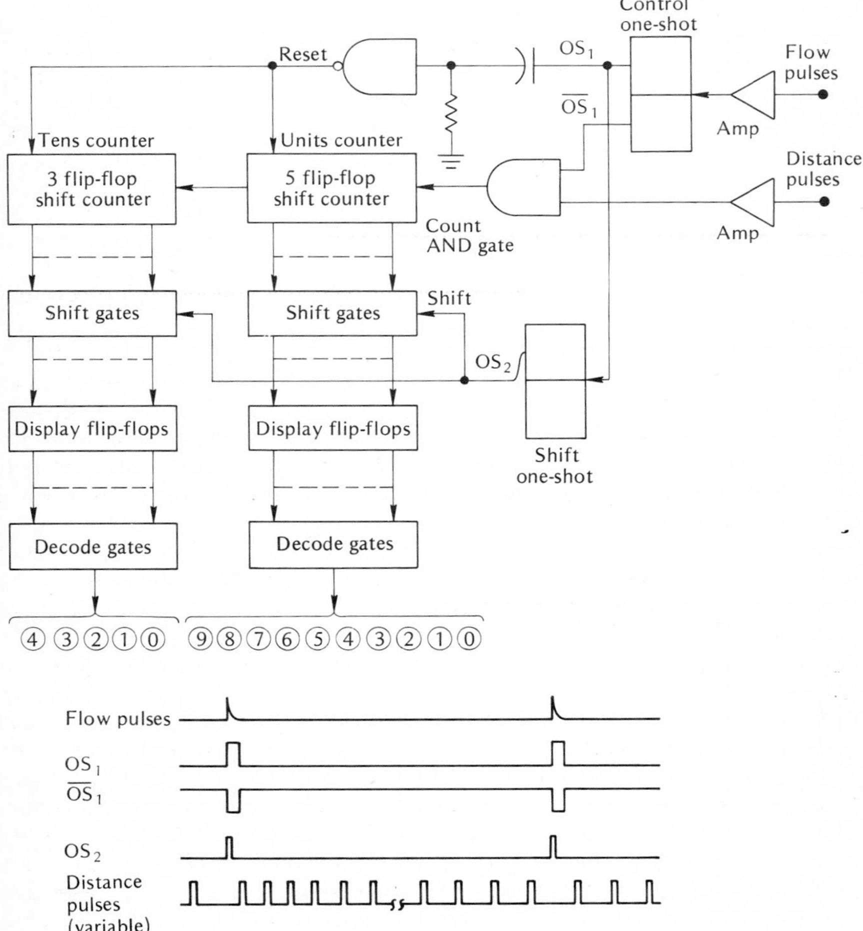

of 10 and 40, we use a five-flip-flop shift counter for the *units* digits, and a three-flip-flop shift counter for the *tens* digits of miles per gallon.

One conversion time is the time between two flow pulses, and we want to shift the accumulated count into the display flip-flops at the end of each conversion cycle. Notice first of all that, when $\overline{OS_1}$ is low, the *count* AND gate is disabled and therefore the *units* and *tens* counters cannot change states. It is during this time that we must shift the contents of these counters into the display flip-flops. We use the leading edge of OS_1 to trigger the *shift* one-shot and develop the shift waveform OS_2. The falling edge of OS_2 is applied to the shift gates, and at this time the count stored in the *units* and *tens* counters is shifted into the display flip-flops. The falling edge of OS_1 is then used to reset all flip-flops in the *units* and *tens* counters. The contents of the display flip-flops are then decoded and used to illuminate the indicator lights. In this system, the distance pulses can be considered to be the basic system clock. The flow pulses form a variable control gate by means of the *control* one-shot which determines the period of time that the *count* AND gate is enabled and therefore the number of distance pulses counted. The output of the *shift* one-shot OS_2 can be considered as a strobe pulse which shifts data from the counters into the display flip-flops in such a way that racing is avoided. The system clearly has an accuracy of $\pm$ one count, which corresponds to ± 1 mile per gallon.

14-4 GENERAL-PURPOSE COMPUTER

The MPG computer discussed in the previous section is considered a special-purpose computer since it is designed and constructed to perform a single function; to alter it so that it could perform another function would require a major change in design. On the other hand, a general-purpose computer is designed so that it can perform a number of fundamental operations—addition, subtraction, multiplication, division, comparison, etc. The computer can then be used in any number of different applications by simply instructing it to perform the appropriate operations in an orderly fashion. The functions to be performed, listed in the order in which they are to be accomplished, is known as a *program* (instruction set). This list of instructions, or program, is normally stored in the computer memory; when the computer is started, it simply performs these instructions in the order stored. Herein lies the difference between an electronic calculator and a general-purpose digital computer—the calculator performs a function (add, subtract, etc.) each time an operator depresses a button, but the stored-program computer performs the complete list of stored instructions without human intervention. Furthermore, the computer is capable of completing the instruction set in a very short period of time (addition in perhaps a few microseconds), and the operation is virtually error free.

The simplified block diagram in Fig. 14-12 shows the basic units to be found in any general-purpose computer system. The *input/output* block represents the interface between man and machine. It could simply be a teletype unit, where input information is typed in on the keyboard and output information is printed on paper. It could also represent any of the other input/output media previously discussed, such as punched paper tape, punched unit-record cards, and magnetic tape. In any case,

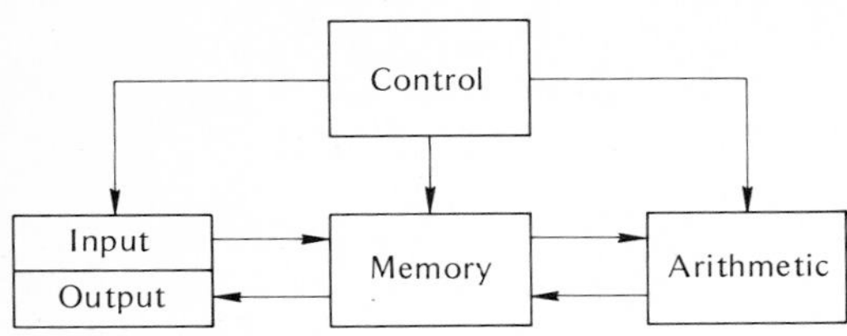

Fig. 14-12. Basic computer unit block diagram.

input data are taken into the system and stored in the memory according to the appropriate signals as generated by the control block. Similarly, the control unit generates the appropriate signals to read data from the memory and move it to the output block.

The arithmetic unit consists of the registers, counters, and logic required for the basic operations, including addition, subtraction, complementation, shifting right or left, comparison, etc. Since the manipulation of data is accomplished in this unit, it is sometimes referred to as the *central processing unit* (CPU). The topics previously covered (number systems, digital arithmetic, etc.) provide an insight into the logic circuits and configurations required in a CPU. Again, the control unit provides the necessary signals to move data from the memory unit to the arithmetic unit, perform the desired data manipulation, and move the resulting data back into memory.

The memory block represents the area used to store the two types of information present in the computer; namely, the list of instructions (program) and the data to be operated on as well as the resulting output data. The memory itself could be constructed using any of the devices previously discussed—magnetic cores, magnetic drums or disks, semiconductor memory units, magnetic tapes, and so on. Reading data from or writing data into the memory is again under the guidance of the control unit.

The control unit generally contains the counters, registers, and logic necessary to develop the control signals required for moving data into and out of the memory, and for performing the necessary data manipulations in the arithmetic unit. The system clock is a part of the control unit, and it is usually the starting point for generating the proper control signals as discussed in the first part of this chapter.

It is interesting to consider an actual general-purpose digital computer in light of the above discussion. For this purpose, a block diagram of the Digital Equipment Corp. PDP-8/E is shown in Fig. 14-13.[1] Note how the system diagram can be broken into the four basic blocks previously discussed—input/output, arithmetic, memory, and control. A table-model PDP-8/E is shown in Fig. 14-14, and the following excerpt gives a general description of the system.[2]

> The PDP-8/E is specially designed as a general perpose computer. It is fast, compact, inexpensive, and easy to interface. The PDP-8/E is designed to meet

[1] "Small Computer Handbook," chap. 1, Digital Equipment Corporation, Maynard, Mass., 1971.
[2] Ibid.

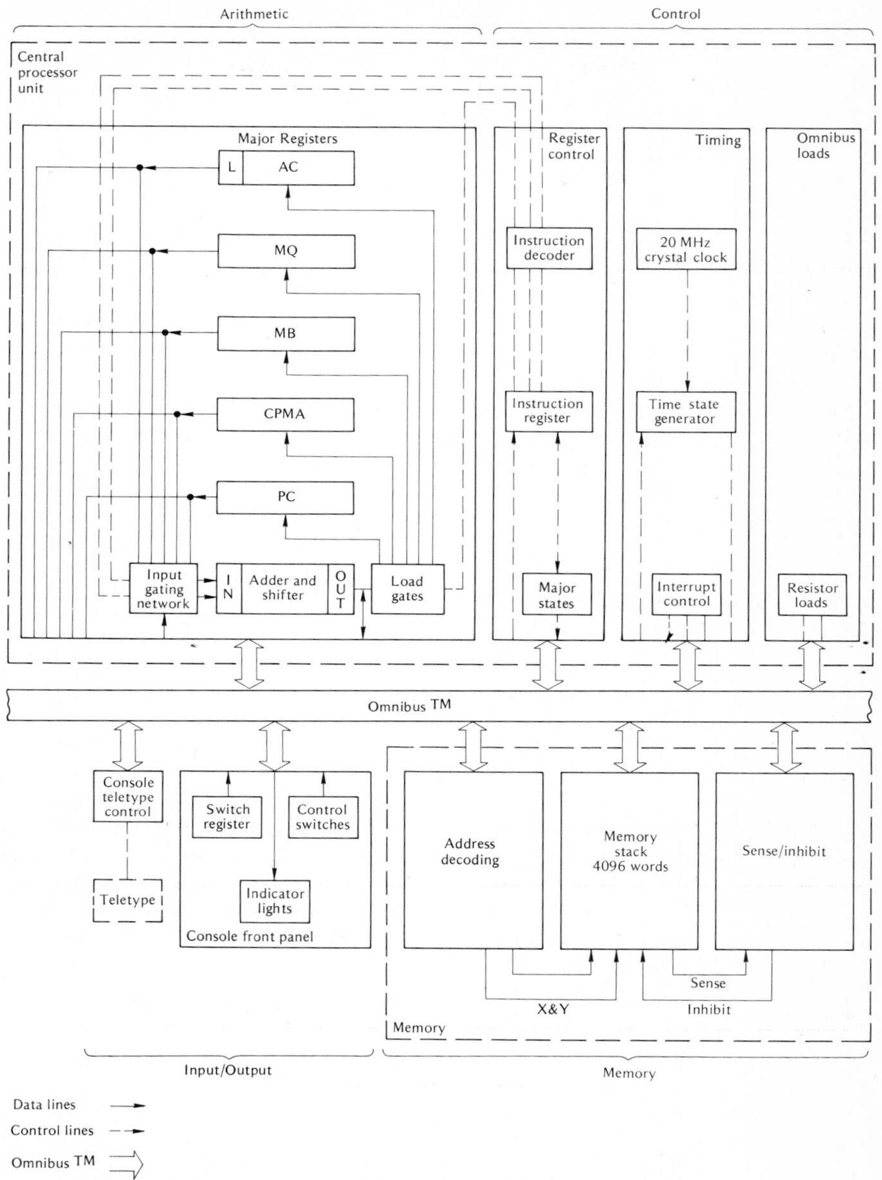

Fig. 14-13. PDP-8/E basic system block diagram.

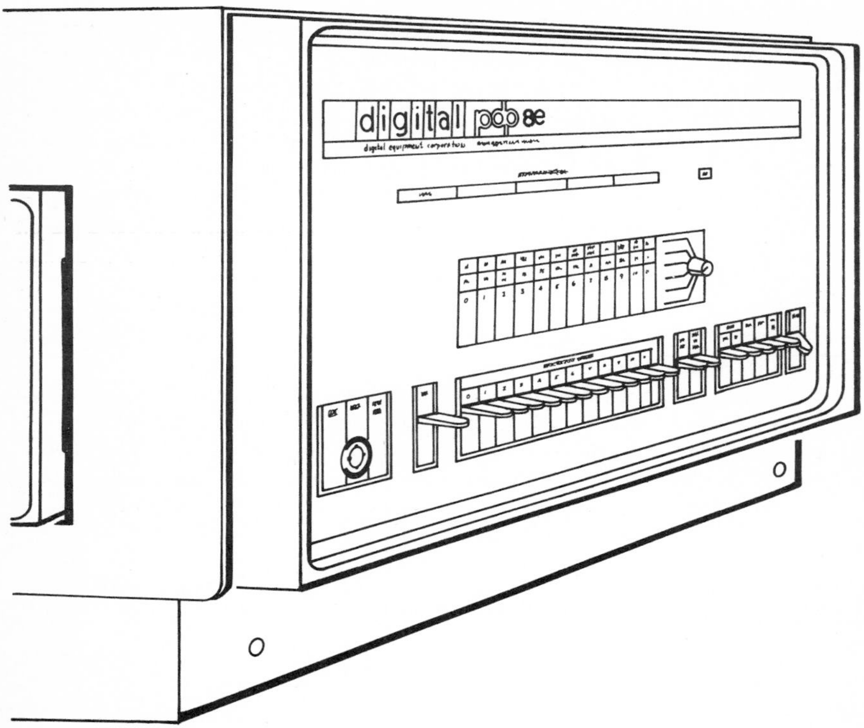

Fig. 14-14. PDP-8/E programmed data processor.

the needs of the average user and is capable of modular expansion to accomodate most individual requirements for a user's specific applications.

The PDP-8/E basic processor is a single-address, fixed word length, parallel-transfer computer using 12-bit, 2's complement arithmetic. The cycle time of the 4096-word random address magnetic core memory is 1.2 microseconds for fetch and defer cycles without autoindex; and 1.4 microseconds for all other cycles. Standard features include indiret addressing and facilities for instruction skip and program interrupt as a function of the input/output device condition.

Five 12-bit registers are used to control computer operations, address memory, operate on data and store data. A Programmer's console provides switches to allow addressing and loading memory and indicators to observe the results. The PDP-8/E may also be programmed using the console Teletype with a reader/punch facility. Thus, programs can be loaded into memory using the switches on the Programmer's console, the Teletype keyboard, or the paper tape reader. Processor operation includes addressing memory, storing data, retrieving data, receiving and transmitting data and mathematical computations.

The 1.2/1.4 microsecond cycle time of the machine provides a computation rate of 385,000 additions per second. Each addition requires 2.6 microseconds (with one number in the accumulator) and subtraction requires 5.0 microseconds (with the subtrahend in the accumulator). Multiplication is performed in 256.5 microseconds or less by a subroutine that operates on two signed 12-bit numbers to produce a 24-bit product, leaving the 12 most significant bits in the accumulator. Division of two signed 12-bit numbers is performed in 342.4 microseconds or less by a subroutine that produces a 12-bit quotient in the accumulator and a 12-bit remainder in core memory. Similar signed multiplication and division operations are performed in approximately 40 microseconds, utilizing the optional Extended Arithmetic Element.

The flexible, high-capacity input/output capabilities of the computer allow it to operate a large variety of peripheral machines. Besides the standard keyboard and paper-tape punch and reader equipment, these computers are capable of operating in conjunction with a number of optional devices (such as high-speed perforated-tape punch and reader equipment, card reader equipment, line printers, analog-to-digital converters, cathode ray tube (CRT) displays, magnetic tape equipment, a 32,764-word random-access disk file, a 262,112-word random-access disk file, etc.).

14-5 COMPUTER ORGANIZATION AND CONTROL

In this short chapter devoted to digital computers, we cannot possibly give an exhaustive treatment of all machines; however, we can discuss in general terms those aspects of computer organization and operation which are common to many different types of digital computers.

The information stored in the computer memory is of two types—either *data words* (numeric information) or *instruction words.* In Sec. 13-1, we considered in some detail the various formats available for storing numbers, including both fixed-point and floating-point numbers. We must now consider an appropriate format for a computer instruction word.

In general, a computer instruction word will have two distinct sections, as shown in Fig. 14-15. In this case the word length is 12 bits; however, the number of bits in a word varies from machine to machine (e.g., 36 in the IBM 7090/7094, 32 in the IBM 360, 36 in the GE 635, and 12 in the PDP-8/E). The first section (the three bits on the left in this case) are used for the *operation code* (op-code) of the instruction to be performed. The op-codes are defined by the computer designer when the machine is initially designed. For example, the op-code for addition might be defined as 001_2. In this case, there are only three bits reserved for op-codes, and a computer using this format would therefore be limited to $2^3 = 8$ op-codes.

The remaining bits in the instruction word shown in Fig. 14-15 are used to specify the address in memory to which the instruction applies. In this case, the nine bits can be used to specify any one of $2^9 = 512$ locations in memory. As an example, the instruction word 001 000001100 means add (001) the contents of the

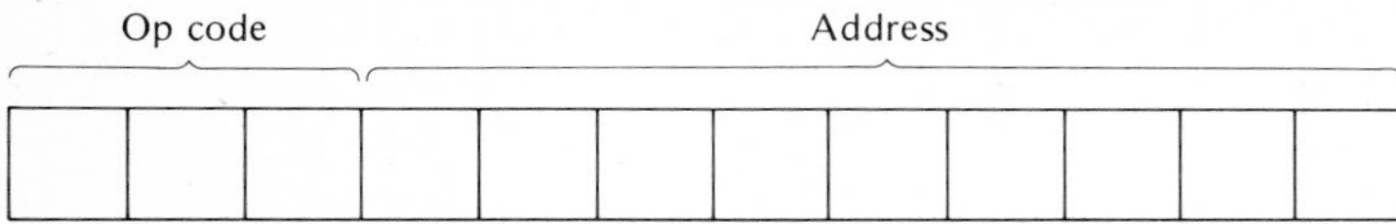

Fig. 14-15. Instruction word format.

memory located at address 12_{10} (000001100) to the contents of the accumulator register in the arithmetic unit.

Frequently the memory is broken up into sections called "pages" in order to provide for more efficient addressing. For example, the PDP-8/E has a basic memory of 4,096 twelve-bit words. The memory is broken up into 32 pages of 128 words on each page. Thus any word on a page can be addressed by means of only seven bits ($2^7 = 128$). The instruction word for the PDP-8/E is then arranged as shown in Fig. 14-16. If the address mode bit (bit 3) is 0, the op-code simply refers to one of the 128 page addresses given by the last seven bits in the word. However, if the address mode bit is 1, *indirect addressing* is indicated. This means the control unit will go either to page 0 or remain on the current page (depending on whether bit 4 is 1 or 0), take the contents of the given address, and treat it as another address. The first five bits of this new address specify which of the 32 pages ($2^5 = 32$), and the remaining seven bits give the address on that page ($2^7 = 128$) containing the data to which the op-code applies.

In this way, the instruction word format need only have seven bits devoted to an address, and only an occasional 12-bit address word is needed to reference data on any one of the other 31 available pages. Clearly this word format is more efficient than simply carrying 12 ($2^{12} = 4{,}096$) bits for address locations in memory.

As an example of indirect addressing, suppose the data being operated on are stored on page 15 of the memory—in order to get to another page, one must use indirect addressing. The instruction word 001 10 0001110 means add (001) the contents of the data located in address 14_{10} (0001110) on page 0 to the contents of the accumulator register in the arithmetic unit. Note that the 1 in the fourth bit position specifies indirect addressing, and the 0 in the fifth bit position refers to page 0. Now, if the contents of memory location 14_{10} on page 0 is 00101 0001111, the data to be added to the accumulator will be found on page 5_{10} (00101) in location 15_{10} (0001111).

Fig. 14-16. PDP-8/E instruction word format.

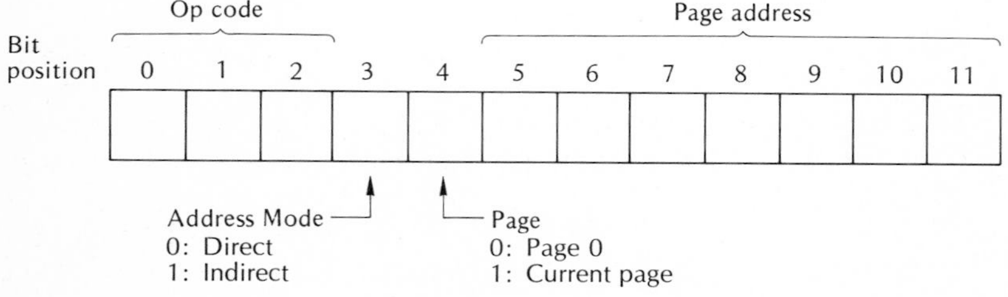

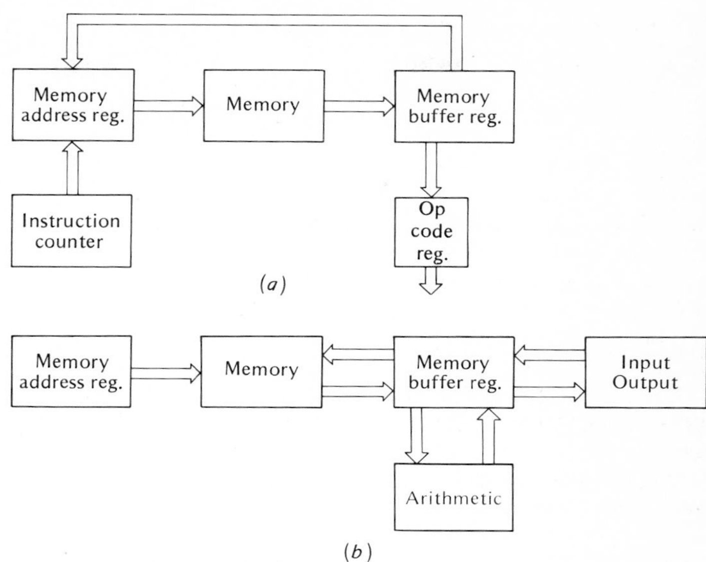

Fig. 14-17. Basic computer operating cycles. (a) *Fetch*. (b) *Execute*.

The instructions to be executed by the computer are normally stored in the memory in the order in which they are to be performed. To begin an operation, the address in the memory of the first instruction to be executed is entered into the machine by an operator. The control unit then *fetches* this instruction from memory, *executes* the proper operation, and proceeds to the next instruction stored in the memory. This basic two-cycle process continues until all the instructions have been completed and the machine stops. Thus the operation of a computer can be explained in terms of two fundamental cycles—*fetch* and *execute*. Let's examine these two cycles and determine the tasks to be accomplished by the control unit during each cycle.

The computer units involved during a fetch cycle are shown in Fig. 14-17*a*. During a *fetch* cycle, the following operations are performed:

1. The address in memory of the first instruction to be executed is placed in the instruction counter. This address is read into the memory address register (MAR) and a *read/write* cycle is initiated in the memory.
2. The instruction stored at the given address in memory is read into the memory buffer register (MBR).
3. The op-code portion of the instruction in the MBR is then stored in the op-code register, and the address portion is placed in the MAR (in place of the previous address) in preparation for the following *execute* cycle.
4. The instruction counter is increased by one in order to be ready for the next *fetch* cycle.

The computer units active during an *execute* cycle are shown in Fig. 14-17*b*, and the following operations are performed:

1. The address in memory containing data to be read out, or where data is to be stored, is contained in the MAR as a result of the previous *fetch* cycle. Similarly, the op-code is contained in the op-code register.
2. The contents of the op-code register are decoded and the control unit provides the necessary control signals to perform the operation called for—e.g. read data from an input TTY, into the MBR and store it at the address in memory according to the contents of the MAR; or, read data from the address in memory as given by the MAR, and move it to the arithmetic unit via the MBR; or, read data from the memory via the MBR and print the data on a TTY: or, read data from the arithmetic unit via the MBR and store it in the memory at the address specified by the MAR.
3. At the completion of the *execute* cycle, return to the next *fetch* cycle.

The *fetch/execute* method of operation is quite common to most general-purpose digital computers, even though the two states might be referred to by different names. When an operation is begun, the control unit first places the computer in the *fetch* mode, and thereafter alternates *execute* and *fetch* modes until the desired operation is complete. A series of clock pulses (perhaps four or five, or even ten) during each *fetch* cycle is used to time the various operations. A similar sequence of clock pulses is utilized during the *execute* cycle.

14-6 COMPUTER INSTRUCTIONS

Every general-purpose computer must have an instruction set. There may be only a few (10 or so) for a small computer, while a large computer may have hundreds of instructions. The set of instructions used with any particular computer is of course devised during the initial design phases, and anyone who uses that computer must become intimately familiar with its instruction set. Incidently, an individual who specializes in efficiently arranging computer instructions for the purpose of solving problems is known as a *computer programmer*.

Inside the computer, every instruction must be represented as a group of binary numbers (e.g., 001 for addition), but to ease the burden of the programmer, the op-codes are frequently assigned *mnemonic* titles. For example, the op-code for addition might be 001, but we could code it as ADD. The programmer could then use ADD in arranging his list of instructions, and when the alphanumeric input ADD appeared at the computer input, it would simply be encoded as the instruction 001.

In general, there are four different types of instructions—arithmetic, data manipulation, transfer, and input/output. Let's list a ficticious set of instructions and then see how they might be arranged as a program to solve a problem. Even though this instruction set is ficticious, it is quite similar to those found in actual computer systems. Each instruction is given in mnemonic form, with its binary code in parenthesis, and a description of the operation it requires.

HLT (*0000*) Halts computer operation. Operator may restart by depressing the start button.
ADDX (*0001*) The content of memory location X is added to the content of the accumulator register in the arithmetic unit.
SUBX (*0010*) The content of memory location X is subtracted from the content of the accumulator register in the arithmetic unit.
MPYX (*0011*) The content of memory location X is multiplied by the content of the MQ register in the arithmetic unit, and the product is stored in the MQ register.
DIVX (*0100*) The content of memory location X is divided into the content of the MQ register, and the quotient is stored in the MQ register.
DCAX (*0101*) The content of the accumulator is stored in memory location X, and the accumulator is cleared to all zeros.
DCQX (*0110*) The content of the MQ register is stored in memory location X, and the MQ register is cleared to all zeros.
JMPX (*0111*) The next instruction is taken from memory location X.
LDQX (*1000*) The content of memory location X is entered into the MQ register.
REDX (*1001*) One word of data is read at the input device and stored in memory at address X.
PRTX (*1010*) One word of data is read from memory at address X and printed on the output device.

This list of instructions is of course not complete enough to allow every possible operation, but it allows us to illustrate basic machine-language programming. Notice that there are four bits in each op-code; this is necessary since we want to include more than eight but fewer than 16 instructions. Further, suppose these instructions are used in a small general-purpose computer having only 128 memory

Table 14-1

Operation	Instruction	Memory location	Instruction as stored in memory
Read *R* and store at memory address 50.	RED 50	0	1001 0110010
Read *A* and store at memory address 51.	RED 51	1	1001 0110011
Read *Y* and store at memory address 52.	RED 52	2	1001 0110100
Clear MQ register	DCQ 127	3	0110 1111111
Clear accumulator	DCA 127	4	0101 1111111
Put *A* in MQ	LDQ 51	5	1000 0110011
Multiply *A* by *Y*	MPY 52	6	0011 0110100
Store *AY* in 53	DCQ 53	7	0110 0110101
Put *R* in accumulator	ADD 50	8	0001 0110010
Add *AY* to *R* in accumulator	ADD 53	9	0001 0110101
Store *Z* in 54	DCA 54	10	0101 0110110
Print out *Z*	PRT 54	11	1010 0110110
Halt	HLT	12	0000 0000000

locations so that an instruction word is composed of 11_{10} bits—four bits of op-code and seven bits for memory address.

Now, let's utilize the instructions for our fictitious computer to solve the problem $Z = R + AY$. The program will read the values of R, A, and Y, perform the necessary calculations, and print out the value of Z. The complete program, as written in machine language (mnemonic code) and as stored in memory, would appear as in Table 14-1.

To initiate the program, the operator sets the instruction counter at 0 and depresses the start button. The computer initiates a *fetch* cycle and obtains the first instruction (RED 50) from memory address 0. This is followed by an *execute* cycle. The next *fetch* cycle obtains the instruction in memory address 1, and so on. The program ends after the computed value for Z is printed out and the HLT instruction is obtained in memory address 12_{10}.

STUDY AIDS

Summary

There are basically two types of digital computers—special purpose and general purpose. Special-purpose computers are designed for a single purpose only, while general-purpose machines can be used in any number of different applications. A general-purpose machine is designed with a basic set of instructions, and a programmer can use such a computer to solve specific problems. The computer solves problems by executing a set of instructions which have been ordered and placed in the computer memory by a programmer. Most computers operate in a basic two-cycle *fetch/execute* mode, and the appropriate control signals are generated in the control unit in synchronism with the system clock.

Glossary

asynchronous system A system in which logic operations and level changes occur at random times.

clock cycle time One clock period; the reciprocal of clock frequency.

computer program A list of specific instructions which a computer executes to solve a given problem.

fetch/execute The two alternating modes of operation in a general-purpose computer.

general-purpose computer A computer designed to accomplish a number of tasks. For example, all the arithmetic operations as well as decision making (i.e., equal to, greater than, less than, go, no go).

instruction word A computer word having two sections, the op-code section and the address section.

mnemonic Intended to assist the memory.

op-code-Operation code. The code which defines a specific computer operation.

oscillator stability The stability of the frequency of oscillation; usually expressed in parts per thousand or parts per million for a period of time.

secondary clock A clock of frequency lower than the basic system clock which is derived from the basic system clock.

special-purpose computer A computer designed to accomplish only one task, for example, the MPG computer in this chapter.

strobe pulse A pulse developed to interrogate gates or to shift data at a time such that racing is avoided.

synchronous system A system in which logic operations and level changes occur in synchronism with a system clock.

two-phase clock The use of two clock waveforms of the same frequency which are 180° out of phase with one another, for example, the 1 and 0 outputs of a flip-flop.

Review Questions

1. Explain why a clock must be perfectly periodic.
2. How can the clock cycle time be found from the clock frequency?
3. Why must flip-flops have a delay time less than one clock cycle time?
4. What factors affect the oscillating frequency of the multivibrator in Fig. 14-2?
5. What is the purpose of the Schmitt trigger in Fig. 14-4?
6. Explain one method for obtaining a two-phase clock.
7. What is the main purpose for developing a strobe pulse?
8. Why is it advantageous to develop the strobe pulse in Fig. 14-9 by turning the transistor on rather than off?
9. Explain the difference between special- and general-purpose computers.
10. What is a computer program?
11. Explain what is meant by fetch and execute in terms of computer operation.

Problems

14-1. Beginning with a symmetrical square wave, show a method for developing a clock consisting of a series of positive pulses. A series of negative pulses.

14-2. What is the clock cycle time for a system using a 1-MHz clock? A 250-kHz clock?

14-3. What is the maximum delay time for a flip-flop if it is to be used in a system having an 8-MHz clock?

14-4. At what frequency will the multivibrator in Fig. 14-2*a* oscillate if $R = 100$ kΩ, $C = 100$ pF, $V_c = 20$ v dc, and $V_B = 10$ v dc?

14-5. What will be the frequency of the multivibrator in Prob. 14-4 if V_B is changed to 20 V dc?

14-6. What value of C is required for the multivibrator in Fig. 14-2*a* if $V_c = V_B$, $R = 47\ \text{k}\Omega$, and the desired frequency is 100 kHz?

14-7. What is the oscillating frequency of the Wien-bridge oscillator in Fig. 14-2*b* is $R = 47\ \text{k}\Omega$, and $C = 100$ pF?

14-8. If the crystal oscillator in Fig. 14-3 has a stability of ± 3 parts in 10^7 per day, what are the maximum and minimum frequencies of the oscillator?

14-9. Show the logic necessary to develop clock frequencies of 5 MHz, 2.5 MHz, 1 MHz, and 200 kHz.

14-10. The 5-MHz oscillator in Prob. 14-9 has a stability of ± 1 part in 10^6 per day. What will be the maximum and minimum frequency of the 1-MHz clock?

14-11. What would be the maximum and minimum frequency of the 200-kHz clock in Prob. 14-10?

14-12. Draw the waveforms for a parallel binary counter being driven by a two-phase clock. Show that this will result in a solution to the race problem. Remember that each flip-flop has a finite delay time.

14-13. How could the MPG computer be modified to give a solution to the nearest 1/10 mile per gallon?

14-14. Draw a block diagram showing the four major blocks in a general-purpose computer system.

14-15. How many op-code bits would be required in a machine having 35 instructions?

14-16. How many address bits would be required to handle 1,000 words of memory?

14-17. How many page address bits would be required to form a 16-page memory having 64 words per page?

14-18. Write a machine-language program to solve the problem $Z = 3R/(A + B)$.

Appendix A

States and Resolution for Binary Numbers

Word length in bits n	Max number of combinations 2^n	Resolution of a binary ladder ppm
1	2	500 000.
2	4	250 000.
3	8	125 000.
4	16	62 500.
5	32	31 250.
6	64	15 625.
7	128	7 812.5
8	256	3 906.25
9	512	1 953.13
10	1 024	976.56
11	2 048	488.28
12	4 096	244.14
13	8 192	122.07
14	16 384	61.04
15	32 768	30.52
16	65 536	15.26
17	131 072	7.63
18	262 144	3.81
19	524 288	1.91
20	1 048 576	0.95
21	2 097 152	0.48
22	4 194 304	0.24
23	8 388 608	0.12
24	16 777 216	0.06

Appendix B

Decimal-Octal-Binary Number Conversion Table

Decimal	Octal	Binary	Decimal	Octal	Binary
0	00	000 000	32	40	100 000
1	01	000 001	33	41	100 001
2	02	000 010	34	42	100 010
3	03	000 011	35	43	100 011
4	04	000 100	36	44	100 100
5	05	000 101	37	45	100 101
6	06	000 110	38	46	100 110
7	07	000 111	39	47	100 111
8	10	001 000	40	50	101 000
9	11	001 001	41	51	101 001
10	12	001 010	42	52	101 010
11	13	001 011	43	53	101 011
12	14	001 100	44	54	101 100
13	15	001 101	45	55	101 101
14	16	001 110	46	56	101 110
15	17	001 111	47	57	101 111
16	20	010 000	48	60	110 000
17	21	010 001	49	61	110 001
18	22	010 010	50	62	110 010
19	23	010 011	51	63	110 011
20	24	010 100	52	64	110 100
21	25	010 101	53	65	110 101
22	26	010 110	54	66	110 110
23	27	010 111	55	67	110 111
24	30	011 000	56	70	111 000
25	31	011 001	57	71	111 001
26	32	011 010	58	72	111 010
27	33	011 011	59	73	111 011
28	34	011 100	60	74	111 100
29	35	011 101	61	75	111 101
30	36	011 110	62	76	111 110
31	37	011 111	63	77	111 111

Answers to Selected Problems

CHAP. 1

1-1 50 μA, 0.5 mA **1-3** 20 nW

CHAP 2

2-1 1, 10, 11, 100, 101, 110, 111, 1000, 1001, 1010, 1011, 1100, 1101, 1110, 1111, 10000, 10001, 10010, 10011, 10100, 10101, 10110, 10111, 11000, 11001, 11010, 11011 **2-3** (*a*) 10100 (*b*) 11000110 (*c*) 11000111 **2-5** 11000, 1000001, 1101010 **2-7** 10111.01110 plus a remainder **2-9** 25.375 **2-11** (*a*) −011 (*b*) 10000 (*c*) −10111 **2-13** 1000. No **2-15** (*a*) 0100 (*b*) 001110 (*c*) −010111 **2-17** (*a*) 100011 (*b*) 10000.1110 (*c*) −10000010.1 **2-19** 110.11 **2-21** 53 **2-23** (*a*) 1216 (*b*) 215 **2-25** (*a*) 257 (*b*) 15.334 (*c*) 123.55 2-27 (*a*) 11100101 (*b*) 101101001101 (*c*) 0111101011110100

CHAP 3

3-1 (*a*) 0101 1001 (*b*) 0011 1001 0101 1000 0100 **3-3** (*a*) 387 (*b*) 967,873 **3-5** (*a*) 0111 1001 1010 (*b*) 1000 1011 0110 1100 **3-7** (*a*) 846 (*b*) 75,320 **3-9** 1010 1000 0011 **3-11** (*a*) 0100 0011 1000 (*b*) 0010 1011 1000 **3-13** (*a*) 952 (*b*) 047 **3-15** (*a*) 10110010 (*b*) 1100011000110 (*c*) 101011100111000 **3-17** 0, 1, 1, 0, 0, 0, 0, 1, 1, 0 **3-19** (*a*) 11000 01111 (*b*) 10000 00111 00011 (*c*) 11111 01111 11000 00011 **3-21** (*a*) 01 00100 10 00001 (*b*) 01 10000 10 00001 01 00100 **3-23** (*a*) 11101 (*b*) 110100110101 (*c*) 11111011001010 **3-25** 850 Ω

CHAP 4

4-1 The *ABy* entries are 3-3-3, 3-12-12, 12-3-12, 12-12-12 **4-3** The *ABCy* entries are 1-1-1-1, 1-1-10-1, 1-10-1-1, 1-10-10-1, 10-1-1-1, 10-1-10-1, 10-10-1-1, 10-10-10-10 **4-5** The *ABCy* entries are 0000, 0010, 0100, 0111, 1001, 1011, 1101, 1111 **4-7** $y = (\overline{A + B})B$. The *ABy* entries are 000, 010, 100, 110 **4-9** The *ABy* entries are 000, 011, 101, 110 **4-11** Construct the truth tables to show equality **4-15** $A \cdot \overline{B}$ **4-23** AB **4-25** 40 ns **4-27** 0

CHAP 5

5-1 Use an exclusive-OR gate to perform mod-2 addition of the sign bits. When the signs are the same, a 0 comes out; when the signs are different, a 1 comes out **5-3** Draw a circuit like Fig. 5-6*a*, except use nine exclusive-OR gates **5-5** There are four input possibilities: $A\ B = 0\ 0,\ 0\ 1,\ 1\ 0,\ 1\ 1$. Substitute the corresponding values of A and B into the left side to get the value of $A\overline{B} + \overline{A}B$ and into the right side to get the value of $(A + B)\overline{AB}$. In each case, the left side equals the right side **5-7** This is straightforward substitution of the given values in Fig. 5-14 **5-9** You have to add 0011 to the four-bit 8421 number. This requires the following adders in the given order: half-adder, half-adder, full-adder, and half-adder. Other logic circuits are also possible **5-11** One half-subtractor and 34 full-subtractors.

CHAP 6

6-1 (*a*) $\overline{A}B\overline{C}$ (*b*) $A\overline{B}\overline{C}D$ **6-3** $y = \overline{A}BC + A\overline{B}C$ **6-5** Two two-input AND gates working into a two-input OR gate; $\overline{A}, B, C$ are the inputs to one AND gate, and $A, \overline{B}, C$ are the inputs to another AND gate **6-7** Four three-input AND gates working into a four-input OR gate; inputs to AND gates are the factors in each logical product **6-9** Simplified equation is $y = AC$; circuit has two input gate leads **6-11** The Karnaugh map is the complement of Fig. 6-16*a*; that is, replace each 0 by a 1, and each 1 by a 0, in Fig. 6-16*a* **6-13** $y = A\overline{C}\overline{D} + \overline{A}\overline{B}CD + \overline{A}\overline{B}C\overline{D}$ **6-15** $Y = D$ **6-17** Taking advantage of *don't cares* results in an equation $y = A\overline{D}$; the corresponding logic circuit is an AND gate with inputs of A and $\overline{D}$ **6-19** Three three-input NAND gates working into a three-input NAND gate; the inputs to the logic circuit are the terms in each logical product **6-21** The Karnaugh map shows isolated 1s; therefore, no simplification is possible and the Boolean equation is $y_{\text{sum}} = \overline{A}\overline{B}C + \overline{A}B\overline{C} + A\overline{B}\overline{C} + ABC$; the logic circuit contains four three-input NAND gates working into a four-input NAND gate

CHAP 7

7-1 250 kHz **7-3** 4 Hz **7-5** There are 2^5 or 32 output states; these states follow the binary number progression from 00000 through 11111 **7-7** 999,999; the BCD number is 0000 0000 0100 0111 1001 0011 **7-9** 32,000; cascade three decade counters, one four-bit binary counter, and one T flip-flop **7-11** (*a*) 11, 01, 00, 10, 11, . . . (*b*) 00, 10, 11, 01, 00, . . . **7-13** 2.8 kHz **7-15** You get a train of 3-μs wide pulses, with the leading edges coincident with the triggers shown in Fig. 7-35

CHAP. 8

8-1 (*a*) 3 (*b*) 4 (*c*) 4 (*d*) 5 (*e*) 5 **8-3** Mod-7
8-5

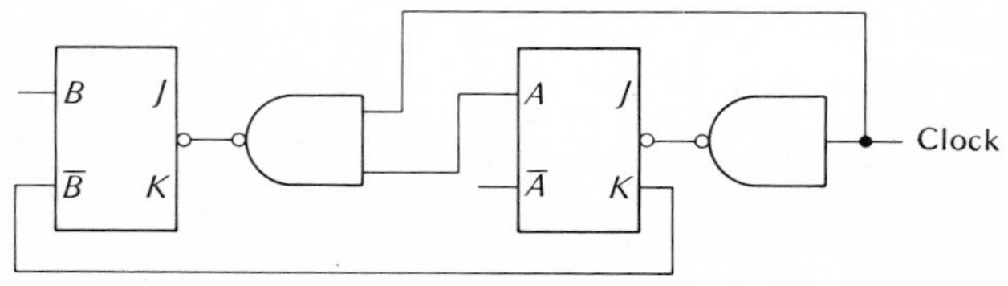

8-5 (cont.)

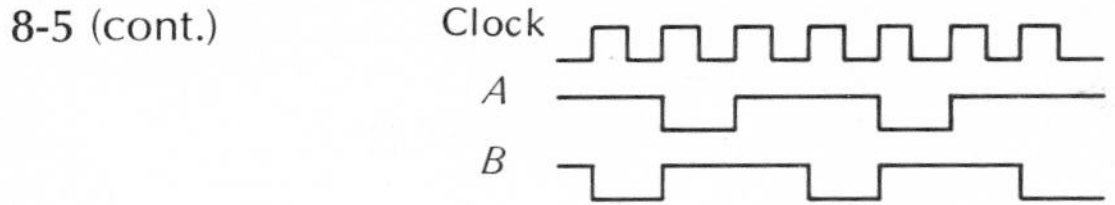

8-7 Yes, cured

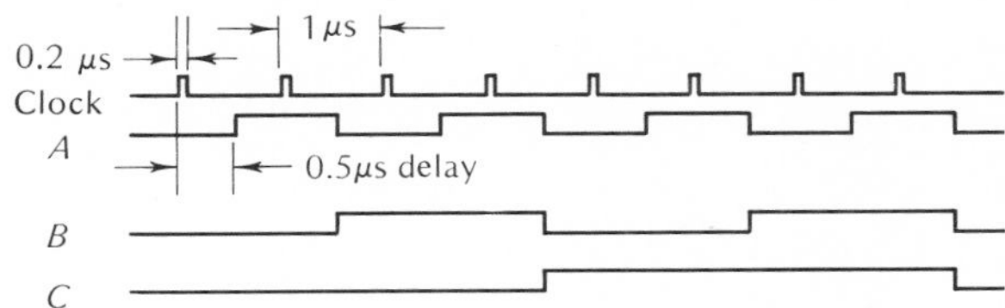

8-9

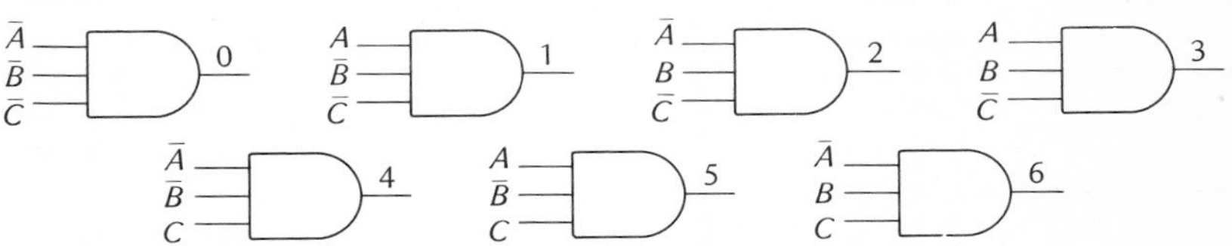

8-11 The difficulty is in decoding

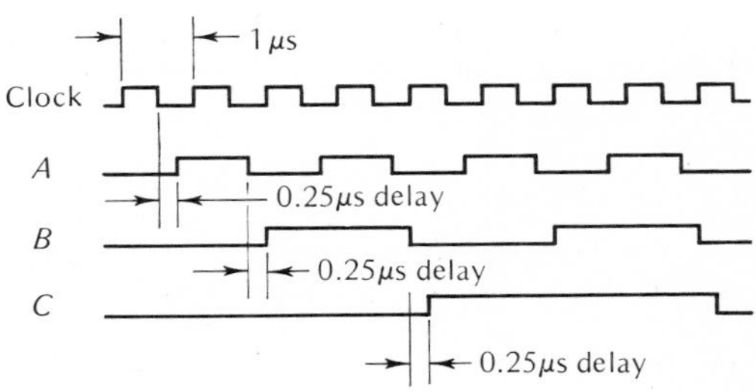

8-13 $2 \times 3 \times 3, 2 \times 9, 3 \times 6$

CHAP. 9

9-1 The only restriction on *RS* flip-flops is that both the *R* and *S* inputs cannot be high at the same time. Since it cannot possibly occur in Fig. 9-2, the logic diagram and waveforms are exactly as shown in Fig. 9-2. **9-3** Apply a positive pulse to *direct reset* of flip-flops *A, B, C, D,* and *E* and *direct set* of *F* **9-5** Set 1s in any two adjacent flip-flops **9-7** Yes **9-9** Yes, 5 **9-11** Corrected on counts 2 or 10 or 18 or 26. Worst case is 9 clock periods. $\overline{C}$ input to gate is not necessary **9-15** Skips counts 16, 17, 48, and 49. *E* is symmetrical. Would not matter if *E* were not symmetrical

9-17

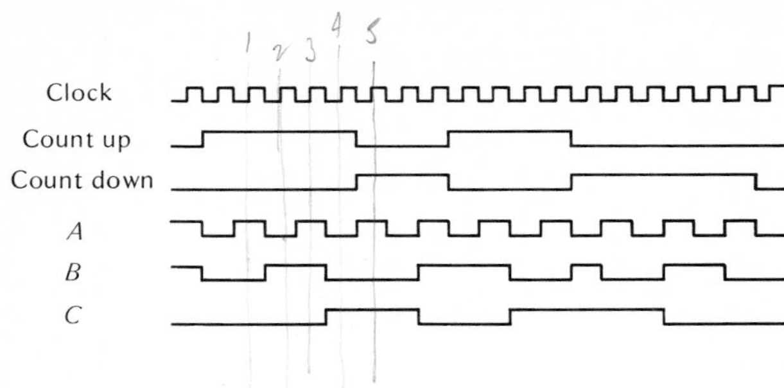

CHAP. 10

10-7 120 in **10-9** 1.37×10^6 **10-11** $F = 5/11$, $L = 275$ in **10-13** No savings in diodes **10-15** Eight three-input AND gates using the three LSBs of the binary number will form the units digit of the octal number. The fourth bit of the binary number will form the other digit of the octal number (1 or 0)

CHAP. 11

11-1 $^1/_{63}$, $^2/_{63}$, $^4/_{63}$, $^8/_{63}$, $^{16}/_{63}$, $^{32}/_{63}$, **11-5** 51.2 mA **11-7** (a) 0.641 V (b) 0.923 V (c) 0.766 V **11-9** 1 part in 4,096; 2.44 mV **11-11** 31 **11-13** (a) 4.096 ms (b) 2.048 ms (c) approximately 500 conversions per second **11-15** 12 μs, not counting delay and control times **11-17** Resolution = 0.0244 percent; accuracy of about 0.05 percent

CHAP. 12

12-1

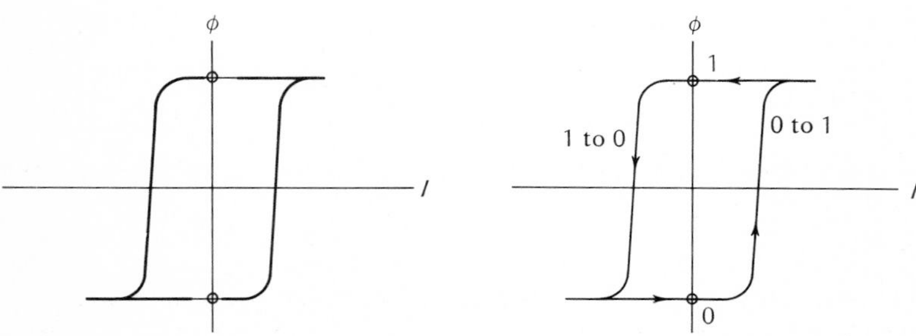

12-3 Symbol shown in Fig. 12-7*b*. Pulse at 0 input will reset; pulse at 1 will set; pulse at advance will cause a 1 output pulse if core previously held a 1 **12-5** Fig. 12-9

12-7

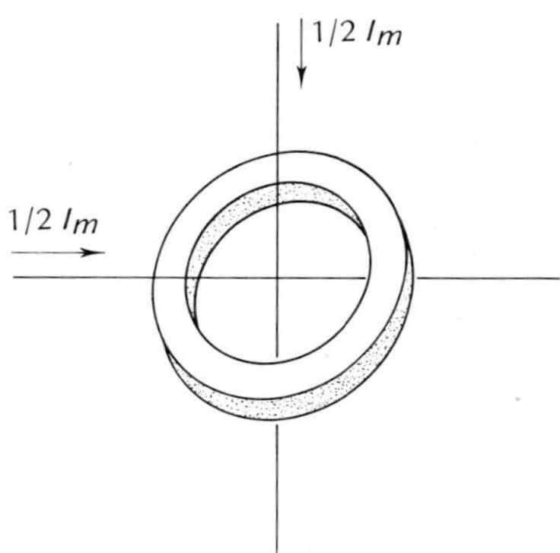

12-9 36 planes; each plane a square array of 4,096 cores, 64 cores on each edge **12-11** 64 *X* lines, 64 *Y* lines, 36 *inhibit* lines **12-13** 12 bits in the address word, six bits for *X* and six bits for *Y* **12-17** 125,600 bits **12-19 0.833** ms; use two sets of *read/write* heads

CHAP. 13

13-1 0 to 0.996094 **13-3** (*a*) −0.156 (*b*) +0.0938 (*c*) −0.312 (*d*) +0.656
13-5 10^4 **13-7** $X_0.X_1X_2X_3X_4\ X_5X_6X_7X_8X_9X_{10}X_{11}X_{12}$
13-9 $(+x) - (+y) = \ x - y,\ (+x) - (-y) = \ x + y$
$(-x) - (+y) = -x - y,\ (-x) - (-y) = -x + y$
13-11 Approximately 1.2 μs
13-13

8421	True	9s	10s
67_{10}	0.0110 0111		
-34_{10}	1.0011 0100	1.0110 0101	1.0110 0110
-93_{10}	1.1001 0011	1.0000 0110	1.0000 0111
67_{10}	0.1001 1010		
-34_{10}	1.0110 0111	1.1001 1000	1.1001 1001
-93_{10}	1.1100 0110	1.0011 1001	1.0011 1010

13-15 Same as Fig. 13-12 with four more bits in each register and one more BCD adder **13-17** 19μs; 46 μs

CHAP. 14

14-1

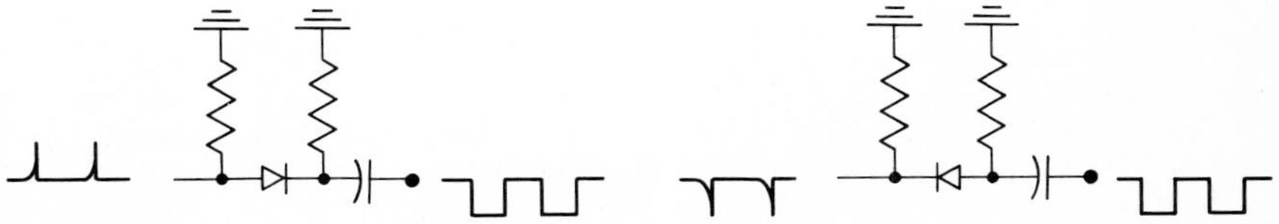

14-3 125 ns **14-5** 722 kHz **14-7** 33.9 kHz
14-9

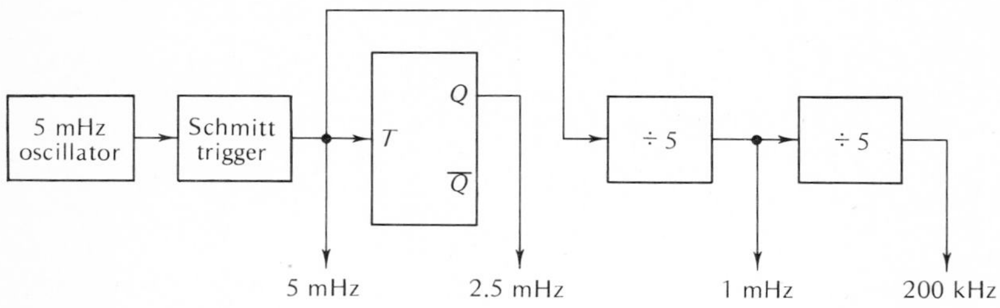

14-11 200 kHz ±0.2 Hz **14-13** Use transducers having 10^4 pulses per mile and 10^4 pulses per gallon **14-15** Six

Index